THE FRACTAL GEOMETRY OF NATURE

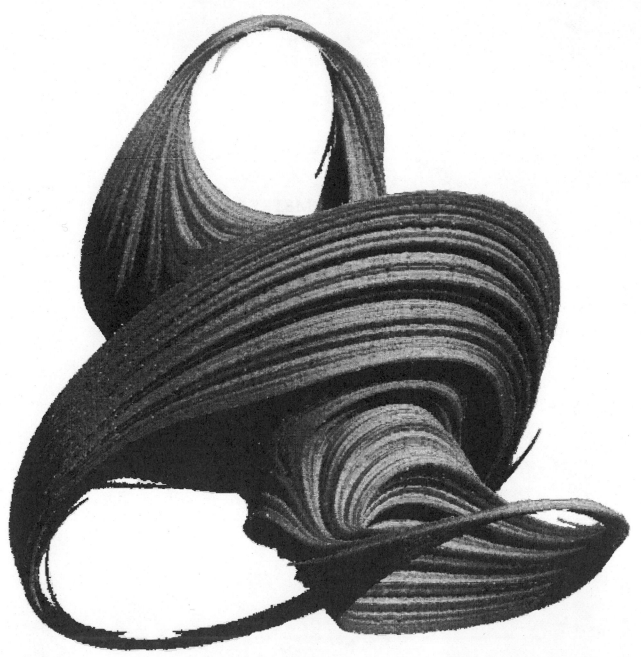

Plate II ¤ ILLUSTRATES PAGE 464

THE FRACTAL GEOMETRY OF NATURE

Updated and Augmented

Benoit B. Mandelbrot

INTERNATIONAL BUSINESS MACHINES
THOMAS J. WATSON RESEARCH CENTER

W. H. FREEMAN AND COMPANY
New York

ABOUT THE AUTHOR ◻◻ IBM Fellow at the IBM Thomas J. Watson Research Center. A Fellow of the American Academy of Arts and Sciences. Graduate, Ecole Polytechnique; M.S. and Ae.E. in Aeronautics, Caltech; Docteur ès Sciences Mathématiques, University of Paris. Dr. Mandelbrot's first positions were with the French Research Council (CNRS), the School of Mathematics at the Institute for Advanced Study (under J. von Neumann), and the University of Geneva. Immediately before joining IBM, he was a junior Professor of Applied Mathematics at the University of Lille and of Mathematical Analysis at Ecole Polytechnique. On leave from IBM, he has been a Visiting Professor of Economics, later of Applied Mathematics, and then of Mathematics at Harvard, of Engineering at Yale, of Physiology at the Albert Einstein College of Medicine, and of Mathematics at the University of Paris-Sud. He has visited M.I.T. several times, first in the Electrical Engineering Department, and most recently as an Institute Lecturer. He was a Visitor at the Institut des Hautes Etudes Scientifiques (Bures). He has been a Fellow of the Guggenheim Foundation, Trumbull Lecturer at Yale, Samuel Wilks Lecturer at Princeton, Abraham Wald Lecturer at Columbia, Goodwin-Richards Lecturer at the University of Pennsylvania, and National Lecturer of Sigma Xi, the Scientific Research Society. Many times a Lecturer at Collège de France since 1973, he delivered there the *leçons* that eventually developed into the present Essay.

AMS Classifications: Primary, 5001, 2875, 60D05; Secondary, 42A36, 54F45, 60J65, 76F05, 85A35, 86A05.

Current Physics Index Classification (ICSU): o5.40. + j, 45.30. − b, 98.20.Qw, 91.65.Br.

Library of Congress Cataloging in Publication Data

Mandelbrot, Benoit B.
   The fractal geometry of nature.

   Rev. ed. of: Fractals. c1977.
   Bibliography: p.
   Includes index.
   1. Geometry.   2. Mathematical models.   3. Stochastic processes. I. Title.
QA447.M357 1982   516'.15   81-15085
ISBN 0-7167-1186-9   AACR2

Printed in the United States of America

7 8 9 10 11 KP 4 3 2 1 0 8 9 8 7 6

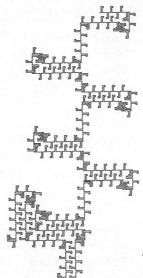

In Memoriam, B. et C.

Pour Aliette

CONTENTS

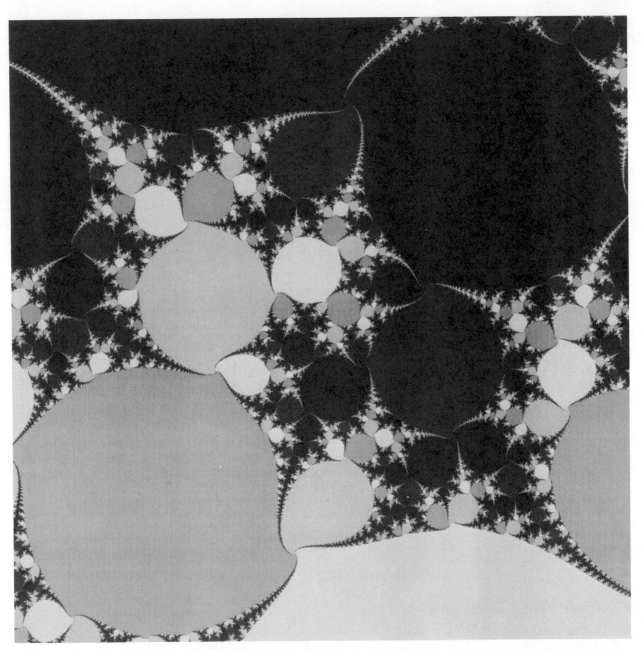

Plate VIII ¤ ILLUSTRATES PAGE 465

FOREWORD

This work follows and largely replaces my 1977 Essay, FRACTALS: FORM, CHANCE AND DIMENSION, which had followed and largely replaced my 1975 Essay in French, LES OBJETS FRACTALS: FORME, HASARD ET DIMENSION. Each stage involved new art, a few deletions, extensive rewriting that affected nearly every section, additions devoted to my older work, and—most important—extensive additions devoted to new developments.

Richard F. Voss made an essential contribution to the 1977 Essay and to this work, especially by designing and now redesigning the fractal flakes, most landscapes, and the planets. The programs for many striking illustrations new to this Essay are by V. Alan Norton.

Other invaluable, close, long-term associates were Sigmund W. Handelman, then Mark R. Laff, for computation and graphics, and H. Catharine Dietrich, then Janis T. Riznychok, for editing and typing.

Individual acknowledgments for the programs behind the illustrations and for other specific assistance are found after the list of references at the end of the volume.

For their backing of my research and my books, I am deeply indebted to the Thomas J. Watson Research Center of the International Business Machines Corporation. As Group Manager, Department Director, and now Director of Research, IBM Vice President Ralph E. Gomory imagined ways of sheltering and underwriting my work when it was a gamble, and now of giving it all the support I could use.

My first scientific publication came out on April 30, 1951. Over the years, it had seemed to many that each of my investigations was aimed in a different direction. But this apparent disorder was misleading: it hid a strong unity of purpose, which the present Essay, like its two predecessors, is intended to reveal. Against odds, most of my works turn out to have been the birth pangs of a new scientific discipline.

Plate X  ILLUSTRATES PAGE 465

THE FRACTAL GEOMETRY OF NATURE

1 ⌑ Theme

Why is geometry often described as "cold" and "dry?" One reason lies in its inability to describe the shape of a cloud, a mountain, a coastline, or a tree. Clouds are not spheres, mountains are not cones, coastlines are not circles, and bark is not smooth, nor does lightning travel in a straight line.

More generally, I claim that many patterns of Nature are so irregular and fragmented, that, compared with *Euclid*—a term used in this work to denote all of standard geometry—Nature exhibits not simply a higher degree but an altogether different level of complexity. The number of distinct scales of length of natural patterns is for all practical purposes infinite.

The existence of these patterns challenges us to study those forms that Euclid leaves aside as being "formless," to investigate the morphology of the "amorphous." Mathematicians have disdained this challenge, however, and have increasingly chosen to flee from nature by devising theories unrelated to anything we can see or feel.

Responding to this challenge, I conceived and developed a new geometry of nature and implemented its use in a number of diverse fields. It describes many of the irregular and fragmented patterns around us, and leads to full-fledged theories, by identifying a family of shapes I call *fractals*. The most useful fractals involve *chance* and both their regularities and their irregularities are statistical. Also, the shapes described here tend to be *scaling*, implying that the degree of their irregularity and/or fragmentation is identical at all scales. The concept of *fractal* (Hausdorff) *dimension* plays a central role in this work.

Some fractal sets are curves or surfaces, others are disconnected "dusts," and yet others are so oddly shaped that there are no good terms for them in either the sciences or the arts. The reader is urged to sample them now, by browsing through the book's illustrations.

Many of these illustrations are of shapes that had never been considered previously, but others represent known constructs, often for the first time. Indeed, while fractal geometry as such dates from 1975, many of its tools and concepts had been previously developed, for diverse purposes altogether different from mine. Through old stones inserted in the newly built structure, fractal geometry was able to "borrow" exceptionally extensive rigorous foundations, and soon led to many compelling new questions in mathematics.

Nevertheless, this work pursues neither abstraction nor generality for its own sake, and is neither a textbook nor a treatise in mathematics. Despite its length, I describe it as a scientific Essay because it is written from a personal point of view and without attempting completeness. Also, like many Essays, it tends to digressions and interruptions.

This informality should help the reader avoid the portions lying outside his interest or beyond his competence. There are many mathematically "easy" portions throughout, especially toward the very end. *Browse and skip*, at least at first and second reading.

PRESENTATION OF GOALS

This Essay brings together a number of analyses in diverse sciences, and it promotes a new mathematical and philosophical synthesis. Thus, it serves as both a *casebook* and a *manifesto*. Furthermore, it reveals a totally new world of plastic beauty.

A SCIENTIFIC CASEBOOK

Physicians and lawyers use "casebook" to denote a compilation concerning actual cases linked by a common theme. This term has no counterpart in science, and I suggest we appropriate it. The major cases require repeated attention, but less important cases also deserve comment; often, their discussion is shortened by the availability of "precedents."

One case study concerns the widely known application of widely known mathematics to a widely known natural problem: Wiener's geometric model of physical Brownian motion. Surprisingly, we encounter no fresh direct application of Wiener's process, which suggests that, among the phenomena of higher complexity with which we deal, Brownian motion is a special case, an exceptionally simple and unstructured one. Nevertheless, it is included because many useful fractals are careful modifications of Brownian motion.

The other case studies report primarily upon my own work, its pre-fractal antecedents, and its extensions due to scholars who reacted to this Essay's 1975 and 1977 predecessors. Some cases relate to the highly visible worlds of mountains and the like, thus fulfilling at long last the promise of the term *geometry*. But other cases concern submicroscopic assemblies, the prime object of physics.

The substantive topic is occasionally esoteric. In other instances, the topic is a familiar one, but its geometric aspects had not been attacked adequately. One is reminded on this account of Poincaré's remark that there are questions that one chooses to ask and other questions that ask themselves. And a question that had long asked itself without response tends to be abandoned to children.

Due to this difficulty, my previous Essays stressed relentlessly the fact that the fractal approach is both effective and "natural." Not only should it not be resisted, but one ought to wonder how one could have gone so long without it. Also, in order to avoid needless controversy, these earlier texts minimized the discontinuities between exposition of standard and other published material, exposition with a new twist, and presentation of my own ideas and results. In the present Essay, to the contrary, I am precise in claiming credit.

Most emphatically, I do not consider the fractal point of view as a panacea, and each case analysis should be assessed by the criteria holding in its field, that is, mostly upon the basis of its powers of organization, explanation, and prediction, and not as example of a mathematical structure. Since each case study must be cut short before it becomes truly technical, the reader is referred elsewhere for detailed developments. As a result (to echo d'Arcy Thompson 1917), this Essay is preface from beginning to end. Any specialist who expects more will be disappointed.

A MANIFESTO: THERE IS A FRACTAL FACE TO THE GEOMETRY OF NATURE

Now, the reason for bringing these prefaces together is that each helps one to understand the others because they share a common mathematical structure. F. J. Dyson has given an eloquent summary of this theme of mine.

"*Fractal* is a word invented by Mandelbrot to bring together under one heading a large class of objects that have [played]...an historical role...in the development of pure mathematics. A great revolution of ideas separates the classical mathematics of the 19th century from the modern mathematics of the 20th. Classical mathematics had its roots in the regular geometric structures of Euclid and the continuously evolving dynamics of Newton. Modern mathematics began with Cantor's set theory and Peano's space-filling curve. Historically, the revolution was forced by the discovery of mathematical structures that did not fit the patterns of Euclid and Newton. These new structures were regarded...as 'pathological,'... as a 'gallery of monsters,' kin to the cubist painting and atonal music that were upsetting established standards of taste in the arts at about the same time. The mathematicians who created the monsters regarded them as important in showing that the world of pure mathematics contains a richness of possibilities going far beyond the simple structures that they saw in Nature. Twentieth-century mathematics flowered in the belief that it had transcended completely the limitations imposed by its natural origins.

"Now, as Mandelbrot points out,...Nature has played a joke on the mathematicians. The 19th-century mathematicians may have been lacking in imagination, but Nature was not. The same pathological structures that the mathematicians invented to break loose from 19th-century naturalism turn out to be inherent in familiar objects all around us." *

In brief, I have confirmed Blaise Pascal's observation that imagination tires before Nature. ("L'imagination se lassera plutôt de concevoir que la nature de fournir.")

Nevertheless, fractal geometry is *not* a straight "application" of 20th century mathematics. It is a new branch born belatedly of the crisis of mathematics that started when duBois Reymond 1875 first reported on a continuous nondifferentiable function constructed by Weierstrass (Chapters 3, 39, and 41). The crisis lasted approximately to 1925, major actors being Cantor, Peano, Lebesgue, and Hausdorff. These names, and those of Besicovitch, Bolzano, Cesàro, Koch, Osgood, Sierpiński, and Urysohn, are not ordinarily encountered in the empirical study of Nature, but I claim that the impact of the work of these giants far transcends its intended scope.

I show that behind their very wildest creations, and unknown to them and to several generations of followers, lie worlds of interest to all those who celebrate Nature by trying to imitate it.

Once again, we are surprised by what several past occurrences should have led us to expect, that "the language of mathematics reveals itself unreasonably effective in the natural sciences..., a wonderful gift which we neither understand nor deserve. We should be grateful for it and hope that it will remain

valid in future research and that it will extend, for better or for worse, to our pleasure even though perhaps also to our bafflement, to wide branches of learning" (Wigner 1960).

MATHEMATICS, NATURE, ESTHETICS

In addition, fractal geometry reveals that some of the most austerely formal chapters of mathematics had a hidden face: a world of pure plastic beauty unsuspected till now.

"FRACTAL" AND OTHER NEOLOGISMS

There is a saying in Latin that "to name is to know:" *Nomen est numen.* Until I took up their study, the sets alluded to in the preceding sections were not important enough to require a term to denote them. However, as the classical monsters were defanged and harnessed through my efforts, and as many new "monsters" began to arise, the need for a term became increasingly apparent. It became acute when the first predecessor of this Essay had to be given a title.

I coined *fractal* from the Latin adjective *fractus.* The corresponding Latin verb *frangere* means "to break:" to create irregular fragments. It is therefore sensible—and how appropriate for our needs!—that, in addition to "fragmented" (as in *fraction* or *refraction*), *fractus* should also mean "irregular," both meanings being preserved in *fragment.*

The proper pronunciation is *frac'tal*, the stress being placed as in *frac'tion.*

The combination *fractal set* will be defined

*From "Characterizing Irregularity" by Freeman Dyson, *Science*, May 12, 1978, vol. 200, no. 4342, pp. 677–678. Copyright © 1978 by the American Association for the Advancement of Science.

rigorously, but the combination *natural fractal* will serve loosely to designate a natural pattern that is usefully representable by a fractal set. For example, Brownian curves are fractal sets, and physical Brownian motion is a natural fractal.

(Since *algebra* derives from the Arabic *jabara* = to bind together, *fractal* and *algebra* are etymological opposites!)

More generally, in my travels through newly opened or newly settled territory, I was often moved to exert the right of naming its landmarks. Usually, to coin a careful neologism seemed better than to add a new wrinkle to an already overused term.

And one must remember that a word's common meaning is often so entrenched, that it is not erased by any amount of redefinition. As Voltaire noted in 1730, "if Newton had not used the word *attraction,* everyone in [the French] Academy would have opened his eyes to the light; but unfortunately he used in London a word to which an idea of ridicule was attached in Paris." And phrases like "the probability distribution of the Schwartz distribution in space relative to the distribution of galaxies" are dreadful.

The terms coined in this Essay avoid this pitfall by tapping underutilized Latin or Greek roots, like *trema,* and the rarely borrowed robust vocabularies of the shop, the home, and the farm. Homely names make the monsters easier to tame! For example, I give technical meanings to *dust*, *curd*, and *whey*. I also advocate *pertiling* for a thorough (= *per*) form of tiling.

RESTATEMENT OF GOALS

In sum, the present Essay describes the solutions I propose to a host of concrete problems, including very old ones, with the help of mathematics that is, in part, likewise very old, but that (aside from applications to Brownian motion) had never been used in this fashion. The cases this mathematics allows us to tackle, and the extensions these cases require, lay the foundation of a new discipline.

Scientists will (I am sure) be surprised and delighted to find that not a few shapes they had to call *grainy, hydralike, in between, pimply, pocky, ramified, seaweedy, strange, tangled, tortuous, wiggly, wispy, wrinkled,* and the like, can henceforth be approached in rigorous and vigorous quantitative fashion.

Mathematicians will (I hope) be surprised and delighted to find that sets thus far reputed exceptional (Carleson 1967) should in a sense be the rule, that constructions deemed pathological should evolve naturally from very concrete problems, and that the study of Nature should help solve old problems and yield so many new ones.

Nevertheless, this Essay avoids all purely technical difficulties. It is addressed primarily to a mixed group of scientists. The presentation of each theme begins with concrete and specific cases. The nature of fractals is meant to be gradually discovered by the reader, not revealed in a flash by the author.

And the art can be enjoyed for itself. ▪▬

2 ¤ The Irregular and Fragmented in Nature

"All pulchritude is relative.... We ought not...to believe that the banks of the ocean are really deformed, because they have not the form of a regular bulwark; nor that the mountains are out of shape, because they are not exact pyramids or cones; nor that the stars are unskillfully placed, because they are not all situated at uniform distance. These are not natural irregularities, but with respect to our fancies only; nor are they incommodious to the true uses of life and the designs of man's being on earth." This opinion of the seventeenth century English scholar Richard Bentley (echoed in the opening words of this Essay) shows that to bring coastline, mountain, and sky patterns together, and to contrast them with Euclid, is an ancient idea.

FROM THE PEN OF JEAN PERRIN

Next we tune to a voice nearer in time and profession. To elaborate upon the irregular or fragmented character of coastlines, Brownian trajectories, and other patterns of Nature to be investigated in this Essay, let me present in free translation some excerpts from Perrin 1906. Jean Perrin's subsequent work on Brownian motion won him the Nobel Prize and spurred the development of probability theory. But here I quote from an early philosophical manifesto. Although it was later paraphrased in the preface to Perrin 1913, this text *failed to gain attention until quoted in this Essay's first (French) version*. It had come to my notice too late to have a substantive effect on my work, but it spurred me on at a time of need, and its eloquence remains unmatched.

"It is well known that, before giving a rigorous definition of continuity, a good teacher shows that beginners already possess the idea which underlies this concept. He draws a well-defined curve and says, holding a ruler, 'You see that there is a tangent at every point.' Or again, in order to impart the notion of the true velocity of a moving object at a point in its trajectory, he says, 'You see, of course, that the mean velocity between two neighboring points does not vary appreciably as these points approach infinitely near to each other.' And many minds, aware that for certain familiar motions this view appears true enough, do not see that it involves considerable difficulties.

"Mathematicians, however, are well aware that it is childish to try to show by drawing curves that every continuous function has a derivative. Though differentiable functions are the simplest and the easiest to deal with, they are exceptional. Using geometrical language, curves that have no tangents are the rule, and regular curves, such as the circle, are interesting but quite special.

"At first sight the consideration of the general case seems merely an intellectual exercise, ingenious but artificial, the desire for absolute accuracy carried to a ridiculous length. Those who hear of curves without tangents, or of functions without derivatives, often think at first that Nature presents no such complications, nor even suggests them.

"The contrary, however, is true, and the logic of the mathematicians has kept them nearer to reality than the practical representations employed by physicists. This assertion may be illustrated by considering certain experimental data without preconception.

"Consider, for instance, one of the white flakes that are obtained by salting a solution of soap. At a distance its contour may appear sharply defined, but as we draw nearer its sharpness disappears. The eye can no longer draw a tangent at any point. A line that at first sight would seem to be satisfactory appears on close scrutiny to be perpendicular or oblique. The use of a magnifying glass or microscope leaves us just as uncertain, for fresh irregularities appear every time we increase the magnification, and we never succeed in getting a sharp, smooth impression, as given, for example, by a steel ball. So, if we accept the latter as illustrating the classical form of continuity, our flake could just as logically suggest the more general notion of a continuous function without a derivative."

An interruption is necessary to draw attention to Plates 10 and 11.

The black-and-white plates first mentioned in a given chapter are collected on pages that follow immediately, and are numbered as the pages on which they occur. The color plates form a special signature, whose captions are written to be fairly independent of the rest of the book.

The quote resumes.

"We must bear in mind that the uncertainty as to the position of the tangent at a point on the contour is by no means the same as the uncertainty observed on a map of Brittany. Although it would differ according to the map's scale, a tangent can always be found, for a map is a conventional diagram. On the contrary, an essential characteristic of our flake and of the coast is that we *suspect,* without seeing them clearly, that any scale involves details that absolutely prohibit the fixing of a tangent.

"We are still in the realm of experimental reality when we observe under the microscope the Brownian motion agitating a small particle suspended in a fluid [this Essay's Plate 13]. The direction of the straight line joining the positions occupied at two instants very close in time is found to vary absolutely irregularly as the time between the two instants is decreased. An unprejudiced observer would therefore conclude that he is dealing with a function without derivative, instead of a curve to which a tangent could be drawn.

"It must be borne in mind that, although closer observation of any object generally leads to the discovery of a highly irregular structure, we often can with advantage approximate its properties by continuous functions. Although wood may be indefinitely porous, it is useful to speak of a beam that has been sawed and planed as having a finite area. In other words, at certain scales and for certain methods of investigation, many phenomena may be represented by regular continuous functions, somewhat in the same way that a sheet of tinfoil may be wrapped round a sponge without following accurately the latter's complicated contour.

"If, to go further, we... attribute to matter the *infinitely* granular structure that is in the spirit of atomic theory, our power to apply to reality the *rigorous* mathematical concept of continuity will greatly decrease.

"Consider, for instance, the way in which we define the density of air at a given point and at a given moment. We picture a sphere of volume v centered at that point and including the mass m. The quotient m/v is the mean density within the sphere, and by *true* density we denote some limiting value of this quotient. This notion, however, implies that at the given moment the mean density is practically constant for spheres below a certain volume. This mean density may be notably different for spheres containing 1,000 cubic meters and 1 cubic centimeter respectively, but it is expected to vary only by 1 in 1,000,000 when comparing 1 cubic centimeter to one-thousandth of a cubic millimeter.

"Suppose the volume becomes continually smaller. Instead of becoming less and less important, these fluctuations come to increase. For scales at which the Brownian motion shows great activity, fluctuations may attain 1 part in 1,000, and they become of the order of 1 part in 5 when the radius of the hypothetical spherule becomes of the order of a hundredth of a micron.

"One step further and our spherule becomes of the order of a molecule radius. In a gas, it will generally lie in intermolecular space, where its mean density will henceforth *vanish*. At our point the *true* density will also *vanish*. But about once in a thousand times that point will lie within a molecule, and the mean density will be a thousand times higher than the value we usually take to be the true density of the gas.

"Let our spherule grow steadily smaller. Soon, except under exceptional circumstances, it will become empty and remain so henceforth owing to the emptiness of intra-atomic space; the true density *vanishes* almost everywhere, except at an infinite number of isolated points, where it reaches an infinite value.

"Analogous considerations are applicable to properties such as velocity, pressure, or temperature. We find them growing more and more irregular as we increase the magnification of our necessarily imperfect image of the universe. The function that represents any physical property will form in intermaterial space a *continuum* with an infinite number of singular points.

"Infinitely discontinuous matter, a continuous ether studded with minute stars, also appears in the cosmic universe. Indeed, the conclusion we have reached above can also be arrived at by imagining a sphere that successively embraces planets, solar system, stars, and nebulae....

"Allow us now a hypothesis that is arbitrary but not self-contradictory. One might encounter instances where using a function without a derivative would be simpler than using one that can be differentiated. When this happens, the mathematical study of irregular continua will prove its practical value."

Then, starting a new section for emphasis. "However, this hope is nothing but a daydream, as yet."

WHEN A "GALLERY OF MONSTERS" BECOMES A MUSEUM OF SCIENCE

Part of this daydream, relative to Brownian motion, did become reality in Perrin's own lifetime. Perrin 1909 chanced to catch the attention of Norbert Wiener (Wiener 1956, pp. 38–39, or 1964, pp. 2–3), who, to his own "surprise and delight" was moved to define and study rigorously a nondifferentiable first model of Brownian motion.

This model remains essential, but physicists stress that its nondifferentiability is traceable to abusive idealization, namely the neglect of inertia. In doing so, physicists turn their back to the feature of Wiener's model that is most significant for the present work.

As to the other applications of mathematics to physics that Perrin foresaw, they were not even attempted until the present work. The collection of sets to which Perrin was alluding (Weierstrass curves, Cantor dusts, and the like) continued to remain a part of "pure mathematics."

Some writers, for example Vilenkin 1965, call this collection a "Mathematical Art Museum," without suspecting (I am sure) how accurate those words were to be proven by the present work. We know from Chapter 1 that other writers (beginning with Henri Poincaré) call it a "Gallery of Monsters," echoing the *Treatise of Algebra* of John Wallis (1685), where the fourth dimension is described as "a Monster in Nature, and less possible than a *Chimera* or *Centaure*."

One of the aims of the present Essay is to show, through relentless hammering at diverse explicit "cases," that the same Gallery may also be visited as a "Museum of Science."

Mathematicians are to be praised for having devised the first of these sets long ago, and scolded for having discouraged us from using them. ■

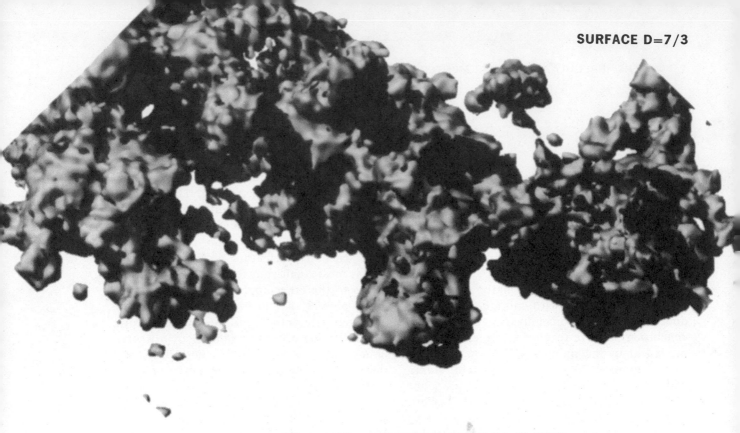

Plates 10 and 11 ¤ ARTIFICIAL FRACTAL FLAKES

In an inspiring text quoted in Chapter 2, Jean Perrin comments on the form of the "white flakes that are obtained by salting a solution of soap." These illustrations are meant to accompany Perrin's remarks.

One must hasten to state that they are neither photographs nor computer reconstitutions of any real object, be it a soap flake, a rain cloud, a volcanic cloud, a small asteroid, or a piece of virgin copper.

Nor do they claim to result from a theory embodying the diverse aspects of a real flake's formation—chemical, physico-chemical, and hydrodynamical.

A fortiori, they do not claim to be directly related to scientific principles.

They are computer-generated shapes meant to illustrate as simply as I can manage certain geometric characteristics that seem to be embodied in Perrin's description, and that

I propose to model using the notion of fractal.

These flakes exist only in a computer's memory. They were never made into hard models, and the shading too was implemented by computation.

The flakes' construction is explained in Chapter 30. The obvious perceptual differences between them are due to differences in the value of a parameter D written next to each. It is called fractal dimension, is basic to the present work, and is introduced in Chapter 3. The overall shapes being the same in all 3 cases is due to bias introduced by the use of an approximation, and is discussed in the caption of Plates 266 and 267.

An earlier version was oddly reminiscent of a presumed photograph of the Loch Ness monster. Could this convergence of form be coincidental? ■

SURFACE D=5/2

SURFACE D=8/3

Plate 13 ¤ JEAN PERRIN'S CLASSIC DRAWINGS OF PHYSICAL BROWNIAN MOTION

Physical Brownian motion is described in Perrin 1909 as follows: "In a fluid mass in equilibrium, such as water in a glass, all the parts appear completely motionless. If we put into it an object of greater density, it falls. The fall, it is true, is the slower the smaller the object; but a visible object always ends at the bottom of the vessel and does not tend again to rise. However, it would be difficult to examine for long a preparation of very fine particles in a liquid without observing a perfectly irregular motion. They go, stop, start again, *mount*, descend, *mount again*, without in the least tending toward immobility."

The present plate, the only one in this book to picture a natural phenomenon, is reproduced from Perrin's *Atoms*. We see four separate tracings of the motion of a colloidal particle of radius 0.53μ, as seen under the microscope. The successive positions were marked every 30 seconds (the grid size being 3.2μ), then joined by straight intervals having no physical reality whatsoever.

To resume our free translation from Perrin 1909, "One may be tempted to define an 'average velocity of agitation' by following a particle as accurately as possible. But such evaluations are *grossly wrong*. The apparent average velocity varies crazily in magnitude and direction. This plate gives only a weak idea of the prodigious entanglement of the real trajectory. If indeed this particle's positions were marked down 100 times more frequently, each interval would be replaced by a polygon smaller than the whole drawing but just as complicated, and so on. It is easy to see that in practice the notion of tangent is meaningless for such curves."

This Essay shares Perrin's concern, but attacks irregularity from a different angle. We stress the fact that when a Brownian trajectory is examined increasingly closely, Chapter 25, its length increases without bound.

Furthermore, the trail left behind by Brownian motion ends up by nearly filling the whole plane. Is it not tempting to conclude that in some sense still to be defined, this peculiar curve has the same dimension as the plane? Indeed, it does. A principal aim of this Essay will be to show that the loose notion of dimension splits into several distinct components. The Brownian motion's trail is *topologically* a curve, of dimension 1. However, being practically plane filling, it is *fractally* of dimension 2. The discrepancy between these two values will, in the terminology introduced in this Essay, qualify Brownian motion as being a fractal. ■

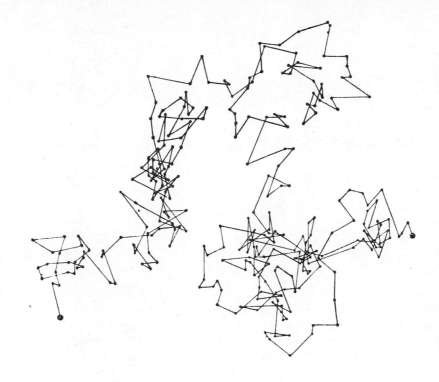

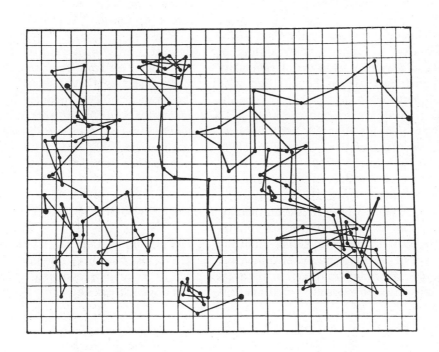

3 ¤ Dimension, Symmetry, Divergence

A central role is played in this Essay by the ancient notions of *dimension* (meaning *number of dimensions* or *dimensionality*) and of *symmetry*. Furthermore, we constantly encounter diverse symptoms of *divergence*.

THE IDEA OF DIMENSION

Mathematicians recognized during their 1875-1925 crisis that a proper understanding of irregularity or fragmentation (as of regularity and connectedness) cannot be satisfied with defining dimension as a number of coordinates. The first step of a rigorous analysis is taken by Cantor in his June 20, 1877, letter to Dedekind, the next step by Peano in 1890, and the final steps in the 1920's.

Like all major intellectual developments, the outcome of this story can be interpreted in diverse ways. Anyone who writes a mathematical book on *the* theory of dimension implies that this theory is unique. But to my mind the main fact is that the loose notion of dimension turns out to have many mathematical facets that not only are conceptually distinct but may lead to different numerical values. Just as William of Occam says of entities, dimensions must not be multiplied beyond necessity, but a multiplicity of dimensions is absolutely unavoidable. Euclid is limited to sets for which all the useful dimensions coincide, so that one may call them *dimensionally concordant* sets. On the other hand, the different dimensions of the sets to which the bulk of this Essay is devoted *fail* to coincide; these sets are *dimensionally discordant*.

Moving on from the dimensions of mathematical sets to the "effective" dimensions of the physical objects modeled by these sets, we encounter a different sort of inevitable and concretely essential ambiguity. Both the mathematical and the physical aspects of dimension are previewed in this chapter.

DEFINITION OF THE TERM FRACTAL

The present section uses undefined mathematical terms, but many readers may find it helpful, or at least reassuring, to scan this text, and anybody can skip it.

This and later digressions in this Essay are delimited by the new brackets ◁ and ►. The latter is very bold, so as to be readily found by anyone who becomes lost in a digression and wants to skip ahead. But the "open bracket" symbol avoids attracting attention, so as to prevent digressions from receiving excessive attention. Material discussed later often receives advance mention in digressions.

◁ The fact that the basic fractals are dimensionally discordant can serve to transform the concept of fractal from an intuitive to a mathematical one. I chose to focus on two definitions, each of which assigns to every set $\mathbb{R}^E$ in Euclidean space, no matter how "pathological," a real number which on intuitive and formal grounds strongly deserves to be called its dimension. The more intuitive of the two is the topological dimension according to Brouwer, Lebesgue, Menger, and Urysohn. We denote it by D_T. It is described in an entry in Chapter 41. The second dimension was formulated in Hausdorff 1919 and put in final form by Besicovitch. It is discussed in Chapter 39. We denote it by D.

◁ Whenever (as is usually the case) we work in the Euclidean span $\mathbb{R}^E$, both D_T and D are at least 0 and at most E. But the resemblance ends here. The dimension D_T is *always an integer*, but D *need not be an integer*. And the two dimensions need not coincide; they only satisfy the Szpilrajn inequality (Hurewicz & Wallman 1941, Chapter 4)

$$D \geq D_T.$$

For all of Euclid, $D = D_T$. But nearly all sets in this Essay satisfy $D > D_T$. There was no term to denote such sets, which led me to coin the term *fractal*, and to define it as follows:

◁ *A fractal is by definition a set for which the Hausdorff Besicovitch dimension strictly exceeds the topological dimension.*

◁ Every set with a noninteger D is a fractal. For example, the original Cantor set is a fractal because we see in Chapter 8 that

$$D = \log 2 / \log 3 \sim 0.6309 > 0, \text{ while } D_T = 0.$$

And a Cantor set in $\mathbb{R}^E$ can be tailored and generalized so that $D_T = 0$, while D takes on any desired value between 0 and E (included).

◁ Furthermore, the original Koch curve is a fractal because we see in Chapter 6 that

$$D = \log 4 / \log 3 \sim 1.2618 > 1, \text{ while } D_T = 1.$$

◁ However, a fractal may have an integer D. For example, Chapter 25 shows that the trail of Brownian motion is a fractal because

$$D = 2, \text{ while } D_T = 1.$$

◁ The striking fact that D need not be an integer deserves a terminological aside. If one uses *fraction* broadly, as synonymous with a noninteger real number, several of the above-listed values of D are fractional, and indeed the Hausdorff Besicovitch dimension is often called *fractional dimension*. But D may be an integer (not greater than E but strictly greater than D_T). I call D a *fractal dimension*. ►

FRACTALS IN HARMONIC ANALYSIS

◁ Part of the study of fractals is the geometric face of harmonic analysis, but this fact is not stressed in the present work. Harmonic (= spectral or Fourier) analysis is not known to most readers, and many who use it effectively are not acquainted with its basic structures.

Also, both the fractal and the spectral approach have their own strong flavor and personality, which are better appreciated by first investigating each for its own sake. Finally, compared to harmonic analysis, the study of fractals is easy and intuitive. ►

OF "NOTIONS THAT ARE NEW,... BUT"

Lebesgue made fun of certain "notions that are new, to be sure, but of which no use can be made after they have been defined." This comment never applied to D, but the use of D remained concentrated in few areas, all of them in pure mathematics. I was the first to use D successfully in the description of Nature. And one of the central goals of this work is to establish D in a central position in empirical science, thereby showing it to be of far broader import than anyone imagined.

Several areas of physics accepted my claim concerning D with exceptional promptness. In fact, having recognized the inadequacies of standard dimension, numerous scholars in these areas had already been groping towards *broken, anomalous* or *continuous* dimensions of all kind. These approaches had remained mutually unrelated, however. Furthermore, few definitions of dimension were used more than once, none had the backing of a mathematical theory, and none was developed far enough for the lack of mathematical backing to make a difference. For the developments to be described here, to the contrary, the existence of a mathematical theory is vital.

A MATHEMATICAL STUDY OF FORM MUST GO BEYOND TOPOLOGY

A mathematician, if asked which well-defined branch of mathematics studies form, is very likely to mention topology. This field is important to our purposes and is referred to in the preceding section, but the present Essay advances and defends the claim that the loose notion of form possesses mathematical aspects other than topological ones.

Topology, which used to be called *geometry of situation* or *analysis situs* (Τοπος means position, situation in Greek), considers that all pots with two handles are of the same form because, if both are infinitely flexible and compressible, they can be molded into any other continuously, without tearing any new opening or closing up any old one. It also teaches that all single island coastlines are of the same form, because they are topologically identical to a circle. And that the topological dimension is the same for coastlines and circles: equal to 1. If one adds offshore "satellite islands," the cumulative coastline is topologically identical to "many" circles. Thus, topology *fails* to discriminate between different coastlines.

By way of contrast, Chapter 5 shows that different coastlines tend to have different fractal dimensions. Differences in fractal dimension express differences in a *nontopological aspect of form,* which I propose to call *fractal form.*

Most problems of real interest combine fractal and topological features in increasingly subtle fashion.

Observe that in the case of topology, the definitions of the field itself and of D_T were refined in parallel, while the notion of D preceded the present study of fractal form by half a century.

Incidentally, Felix Hausdorff's name being given to a class of topological spaces, the widely used contracted term for D, Hausdorff dimension, seems to have undertones of "dimension of a Hausdorff space," thus suggesting it is a topological concept—which emphatically is *not* the case. This is yet another reason for preferring *fractal dimension.*

EFFECTIVE DIMENSION

In addition to the mathematical notions underlying D_T and D, this Essay often invokes *effective dimension*, a notion that *should not* be defined precisely. It is an intuitive and potent throwback to the Pythagoreans' archaic Greek geometry. A novelty of this Essay is that it allows the value of effective dimension to be a fraction.

Effective dimension concerns the relation between mathematical sets and natural objects. Strictly speaking, physical objects such as a veil, a thread, or a tiny ball should all be represented by three-dimensional shapes. However, physicists prefer to think of a veil, a thread, or a ball—if they are fine enough—as being "in effect" of dimensions 2, 1, and 0, respectively. For example, to describe a thread, the theories relating to sets of dimension 1 or 3 must be modified by corrective terms. And the better geometrical model is determined after the fact, as involving the smaller corrections. If luck holds, this model continues to be helpful even when corrections are omitted. In other words, effective dimension inevitably has a subjective basis. It is a matter of approximation and therefore of degree of resolution.

DIFFERENT EFFECTIVE DIMENSIONS IMPLICIT IN A BALL OF THREAD

To confirm this last hunch, a ball of 10 cm diameter made of a thick thread of 1 mm diameter possesses (in latent fashion) several distinct effective dimensions.

To an observer placed far away, the ball appears as a zero-dimensional figure: a point. (Anyhow, it is asserted by Blaise Pascal and by medieval philosophers that on a cosmic scale our whole world is but a point!) As seen from a distance of 10 cm resolution, the ball of thread is a three-dimensional figure. At 10 mm, it is a mess of one-dimensional threads.

At 0.1 mm, each thread becomes a column and the whole becomes a three-dimensional figure again. At 0.01 mm, each column dissolves into fibers, and the ball again becomes one-dimensional, and so on, with the dimension crossing over repeatedly from one value to another. When the ball is represented by a finite number of atomlike pinpoints, it becomes zero-dimensional again. An analogous sequence of dimensions and crossovers is encountered in a sheet of paper.

The notion that a numerical result should depend on the relation of object to observer is in the spirit of physics in this century and is even an exemplary illustration of it.

Most of the objects considered in this Essay are like our ball of thread: they exhibit a succession of different effective dimensions. But a vital new element is added: certain ill-defined transitions between zones of well-defined dimension are reinterpreted as being fractal zones within which $D > D_T$.

SPATIAL HOMOGENEITY, SCALING, AND SELF–SIMILARITY

Having finished with dimensions for the time being, let us prepare for the theme of symmetry by recalling that Euclid begins with the simplest shapes, such as lines, planes, or spaces. And the simplest physics arises when some quantity such as density, temperature, pressure, or velocity is distributed in a homogeneous manner.

The homogeneous distribution on a line, plane, or space has two very desirable properties. It is *invariant under displacement*, and it is *invariant under change of scale*. When we move on to fractals, either invariance *must* be modified and/or restricted in its scope. Hence, the best fractals are those that exhibit the maximum of invariance.

Concerning displacement: different parts of the trail of Brownian motion can never be precisely superposed on each other—as can be done with equal parts of a straight line. Nevertheless, the parts can be made to be superposable in a statistical sense. Nearly all the fractals in the present Essay are to some extent invariant under displacement.

Furthermore, most fractals in this Essay are invariant under certain transformations of scale. They are called *scaling*. A fractal invariant under ordinary geometric similarity is called *self-similar*.

In the compound term *scaling fractals*, the adjective serves to mitigate the noun. While the primary term *fractal* points to disorder and covers cases of intractable irregularity, the modifier *scaling* points to a kind of order. Alternatively, taking *scaling* as the primary term pointing to strict order, *fractal* is a modifier meant to exclude lines and planes.

The motivation for assuming homogeneity and scaling must not be misinterpreted. Here as in standard geometry of nature, no one believes that the world is strictly homogeneous or scaling. Standard geometry investigates straight lines as a preliminary. And mechanics also views uniform rectilinear motion as merely a first step.

The same is true of the study of scaling fractals, but the first step takes much longer in this case because the straight line is replaced by a wealth of diverse possibilities, which this book can merely sample. One should not be surprised that scaling fractals should be limited to providing first approximations of the natural shapes to be tackled. One must rather marvel that these first approximations are so strikingly reasonable.

It is good to point out that self-similarity is an old idea. In the case of the line, it occurred to Leibniz circa 1700 (see under SCALING IN LEIBNIZ AND LAPLACE in Chapter 41). And its generalization beyond lines and planes is almost a hundred years old in mathematics, though its concrete importance was not appreciated until this Essay. Also, it is not new in science, since Lewis F. Richardson postulated in 1926 that over a wide range of scales turbulence is decomposable into self-similar eddies. Furthermore, striking *analytical* consequences of this idea in mechanics are drawn in Kolmogorov 1941. And the analytic aspects of scaling in physics are associated with the notion of renormalization group, Chapter 36.

However, this Essay's 1975 predecessor was the first to address itself to the *geometric* aspects of nonstandard scaling in *Nature*.

"SYMMETRIES" BEYOND SCALING

After it finishes with lines, Euclid tackles shapes with richer properties of invariance, usually called "symmetries." And this Essay also makes a fairly lengthy excursion into nonscaling fractals, in Chapters 15 to 20.

Self-mapping but nonscaling fractals are intimately linked with some of the most refined and difficult areas of "hard" classical mathematical analysis. Contrary to rumors that analysis is a dry subject, these fractals tend to be astoundingly beautiful.

DIVERGENCE SYNDROMES

Almost every case study we perform involves a divergence syndrome. That is, some quantity that is commonly expected to be positive and finite turns out either to be infinite or to vanish. At first blush, such misbehavior looks most bizarre and even terrifying, but a careful reexamination shows it to be quite acceptable..., as long as one is willing to use new methods of thought.

Cases where a symmetry is accompanied by a divergence are also a familiar fixture of quantum physics, within which diverse divergence eliminating arguments take a place of honor. Luckily, the various fractal divergences are much easier to handle. ► ■

4 ¤ Variations and Disclaimers

Now that the diverse objectives of this Essay are outlined, we examine its manner. It too attempts to integrate several distinct facets.

OBSCURITY IS NOT A VIRTUE

To be accessible to scholars and students not necessarily specializing in the various subjects tackled, many of which are esoteric, this work incorporates much exposition.

*But exposition is **not** its principal purpose.*

Further, an attempt is made not to frighten away those who are not interested in mathematical precision, but who ought to be interested in my main conclusions. Rigorous mathematical backup is available throughout (and is sounder than in much of physics), but the book's style is informal (though precise). All detail is set aside to Chapter 39, to the references, and to diverse works to come.

Since original work is not expected to show such concerns, this Essay is to some extent a *work of popularization.*

*But popularization is **not** its main purpose.*

ERUDITION IS GOOD FOR THE SOUL

As exemplified in Chapter 2, this Essay includes many old and obscure references. Most did not attract my attention until well after my own work in related areas was essentially complete. They did not influence my thinking. However, during the long years when my interests were not shared by anyone, I rejoiced in discovering analogous concerns in ancient works, however fleetingly and ineffectually expressed, witness their failure to be developed. In this fashion, an interest in "classics," which the usual practice of science destroys, was nurtured in my case.

In other words, I rejoiced in finding that the stones I needed—as the architect and builder of the theory of fractals—included many that had been considered by others. But why continue to dwell on this fact today? Casual footnotes would satisfy the prevailing custom, while an excessive stress on distant roots or origins risks fostering the absurd impression that my building is largely a pile of old stones with new names on them.

Thus, my antiquarian curiosity would require a justification, but I shall not attempt one. It is enough to say that, in my opinion, an interest in the history of ideas is good for the scientist's soul.

However, whenever we read a great man's writings in a light with which he was not blessed, we may ponder the delightful preface Lebesgue wrote to a book by Lusin. He disclaimed many profound thoughts with which said book credited him, saying he might have, or should have, had these thoughts, but had not, and that they originated with Lusin. A related item is Whittaker 1953, wherein quotes from Poincaré and Lorentz are marshalled in favor of a thesis both had pointedly disclaimed: that the physical theory of relativity was their creation and not Einstein's.

Also, for each author jotting down years ago an idea which we can now develop but he did not, we run the risk of finding a second author to declare that the idea is absurd. And should we credit the young Henri Poincaré with ideas he failed to develop, and the mature Henri Poincaré rejected? Stent 1972 might lead us to the conclusion that prematurity, being too much ahead of one's time, deserves nothing but compassionate oblivion.

While excessive erudition in relation to the history of ideas is self-defeating. I do wish to assert the echoes from the past, stressing them further in the biographical and historical sketches in Chapters 40 and 41.

*Yet, a display of erudition is certainly **not** the main purpose of this Essay.*

"TO SEE IS TO BELIEVE"

In a letter to Dedekind, at the very beginning of the 1875–1925 crisis in mathematics, Cantor is overwhelmed by amazement at his own findings, and slips from German to French to exclaim that "to see is not to believe" ("Je le vois, mais je ne le crois pas"). And, as if on cue, mathematics seeks to avoid being misled by the graven images of monsters. What a contrast between the rococo exuberance of pre- or counterrevolutionary geometry, and the near-total visual barrenness of the works of Weierstrass, Cantor, or Peano! In physics, an analogous development threatened since about 1800, since Laplace's *Celestial Mechanics* avoided all illustration. And it is exemplified by the statement by P. A. M. Dirac (in the preface of his 1930 *Quantum Mechanics*) that nature's "fundamental laws do not govern the world as it appears in our mental picture in any very direct way, but instead they control a substratum of which we cannot form a mental picture without introducing irrelevancies."

The wide and uncritical acceptance of this view has become destructive. In particular, in the theory of fractals "to see *is* to believe." Therefore, before he proceeds further, the reader is again advised to browse through my picture book. This Essay was designed to help make its contents accessible in various degrees to a wide range of readers, and to try and convince even the purest among mathematicians that the understanding of known concepts and the search for new concepts and conjectures are both helped by fine graphics. Rarely does contemporary scientific literature show such trust in the usefulness of graphics.

*However, showing pretty pictures is **not** the main purpose in this Essay; they are an essential tool, but only a tool.*

One must also recognize that any attempt to illustrate geometry involves a basic fallacy. For example, a straight line is unbounded and infinitely thin and smooth, while any illustration is unavoidably of finite length, of positive thickness, and rough edged. Nevertheless, a rough evocative drawing of a line is felt by many to be useful, and by some to be necessary, to develop intuition and help in the search for proof. And a rough drawing is a more adequate geometric model of a thread than the mathematical line itself. In other words, it suffices for all practical purposes that a geometric concept and its image should fit within a certain range of characteristic sizes, ranging between a large but finite size to be called outer cutoff and a small but positive inner cutoff.

Today, thanks to computer-controlled graphics, the same kind of evocative illustration is practical in the case of fractals. For example, all self-similar fractal curves are also unbounded and infinitely thin. Also, each has a very specific lack of smoothness, which makes it more complicated than anything in Euclid. The best representation, therefore, can only hold within a limited range, on the principles we have already encountered. However, cutting off the very large and the very small detail is not only quite acceptable but even

eminently appropriate, because both cutoffs are either present or suspected in Nature. Thus the typical fractal curve can be evoked satisfactorily by elementary strokes in large but finite number.

The larger the number of strokes and the greater the accuracy of the process, the more useful the representation, because fractal concepts refer to the mutual placement of strokes in space, and it is vital in illustrating them to keep to precise scale. Hand drawing would be prohibitive, but computer graphics serves beautifully. My successive Essays have been very much influenced by the availability of increasingly sophisticated systems—and of increasingly sophisticated programmer-artists to run them! Also, I am fortunate in having access to a device that produces camera ready illustrations. This Essay provides a sample of its output.

Graphics is wonderful for matching models with reality. When a chance mechanism agrees with the data from some analytic viewpoint but simulations of the model do not look at all "real," the analytic agreement should be suspect. A formula can relate to only a small aspect of the relationship between model and reality, while the eye has enormous powers of integration and discrimination. True, the eye sometimes sees spurious relationships which statistical analysis later negates, but this problem arises mostly in areas of science where samples are very small. In the areas we shall explore, samples are huge.

In addition, graphics helps find new uses for existing models. I first experienced this possibility with the random walk illustration in Feller 1950—the curve looked like a mountain's profile or cross section, and the points where it intersects the time axis reminded me of certain records I was then investigating, relative to telephone errors. The ensuing hunches eventually led to the theories presented in Chapters 28 and 31, respectively. My own computer-generated illustrations provided similar inspiration, both to me and to others kind enough to "scout" for me in more sciences than I knew existed.

Naturally, graphics is extended by cinematography: films concerned with some classical fractals have been provided by Max 1971.

THE STANDARD FORM, AND THE NEW FRACTAL FORM, OF GEOMETRIC "ART"

As to this book's endpapers and diverse patterns scattered around, they were the unintended result of faulty computer programming. I hear and read of both the intended and the unintended illustrations being described as a "New Form of Art."

Clearly, competing with artists is not at all a purpose of this Essay. Nevertheless, one must address this issue. The question is *not* whether the illustrations are neatly drawn and printed, and the originals being drawn by computer is not essential either, except in terms of economics. But we do deal with a new form of the controversial but ancient theme that all graphical representations of mathematical concepts are a form of art, one that is best when it is simplest, when (to borrow a painter's term) it can be called "minimal art."

It is widely held that minimal art is restricted to limited combinations of standard shapes: lines, circles, spirals, and the like. But such need not be the case. The fractals used in scientific models are also very simple (because science puts a premium on simplicity). And I agree that many may be viewed as a new form of minimal geometric art.

Is some of it reminiscent of M. C. Escher? It should be, because Escher had the merit of letting himself be inspired by the hyperbolic tilings in Fricke & Klein 1897, which (see Chapter 18) relate closely to shapes that are being incorporated into the fractal realm.

The fractal "new geometric art" shows surprising kinship to Grand Masters paintings or Beaux Arts architecture. An obvious reason is that classical visual arts, like fractals, involve very many scales of length and favor self-similarity (Mandelbrot 1981l). For all these reasons, and also because it came in through an effort to imitate Nature in order to guess its laws, it may well be that fractal art is readily accepted because it is not truly unfamiliar. Abstract paintings vary on this account: those I like also tend to be close to fractal geometric art, but many are closer to standard geometric art—too close for my own comfort and enjoyment.

A paradox emerges here: As observed in Dyson's quote in Chapter 1, modern mathematics, music, painting, and architecture may seem to be related to one another. But this is a superficial impression, notably in the context of architecture: A Mies van der Rohe building is a scalebound throwback to Euclid, while a high period Beaux Arts building is rich in fractal aspects.

POINTS OF LOGISTICS

Successive chapters take up diverse topics by increasing complexity, in order to introduce the basic ideas gradually. The fact that this approach seems feasible is a great asset for the theory of fractals. The amount of built-in repetition is such that the reader is unlikely to lose the main thrust of the argument if he skips the passages he feels to be either repetitious or too complicated (in particular, those that go beyond the most elementary mathematics). Much information is included in the captions of the plates.

As already mentioned, the plates are grouped after the chapters where they are first examined. Also this writer feels every so often the need to engage in private conversation, so to speak, with specific groups of readers who might be overly troubled if some point were left unmentioned or unexplained. The digressions are left in the text but marked by the newfangled brackets ◁ and ▶, which should make them easier to skip. Other digressions are devoted to incidental remarks I have no time to explore fully. But this Essay is less digressive than the 1977 *Fractals*.

An attempt is made to show at a glance whether the discussion is concerned with theoretical or empirical dimensions D. The latter are mostly known to one or two decimals, and are therefore written as 1.2 or 1.37. The former are written as integers, ratios of integers, ratios of logarithms of integers, or in decimal form to at least *four* decimals.

BACK TO THE BASIC THEME

Having disclaimed diverse goals that are peripheral to this Essay, let me echo Chapter 1. This work is *a manifesto and a casebook*, devoted nearly exclusively to theories and theses which I initiated but which often led to the revival and the reinterpretation of diverse old works.

None of these theories stopped growing, and a few are still at the seed stage. Some are published here for the first time, while others had been described in my earlier articles. In addition, I mention numerous developments my earlier Essays had inspired, and which in turn stimulated me. However, I do not attempt to list all the fields where fractals prove useful, for fear of destroying the style of an Essay and the flavor of a manifesto.

Last reminder: I do not propose to develop any case study in the full detail desired by the specialists. But many topics are touched upon repeatedly; don't forget to use the index.   ■■

5 ¤ How Long Is the Coast of Britain?

To introduce a first category of fractals, namely curves whose fractal dimension is greater than 1, consider a stretch of coastline. It is evident that its length is at least equal to the distance measured along a straight line between its beginning and its end. However, the typical coastline is irregular and winding, and there is no question it is much longer than the straight line between its end points.

There are various ways of evaluating its length more accurately, and this chapter analyzes several of them. The result is most peculiar: coastline length turns out to be an elusive notion that slips between the fingers of one who wants to grasp it. All measurement methods ultimately lead to the conclusion that the typical coastline's length is very large and so ill determined that it is best considered infinite. Hence, if one wishes to compare different coastlines from the viewpoint of their "extent," length is an inadequate concept.

This chapter seeks an improved substitute, and in doing so finds it impossible to avoid introducing various forms of the fractal concepts of dimension, measure, and curve.

MULTIPLICITY OF ALTERNATIVE METHODS OF MEASUREMENT

METHOD A: Set dividers to a prescribed opening ϵ, to be called the yardstick length, and walk these dividers along the coastline, each new step starting where the previous step leaves off. The number of steps multiplied by ϵ is an approximate length $L(\epsilon)$. As the dividers' opening becomes smaller and smaller, and as we repeat the operation, we have been taught to expect $L(\epsilon)$ to settle rapidly to a well-defined value called the *true length*. But in fact what we expect does not happen. In the typical case, the observed $L(\epsilon)$ tends to increase without limit.

The reason for this behavior is obvious:

When a bay or peninsula noticed on a map scaled to 1/100,000 is reexamined on a map at 1/10,000, subbays and subpeninsulas become visible. On a 1/1,000 scale map, sub-subbays and sub-subpeninsulas appear, and so forth. Each adds to the measured length.

Our procedure acknowledges that a coastline is too irregular to be measured directly by reading it off in a catalog of lengths of simple geometric curves. Therefore, METHOD A replaces the coastline by a sequence of broken lines made of straight intervals, which are curves we know how to handle.

METHOD B: Such "smoothing out" can also be accomplished in other ways. Imagine a man walking along the coastline, taking the shortest path that stays no farther from the water than the prescribed distance ϵ. Then he resumes his walk after reducing his yardstick, then again, after another reduction; and so on, until ϵ reaches, say, 50 cm. Man is too big and clumsy to follow any finer detail. One may further argue that this unreachable fine detail (a) is of no direct interest to Man and (b) varies with the seasons and the tides so much that it is altogether meaningless. We take up argument (a) later on in this chapter. In the meantime, we can neutralize argument (b) by restricting our attention to a rocky coastline observed when the tide is low and the waves are negligible. In principle, Man could follow such a curve down to finer details by harnessing a mouse, then an ant, and so forth. Again, as our walker stays increasingly closer to the coastline, the distance to be covered continues to increase with no limit.

METHOD C: An asymmetry between land and water is implied in METHOD B. To avoid it, Cantor suggests, in effect, that one should view the coastline with an out-of-focus camera that transforms every point into a circular blotch of radius ϵ. In other words, Cantor considers all the points of both land and water for which the distance to the coastline is no more than ϵ. These points form a kind of sausage or tape of width 2ϵ, as seen in a different context on Plate 32. Measure the area of the tape and divide it by 2ϵ. If the coastline were straight, the tape would be a rectangle, and the above quotient would be the actual length. With actual coastlines, we have an estimated length $L(\epsilon)$. As ϵ decreases, this estimate increases without limit.

METHOD D: Imagine a map drawn in the manner of pointillist painters using circular blotches of radius ϵ. Instead of using circles centered on the coastline, as in METHOD C, let us require that the blotches that cover the entire coastline be as few in number as possible. As a result, they may well lie mostly inland near the capes and mostly in the sea near the bays. Such a map's area, divided by 2ϵ, is an estimate of the length. This estimate also "misbehaves."

ARBITRARINESS OF
THE RESULTS OF MEASUREMENT

To summarize the preceding section, the main finding is always the same. As ϵ is made smaller and smaller, every approximate length tends to increase steadily without bound.

In order to ascertain the meaning of this result, let us perform analogous measurements on a standard curve from Euclid. For an interval of straight line, the approximate measurements are essentially identical and define the length. For a circle, the approximate measurements increase but converge rapidly to a limit. The curves for which a length is thus defined are called *rectifiable*.

An even more interesting contrast is provided by the results of measurement on a coastline that Man has tamed, say the coast at Chelsea as it is today. Since very large features are unaffected by Man, a very large yardstick again yields results that increase as ε decreases.

However, there is an intermediate zone of ε's in which L(ε) varies little. This zone may go from 20 meters down to 20 centimeters (but do not take these values too strictly). But L(ε) increases again after ε becomes less than 20 centimeters and measurements become affected by the irregularity of the stones. Thus, if we trace the curves representing L(ε) as a function of ε, there is little doubt that the length exhibits, in the zone of ε's between ε=20 meters and ε=20 centimeters, a flat portion that was not observable before the coast was tamed.

Measurements made in this zone are obviously of great practical use. Since boundaries between different scientific disciplines are largely a matter of conventional division of labor between scientists, one might restrict geography to phenomena above Man's reach, for example, on scales above 20 meters. This restriction would yield a well-defined value of geographical length. The Coast Guard may well choose to use the same ε for untamed coasts, and encyclopedias and almanacs could adopt the corresponding L(ε).

However, the adoption of the same ε by all the agencies of a government is hard to imagine, and its adoption by all countries is all but inconceivable. For example, Richardson 1961, the lengths of the common frontiers between Spain and Portugal, or Belgium and Netherlands, as reported in these neighbors' encyclopedias, differ by 20%. The discrepancy must in part result from different choices of ε. An empirical finding to be discussed soon shows that it suffices that the ε differ by a factor of 2, and one should not be surprised that a small country (Portugal) measures its borders more accurately than its big neighbor.

The second and more significant reason against deciding on an arbitrary ε is philosophical and scientific. Nature does exist apart from Man, and anyone who gives too much weight to any specific ε and L(ε) lets the study of Nature be dominated by Man, either through his typical yardstick size or his highly variable technical reach. If coastlines are ever to become an object of scientific inquiry, the uncertainty concerning their lengths cannot be legislated away. In one manner or another, the concept of geographic length is not as inoffensive as it seems. It is not entirely "objective." The observer inevitably intervenes in its definition.

IS THIS ARBITRARINESS GENERALLY RECOGNIZED, AND DOES IT MATTER?

The view that coastline lengths are nonrectifiable is doubtless held true by many people, and I for one do not recall ever thinking otherwise. But my search for written statements to this effect is a near fiasco. Aside from the Perrin quote in Chapter 2, there is the observation in Steinhaus 1954 that "the left bank of the Vistula, when measured with increasing precision, would furnish lengths ten, hundred or even thousand times as great as the length read off the school map...[A] statement nearly approaching reality would be to call most arcs encountered in nature nonrectifiable. This statement is contrary to the belief that nonrectifiable arcs are an invention of mathematicians and that natural arcs are rectifiable: it is the opposite that is true." But neither Perrin nor Steinhaus follow up on this insight.

Let me also retell a story reported by C. Fadiman. His friend Edward Kasner would ask small tots "to guess the length of the eastern coast line of the United States. After a 'sensible' guess had been made...he would...point out that this figure increased enormously if you measured the perimeter of each bay and inlet, then that of every projection and curve of each of these, then the distance separating every small particle of coastline matter, each molecule, atom, etc. Obviously the coast line is as long as you want to make it. The children understood this at once; Kasner had more trouble with grownups." The story is nice, but it is not relevant here:

Kasner's goal was *not* to point out an aspect of Nature worthy of further exploration.

Therefore, Mandelbrot 1967s and the present Essay are effectively the first works on this subject.

One is reminded of William James writing in *The Will to Believe* that "The great field for new discoveries...is always the unclassified residuum. Round about the accredited and orderly facts of every science there ever floats a sort of dust-cloud of exceptional observations, of occurrences minute and irregular and seldom met with, which it always proves more easy to ignore than to attend to. The ideal of every science is that of a closed and completed system of truth... Phenomena unclassifiable within the system are paradoxical absurdities, and must be held untrue...—one neglects or denies them with the best of scientific consciences... Any one will renovate his science who will steadily look after the irregular phenomena. And when the science is renewed, its new formulas often have more of the voice of the exception in them than of what were supposed to be the rules."

This Essay, whose ambition is indeed to renew the Geometry of Nature, relies upon many puzzles so unclassified that they are only published when the censors nod. The next section discusses a first example.

THE RICHARDSON EFFECT

The variation of the approximate length $L(\epsilon)$ obtained by Method A has been studied empirically in Richardson 1961, a reference that chance (or fate) put in my way. I paid attention because (Chapter 40) I knew of Lewis Fry Richardson as a great scientist whose originality mixed with eccentricity. As we shall learn in Chapter 10, we are indebted to him for some of the most profound and most durable ideas regarding the nature of turbulence, notably the notion that turbulence involves a self-similar cascade. He also concerned himself with other difficult problems, such as the nature of armed conflict between states. His experiments were of classic simplicity, but he never hesitated to use refined concepts when he deemed them necessary.

The diagrams reproduced in Plate 33, found among his papers after he died, were published in a near confidential (and totally inappropriate) *Yearbook*. They all lead to the conclusion that there are two constants, which we shall call λ and D, such that—to approximate a coastline by a broken line—one needs roughly $F\epsilon^{-D}$ intervals of length ϵ, adding up to the length

$$L(\epsilon) \sim F\epsilon^{1-D}.$$

The value of the exponent D seems to depend upon the coastline that is chosen, and different pieces of the same coastline, if considered separately, may produce different values of D. To Richardson, the D in question was a simple exponent of no particular significance. However, its value seems to be independent of the method chosen to estimate the length of a coastline. Thus D seems to warrant attention.

A COASTLINE'S FRACTAL DIMENSION (MANDELBROT 1967s)

Having unearthed Richardson's work, I proposed (Mandelbrot 1967s) that, despite the fact that the exponent D is not an integer, it can and should be interpreted as a dimension, namely, as a fractal dimension. Indeed, I recognized that all the above listed methods of measuring $L(\epsilon)$ correspond to nonstandard generalized definitions of dimension already used in pure mathematics. The definition of length based on the coastline being covered by the smallest number of blotches of radius ϵ is used in Pontrjagin & Schnirelman 1932 to define the covering dimension. The definition of length based on the coastline being covered by a tape of width 2ϵ implements an idea of Cantor and Minkowski (Plate 32), and the corresponding dimension is due to Bouligand. Yet these two examples only hint at the many dimensions (most of them known only to a few specialists) that star in diverse specialized chapters of mathematics. A certain number of them are discussed further in Chapter 39.

Why did mathematicians introduce this plethora of distinct definitions? Because in some cases they yield distinct values. Luckily, however, such cases are never encountered in this Essay, and the list of possible alternative dimensions can be reduced to two that I have not yet mentioned. The older and best investigated one dates back to Hausdorff and serves to define fractal dimension; we come to it momentarily. The simpler one is similarity dimension: it is less general, but in many cases is more than adequate; it is explored in the following chapter.

Clearly, I do not propose to present a mathematical proof that Richardson's D is a dimension. No such proof is conceivable in any natural science. The goal is merely to convince the reader that the notion of length poses a conceptual problem, and that D provides a manageable and convenient answer. Now that fractal dimension is injected into the study of coastlines, even if specific reasons come to be challenged, I think we shall never return to the stage when D=1 was accepted thoughtlessly and naively. He who continues to think that D=1 has to argue his case.

The next step, to explain the shape of the coastlines and to deduce the value of D from other more basic considerations, is put off until Chapter 28. Suffice at this point to announce that to a first approximation D=3/2. This value is much too large to describe the facts but more than sufficient to establish that it is natural, proper, and expected for a coastline's dimension to exceed the standard Euclidean value D=1.

HAUSDORFF FRACTAL DIMENSION

If we accept that various natural coasts are really of infinite length and that the length based on an anthropocentric value of ϵ gives only a partial idea of reality, how can different coastlines be compared to each other? Since infinity equals four times infinity, every coastline is four times longer than each of its quarters, but this is not a useful conclusion. We need a better way to express the sound idea that the entire curve must have a "measure" that is four times greater than each of its fourths.

A most ingenious method of reaching this goal has been provided by Felix Hausdorff. It is intuitively motivated by the fact that the linear measure of a polygon is calculated by adding its sides' lengths without transforming them in any way. One may say (the reason for doing so will soon become apparent) that these lengths are raised to the power D=1, the Euclidean dimension of a straight line. The surface measure of a closed polygon's interior is similarly calculated by paving it with squares, and adding the squares' sides raised to the power D=2, the Euclidean dimension of a plane. When, on the other hand, the "wrong" power is used, the result gives no specific information: the area of every closed polygon is zero, and the length of its interior is infinite.

Let us proceed likewise for a polygonal approximation of a coastline made up of small intervals of length ϵ. If their lengths are raised to the power D, we obtain a quantity we may call tentatively an "approximate measure in the dimension D." Since according to Richardson the number of sides is $N = F\epsilon^{-D}$, said approximate measure takes the value $F\epsilon^{D}\epsilon^{-D} = F$.

Thus, *the approximate measure in the dimension D is independent of* ϵ. With actual data, we simply find that this approximate measure varies little with ϵ.

In addition, the fact that the length of a square is infinite has a simple counterpart and generalization: a coastline's approximate measure evaluated in any dimension d smaller than D tends to ∞ as $\epsilon \to 0$. Similarly, the area and the volume of a straight line are zero. And when d takes any value larger than D, the corresponding approximate measure of a coastline tends to 0 as $\epsilon \to 0$. The approximate measure behaves reasonably if and only if d=D.

A CURVE'S FRACTAL DIMENSION MAY EXCEED 1; FRACTAL CURVES

By design, the Hausdorff dimension preserves the ordinary dimension's role as exponent in defining a *measure*.

But from another viewpoint, D is very odd indeed: it is a fraction! In particular, it exceeds 1, which is the intuitive dimension of curves and which may be shown rigorously to be their topological dimension D_T.

I propose that curves for which the fractal dimension exceeds the topological dimension 1 be called *fractal curves*. And the present chapter can be summarized by asserting that, within the scales of interest to the geographer, coastlines can be modeled by fractal curves. Coastlines are *fractal patterns*. ▬

Plate 31 ⬚ **MONKEYS TREE**

At this point, the present small incidental plate should be viewed as merely a decorative drawing, filling a gap.

However, when the reader has finished Chapter 14, he will find in this drawing a hint to help unscramble the "architecture" in Plate 146. A more sober hint resides in the following generator.

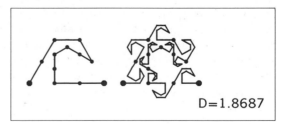

$D=1.8687$

▬

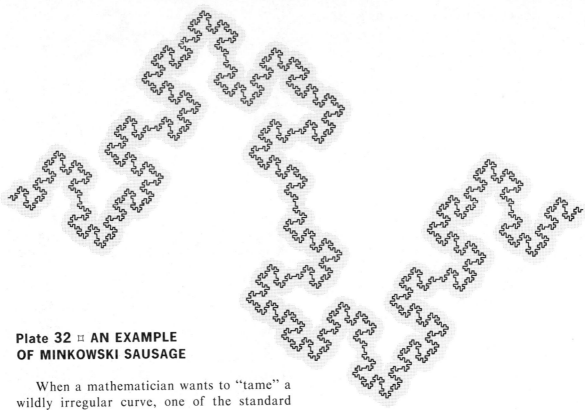

Plate 32 ◻ AN EXAMPLE OF MINKOWSKI SAUSAGE

When a mathematician wants to "tame" a wildly irregular curve, one of the standard procedures is to select a radius ϵ and to draw around each point of the curve a disc of radius ϵ. This procedure, dating back at least to Hermann Minkowski and possibly to Georg Cantor, is brutal but very effective. (As to the term *sausage,* unverifiable rumor claims it is a leftover of an application of this procedure to the Brownian curves of Norbert Wiener.)

In the present illustration such smoothing is not applied to an actual coastline but to a theoretical curve that will be constructed later (Plate 49) by continual addition of ever smaller detail. Comparing the piece of sausage drawn to the right with the rightmost end of the sausage drawn above it, we see that the construction of the curve passes a critical stage when it begins to involve details of size smaller than ϵ. Later stages of construction leave the sausage essentially unaffected. ◼◻

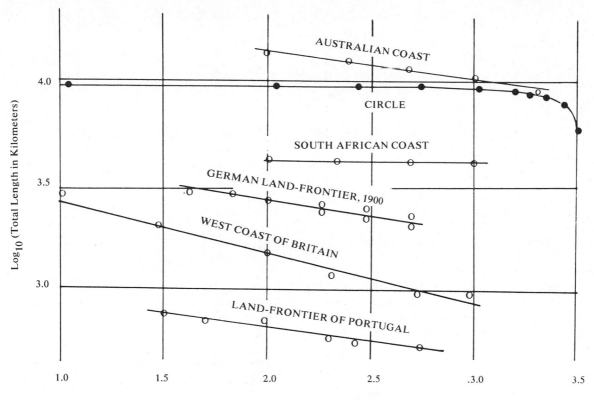

Plate 33 ¤ **RICHARDSON'S EMPIRICAL DATA**
CONCERNING THE RATE OF INCREASE OF COASTLINES' LENGTHS

This Figure reproduces Richardson's experimental measurements of length performed on various curves using equal-sided polygons of increasingly short side ϵ. As expected, increasingly precise measurements made on a circle stabilize very rapidly near a well-determined value.

In the case of coastlines, on the contrary, the approximate lengths do not stabilize at all. As the yardstick length ϵ tends to zero, the approximate lengths, as plotted on doubly logarithmic paper, fall on a straight line of negative slope. The same is true of boundaries between countries. Richardson's search in en-

cyclopedias reveals notable differences in the lengths of the common land frontiers claimed by Spain and Portugal (987 versus 1214 km), and by the Netherlands and Belgium (380 versus 449 km). With a slope of −0.25, the 20% differences between these claims can be accounted for by assuming that the ϵ's differ by a factor of 2, which is not unlikely.

To Richardson, his lines' slopes had no theoretical interpretation. The present Essay, on the other hand, interprets coastlines as approximate fractal curves, and uses the slope of each line as an estimate of 1−D, where D is the fractal dimension. ■

6 ⌗ Snowflakes and Other Koch Curves

In order to understand fully my interpretation of Richardson's D as a fractal dimension, we move from natural phenomena over which we have no control, to geometric constructs we can design at will.

SELF-SIMILARITY AND CASCADES

Until now we stressed that coastlines' geometry is complicated, but there is also a great degree of order in their structure.

Although maps drawn at different scales differ in their specific details, they have the same generic features. In a rough approximation, the small and large details of coastlines are geometrically identical except for scale.

One may think of such a shape as drawn by a sort of fireworks, with each stage creating details smaller than those of the preceding stages. However, a better term is suggested by our Lewis Richardson's noted work on turbulence: the generating mechanism may be called a *cascade*.

When each piece of a shape is geometrically similar to the whole, both the shape and the cascade that generate it are called *self-similar*. This chapter probes self-similarity using very regular figures.

The most extreme contrasts to self-similar shapes are provided by curves that (a) have a single scale, like the circle, or (b) have two clearly separated scales, like a circle adorned with "scallops." Such shapes can be described as *scalebound*.

COASTLIKE TERAGONS AND THE TRIADIC KOCH CURVE $\mathcal{K}$

To insure that an infinite number of scales of length are present in a curve, the safest is to put them in deliberately and separately. A regular triangle of side 1 has a single scale, triangles of side ⅓ have a smaller scale, and triangles of side $(⅓)^k$ are of increasingly small scale. And by piling these triangles on top of each other, as in Plate 42, one is left with a shape combining all scales below 1.

In effect, we assume that a bit of coastline drawn to a scale of 1/1,000,000 is a straight interval of length 1, to be called *initiator*.

Then we assume that the detail that becomes visible on a map at 3/1,000,000 replaces the earlier interval's middle third by a promontory in the shape of an equilateral triangle. The resulting second approximation is an broken line formed of four intervals of equal lengths, to be called *generator*. We further assume that the new detail that appears at 9/1,000,000 results from the replacement of each of the generator's four intervals by the generator reduced in a ratio of one-third, forming subpromontories.

Proceeding in this fashion, we break each straight line interval, replacing the initiator by an increasing broken curve. Since we deal with them throughout this Essay, let me coin for such curves the term *teragon*, from the Greek $\tau\epsilon\rho\alpha\varsigma$, meaning "monster, strange creature," and $\gamma\omega\nu\iota\alpha$, meaning "corner, angle." Very appropriately, the metric system uses *tera* as prefix for the factor 10^{12}.

And, if the same cascade process is made to continue to infinity, our teragons converge to a limit first considered by von Koch 1904, Plate 45. We must be specific, and shall call it the triadic Koch curve and denote it by $\mathcal{K}$.

This curve's area vanishes, as is obvious on Plate 43. On the other hand, each stage of construction increases its total length in a ratio of 4/3, hence the limit curve is of infinite length. Furthermore, it is continuous, but it has no definite tangent anywhere—like the graph of a continuous function without a derivative.

As a model of a coastline, $\mathcal{K}$ is only a suggestive approximation, but not because it is too irregular, rather because, in comparison with a coastline, its irregularity is far too systematic. Chapters 24 and 28 "loosen it up" to make it fit better.

THE KOCH CURVE AS MONSTER

As introduced in the preceding section, the Koch curve must seem the most intuitive thing in geometry. But the conventional motivation for it is totally different. So is the conventional attitude towards it on the part of mathematicians. They are all but unanimous in proclaiming that $\mathcal{K}$ is a monstrous curve! For elaboration, let us look up *The Crisis of Intuition*, Hahn 1956, which will serve us repeatedly. We read that "the character of [a *non*rectifiable curve or of a curve without a tangent] entirely eludes intuition; indeed after a few repetitions of the segmenting process the evolving figure has grown so intricate that intuition can scarcely follow; and it forsakes us completely as regards the curve that is approached as a limit. Only thought, or logical analysis, can pursue this strange object to its final form. Thus, had we relied on intuition in this instance, we should have remained in error, for intuition seems to force the conclusion that there cannot be curves lacking a tangent at any point. This first example of the failure of intuition involves the fundamental concepts of differentiation."

The best one can say of these words is that they stop short of a celebrated exclamation of Charles Hermite, writing on May 20, 1893, to

T. Stieltjes of "turning away in fear and horror from this lamentable plague of functions with no derivatives." (Hermite & Stieltjes 1905, II, p. 318.) One likes to believe that great men are perfect, and that Hermite was being ironic, but Lebesgue's 1922 *Notice* (Lebesgue 1972–, I) suggests otherwise. Having written a paper concerned with surfaces devoid of tangent planes, "thoroughly crumpled handkerchiefs," Lebesgue wanted it published by the Académie des Sciences, but "Hermite for a moment opposed its inclusion in the *Comptes Rendus*; this was about the time when he wrote to Stieltjes...."

We recall that Perrin and Steinhaus knew better, but the only mathematician to argue otherwise on the basis of intuition alone (Steinhaus argues on the basis of fact) is Paul Lévy (Lévy 1970): "[I have] always been surprised to hear it said that geometric intuition inevitably leads one to think that all continuous functions are differentiable. From my first encounter with the notion of derivative, my experience proved that the contrary is true."

These voices had not been heard, however. Not only near every book but every science museum proclaims that nondifferentiable curves are counter-intuitive, "monstrous," "pathological," or even "psychopathic."

THE KOCH CURVE, TAMED.
THE DIMENSION D=log 4/log 3=1.2618

I claim that a Koch curve is a rough but vigorous model of a coastline. For a first quantitative test, let us investigate the length $L(\epsilon)$ of the triadic Koch teragon whose sides are of length ϵ. This lengths can be measured exactly, and the result is extraordinarily satisfying:

$$L(\epsilon)=\epsilon^{1-D}.$$

This *exact* formula is identical with Richardson's *empiric* law relative to the coast of Britain. For the triadic Koch curve,

$$D=\log 4/\log 3 \sim 1.2618,$$

hence D lies in the range of values observed by Richardson!

◁ PROOF: Clearly, $L(1)=1$ and

$$L(\epsilon/3)=(4/3)L(\epsilon).$$

This equation has a solution of the form $L(\epsilon)=\epsilon^{1-D}$ if D satisfies

$$3^{D-1}=4/3.$$

Hence $D=\log 4/\log 3$, as asserted. ▶

Naturally, the Koch D is not an empirical but a mathematical constant. Therefore the argument for calling D a dimension becomes even more persuasive in the case of the Koch curve than in the case of coastlines.

On the other hand, the approximate Hausdorff measure in the dimension D (a notion introduced in the preceding chapter) equals ϵ^D multiplied by the number of legs of length ϵ, that is, equals $\epsilon^D \cdot \epsilon^{-D}=1$. This is a good indication that the Hausdorff dimension is D. Unfortunately, the Hausdorff definition is disappointingly difficult to handle rigorously. Moreover, even if it had been easy to handle, the generalization of dimension beyond integers is so far-reaching an idea that one should welcome further motivation for it.

THE SIMILARITY DIMENSION

It happens that in the case of self-similar shapes a very easy further motivation is available in the notion of *similarity dimension*. One often *hears* mathematicians use the similarity dimension to guess the Hausdorff dimension, and the bulk of the present Essay encounters only cases where this guess is correct. In their context, there can be no harm in thinking of fractal dimension as being synonymous with similarity dimension. ◁ We have here a counterpart to the use of topological dimension as synonymous with "intuitive" dimension. ▶

As a motivating prelude, let us examine the standard self-similar shapes: intervals in the line, rectangles in the plane, and the like; see Plate 44. Because a straight line's Euclidean dimension is 1, it follows for every integer "base" b that the "whole" interval $0 \leq x < X$ may be "paved" (each point being covered once and only once) by $N=b$ "parts." These "parts" are the intervals $(k-1)X/b \leq x < kX/b$, where k goes from 1 to b. Each part can be deduced from the whole by a similarity of ratio $r(N)=1/b=1/N$.

Likewise, because a plane's Euclidean dimension is 2, it follows that whatever the value of b, the "whole" made up of a rectangle $0 \leq x < X$; $0 \leq y < Y$ can be "paved" exactly by $N=b^2$ parts. These parts are rectangles defined by the combined inequalities

$$(k-1)X/b \leq x < kX/b,$$

$$\text{and } (h-1)Y/b \leq y < hY/b,$$

wherein k and h go from 1 to b. Each part can now be deduced from the whole by a similarity of ratio $r(N)=1/b=1/N^{1/2}$.

For a rectangular parallelepiped, the same argument gives us $r(N)=1/N^{1/3}$.

And there is no problem in defining spaces whose Euclidean dimension is $E > 3$. (The Euclidean—or Cartesian—dimension is denoted by E in this book.) All D-dimensional parallelepipeds defined for $D \leq E$ satisfy

$$r(N)=1/N^{1/D}.$$

Thus,

$$Nr^D=1.$$

Equivalent alternative expressions are

$$\log r(N)=\log(1/N^{1/D})=-(\log N)/D,$$

$$D=-\log N/\log r(N)=\log N/\log(1/r).$$

Now let us move on to nonstandard shapes. In order for the exponent of self-similarity to have formal meaning, the sole requirement is that the shape be self-similar, i.e., that the whole may be split up into N parts, obtainable from it by a similarity of ratio r (followed by displacement or by symmetry). The D obtained in this fashion always satisfies

$$0 \leq D \leq E.$$

In the example of the triadic Koch curve, N=4 and r=⅓, hence D=log 4/log 3, identical to the Hausdorff dimension.

CURVES; TOPOLOGICAL DIMENSION

Thus far, we have been casual in calling Koch's $\mathcal{K}$ a curve, but we must return to this notion. Intuitively, a standard arc is a connected set that becomes disconnected if any single point is removed. And a closed standard curve is a connected set that separates into standard arcs if 2 points are removed. For the same reason, Koch's $\mathcal{K}$ is a curve.

The mathematician says that all the shapes with the above property, e.g., $\mathcal{K}$, [0,1] or a circle, are of topological dimension $D_T=1$. Thus, yet another notion of dimension has to be considered! Being disciples of William of Ockham, all scientists know that "entities must not be multiplied beyond necessity." It must therefore be confessed that our switching back and forth between several near equivalent forms of fractal dimension is a matter of convenience. However, the coexistence of a fractal and a topological dimension *is a matter of necessity*. Readers who skipped the digressive definition of fractal in Chapter 3 are advised to scan it now, and everyone is advised to read the entry devoted to DIMENSION in Chapter 41.

INTUITIVE MEANING OF D IN THE PRESENCE OF CUTOFFS Λ AND λ

Cesàro 1905 begins with the motto,

The will is infinite
and the execution confined,
the desire is boundless
and the act a slave to limit.

Indeed, limits apply to scientists no less than to Shakespeare's Troilus and Cressida. To obtain a Koch curve, the cascade of smaller and smaller new promontories is pushed to infinity, but in Nature every cascade must stop or change character. While endless promontories may exist, the notion that they are self-similar can only apply between certain limits. Below the lower limit, the concept of coastline ceases to belong to geography.

It is therefore reasonable to view the real coastline as involving two *cutoff scales*. Its *outer cutoff* Ω might be the diameter of the smallest circle encompassing an island, or perhaps a continent, and the *inner cutoff* ϵ might be the 20 meters mentioned in Chapter 5. Actual numerical values are hard to pinpoint, but the need for cutoffs is unquestionable.

Yet, after the very big and the very small details are cut off, D continues to stand for an *effective dimension* as described in Chapter 3.

Strictly speaking, the triangle, the Star of David, and the finite Koch teragons are of dimension 1. However, both intuitively and from the pragmatic point of view of the simplicity and naturalness of the corrective terms required, it is reasonable to consider an advanced Koch teragon as being closer to a curve of dimension $\log 4 / \log 3$ than to a curve of dimension 1.

As for a coastline, it is likely to have several separate dimensions (remember the balls of thread in Chapter 3). Its geographic dimension is Richardson's D. But in the range of sizes of interest in physics, the coastline may have a different dimension—associated with the concept of interface between water, air, and sand.

ALTERNATIVE KOCH GENERATORS AND SELF-AVOIDING KOCH CURVES

Let us restate the basic principle of construction of the triadic Koch curve. One begins with *two shapes,* an *initiator* and a *generator.* The latter is an oriented broken line made up of N equal sides of length r. Thus each stage of the construction begins with a broken line and consists in replacing each straight interval with a copy of the generator, reduced and displaced so as to have the same end points as those of the interval being replaced. In all cases, $D = \log N / \log (1/r)$.

It is easy to change this construction by modifying the generator, in particular by combining promontories with bays, as exem-plified in upcoming plates. In this way we obtain Koch teragons that converge to curves whose dimensions are between 1 and 2.

All these Koch curves are self-avoiding: have no self-intersection. This is why their wholes can be divided into disjoint parts with no ambiguity, in order to define D. However, a Koch construction using carelessly chosen generators risks self-contact or self-intersection, or even self-overlap. When the desired D is small, it is easy to avoid double points by careful choice of the generator. The task becomes increasingly difficult as D increases, but remains possible as long as $D < 2$.

However, any Koch construction that attempts to reach a dimension $D > 2$ leads inevitably to curves that cover the plane infinitely many times. The case $D = 2$ deserves a special discussion to be provided in Chapter 7.

KOCH ARCS AND HALF LINES

In some cases, the term *Koch curve* must be replaced by more precise, and pedantic, terminology. The shape at the bottom of Plate 44 is technically the *Koch map* of a line interval, and can be called a *Koch arc.* Thus the boundary in Plate 45 is made of three Koch arcs. And it is often useful to extrapolate an arc into a *Koch half line:* The extrapolation enlarges the original arc, using its left end point as focus, in the ratio $1/r = 3$, then in the ratio 3^2 and so on. Each successive extrapolate contains the preceding one, and the limit curve contain all the intermediate finite stages.

DEPENDENCE OF MEASURE ON THE RADIUS, WHEN D IS A FRACTION

Let us now extend from Euclidean to fractal dimensions another standard result in Euclid. For idealized physical objects of uniform density ρ, the weight $M(R)$ of a rod of length $2R$, of a disc of radius R or of a ball of radius R is proportional to ρR^E. For $E=1$, 2, and 3, the proportionality constants are respectively equal to 2, 2π, and $4\pi/3$.

The rule $M(R) \propto R^D$ *also applies to fractals when they are self-similar.*

In the triadic Koch case, the proof is easiest when the origin is the end point of a Koch half line. When a circle of radius $R_0 = 3^k$ (with $k \geq 0$) contains the mass $M(R_0)$, the circle of radius $R = R_0/3$ contains the mass $M(R) = M(R_0)/4$. Hence,

$$M(R) = M(R_0)(R/R_0)^D = [M(R_0)R_0^{-D}]R^D.$$

Consequently, the ratio $M(R)/R^D$ is independent of R, and can serve to define a "density" ρ.

KOCH MOTION

Imagine a point moving along a Koch half line, taking equal time to cover arcs of equal measure. If we then invert the function giving time as function of position, we obtain a posi-

tion as function of time, that is, a motion. Of course its velocity is infinite.

PREVIEW OF RANDOM COASTLINES

The Koch curve reminds us of real maps, but has major defects one encounters almost unchanged in the early models of every case study in this Essay. Its parts are identical to each other, and the self-similarity ratio r must be part of a strict scale of the form b^{-k}, where b is an integer, namely, $\frac{1}{3}$, $(\frac{1}{3})^2$, and so on. Thus, a Koch curve is a very preliminary model of a coastline.

I have developed diverse ways of avoiding both defects, but all involve probabilistic complications which are better tackled after we settle many issues concerning nonrandom fractals. However, curious readers familiar with probability may peek ahead to the models based on my "squig curves" (Chapter 24), and, more important, on level curves of fractional Brown surfaces (Chapter 28).

The same method of exposition is followed later in this Part. Numerous patterns of Nature are discussed against the background of systematic fractals that provide a very preliminary model, while the random models I advocate are postponed to later chapters.

REMINDER. In all cases where D is known precisely, is not an integer, and is written in decimal form to enable comparisons, it is carried to *four* decimals. This number 4 is chosen to make obvious that D is *neither* an empirical value (all empirical values are known at present to 1 or 2 decimals), *nor* an incompletely determined geometric value (at present, the latter are known either to 1 or 2 decimals, or to 6 decimals and more.)

COMPLEX, OR SIMPLE AND REGULAR?

Koch curves exhibit a novel and most interesting combination of complexity and simplicity. At first blush, they are enormously more complicated than the standard curves of Euclid. However, the Kolmogorov and Chaitin theory of mathematical algorithms suggests the contrary conclusion, that a Koch curve is *not* significantly more complicated than a circle! This theory starts with a collection of "letters" or "atomic operations," and takes the length of the shortest known algorithm that yields a desired function as an objective upper bound to the function's complexity.

To apply this way of thinking to the construction of curves, let the letters or "atoms" of the graphic process be straight "strokes." In this alphabet, tracing a regular polygon requires a finite number of strokes, each described by a finite number of lines of instruction, hence it is a task of finite complexity. By contrast, a circle involves an "infinite number of infinitely short strokes," hence seems a curve of infinite complexity. However, if the construction of the circle is made to proceed recursively, it is seen to involve only a finite number of instructions, hence to be also a task of finite complexity. For example, starting with a regular polygon of 2^m sides (m > 2),

one replaces each stroke of length $2 \sin(\pi/2^m)$ by two strokes of length $2 \sin(\pi/2^{m+1})$; then the loop starts again. To construct Koch curves, the same approach is used, but with simpler operations, since the stroke length has simply to be multiplied by r, and the replacement strokes' relative positions are the same throughout. Hence this punchline: When complexity is measured by the presently best algorithm's length in this particular alphabet, a *Koch curve is actually simpler than a circle.*

This peculiar ranking of curves by relative simplicity should not be taken seriously. Most notably, the contrary conclusion is reached if the alphabet is based on the compass and ruler—meaning that the circle is relabeled as "atomic." Nevertheless, as long as a sensible alphabet is used, any Koch curve is not only of finite complexity but simpler than most curves in Euclid.

Being fascinated with etymology, I cannot leave this discussion without confessing that I hate to call a Koch curve "irregular." This term is akin to *ruler,* and is satisfactory as long as one keeps to the meaning of *ruler* as an instrument used to trace straight lines: Koch curves are far from straight. But when thinking of a ruler as a king (= *rex,* same Latin root), that is, as one who hands down a set of detailed rules to be followed slavishly, I protest silently that nothing is more "regular" than a Koch curve. ▬

Plate 42 ◻ TRIADIC KOCH ISLAND OR SNOWFLAKE $\mathcal{K}$. ORIGINAL CONSTRUCTION BY HELGE VON KOCH (COASTLINE DIMENSION D=log 4/log 3~1.2618)

The construction begins with an "initiator," namely, a black △ (equilateral triangle) with sides of unit length. Then one pastes upon the midthird of each side a △-shaped peninsula with sides of length ⅓. This second stage ends with a star hexagon, or Star of David. The same process of addition of peninsulas is repeated with the Star's sides, and then again and again, ad infinitum.

Each addition displaces the points in an interval's midthird in a perpendicular direction. The triangular initiator vertices never move. The other 9 vertices of the Star of David achieve their final positions after a finite number of stages. Still other points are displaced without end, but move by decreasing amounts and eventually converge to limits, which define the coastline.

The island itself is the limit of a sequence of domains bounded by polygons, each of which contains the domain bounded by the

preceding polygon. A photographic negative of this limit is part of Plate 45.

Observe that this and many other plates in the book represent islands or lakes rather than coastlines, and in general represent "solid areas" rather than their contours. This method takes fullest advantage of the fine resolution of our graphics system.

WHY A TANGENT CANNOT BE DEFINED HERE. Take as fixed point a vertex of the original △ and draw a cord to a point on the limit coastline. As this point converges clockwise to the vertex, the connecting cord oscillates within a 30° angle, and never tends to a limit one could call a clockwise tangent. The counterclockwise tangent is not defined either. A point where there is no tangent because clockwise and counterclockwise chords oscillate in well-defined angles is called *hyperbolic*. The points that $\mathcal{K}$ attains asymptotically fail to have a tangent for a different reason. ■

Plate 43 ¤ TRIADIC KOCH ISLAND OR SNOWFLAKE 𝒦. ALTERNATIVE CONSTRUCTION BY ERNEST CESÀRO (COASTLINE DIMENSION D=log 4/log 3~1.2618)

An alternative construction of the Koch island is given in Cesàro 1905, a work of such charm as to make me forget the hard search for the original (and the irritation at later finding it reprinted in Cesàro 1964). Here is a free translation of a few ecstatic lines. "This endless imbedding of this shape into itself gives us an idea of what Tennyson describes somewhere as the *inner* infinity, which is after all the only one we could conceive in Nature. Such similarity between the whole and its parts, even its infinitesimal parts, leads us to consider the triadic Koch curve as truly marvelous. Had it been given life, it would not be possible to do away with it without destroying it altogether for it would rise again and again from the depths of its triangles, as life does in the Universe."

Cesàro's initiator is a regular hexagon with sides of length √3/3. The surrounding ocean is in gray. Increasingly small △-shaped bays are squeezed in ad infinitum, the Koch island being the limit of *decreasing* approximations.

This method of construction and Koch's method described in Plate 42 are carried out in parallel in the present plate. In this way, the Koch coastline is squeezed between an inner and an outer teragon that grow increasingly close to each other. One can think of a cascade process starting with three successive rings: solid land (in black), swamp (in white), and water (in gray). Each cascade stage transfers chunks of swamp to either solid land or water. At the limit the swamp exhausts itself from a "surface" down to a curve.

MIDPOINT DISPLACEMENT INTERPRETATION. It involves the following generator and next step (the angle here is 120°)

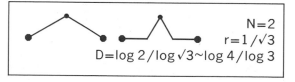

$$N=2$$
$$r=1/\sqrt{3}$$
$$D=\log 2/\log \sqrt{3}\sim\log 4/\log 3$$

When placed outside the inner kth teragon, it yields the outer kth teragon; when placed inside the outer kth teragon, it yields the inner (k+1)st teragon. This approach is useful in Plates 64 and 65, and in Chapter 25. ∎

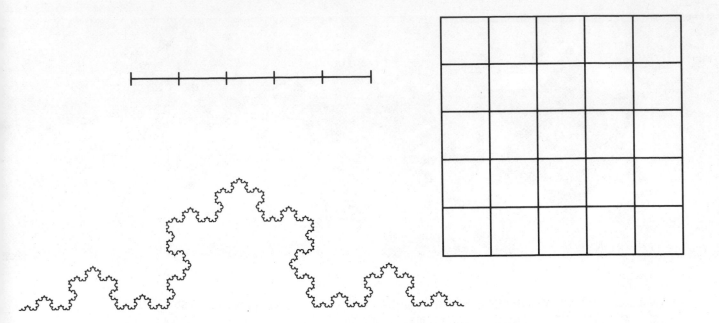

Plate 44 ¤ TWO KINDS OF SELF-SIMILARITY: STANDARD AND FRACTAL

The top Figures recall how, given an integer (here, b=5), a straight interval of unit length may be divided into N=b subintervals of length r=1/b. Similarly, a unit square can be divided into N=b² squares of side r=1/b. In either case, log N/log(1/r) is the shape's similarity dimension—a notion school geometry feels no need of pinpointing, since its value reduces to the Euclidean dimension.

The bottom Figure is a triadic Koch curve,

one-third of a Koch coastline. It too can be decomposed into reduced-size pieces, with N=4 and r=⅓. The resulting similarity dimension D=log N/log(1/r) is not an integer (its value is ~1.2618), and it corresponds to nothing in standard geometry.

Hausdorff showed that D is of use in mathematics, and that it is identical to the Hausdorff, or fractal, dimension. My claim is that D is also vital in natural science. ■

Plate 45 ¤ TRIADIC KOCH LAKE $\mathcal{K}$ (COASTLINE DIMENSION D=log 4/log 3~1.2618)

The construction described in the captions of Plates 42 and 43 has been carried much further, and a photographic negative taken, yielding a lake rather than an island.

The peculiar pattern of gray "waves" that fills this lake is not haphazard. It is explained

in Plates 68 and 69.

The coastline on this Plate is *not* self-similar, because a loop cannot be decomposed into the union of other loops. ◁ However, Chapter 13 uses the notion of self-similarity within an infinite collection of islands. ► ■

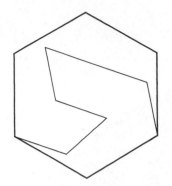

Plates 46 and 47 ¤ ALTERNATIVE KOCH ISLAND AND LAKE
(COASTLINE DIMENSION D=log 9/log 7~1.1291)

This variant of the Koch island is due to W. Gosper (Gardner 1976): the initiator is a regular hexagon, and the generator is

$$N=3$$
$$1/r=\sqrt{7}$$
$$D=\log 3/\log (\sqrt{7})\sim 1.1291$$

PLATE 46. In this plate, several stages of construction of the "Gosper island" are drawn as a bold line "wrapping." The corresponding thin line "filling" is explained in Plate 70.

PLATE 47. This is an advanced construction stage of the "wrapping." The variable thickness "filling" is, again, explained in Plate 70.

Observe that, contrary to Koch's original, the present generator is symmetric with respect to its center point. It combines peninsulas and bays in such a way that the island's area remains constant throughout the construction. The same is true of the Koch curves up to Plate 57.

TILING. The plane can be covered using Gosper islands. This property is called *tiling*.

PERTILING. Moreover, the present island is self-similar, as is made obvious by using variable-widths hatching. That is, each island divides into seven "provinces" deducible from the whole by a similarity of ratio $r=1/\sqrt{7}$. I denote this property by the neologism *pertiling*, coined with the Latin prefix *per-*, as used for example in "to perfume" = "to penetrate thoroughly with fumes."

Most tiles *cannot* be subdivided into equal tiles similar to the whole. For example, it is a widespread source of irritation that hexagons put together do not quite make up a bigger hexagon. The Gosper flake fudges the hexagon just enough to allow exact subdivision into 7. Other fractal tiles allow subdivision into different numbers of parts.

FRANCE. A geographical outline of unusual regularity often described as *the* Hexagon, namely the outline of France, resembles a hexagon less than it resembles Plate 47 (although Brittany is undernourished here.)

◄ REASON WHY A TANGENT CANNOT BE DEFINED AT ANY POINT OF THESE COASTLINES. Fix any point that the coastline attains after a finite number of stages of construction, and join it by a cord to a moving point on the limit coastline. As the moving point approaches the fixed point along the limit coastline, either clockwise or counter-clockwise, the cord's direction winds without end around the fixed point. Such a point is called *loxodromic*. ►

■

Plate 49 ¤ ALTERNATIVE KOCH ISLANDS AND LAKES
(COASTLINE DIMENSIONS FROM 1 TO D=log 3/log √~1.3652)

Throughout this sequence of fractal curves, the initiator is a regular polygon with M sides, and the generator is such that N=3 and that the angles between the first and second and second and third legs are both $\theta=2\pi/M$. Plates 46 and 47 had involved the special value M=6 (not repeated here), and the value M=3 is discussed in Plate 72. The present plate exhibits advanced teragons for the values M=4, 8, 16, and 32, in the form of nested lakes and islands. For example, M=4 corresponds to the generator

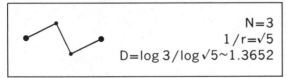

$$N=3$$
$$1/r=\sqrt{5}$$
$$D=\log 3/\log \sqrt{5}\sim 1.3652$$

The shading on the central island (M=4) is explained in Plates 72 and 73.

Were this pattern extended to M=∞, it would converge to a circle. As we move in, the figures "shrivel," first gradually, then by rapid jumps. The next stage of shriveling would lead to M=3, but the corresponding curve is

no longer self-avoiding. We meet it later, in Plates 72 and 73.

A CRITICAL DIMENSION. When the initiator is [0,1], the angle θ may take any value from 180° down to 60°. There is a critical angle θ_{crit}, such that the "coastline" is self-avoiding if, and only if, $\theta>\theta_{crit}$. The corresponding D_{crit} is a *critical dimension* for self-intersection. The angle θ_{crit} is close to 60°.

GENERALIZATION. The constructions of Plates 46 to 57 are easily generalized as follows. Let the generators that are shown be called straight (S), and define the flipped generator (F) as the mirror image of the straight generator in the line y=0. Each stage of the construction must use the same generator throughout, either S or F, but different stages may select different generators. These plates, and more which follow, use S throughout, but other infinite sequences of S and F yield immediate variants.

◁ If F and S alternate, the formerly loxodromic points become hyperbolic, as in the Koch curve. ►■

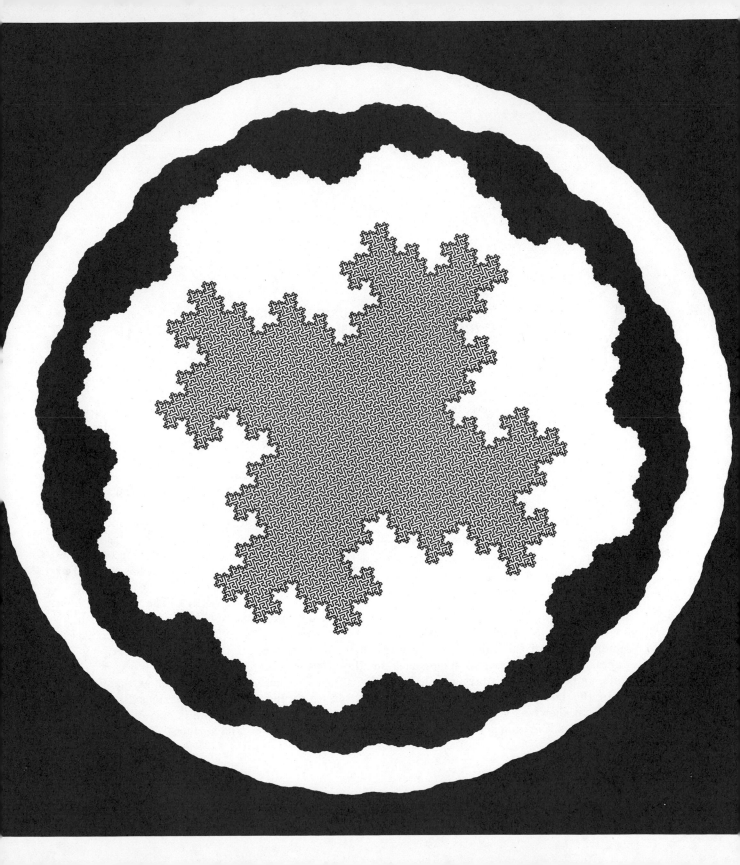

Plate 51 ◻ A QUADRIC KOCH ISLAND
(COASTLINE DIMENSION D=3/2=1.5000)

Plates 49 to 55 show several Koch constructions initiated with a square (hence the term *quadric*). One advantage is that one can experiment with these constructions even when the available graphic systems are crude. ◅ Another advantage is that quadric fractal curves lead on directly to the original Peano curve described on Plate 63. ▶

PLATE 51. Here, the initiator is a square, and the generator is

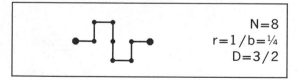

$$N=8$$
$$r=1/b=\tfrac{1}{4}$$
$$D=3/2$$

As in Plates 46 to 49, the total island area remains constant throughout the succession of stages. Plate 51 shows two stages on a small scale, and the next on a larger scale.

In the last stage, enlarged even further, the detail shows as very thin and barely visible whiskers, but much would be lost perceptually if the graphics were less excellent, forcing us to omit this detail.

Both the teragons and the limit curve involve no self-overlap, no self-intersection, and no self-contact. The same is true through Plate 55.

◅ One should not forget that the fractal in Plates 51 to 55 is the coastline; the land and sea are conventional shapes that have positive and finite areas. Page 144 mentions a case in which the "sea" alone has a well-defined area, being again the union of simple-shaped tremas, while the land has no interior point. ▶

TILING AND PERTILING. The present island is decomposable into 16 islands reduced by the ratio of $r=\tfrac{1}{4}$. Each is the Koch island built on one of the 16 squares forming the first stage of the construction.

◅ Chapters 25 and 29 show that $D=3/2$ is also encountered for various Brown functions. Hence this value is easy to obtain with random curves and surfaces. ∎

Plate 53 ¤ A QUADRIC KOCH ISLAND
(COASTLINE DIMENSION D=log 18/log 6~1.6131)

The initiator is again a square, and the generator is

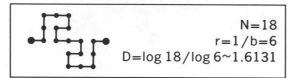

$$N=18$$
$$r=1/b=6$$
$$D=\log 18/\log 6 \sim 1.6131$$

The fact that the form of the quadric Koch islands in the present portfolio of illustrations depends very markedly upon D is significant. However, their having roughly the same overall outline is due to the initiator's being a square. When the initiator is an M-sided regular polygon (M>4), the overall shape looks smoother, increasingly so as M increases. A genuine link between overall form and the value of D will not enter until Chapter 28, which deals with random coastlines that effectively determine the generator and the initiator at the same time.

◁ MAXIMALITY. Another fact that contributes to the similarity of overall outline is that the quadric Koch curves in Plates 49 to 55 possess an interesting property of maximality. Consider all Koch generators that yield self-avoiding curves are traced on a square lattice made by straight lines parallel and perpendicular to [0,1], and in addition can be used with any initiator on the square lattice. We denote as *maximal* the generators that attain the highest possible value of N and hence of D. One finds that $N_{max}=b^2/2$ when b is even, while $N_{max}=(b^2+1)/2$ when b is odd.

◁ As the value of b increases, so does the maximal N, and so also does the number of alternative maximal polygons. Therefore, the limit Koch curve becomes increasingly influenced by the original generator. It also looks increasingly contrived, because the wish to achieve a maximal dimension without contact points imposes a degree of discipline that increases with D. It reaches its paroxysm in the next chapter, for the Peano limit D=2.

◁ LACUNARITY. Fractal curves sharing D but with different N and r may differ qualitatively from each other. The resulting parameter beyond D is discussed in Chapter 34. ■

CAPTION OF PLATE 55, CONTINUED

◁ In fact, the value of D is likely to depend on the fluid's initial energy, and on the size of the vessel in which dispersion is contained. A low initial energy would wither a disc-shaped blob into a curve with D close to 1 (Plate 49). A high initial energy in a small vessel might lead to more thorough dispersion, with planar sections more reminiscent of Plate 54 (D~1.7373) or even of the dimension D=2 (Chapter 8). See Mandelbrot 1976c.

◁ If this last inference is valid, the next step would be to investigate the relation between initial energy and D, and to seek the lowest energy that yields D=2 in the plane, i.e., D=3 in space. When we examine the limit case D=2 (Chapter 7), we shall see that it differs qualitatively from D<2 because it allows ink particles that start far apart to come into asymptotic contact. ◁ Thus, I would not be at all surprised if it turns out that the turbulent dispersion is a single term representing two sharply distinct phenomena.

◁ POSTSCRIPT. Well after this plate had first appeared in the 1977 *Fractals*, Paul Dimotakis photographed thin sections of a turbulent jet dispersing in a laminar medium. The resemblance with the present plate is most gratifying. ► ■

Now the same construction as in Plate 49 is carried out with the following generators. In Plate 55,

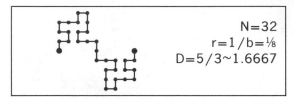

N=32
r=1/b=⅛
D=5/3~1.6667

and in Plate 54,

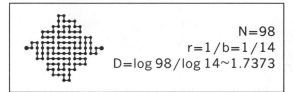

N=98
r=1/b=1/14
D=log 98/log 14~1.7373

The causeways and the channels in these nightmarish marinas become increasingly nar-row as one proceeds toward the peninsulas' tips or the bays' deepest points. In addition, these widths tend to narrow down as the fractal dimension increases, and "wasp waists" appear around D~5/3.

◁ DIGRESSION CONCERNING TURBULENT DISPERSION. I see an uncanny resemblance between the sequence of approximate fractals drawn in Plate 55, and the successive stages of turbulent dispersion of ink in water. Actual dispersion is of course less systematic, a feature one can mimic by invoking chance.

◁ One can almost see a Richardsonian cascade at work. A finite pinch of energy spreads a square ink blob around. Then the original eddy splits into smaller scale eddies, the effects of which are more local. The initial energy cascades down to ever smaller typical sizes, eventually contributing nothing but slight fuzziness to the outline of the final ink blob, just as in the following diagram from Corrsin 1959b.

◁ The conclusion that a Richardsonian cascade leads to a shape bounded by a fractal is inescapable, but the conclusion that D=5/3 is shaky. This value of D corresponds to planar sections of spatial surfaces with D=8/3, which occur often in turbulence. In the case of isosurfaces of scalars (studied in Chapter 30), D=8/3 is reducible to the Kolmogorov theory. Nevertheless, numerological analogies are not to be trusted.

THIS CAPTION CONTINUES ON PAGE 52

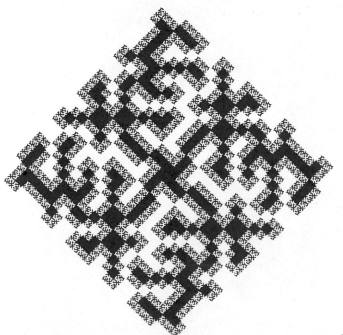

Plates 56 and 57 ¤ GENERALIZED KOCH CURVES AND SELF-SIMILARITY WITH UNEQUAL PARTS (D~1.4490, D~1.8797, D~1+ε)

These plates are constructed in the manner of Koch, except that the lengths of the generators' sides take different values r_m. Until now, we assume that the N "parts" into which our "whole" is divided all involve the same similarity ratio r. Using unequal r_m, the Koch curve becomes less relentlessly regular. Thus Plate 56 adds variety to the triadic Koch curve.

Note that in all this series of plates, the construction continues until it reaches details of a predetermined small size. When $r_m \equiv r$, this goal is reached after a predetermined number of construction stages, but here we need a variable number of stages.

The next task is to extend the notion of similarity dimension to this generalization of the Koch recursion. In a search for suggestions, let ordinary Euclidean shapes be paved with parts reduced in the respective ratios r_m. When D=1, the r_m must satisfy $\Sigma r_m = 1$, and, more generally, Euclidean shapes require

$\Sigma r_m{}^D = 1$. Furthermore, in the case of fractals that can be split into equal parts, the familiar condition $Nr^D = 1$ can be rewritten as $\Sigma r_m{}^D = 1$. These precedents suggest forming the dimension-generating function, namely $G(D) = \Sigma r_m{}^D$, and defining D as its unique real root of $G(D) = 1$. It remains to investigate whether or not said D coincides with the Hausdorff Besicovitch dimension. In every case I know of, it does.

EXAMPLES. Plate 56 has a D above Koch's original $\log 4/\log 3$. The top of Plate 57 has a D slightly below 2. As $D \to 2$, the coastline on this Figure tends toward the Peano-Pólya curve, a variant of the Peano curves examined in the next chapter. The resemblance between this Figure and a row of trees is not accidental, as seen in Chapter 17. Finally, the bottom of Plate 57 has a D slightly above 1. ■

7 ◻ Harnessing the Peano Monster Curves

When the end of Chapter 6 tackles generalized Koch curves that do not self-intersect, there is good reason for stopping short of D=2. When D reaches D=2, a profound qualitative change occurs.

We shall assume that the teragons do not self-intersect, although they may self-contact. Then one symptom of reaching D=2 is that points of self-contact become *inevitable* asymptotically. The major symptom is that it is *inevitable* that the limit should fill a "domain" of the plane, that is, a set that contains discs (filled in circles).

This double conclusion is *not* due to a corrigible lack of imagination on the part of mathematicians. It involves a fundamental principle, central to the 1875–1925 crisis in mathematics.

PEANO "CURVES," MOTIONS, SWEEPS

The corresponding limits, exemplified in upcoming plates, are called *Peano curves*, because the first is found in Peano 1890. They are also called *plane-filling curves*. For them, the formal definition of dimension by $\log N/\log(1/r)=2$ is justified, but for a disappointing reason. From the mathematical viewpoint, a Peano curve is merely an unusual way of looking at a domain or piece of plane, a set for which all the classical definitions yield the dimension 2. In other words, the term *plane-filling curve* should be avoided by careful writers.

Fortunately, most Peano "curves," including those obtained by a recursive Koch construction, are parametrized naturally by a scalar t, which may be called "time." In their case, we can (with no fear of the guardians of rigor) use the terms *Peano motions*, *plane-filling motions*, *tile sweeping motions*, or *tile sweeps* (tiles are discussed later in the chapter). We shall do so when it seems appropriate, but Essays need not attempt full consistency on any account.

THE PEANO CURVES AS MONSTERS

"Everything had come unstrung! It is difficult to put into words the effect that [Giuseppe]

Peano's result had on the mathematical world. It seemed that everything was in ruins, that all the basic mathematical concepts had lost their meaning" (Vilenkin 1965). "[Peano motion] cannot possibly be grasped by intuition; it can only be understood by logical analysis" (Hahn 1956). "Some mathematical objects, like the Peano curve, are totally non-intuitive..., extravagant" (Dieudonné 1975).

THE PEANO CURVES' TRUE NATURE

I claim that the preceding quotes merely prove that no mathematician ever examined a good Peano graph with care. An unkind observer could say these quotes demonstrate a lack of geometric imagination.

I assert to the contrary that, after Peano teragons are observed attentively, letting one's thoughts wander about, it becomes very difficult *not* to associate them with diverse aspects of Nature. This chapter takes up the self-avoiding curves, those whose teragons *avoid* self-contact. Chapter 13 takes up teragons that self-contact *moderately*. Teragons that fill a lattice (e.g., lines parallel to the axes and having integer coordinates) must first be processed to eliminate the self-contacts.

RIVER AND WATERSHED TREES

Examining diverse Peano teragons, I saw in each case a set of two *trees* (or sets of trees) possessing an endless variety of concrete interpretations. They are particularly conspicuous on the "snowflake sweep" Peano curve I designed, Plate 69. It is, for example, easy to visualize this Plate as a collection of bushes rooted side by side along the bottom third of a Koch snowflake, and creeping up a wall. Alternatively, one may choose to be reminded of the boldly emphasized outline of a collection of rivers meandering around, and eventually flowing into a river that follows the snowflake's bottom. This last interpretation suggests immediately that the curves that separate the rivers from each other combine into watershed trees. And of course, the labels *river* and *watershed* can be interchanged.

This *new* rivers-watersheds analogy is so obvious *after the fact* that it lays to rest any notion that the Peano curve is necessarily pathological. As a matter of fact if a tree made of rivers of vanishing width is to drain an area thoroughly, it *must* penetrate everywhere. One who follows the rivers' combined bank performs a plane-filling motion. Ask any child for confirmation!

Helped by the intuition garnered from Plate 68, it would be difficult not to see analogous conjugate networks in every Peano teragon. Even the crude island of Plate 63 begins to make intuitive sense. The thin fingers of water that penetrate it cannot be viewed as a marina, however exaggerated, but can be viewed as branching rivers.

When rivers give rise to a proper science, it should be called *potamology*—Maurice Pardé's coinage from ποταμος (= river) and λογος. But sober usage merges the study of rivers into the science of water, hydrology, into which this Essay makes many incursions.

MULTIPLE POINTS ARE UNAVOIDABLE IN TREES, HENCE IN PEANO MOTIONS

Suddenly many mathematical properties of Peano curves become obvious too. To account for double points, assume one starts on a river's shore in a Peano river tree, and moves upstream or downstream, making a detour for the slightest branch (moving ever faster as one gets to finer branches). It is clear that one will eventually face the point of departure from across the river. And since the limit river is infinitely narrow, one will effectively return to the starting point. Thus, double points in a Peano curve are *inevitable*, not only from a logical but also from an intuitive viewpoint. Furthermore, they are *everywhere dense*.

Also, it is inevitable that some points be visited more than twice, because a point where rivers join is one where at least three points of the bank coincide. When all points of confluence involve only two rivers, there is no point of multiplicity above three. On the other hand, one can do without points of multiplicity of three if one agrees to have points of higher multiplicity.

All the assertions in the preceding paragraphs have been proven, and, since the proofs are delicate and led to controversy, the properties themselves seem "technical." But the contrary is the case. Who would continue to argue that a purely logical approach toward them is preferable to my own intuitive one?

Typically, a Peano curve's rivers are not standard shapes but fractal curves. This is fortunate for the needs of modeling, because every argument in Chapter 5 to the effect that geographic curves are nonrectifiable applies equally well to river banks. In fact, the Richardson data include frontiers that follow rivers or watersheds. And rivers are involved in the quote from Steinhaus 1954. As to rivers' drainage basins, they are surrounded by closed curves akin to island coastlines, made of portions of watershed. Each basin is the juxtaposition of partial basins and is crisscrossed by the rivers themselves, but plane-filling curves that are bounded by fractal curves display all the structure we need.

PEANO MOTION AND PERTILING

Taking the original Peano curve (Plate 63), develop t in the counting base N=9, in the form $0.\tau_1\tau_2....$ Times sharing the same first "digit" are mapped on the same ninth of the initial square, those with the same second digit on the same 9^2-th, etc. Thus, the tiling of [0,1] into 9-th maps on a tiling of the square. Successive 9-ths of the linear tiles map on successive planar subtiles. And the interval's property of being pertiling (page 46), i.e. subdivisible recursively and ad infinitum into smaller tiles similar to [0,1], is mapped on the square. Alternative Peano motions, due to E. Cesàro, G. Pólya and others, map this property on diverse pertilings of the triangle.

More generally, most Peano motions generate pertilings of the plane. In the simplest case, there is a base N, and one starts with a linear pertiling that consists of successive divisions into N-th. But the snowflake sweep of Plate 68-69 requires an irregular division of the [0,1] interval of t, into four subintervals of length $1/9$, then four of length $1/9\sqrt{3}$, one of length $1/9$, two of length $1/9\sqrt{3}$, and two of length $1/9$.

ON MEASURING DISTANCE BY AREA

Exquisite relationships, wherein length and area interchange, are a common occurrence in Peano motion, especially if it is *isometric,* meaning that a time interval $[t_1, t_2]$ maps on an *area* equal to the *length* $|t_1-t_2|$. (Most Peano motions are both isometric and pertiling, but these are distinct notions.) Calling the map of the time interval $[t_1,t_2]$ a planar Peano *interval* implies that, instead of measuring distances through a time, one may do so through an area. But we encounter a vital complication, because points that sit across from each other on different banks of a river coincide in space but are visited repeatedly.

The definition of "Peano distance" may involve only the order of the visits. Denoting the instants of first and last visits of P_1 and P_2 by t'_1 and t'_2 and by t''_1 and t''_2, the *left Peano interval* $\mathcal{L}\{P_1, P_2\}$ is defined as the map of $[t'_1, t'_2]$ and the *right Peano interval* $\mathcal{R}\{P_1, P_2\}$ is defined as the map of $[t''_1,t''_2]$. These intervals' lengths define the *left*

distance and the *right distance* as $|\mathcal{L}\{P_1, P_2\}| = |t'_1-t'_2|$ and $|\mathcal{R}\{P_1, P_2\}| = |t''_1-t''_2|$. Each of these distances is additive, meaning for example that if three points P_1, P_2, and P_3 are left ordered according to the order of first visits, one has

$$|\mathcal{L}(P_1,P_3)|=|\mathcal{L}(P_1,P_2)|+|\mathcal{L}(P_2,P_3)|.$$

Alternate definitions of interval and distance distinguish between river and watershed points. Denote by t' and t'' the instants of first and last visit of P. P is a *river point* if the map of [t',t''] is bounded by P and watersheds. Successive visits of P face each other across rivers. P is a *watershed point* if the map of [t', t''] is bounded by P and rivers.

Furthermore, once a Peano curve is represented as the common shore of a river tree and a watershed tree, the paths that link P_1 and P_2 through rivers (resp., along watersheds) include a common minimal path. It is reasonable to follow this path in order to measure the distance between P_1 and P_2. Save for exceptional cases, the rivers' and watersheds' dimension D is strictly below 2 and strictly above 1. Hence the minimal path can be measured neither by length nor by area, but in typical cases it has a nontrivial Hausdorff measure in the dimension D.

MORE. Very important additional considerations on Peano motions are detailed in the captions that follow.   ▄▄

Plate 63 ⌑ A QUADRIC KOCH CONSTRUCTION OF DIMENSION D=2: THE ORIGINAL PEANO CURVE, A SQUARE SWEEP

The *Peano plane-filling curve* in this plate is the original one. Giuseppe Peano's incredibly terse algorithm was graphically implemented in Moore 1900 (which receives undue credit in my 1977 *Fractals*). The present plate rotates Peano's curve by 45°, and by doing so brings it into the fold of Koch curves in the strict sense: the generator is always placed in the same way on the sides of the teragon obtained at the preceding stage.

The initiator here is the unit square (bounding the black box) and the generator is

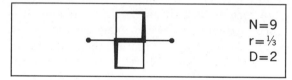

$$N=9 \quad r=\tfrac{1}{3} \quad D=2$$

Because this generator self-contacts, the resulting finite Koch islands are sets of black squares on a chunk from an infinite chessboard. And the nth Koch teragon is a grid of lines, a distance of $\eta=3^{-n}$ apart; they crisscross a square of area equal to 2 that becomes covered increasingly tightly as $k\to\infty$. It suffices to show one example of this dull design (next to the initial black box).

Three illustrations on the top of page 63 avoid ambiguity by cutting off the corners while leaving the total area invariant.

On the same scale, the fourth stage of this sequence would merge into 50% gray, but a larger drawing of one-fourth of the coastline can be followed unambiguously (at some risk of becoming seasick). It shows graphically what is meant by saying that the limit Koch curve fills the plane.

It would have been nice to be able to define a limit island in analogy to the Koch islands of Chapter 6, but in the present case it is impossible. A point chosen at random almost surely flips between being inland and in the ocean, without end. Advanced teragons

are penetrated by bays or rivers so deeply and uniformly that a square of middling side x—such that $\eta\ll x\ll 1$—divides between dry land and water in near equal proportions!

INTERPRETATION. The limit Peano curve establishes a continuous correspondence between the straight line and the plane. The fact that self-contacts are mathematically unavoidable is classical. The fact that they are valuable in modeling Nature is new to this work.

LONG-RANGE ORDER. Without knowing of the descending cascades that built our finite Peano curves, one would be baffled by the extraordinary long-range order that allows these curves to avoid not only self-intersection but also self-contact. Any lapse in discipline would make the latter very likely.

◁ And total breakdown of discipline makes endlessly repeated self-intersection almost certain, since a totally undisciplined Peano curve is Brownian motion, mentioned in Chapter 2 and explored in Chapter 25.

◁ LIOUVILLE THEOREM AND ERGODICITY. Mechanics represents the state of a complex system by a single point in a "phase space." Under the equations of motion, every domain in this space is known to behave as follows: its measure (hyper-volume) remains invariant (Liouville theorem), but its shape changes and it disperses and fills all the space available to it with increasing uniformity. Clearly, both of these characteristics are echoed by the behavior we impose upon the black domain in the present Peano construction. It is interesting, therefore, to dig deeper, by observing that in many simplified "dynamical" systems that allow a detailed study each domain disperses by transforming into an increasingly long and thin ribbon. It would be interesting to see whether other systems' dispersion proceeds through Peano-like trees instead of ribbons. ► ■

Plates **64** and **65** ¤ QUADRIC KOCH CONSTRUCTIONS OF DIMENSION D=2: CESÀRO'S AND POLYA'S TRIANGLE SWEEPS, AND VARIANTS

The simplest generator one could imagine is made of N=2 equal intervals making an angle θ that satisfies $90° \leq \theta \leq 180°$. The limit case $\theta = 180°$ generates a straight interval; the case $\theta = 120°$ (illustrated in the caption of Plate 43) generates the triadic Koch curve (among others). The limit case $\theta = 90°$ is

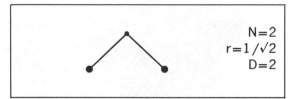

N=2
r=1/√2
D=2

This generator gives rise to an uncanny number of different Peano curves, according to the initiator's shape, and the rule of placement of the generator upon the preceding teragon. Plates 64 to 67 examine a few notable examples.

◁ In addition, Chapter 25 obtains Brownian motion by randomizing the class of all Peano curves with these N and r ►.

PÓLYA'S TRIANGLE SWEEP. The initiator is [0,1], the generator is as above, and it alternates between the right and the left of the teragon. The first position also alternates. The early construction stages yield the following

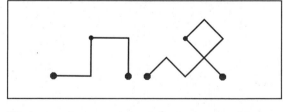

The teragons are pieces of square graph paper contained within a right isosceles triangle

whose *side* is [0,1]. The limit curve sweeps this triangle.

PLATE 64. PÓLYA SWEEP OVER A RIGHT NONISOSCELES TRIANGLE. The generator is changed to be made of two unequal orthogonal intervals. Guessing the processing chosen to avoid self-contact is left to the reader as an exercise.

CESÀRO'S TRIANGLE SWEEP. The initiator is [1,0], the generator is again as above, and the next two construction stages are as follows (for the sake of clarity, the drawing refers to $\theta = 85°$ instead of $\theta = 90°$).

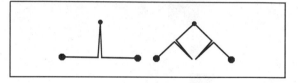

Thus, in all the odd-numbered construction stages, the generator is positioned to the right, yielding as teragon a grid of lines parallel to the initiator's diagonals. And in all the even-numbered stages, the generator is positioned to the left, yielding as teragon a grid of lines parallel to the initiator's sides. Asymptotically, this curve fills a right isosceles triangle whose *hypotenuse* is [0,1].

PLATE 65. This plate represents a square sweep obtained by adding the Cesàro sweeps initiated by [0,1] and [1,0]. (Again, $\theta=85°$ instead of $90°$ for the sake of clarity.)

SELF-OVERLAP. Each interval in the grids covered by the Cesàro teragons is covered *twice*. Not only the construction is self-contacting, but it is self-overlapping.

"EFFICIENCY" OF PLANE FILLING. AN EXTREMAL PROPERTY OF THE PEANO-CESÀRO DISTANCE. The Peano curve of Plate 63 maps [0,1] on the square of diagonal [0,1] and area ½. The same shape is covered by the Pólya curve. But the Cesàro curve fills a right isosceles triangle of hypotenuse [0,1] and area ¼. To cover the whole square, Cesàro must add the maps of [1,0] and [0,1]. Thus, the Cesàro curve in the less "efficient," of the two. As a matter of fact, it is the least efficient non-self-intersecting Peano curve on a square lattice. But this fact endows it with a redeeming virtue: the left or right Peano distance (see p. 61) between two points P_1 and P_2 is at least equal to the square Euclidean distance:

$$|\mathcal{L}(P_1,P_2)| \geq |P_1P_2|^2; \; |\mathcal{R}(P_1,P_2)| \geq |P_1P_2|^2$$

For other Peano curves, the difference between Peano and Euclid distance may take either sign.

KAKUTANI-GOMORY PROBLEM. After selecting M points P_m in the square $[0,1]^2$, Kakutani (private communication) investigates the expression $\inf\Sigma|P_mP_{m+1}|^2$, where the infinitum is taken over all the chains that join the P_m in sequence. He proves that $\inf\leq8$, but conjectures that this bound is not the best one. Indeed, R. E. Gomory (private communication) obtains the improved bound $\inf\leq4$.

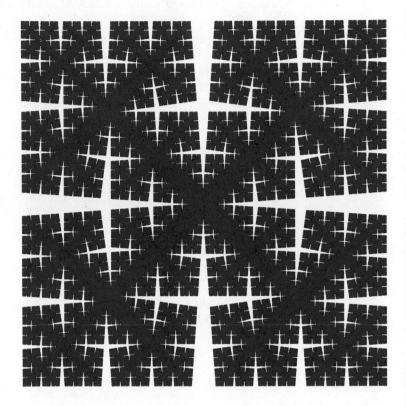

The proof uses the Peano-Cesàro curve, as follows. (A) Add the square's corners if they are not already among the P_m. (B) Rank the M points P_m in the order they are first visited by the string of four Peano-Cesàro curves drawn inside the square, along its sides. (C) Observe that, by lengthening the chain in step (A), we did not decrease $\Sigma|P_mP_{m+1}|^2$. (D) Observe that each addend $|P_mP_{m+1}|^2$ is not decreased if replaced by $|\mathcal{L}(Z_m,Z_{m+1})|$. (E) Observe that $\Sigma|\mathcal{L}(Z_m,Z_{m+1})| = 4$. If different Peano curves were used, steps (B) and (D) would be invalid. ■

The generator is the same here as in Plates 64 and 65, but seemingly slight changes in other rules have lasting consequences.

A LATER SQUARE SWEEP BY PEANO. The initiator is [0,1], but the second, fourth and sixth construction stages are changed to

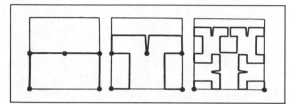

EFFICIENCY. AN EXTREMAL PROPERTY. This curve fills a domain of area equal to 1, while the curves of Plates 64-65 and the dragon curve to be below covers ½ or ¼. When the teragons lie on an orthogonal lattice, the covered area cannot exceed 1. It reaches this maximum whenever the teragons are self-avoiding. In other words, absence of self-contact is more than a matter of esthetics, and a self-contacting curve whose self-contacts are rounded off, as in Plate 63, does *not* become equivalent to a self-avoiding Koch curve.

By taking the odd numbered stages of the present square sweep, then joining the midpoints of the teragons' successive intervals to avoid self-contact, one falls back on a Peano curve due to Hilbert.

PLATE 67. A CURVE SWEEPING A RIGHT TRAPEZOID. The generator is changed to be made of two unequal orthogonal intervals. The processing to avoid self-contact is the same as in the preceding plate.

THE HARTER-HEIGHTWAY DRAGON. (See Gardner 1967, Davis & Knuth 1970.) Here the initiator is [1,0], the generator is as above, and it alternates between the right and the left of the teragon. The only difference with the Polya triangle sweep is that the first

position is always to the right at every stage of construction, early stages being as follows

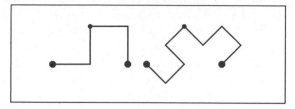

The consequences of this change are dramatic, since a mature stage looks like this

On this illustration, the curve itself has become indistinct, and we see only its boundary, called *dragon curve*. Thus, this Peano curve deserves to be called *dragon sweep*. As any Koch curve initiated by [0,1], the dragon is self-similar. But in addition it is seen to be segmented into portions, which join at wasp waists. The sections are similar to one another, but not to the dragon itself.

TWINDRAGON. The 1977 *Fractals* points out that, with the dragon's rules of construction, a more natural initiator is [0,1] followed by [1,0], and terms the shape that is swept as a result, a *twindragon*. This shape is encountered number representations, Knuth 1980. It looks like this (one component dragon is in black and the other is in gray).

D~1.5236

The short and long intervals here are of
lengths $r_1 = 1/\sqrt{2}$ and $r_2 = (\frac{1}{2})(\sqrt{2}) = r_1^3$,
respectively. Hence, the dimension generating
function is $(1/\sqrt{2})^D + 2(2\sqrt{2})^D = 1$, showing
that the quantity $2^{D/2}$ satisfies $x^3 - x^2 - 2 = 0$.

ALTERNATE DRAGONS. (Davis & Knuth
1970.) Pick any infinite sequence x_1, x_2...,
where each x_k can be either 0 or 1, and use
the value of x_k to determine the first position
of the generator during the k-th stage off construction: when $x_k = 1$, a generator is first positioned to the right, but when $x_k = 0$ it is first
positioned to the left. Each sequence generates a different alternate dragon. ■

TWINDRAGON RIVER. After the streams
near the source are erased (for legibility), the
river tree of a twindragon looks like this.

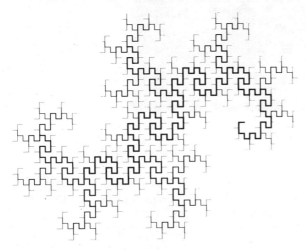

A twindragon can be tiled by reduced size
replicas of itself, like this.

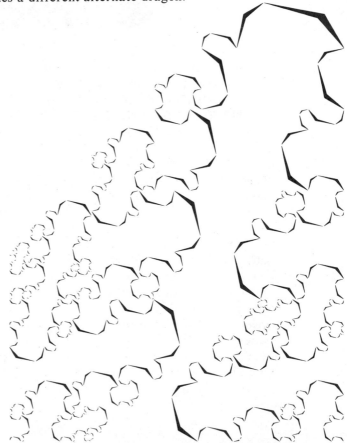

**Plates 68 and 69 ¤ THE SNOWFLAKE
SWEEPS: NEW PEANO CURVES
AND TREES (WATERSHED
AND RIVER DIMENSIONS D~1.2618)**

These plates illustrate a family of Peano curves I designed. They fill the original Koch snowflake (Plate 45), hence two basic monsters of circa 1900 are brought together.

A more important virtue is that a glance suffices here to document a major theme of the present Essay: Peano curves are far from being mathematical monsters with no concrete interpretation. If they fail to self-contact, they involve readily visible and interpretable conjugate trees. These trees are good first-order models of rivers, watersheds, botanical trees, and human vascular systems.

As a by-product, we obtain here a method

for tiling the snowflake with unequal snow-flakes.

SEVEN INTERVAL GENERATOR. Let the initiator be [0,1], and the generator and the second construction stage be

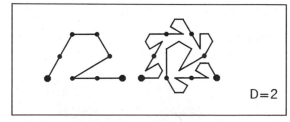

D=2

More precisely, let the above generator be denoted by S and called straight, and define the flipped generator F as the mirror image of S in the line $x=\frac{1}{2}$. At any stage of the construction of the snowflake sweep, one can use either the F, or the S generator, at will. Hence, each infinite sequence of F and S yields a different snowflake sweep.

ROUNDED OFF TERAGONS. Broken lines tend to look raw, and the snowflake sweep's teragons are made to look isotropic and otherwise much more "natural", if each interval is rounded off into one sixth of the circle.

PLATE 45. An advanced teragon of a seven interval snowflake sweep, rounded off, and later filled in, was used long ago in Plate 45 to provide a wavy background shading. Looking at it again, we are reminded of a liquid's flow past a fractal boundary, and of the shear lines between two roughly parallel flows of different velocities.

THIRTEEN INTERVAL GENERATOR. Now change the above 7 interval generator by replacing 5-th leg by a reduced version of the whole. This version can be positioned either in the S or the F position. The latter yields the following generator and second construction stage

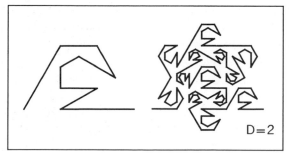

PLATE 68. This advanced teragon, shown as boundary between two fantastically intertwined domains serves better than any number of words to explain what plane-filling means.

PLATE 69. Let the above 13-interval generator be rounded off, and do the same in parallel to the snowflake curve. The resulting first few stages are shown in Plate 69.

RIVER DIMENSIONS. In Peano's original curve, each individual river is of finite length, hence of dimension 1. Here individual rivers are of dimension $\log 4 / \log 3$. To achieve the dimension $D=2$, all rivers have to be taken together. ∎

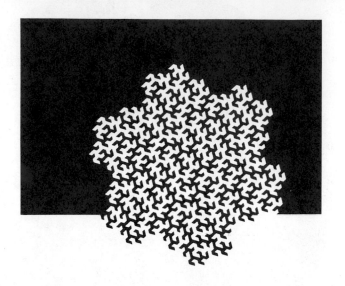
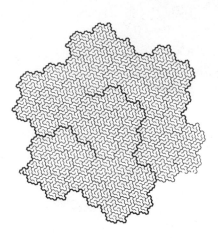

Plates 70 and 71 ¤ THE PEANO-GOSPER CURVE. ITS TREES, AND RELATED KOCH TREES (WATERSHED AND RIVER DIMENSIONS D~1.1291)

BACK TO PLATE 46. The thin broken lines on this plate, unexplained until now, represent the early construction stages 1 to 4 of a curve due to Gosper (Gardner 1976). This was the first self-avoiding Peano curve to be obtained by the Koch method without further processing.

The initiator is [0,1], and the generator is

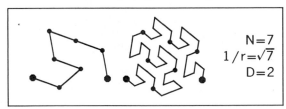

$$N = 7$$
$$1/r = \sqrt{7}$$
$$D = 2$$

By turning the generator counterclockwise until its first link becomes horizontal, one sees that it is drawn as a triangular lattice, on which it occupies 7 out of 3×7 links. This feature extends to triangular lattices a property which page 66 discusses for square lattices.

Now we see that the present Peano curve fills the Koch curve of Plate 46. The variable

width hatching in Plate 46 can be explained now: it represents the fifth stage of the present construction.

LEFT OF PLATE 70. The fourth teragon of the Gosper curve is redrawn as the boundary between a black and a white region.

RIGHT OF PLATE 70. RIVER AND WATERSHED TREES. Rivers and watersheds are drawn along the midlines of the white and black "fingers" of the figure to the left of Plate 70.

TOP OF PLATE 71. Starting with the river and watershed trees to the right of Plate 70, the widths of the links are redrawn according to their relative importance in the Horton-Strahler scheme (Leopold 1962). In this instance, the river or watershed links are given widths proportional to their lengths as the crow flies. The rivers are in black, and the watersheds in gray.

DIMENSIONS. Each Peano curve determines the D of its own boundary. In Plates 63 and 64, said boundary is merely a square. In later plates, it was a dragon's skin, then a snow-

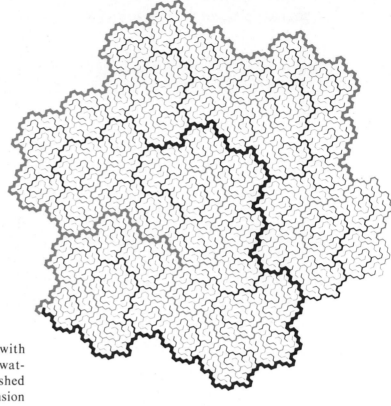

flake curve. Here it is a fractal curve with
D~1.1291, which is part river and part watershed. And every other river and watershed
also converges to a curve of fractal dimension
D~1.1291.

FRANCE. One who as a schoolboy often
gazed on a map showing the rivers Loire and
Garonne does not feel far from home.

BOTTOM OF PLATE 71. A RIVER TREE CON-
STRUCTED DIRECTLY BY A KOCH CASCADE.
When the generator is itself tree-shaped, it
generates a tree. For example, let the generator be

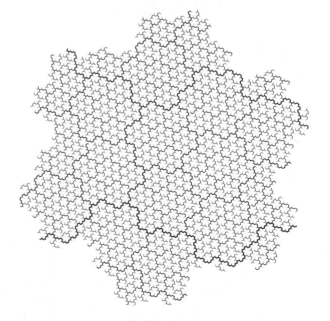

Here we have an alternative method of drain-
ing the Koch curve of Plate 46. (The last
branches near the "sources" have been clip-
ped off.) ■■

The plane-filling "river" trees deduced from some Peano curves can also be obtained by a direct recursive construction. The key is a generator that is itself tree shaped. A dull example is obtained if the tree generator is made up of 4 legs forming a + sign. One obtains the river tree of the Peano Cesàro curve (Plate 65).

FUDGEFLAKE. A better example results from taking [0,1] as initiator, and using the following generator

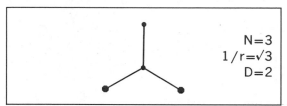

$$N=3$$
$$1/r=\sqrt{3}$$
$$D=2$$

We begin by observing that individual rivers are generated by a midpoint displacement shape like on Plate 43. Hence, every asymptotic river has the dimension $D = \log 2/\log\sqrt{3} = \log 4/\log 3$. This value is very familiar from the snowflake curve, but the curve with which we deal here is not a snowflake, because the positioning of the generator follows a different rule.

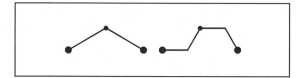

In order to leave room for the rivers, the generator must be made to alternate between the right and the left. Therefore, the snowflake's symmetry is fudged, and the domain these rivers drain is to be called *fudgeflake*.

Now, we turn to the river tree. Its teragons do not self-overlap, but they self-contact badly. This feature's asymptotic variant is unavoidable, and it is also unobjectionable, since it expresses quite properly the fact that several rivers can originate at the same point. But we shall see later in this caption that river teragons *may* avoid self-contact. Due to self-contacts, the present river teragon is an illegible chunk of hexagonal graph paper, bounded by an approximate fractal.

TOP OF PLATE 73. The river tree is made more transparent by erasing all river intervals that touch a source, and using a bolder pen to draw the principal river. The area drained by this tree is $\sqrt{3}/2\sim.8660$.

FUDGEFLAKE SWEEP. Now draw a Peano curve with a Δ shaped initiator, and a generator in the shape of a Z whose legs are equal and make angles of 60°. This is the extreme case for M=3 of the family of generators used in Plates 46 and 47, but it differs profoundly from all the other cases. It is investigated in Davis & Knuth 1970.

One can verify that this Peano curve's river tree is none else than the tree we just drew directly. The initiator's sides are of length 1, and the corresponding Peano curve sweeps an area equal to $\sqrt{3}/6\sim.2886$ (how inefficient!).

QUARTET. Next, we consider a different Koch curve, together with three curves that fill it: one Peano curve and two trees. These shapes, which I designed, illustrate a further theme of interest.

Take [0,1] as initiator, and take the following generator

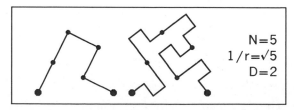

$$N=5$$
$$1/r=\sqrt{5}$$
$$D=2$$

This curves' boundary converges to a Koch curve of dimension $D=\log 3/\log\sqrt{5}=1.3652$. Advanced teragons of the boundary and of the Peano curve are seen in the center of Plate 49,

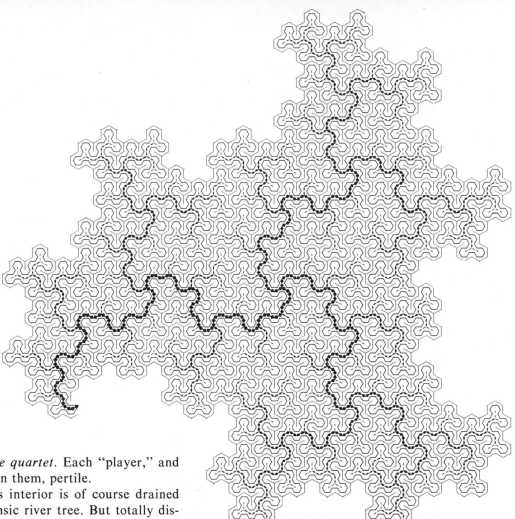

which I term *the quartet*. Each "player," and the table between them, pertile.

The quartet's interior is of course drained by its own intrinsic river tree. But totally distinct patterns of chainage are obtained by using either of the following generators

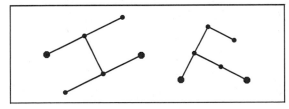

With the generator to the left, the teragons self-contact, as with the first example in this caption. And the drainage area turns out to be ½. With the generator to the right, the teragons *avoid* self-contact. And the drainage area is 1. An advanced teragon is shown in the bottom figure of Plate 73. ■

73

8 ¤ Fractal Events and Cantor Dusts

This chapter's principal goal is to acquaint the reader concretely and painlessly with yet another mathematical object ordinarily viewed as pathological, the Cantor dust, C. This and related dusts we shall describe have fractal dimensions between 0 and 1.

Being formed by points on a straight line, they are easy to study. In addition, they help introduce in simplest form several concepts that are central to fractals but that have been so underutilized in the past that no specific terms were required to denote them. First, the term *dust* is given a technical meaning, as an informal equivalent to a *set of topological dimension* $D_T=0$, just as "curve" and "surface" denote sets of topological dimensions $D_T=1$ and $D_T=2$. Other new terms are *curd, gap,* and *trema,* to be explained.

NOISE

For the layman, a noise is a sound that is too strong, has no pleasing rhythm or purpose, or interferes with more desirable sounds. Partridge 1958 proclaims that the term "derives from the Latin *nausea* (related to *nautes* = sailor), the semantic link being afforded by the noise made by an ancient shipful of passengers groaning and vomiting in bad weather." (The *Oxford English Dictionary* is not so sure.) As to contemporary physics, it is less colorful, and not nearly so precise: it uses *noise* as a synonym of chance fluctuation or error, irrespective of origin and manifestation. This chapter introduces C through the case study of an esoteric but simple noise.

ERRORS IN DATA TRANSMISSION LINES

A transmission channel is a physical system capable of transmitting electricity. However, electric current is subject to spontaneous noise. The quality of transmission depends on the likelihood of error due to noise distortion, which depends, in turn, on the ratio between the intensities of signal and noise.

This chapter is concerned with channels that transmit computer data and involve very strong signals. An interesting fact is that the signal is discrete, hence the distribution of errors simplifies the distribution of noise to the bone, so to speak. Noise involves a function having several possible values, while errors involve a function that has only two possible values. For example, it may be the indicator function, which is 0 when there is no error at time t, and 1 if there is an error.

Physicists have mastered the structure of the noises that predominate in the case of weak signals, e.g., thermal noise. In the problem just described, however, the signal is so strong that the classical noises are negligible.

The nonnegligible *excess noises* are difficult and fascinating because little is known about them. This chapter examines an excess noise that was, around 1962, of practical importance to electrical engineers, so that diverse talents were called upon to investigate it. My contribution to this effort was the first concrete problem in which I experienced the need to use fractals. No one remotely imagined at that time that a careful study of this apparently modest engineering difficulty would get us so far.

BURSTS AND GAPS

Let us subject the errors to increasingly refined analysis. A rough analysis reveals the presence of periods during which no error is encountered. Let these remission periods be called "gaps of rank 0" if their duration exceeds one hour. By contrast, any time interval flanked by gaps of rank 0 is singled out as being a "burst of errors of rank 0." As the

analysis is made three times more accurate, it reveals that the original burst is itself "intermittent." That is, shorter gaps "of rank 1," lasting 20 minutes or more, separate correspondingly shorter bursts "of rank 1." Likewise, each of the latter contains several gaps "of rank 2," lasting 400 seconds, separating bursts "of rank 2," and so on, each stage being based on gaps and bursts that are three times shorter than the previous ones. The process is illustrated very roughly by Plate 80. (Do not pay attention to the caption yet.)

The preceding description suggests something about the relative positions of the bursts of rank k within a burst of rank k−1. The probability distribution of these relative positions seems independent of k. This invariance is obviously an example of self-similarity, and fractal dimension cannot be far behind, but let us not rush. This Essay's diverse case studies are meant, among others, to elicit new themes or refine old ones. With this in mind, it seems best to reverse the historical order, and introduce a new theme through a rough nonrandom variant of the Berger & Mandelbrot stochastic model of errors, Chapter 31.

A ROUGH MODEL OF ERROR BURSTS: THE CANTOR FRACTAL DUST $\mathcal{C}$

The preceding section constructs the set of errors by starting with a straight line, namely the time axis, then cutting out shorter and shorter error-free gaps. This procedure may be unfamiliar in natural science, but pure mathematics has used it at least since Georg Cantor (Hawkins 1970, especially p. 58).

In Cantor 1883, the initiator is the closed interval [0,1]. The term "closed" and the use of brackets indicate that the extreme points are included; this notation was used in Chapter 6, but there was no need until now to make it explicit. The first construction stage consists in dividing [0,1] into 3 pieces, then removing the middle open third, designated]⅓, ⅔[. The term "open" and the use of reversed brackets indicate that the extreme points are excluded. Next, one removes the open middle of each of N=2 remaining thirds. And so on to infinity.

The remainder set C is called either *dyadic,* due to the fact that N=2, or *triadic* or *ternary*, due to the fact that [0,1] is subdivided into 3 pieces.

More generally, the number of pieces, called *base*, is denoted by b, the ratio between each N-th of the set and the whole being r=1/b. C is also called *Cantor discontinuum,* and I shall momentarily suggest the term, *Cantor fractal dust.* Since a point on the time axis marks an "event," C is a fractal sequence of events.

CURDLING, TREMAS, AND WHEY

Cantor's procedure is a *cascade,* to use a term Lewis Richardson had applied to turbulence, and we first borrowed in Chapter 6 to describe coastlines and the Koch curve. "Stuff" that was uniformly distributed over an initiator [0,1] is subjected to a centrifugal eddy which sweeps it into the extreme thirds.

The middle third portion cut out of [0,1] to form a gap is henceforth denoted as *trema generator*. This neologism is being coined in this section from $\tau\rho\eta\mu\alpha$ meaning hole, whose distant relative is the Latin *termes* = termite. It may be the shortest Greek word that has not yet been put to work with a significant scientific meaning.

In this context, tremas coincide with gaps, but in different instances to be encountered later they do not, which is why two different terms are required.

While a "first-order trema" is emptied, the total stuff is conserved and redistributed with uniform density over the outer thirds, to be called *precurds*. Then two centrifugal eddies come in and repeat the same operation, starting with the two intervals [0,⅓] and [⅔,1]. The process continues as a Richardsonian cascade converging at the limit to a set to be called *curd*. If a stage's duration is proportional to the eddy size, the total process is of finite duration.

In parallel, I propose *whey* (a term Miss Muffet should not mind) to denote the space outside the curd.

It is suggested that the above terms be used not only in a mathematical but also in a physical meaning: *curdling* to denote any cascade of instabilities resulting in contraction, and *curd* to denote a volume within which a physical characteristic becomes increasingly concentrated as a result of curdling.

ETYMOLOGY. *Curd* derives from the old

English *crudan,* 'to press, to push hard.' This erudition from Partridge 1958 is not necessarily irrelevant, since the etymological kin of curd doubtless include fractal kin of interest; see Chapter 23.

Note the following free associations: curds → cheese → milk → Milky Way → Galaxy ($\gamma\alpha\lambda\alpha$ = milk) → galaxies. I coined *curdling* while working on galaxies, and the etymological undertones of "galactic curdling" did not escape my notice.

OUTER CUTOFF AND
EXTRAPOLATED CANTOR DUSTS

As a prelude to the extrapolation of C, let us recall a point of history. When Cantor introduced C, he had barely left his original field, the study of trigonometric series. Since such series are concerned with periodic functions, the only extrapolation they involve is endless repetition. Now recall the self-explanatory terms of *inner* and *outer cutoff,* which Chapter 6 borrows from the study of turbulence. These are, respectively, the sizes ϵ and Ω of the smallest and the largest feature present in a set, and one may say that Cantor restricted himself to $\Omega=1$. The k-th construction stage yields $\epsilon=3^{-k}$, but $\epsilon=0$ for C itself. To achieve any other $\Omega<\infty$, for example the value of 2π appropriate in a Fourier series, one enlarges the periodic Cantor dust in the ratio Ω.

However, self-similarity, which this Essay views as valuable, is destroyed by repetition. But it is readily saved, if the initiator is used

only for extrapolation and if extrapolation follows an *inverse* or *upward* cascade. The first stage enlarges C in the ratio $1/r=3$ and positions it on [0,3]. The result is C plus a replica translated to the right and separated from C by a new trema of length 1. The second stage enlarges the outcome of the first stage in the same ratio 3 and positions it on [0,9]. The result is C plus 3 replicas translated to the right and separated by two new tremas of length 1, and one new trema of length 3. The upward cascade continues to enlarge C in the successive ratios of the form 3^k.

If one prefers, one may alternate two stages of interpolation, then a stage of extrapolation, etc. In this fashion, each series of three stages multiplies the outer cutoff Ω by 3 and divides the inner cutoff ϵ by 3.

◁ In this extrapolated dust, the negative axis is empty: an infinite trema. The underlying notion is discussed further in Chapter 13, where we tackle the (infinite) continent and the infinite cluster. ▶

DIMENSIONS D BETWEEN 0 AND 1

The set yielded by infinite interpolation and extrapolation is self-similar, and

$$D=\log N/\log(1/r)=\log 2/\log 3 \sim 0.6309,$$

a fraction between 0 and 1.

By following a different curdling rule, we can achieve other D's, in fact any dimension between 0 and 1. If the first stage trema is of length 1–2r, where 0 < r < ½, the dimension is log 2/log (1/r).

Further variety becomes possible if N≠2. For the sets with N=3 and r=1/5, we find

$$D=\log 3/\log 5 \sim 0.6826.$$

For the sets with N=2 and r=¼, we find

$$D=\log 2/\log 4=\frac{1}{2}.$$

For the sets with N=3 and r=1/9, we also find

$$D=\log 3/\log 9=\frac{1}{2}.$$

Although their D are equal, these last two sets "look" very different. This observation is taken up again and extended in Chapter 34, and leads to the notion of lacunarity.

Observe also that there is at least one Cantor set for every D < 1, but it follows from Nr≤1 that N < 1/r, hence D is never above 1.

$\mathcal{C}$ IS CALLED DUST BECAUSE $D_T=0$

While a Cantor set's D can vary between 0 and 1, from the topological viewpoint all Cantor sets are of dimension $D_T=0$, because any point is by definition cut from the other points, without anything having to be removed to cut it. From this viewpoint, there is no difference between $\mathcal{C}$ and finite sets of points! The fact that $D_T=0$ in this last case is familiar in standard geometry, and Chapter 6 uses

it in arguing that Koch's $\mathcal{K}$ is of topological dimension 1. But $D_T=0$ for all totally disconnected sets.

In the absence of accepted colloquial counterparts to "curve" and "surface" (which are connected sets with $D_T=1$ and $D_T=2$), I propose that sets with $D_T=0$ be called *dusts*.

GAPS' LENGTH DISTRIBUTION

In a Cantor dust, let u be a possible value of a gap's length, and denote by U the length when it is unknown, and by Nr(U > u) the number of gaps or tremas of length U greater than u. ◁ This notation is patterned after the notation Pr(U > u) of probability theory. ► One finds there is a constant prefactor F, such that the graph of the function Nr(U > u) constantly crosses the graph of Fu^{-D}. Here comes dimension again. With log u and log Nr as coordinates, the steps are uniform.

AVERAGE NUMBERS OF ERRORS

As in the case of a coastline, a rough idea of the sequence of errors is obtained if Cantor curdling stops with intervals equal to $\epsilon=3^{-k}$. The ϵ may be the length of time required to transmit a single symbol. One must also use Cantor's periodic extrapolation with a large but finite Ω.

The number of errors between times 0 and R, denoted by M(R), keeps time by counting only those instants that witness something noteworthy. It is an example of *fractal time*.

When the sample begins at t=0 (which is the only case to be considered here), the derivation of M(R) proceeds as in the case of the Koch curve. As long as R is smaller than Ω, the number of errors doubles each time R is multiplied by 3. As a result, $M(R) \propto R^D$.

This expression is like the standard expression for the mass of a disc or ball of radius R in D-dimensional Euclidean space. It is also identical to the expression obtained in Chapter 6 for the Koch curve.

As a corollary, the average number of errors per unit length varies roughly like R^{D-1} as long as R lies between the inner and the outer cutoffs. When Ω is finite, the decrease in the average number of errors continues to the final value of Ω^{D-1}, which is reached with $R=\Omega$. Thereafter, the density remains more or less constant. When Ω is infinite, the average number of errors decreases to zero. Finally, the empirical data often suggest that Ω is finite and very large, but fail to determine its value with any accuracy. If this is the case, the average number of errors has a lower limit that does not vanish but that is so ill-determined as to be of no practical use.

TREMA ENDPOINTS AND THEIR LIMITS

◁ The most conspicuous members of $\mathcal{C}$, the trema endpoints, do *not* exhaust $\mathcal{C}$; in fact they constitute but a tiny portion of it. The other points' physical importance is discussed in Chapter 19. ▶

THE CANTOR DUSTS' TRUE NATURE

The reader who has followed thus far and/or has heard the echo of the rapidly growing literature on Devil's Staircases (caption of Plate 83) must find it hard to believe that, when I started on this topic in 1962, everyone was agreeing that Cantor dusts are *at least* as monstrous as the Koch and Peano curves.

Every self-respecting physicist was automatically "turned off" by a mention of Cantor, ready to run a mile from anyone claiming $\mathcal{C}$ to be interesting in science, and eager to assert that such claims had been advanced, tested, and found wanting. My sole encouragement came from S. Ulam's suggestions, tantalizing despite their failure to be either developed or accepted, concerning the possible role for Cantor sets in the gravitational equilibrium of star aggregates; see Ulam 1974.

To publish on Cantor dusts, I had to erase every mention of Cantor!

But here we were led to $\mathcal{C}$ by Nature's own peculiarities. And Chapter 19 describes a second, very different, physical role for $\mathcal{C}$. All this must mean that the true nature of the Cantor dust is very different.

It is undeniable that in most cases $\mathcal{C}$ itself a very rough model, requiring many improvements. I contend, however, that *the very same properties that cause Cantor discontinua to be viewed as pathological are indispensable in a model of intermittency*, and must be preserved in more realistic substitutes for $\mathcal{C}$. ■

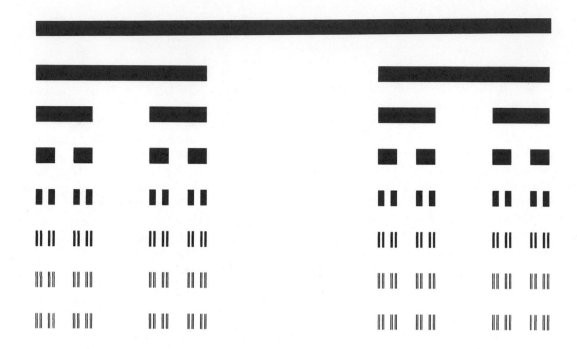

Plates 80 and 81 ◻ CANTORIAN TRIADIC BAR AND CAKE (HORIZONTAL SECTION DIMENSION D=log 2/log 3=0.6309). SATURN'S RINGS. CANTOR CURTAINS.

The Cantor dust uses [0,1] as initiator, and its generator is

N=2
r=⅓
D=log 2/log 3=.6309

PLATE 80. The Cantor dust is extraordinarily difficult to illustrate, because it is thin and spare to the point of being invisible. To help intuition by giving an idea of its form, thicken it into what may be called a Cantor bar. ◁ In technical terms, this is the Cartesian product of a Cantor dust of length 1, by an interval of length 0.03. ►

CURDLING. The construction of the Cantor bar results from the process I call *curdling*. It begins with a round bar (seen in projection as a rectangle in which width/length=0.03). It is best to think of it as having a very low density. Then matter "curdles" out of this bar's middle third into the end thirds, so that the positions of the latter remain unchanged. Next matter curdles out of the middle third of each end third into its end thirds, and so on ad infinitum until one is left with an infinitely large number of infinitely thin slugs of infinitely high density. These slugs are spaced along the line in the very specific fashion induced by the generating process. In this illustration, curdling (which eventually requires hammering!) stops when both the printer's press and our eye cease to follow; the last line is indistinguishable from the last but one: each of its ultimate parts is seen as a gray slug rather than two parallel black slugs.

CANTOR CAKE. When curdling starts with a pancake, much less thick than it is wide, and dough curdles into thinner pancakes (while exuding an appropriate filling), one ends up with an infinitely extrapolated Napoleon, which one might call *Cantor cake*.

SATURN'S RINGS. Saturn was originally believed to have a single ring around it. But eventually a break was discovered, then two, and now Voyager I has identified a very large number of breaks, mostly very thin ones. Voyager also established that the rings are diaphanous: they let sunlight through...as befits a set we called "thin and spare."

Thus, the rings' structure (see Stone & Minen 1981, especially the cover illustration) is suggestive of a collection of near circles, each with a radius corresponding to the distance from some origin to a point in Cantor dust. ◁ The technical term is Cartesian product of a Cantor dust by a circle. Actually, it may be that a closer picture is given by a circle's product with a dust with positive measure, like those examined in Chapter 15. ► Last minute insert: The same idea is stated independently to Avron & Simon 1981, which relates it to Hill's equation; their Note 6 includes many other relevant references.

SPECTRA. Harter 1979–1981 describes some spectra of organic molecules whose resemblance to a Cantor dust is stunning.

PLATE 81. Here, the Cantor dust's shape is clarified by being placed among generalized dusts with $N=2$ and variable r. The vertical coordinate is either r itself, ranging from 0 to ½ (bottom figure), or D ranging from 0 to 1 (top figure). Both theater curtains are topped by the full interval [0,1]. Every horizontal cut of either figure is some Cantor dust, with the arrows pointing out $r = ⅓$ and $D = 0.6309$.

A FAMOUS GREEK PARADOX. Greek philosophers believed that, in order to be indefinitely subdivisible, a body had to be continuous. They had not heard of Cantor dusts. ▪▪

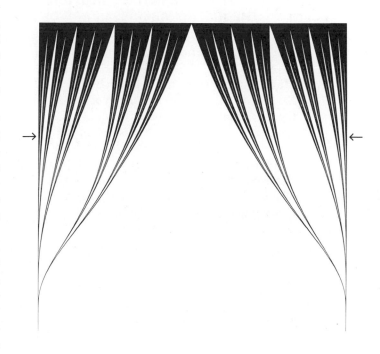

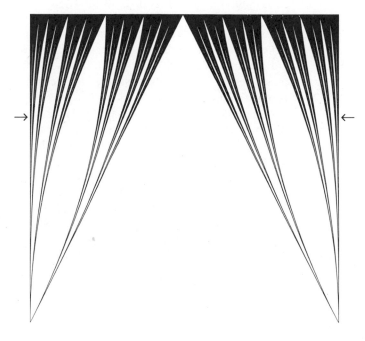

Plate 83 ◻ CANTOR FUNCTION, OR DEVIL'S STAIRCASE (DIMENSION D=1 THE RISERS' ABSCISSAS ARE OF DIMENSION D~0.6309). CANTOR MOTION

The Cantor function describes the distribution of mass along the Cantor bar of Plate 80. Many writers refer to its graph as the *Devil's Staircase*, because it is odd indeed. Set both the bar's length and mass as equal to 1, and for every value of the abscissa R define M(R) as the mass contained between 0 and R. Since there is no mass in the gaps, M(R) remains constant along intervals that add up to the whole length of the bar. However, since hammering does not affect the total mass in the bar, M(R) must manage to increase *somewhere* from the point of coordinates (0,0) to the point of coordinates (1,1). It increases over infinitely many, infinitely small, highly clustered jumps corresponding to the slugs. Hille & Tamarkin 1929 describes this function's odd properties in detail.

REGULARIZING MAPPINGS. The Devil's staircase accomplishes the feat on mapping the drastic nonuniformity of the Cantor bar into something uniform and homogeneous. Starting with two different intervals of the same length on the vertical scale, the inverse function of the Cantor staircase yields two collections of slugs that contain the same mass—even though they usually look very different from each other.

Since science thrives on uniformity, it often happens that such regularizing transformations make fractal irregularity accessible to analysis.

FRACTAL HOMOGENEITY. It is convenient to describe the distribution of mass in the Cantor bar as *fractally homogeneous*.

CANTOR MOTION. As in the case of the Koch curve reinterpreted as Koch motion, or of the Peano motion, it is useful to reinterpret the ordinate M(R) as a time. If so, the inverse function R(M) gives the position of a *Cantor motion* at time t. This motion is most discontinuous. Chapters 31 and 32 describe a ran-domized linear and spatial generalizations.

FRACTAL DIMENSION. The sums of the widths and of the heights of the steps both equal 1, and one finds in addition that this curve has a well-defined length equal to 2. A curve of finite length is called rectifiable and is of dimension D=1. This example demonstrates that the dimension D=1 is compatible with the presence of many irregularities, as long as they remain sufficiently scattered.

◁ One would love to call the present curve a fractal, but to achieve this goal we would have to define *fractals* less stringently, on the basis of notions other than D alone. ▶

SINGULAR FUNCTIONS. The Cantor staircase is a nondecreasing and nonconstant function that is singular, in the sense that it is continuous but *non*differentiable. Its derivative vanishes almost everywhere, and its continuous variation manages to occur over a set whose length—i.e., linear measure—vanishes.

Any nondecreasing function can be written as the sum of a singular function, of a function made of discrete jumps, and of a differentiable function. The last two components are classical in mathematics and of wide use in physics. On the other hand, the singular component is widely regarded in physics as pathological and totally devoid of uses. A principal theme of this Essay is that this last opinion is totally devoid of merit.

DEVIL'S STAIRCASES IN STATISTICAL PHYSICS. The publication of this plate in my 1977 Essay brought the Devil's staircase to the physicists' attention, and stimulated an extensive literature. Diagrams analogous to the "curtains" of Plate 81, or the Fatou curtain of Plate 185, are encountered with growing frequency. See Aubry 1981. Important earlier work (Azbel 1964, Hofstadter 1976), which used to be isolated, merges with this new development. ■

9 ¤ Fractal View of Galaxy Clusters

In Chapters 6 and 7, the Koch and Peano fractals are introduced via geomorphology, but the most significant uses of fractals are rooted elsewhere. Inching toward the mainstream of science, this chapter and the next two tackle two issues of exceptional antiquity, importance and difficulty.

The distribution of the stars, the galaxies, the clusters of galaxies, and so on fascinates the amateur as well as the specialist, yet clustering remains peripheral to astronomy and to astrophysics as a whole. The basic reason is that no one has yet explained why the distribution of matter falls into an irregular hierarchy, at least within a certain range of scales. While there are allusions to clustering in most works on the subject, serious theoretical developments hasten to sweep it under the rug, claiming that on scales beyond some large but unspecified threshold galaxies are uniformly distributed.

Less fundamentally, the hesitation in dealing with the irregular arises from the absence of tools to describe it mathematically. Statistics is asked to decide between two assumptions, only one of which is thoroughly explored (asymptotic uniformity). Is it surprising that the results are inconclusive?

The questions, however, refuse to be set aside. In parallel with efforts to *explain,* I think it indispensable to *describe* clustering, and to mimic reality by purely geometric means. The fractal treatment of this subject, scattered over several chapters of this Essay, proposes to show by explicitly constructed models that the evidence is compatible with a degree of clustering that extends far beyond the limits suggested by existing models.

The present introductory chapter describes an influential theory of the formation of stars and galaxies, due to Hoyle, the principal descriptive model of their distribution, due to Fournier d'Albe (also known as the Charlier model), and, most important, sketches some empirical data. It is shown that both theories and data can be interpreted in terms of a scaling fractal dust. I argue that the distribution of galaxies and of stars includes a zone of self-similarity in which the fractal dimension satisfies $0 < D < 3$. Theoretical reasons for expecting $D=1$ are sketched, raising the question of why the observed D is ~1.23.

PREVIEW. Chapter 22 uses fractal tools to improve our understanding of what the cosmological principle means, how it can and should be modified, and why the modification demands randomness. A discussion of improved model clusters is withheld until Chapters 22, 23, and 32 to 35.

IS THERE A GLOBAL DENSITY OF MATTER?

Let us begin with a close examination of the concept of global density of matter. As with the concept of the length of a coastline, things seem simple, but in fact go awry very quickly and most interestingly. To define and measure density, one starts with the mass $M(R)$ in a sphere of radius R centered on Earth. The approximate density, defined as

$$M(R)/[(4/3)\pi R^3],$$

is evaluated. After that, the value of R is made to tend toward infinity, and the global density is defined as the limit toward which the approximate density converges.

But need the global density converge to a positive and finite limit? If so, the speed of convergence leaves a great deal to be desired. Furthermore, the estimates of the limit density had behaved very oddly in the past. As the depth of the world perceived by telescopes increased, the approximate density diminished in a surprisingly systematic manner. According to de Vaucouleurs 1970, it has remained $\propto R^{D-3}$. The observed exponent D is much smaller than 3, the best estimate, on the basis of indirect evidence, being $D=1.23$.

The thesis of de Vaucouleurs is that the behavior of the approximate density reflects reality, meaning that $M(R) \propto R^D$. This formula recalls the classical result that a ball of radius R in a Euclidean space of dimension E has a volume $\propto R^E$. In Chapter 6 we encounter the same formula for the Koch curve, with the major difference that the exponent is not the Euclidean dimension $E=2$ but a fraction-valued fractal dimension D. And Chapter 8 derives $M(R) \propto R^D$ for the Cantor dust on the time axis (for which $E=1$).

All these precedents suggest very strongly that the de Vaucouleurs exponent D is a fractal dimension.

ARE STARS IN THE SCALING RANGE?

Obviously, the scaling range in which D satisfies $0 < D < 3$ must end before one reaches objects with well-defined edges, such as planets. But does it, or does it not, include stars? According to data by Webbink reported in Faber & Gallagher 1980, the mass of the Milky

Way interior to radius R may very well be represented as $M(R) \propto R^D$, with the D extrapolated from galaxies. But we continue our discussion exclusively in terms of galaxies.

IS THERE AN UPPER CUTOFF TO THE SCALING RANGE?

The question of how far the range in which $0 < D < 3$ extends in the direction of very large scales is controversial and the subject of renewed activity. Many authors either state or imply that the scaling range admits of an outer cutoff corresponding to clusters of galaxies. Other authors disagree. De Vaucouleurs 1970 asserts that "clustering of galaxies, and presumably of all forms of matter, is the dominant characteristic of the structure of the universe on all observable scales with no indication of an approach to uniformity; the average density of matter decreases steadily as even larger volumes of space are considered, and there is no observational basis for the assumption that this trend does not continue out to much greater distances and lower densities."

The debate between these two schools of thought is interesting and important to cosmology—but not for the purposes of this Essay. Even if the range in which $0 < D < 3$ is cut off at both ends, its importance is sufficient in itself to warrant a careful study.

In either case, the Universe (just like the ball of thread discussed in Chapter 3) appears to involve a sequence of several different effective dimensions. Starting with scales of the order of Earth's radius, one first encounters the dimension 3 (due to solid bodies with sharp edges). Then the dimension jumps to 0 (matter being viewed as a collection of isolated points). Next is the range of interest, ruled by some nontrivial dimension satisfying $0 < D < 3$. If scaling clustering continues ad infinitum, so does the applicability of this last value of D. If, on the contrary, there is a finite outer cutoff, a fourth range is added on top, in which points lose their identity and one has a uniform fluid, meaning that the dimension again equals 3.

On the other hand, the most naive idea is to view the galaxies as distributed near uniformly throughout the Universe. Under this untenable assumption, one has the sequence $D=3$, then $D=0$, and again $D=3$.

◁ The general theory of relativity asserts that in the absence of matter, the local geometry of space tends to be flat and Euclidean, with the presence of matter making it locally Riemannian. Here we could speak of a globally flat Universe of dimension 3 with local $D < 3$. This type of disturbance is considered in Selety 1924, an obscure reference which fails to refer to Koch but includes (p. 312) an example of the construction of Chapter 6. ▶

THE FOURNIER UNIVERSE

It remains to construct a fractal that satisfies $M(r) \propto R^D$, and see how it agrees with accepted views concerning the Universe. The first fully described model of this kind is due to E. E.

Fournier d'Albe (Chapter 40). While Fournier 1907 is largely a work of fiction disguised as science, it also contains genuinely interesting considerations to which we come momentarily. It is best, however, to first describe the structure it proposes.

Its construction begins with the centered regular octahedron whose projection is represented near the center of Plate 95. The projection reduces to the four corners of a square whose diagonal is set to be of length 12 "units," and to this square's center. But the octahedron also includes two points above and below our plane, on the perpendicular drawn from the center of the square, and at the same distance of 6 units from this center.

Now, each point is replaced with a ball of radius 1, to be viewed as "stellar aggregate of order 0." And the smallest ball including the basic 7 balls is to be called a "stellar aggregate of order 1." An aggregate of order 2 is achieved by enlarging an aggregate of order 1 in the ratio $1/r=7$ and by replacing each of the resulting balls of radius 7 by a replica of the aggregate of order 1. In the same way, an aggregate of order 3 is achieved by enlarging an aggregate of order 2 in the ratio $1/r=7$ and by replacing each ball by a replica of the aggregate of order 2. And so on.

In sum, between two successive orders of aggregation, the number of points and the radius are enlarged in the ratio $1/r=7$. Consequently, whenever R is the radius of some aggregate, the function $M_0(R)$ expressing the number of points contained in a ball of radius R is $M_0(R)=R$. For intermediate values of R, $M_0(R)$ is smaller (reaching down to $R/7$), but the overall trend is $M_0(R) \propto R$.

Starting from aggregates of order 0, it is also possible to interpolate by successive stages to aggregates of orders −1, −2, and so on. The first stage replaces each aggregate of order 0 with an image of the aggregate of order 1, reduced in the ratio $1/7$, and so forth. If one does so, the validity of the relationship $M_0(R) \propto R$ is extended to ever smaller values of R. After infinite extra- and interpolation, we have a self-similar set with $D = \log 7 / \log 7 = 1$.

We may also note that an object in 3-space for which $D=1$ need not be a straight line nor any other rectifiable curve. It need not even be connected. Each D is compatible with any lesser or equal value of the topological dimension. In particular, since the doubly infinite Fournier universe is a totally disconnected "dust," its topological dimension is 0.

DISTRIBUTION OF MASS; FRACTAL HOMOGENEITY

The step from geometry to the distribution of mass is obvious. If each stellar aggregate of order 0 is loaded with a unit mass, the mass $M(R)$ within a ball of radius $R>1$ is identical to $M_0(R)$, hence $\propto R$. Furthermore, to generate aggregates of order −1 from aggregates of order 0 amounts to breaking up a ball that had been viewed as uniform, and finding it to be made of seven smaller ones. This stage extends the rule $M(R) \propto R$ below $R=1$.

When viewed over the whole 3-space, the resulting mass distribution is grossly inhomogeneous, but over the Fournier fractal it is as homogeneous as can be. (Recall Plate 80.) In particular, any two geometrically identical portions of the Fournier universe carry identical masses. I propose that such a distribution of mass be called *fractally homogeneous*.

◁ The preceding definition is phrased in terms of scaling fractals, but the concept of fractal homogeneity is more general. It applies to any fractal for which the Hausdorff measure for the dimension D is positive and finite. Fractal homogeneity requires the mass carried by a set to be proportional to the set's Hausdorff measure. ▶

FOURNIER UNIVERSE VIEWED AS CANTOR DUST. EXTENSION TO D≠1

I trust the reader was not distracted by the casual use of fractal terminology in the opening sections of this chapter. It is obvious that, without being aware of the fact, Fournier was traveling along a track parallel to that of Cantor, his contemporary. The main difference is that the Fournier construction is imbedded in space instead of the line. To further improve the resemblance, it suffices to change Fournier's aggregates from being balls to being bricks (filled-in cubes). Now, each aggregate of order 0 is a brick of side 1, and it includes 7 aggregates of side 1/7: one of them has the same center as the initial cube, and the other six touch the central subsquares of

the faces of the original cube.

Later we will examine how Fournier obtains the value D=1 from basic physical phenomena, and how Hoyle obtains this same value. Geometrically, however, D=1 is a special case, even if one preserves the overall octahedron and the value N=7. Since the balls do not overlap, 1/r can take any value between 3 and infinity, yielding $M(R) \propto R^D$, with $D = \log 7 / \log (1/r)$ anywhere between 0 and $\log 7 / \log 3 = 1.7712$.

Further, given any D satisfying D<3, it is easy by changing N to construct variants of Fournier's model having this dimension.

THE CHARLIER MODEL AND OTHER FRACTAL UNIVERSES

The above constructs share every one of the characteristic defects of first fractal models. Most conspicuously, just like the Koch curve model in Chapter 6 and the Cantor dust model in Chapter 8, the Fournier model is so regular as to be grotesque. As a corrective, Charlier 1908, 1922 suggests that one allow N and r to vary from one hierarchical level to another, taking on the values N_m and r_m.

The scientific eminence of Charlier was such that, despite the praise he lavished on Fournier, writing in the leading scientific languages of the day, even the simple model soon became credited to its famous expositor instead of its unknown author. It was much discussed in its time, in particular in Selety 1922, 1923a, 1923b, 1924. Furthermore, the model attracted the attention of the very influential Emile Borel, whose comments in Borel 1922, while dry, were perceptive. But from then on, aside from fitful revivals, the model fell into neglect (for not very convincing reasons noted in North 1965, pp. 20–22 and 408–409). Nevertheless, it refuses to die. The basic idea was independently reinvented many times to this day, notably in Lévy 1930. (See the LÉVY entry in Chapter 40.) Most important, the fractal core notion of the Fournier universe is implicit in the considerations about turbulence and galaxies in von Weizsäcker 1950 (see Chapter 10), and in the model of the genesis of the galaxies due to Hoyle 1953, which will be discussed momentarily.

The basic fractal ingredient is also present in my models, Chapters 32 to 35.

In this light, the question arises of whether a model of galaxy distribution can *fail* to be a fractal with one or two cutoffs. I think not. If one agrees that the distribution must be scaling (for reasons to be elaborated in Chapter 11) and that the set on which matter concentrates is not a standard scaling set, it must be a fractal set.

Granted the importance of scaling, Charlier's nonscaling generalization of the Fournier model is ill-inspired. ◁ Incidentally, it allows $\log N_m / \log (1/r_m)$ to vary with m between two bounds, $D_{min} > 0$ and $D_{max} < 3$. We have here yet another theme: effective dimension need not have a single value, and may drift between an upper and a lower limit. This theme is picked up again in Chapter 15. ▶

FOURNIER'S REASON TO EXPECT D=1

We now describe the impressive argument that leads Fournier 1907, p. 103, to conclude that D must be equal to 1. This argument is a strong reason for not forgetting its author.

Consider a galactic aggregate of arbitrary order with mass M and radius R. Using without misgivings a formula applicable to objects with spherical symmetry, assume that the gravitational potential on the surface is GM/R (G being the gravitational constant). A star falling on this universe impacts with the velocity V equal to $(2GM/R)^{1/2}$.

To paraphrase Fournier, an important conclusion may be drawn from the observation that no stellar velocity exceeds 1/300 of the velocity of light. It is that the mass comprised within a world ball increases as its radius, and not as its volume, or in other words, that the density within a world ball varies inversely as the surface of the ball... To make this clearer, the potential at the surface would be always the same, being proportional to the mass and inversely proportional to the distance. And as a consequence, stellar velocities approaching the velocity of light would not prevail in any part of the universe.

HOYLE CURDLING; THE JEANS CRITERION ALSO YIELDS D=1

A hierarchical distribution also arises in a theory advanced in Hoyle 1953, according to which galaxies and stars form by a cascade process starting with a uniform gas.

Consider a gas cloud of temperature T and mass M_0, distributed with a uniform density throughout a ball of radius R. As shown by Jeans a "critical" situation prevails when $M_0/R_0 = JkRT/G$. (Here, k is the Boltzmann constant and J a numerical coefficient.) In this critical case, the primordial gaseous cloud is unstable and must inevitably contract.

Hoyle postulates (a) that M_0/R_0 takes on this critical value at some initial stage, (b) that the resulting contraction stops when the volume of the gas cloud drops to 1/25-th, and (c) that each cloud then splits into five clouds of equal size, mass $M_1 = M_0/5$, and equal radius $R_1 = R_0/5$. Thus the process ends as it started: in an unstable situation followed by a second stage of contraction and subdivision, then a third, and so on. But curdling stops as clouds become so opaque that the heat due to gas collapse can no longer escape.

As in the diverse other fields where the same cascade process is encountered, I propose that the five clouds be called *curds*, and that the cascade process be called *curdling*. As said when I introduced this last term, I could not resist its juxtaposition with *galactic*.

Fournier injects N=7 to facilitate the graphical illustration, but Hoyle claims that N=5 has a physical basis. In another contrast with Fournier, whose geometrical illustration is detailed beyond what is reasonable or needed, Hoyle is vague about the curds' spatial scatter. An explicit implementation has to wait until we describe random curdling in Chapter 23. But these discrepancies do not matter: the main fact is that r=1/N, so that D=1 must be part of the design if curdling is to end as it began, in Jeans instability.

Further, if the duration of the first stage is taken to be 1, gas dynamics shows that the mth stage's duration is 5^{-m}. It follows that the same process could continue to infinity within a total time of 1.2500.

EQUIVALENCE OF THE FOURNIER AND HOYLE DERIVATIONS OF D=1

At the edge of an unstable gas cloud satisfying the Jeans criterion, the velocity and the temperature are linked by $V^2/2 = JkT$, because GM/R is equal to $V^2/2$ (Fournier) and to JkT (Jeans). Now recall that in statistical thermodynamics the temperature of a gas is proportional to the mean square velocity of its molecules. Hence the combination of the Fournier and Jeans criteria suggests that at the edge of a cloud the velocity of the fall of a macroscopic object is proportional to the average velocity of its molecules. A careful analysis of the role of temperature in the Jeans criterion is bound to show the two criteria to be equivalent. ◁ Most likely, the analogy extends to the M(R)∝R relationship *within* galaxies, reported in Wallenquist 1957. ▶

WHY D=1.23 AND NOT D=1?

The disagreement between the empirical D=1.23 and the Fournier and Hoyle theoretical D=1 raises an important issue. P. J. E. Peebles tackled it in 1974 by relativity theory. See Peebles 1980, a full treatment of the physics and of the statistics (but not of the geometry) of this topic.

THE SKY'S FRACTAL DIMENSION

The sky is a projection of a universe, in which every point is first described by its spherical coordinates ρ, θ, and ϕ and then replaced by the point of spherical coordinates 1, θ, and ϕ. When the universe is a fractal of dimension D, and the origin of the frame of references belongs to the universe (see Chapter 22), the structure of this projection is "typically" ruled by the following alternative: D>2 implies that the projection covers a nonzero proportion of the sky, while D<2 implies that the projection is itself of dimension D. ◁ As exemplified in Plates 95 and 96, *typical* allows for exceptions, due to the structure of the fractal and/or the choice of origin. It often means "true with probability 1." ▶

ASIDE ON THE BLAZING SKY EFFECT (WRONGLY CALLED OLBERS PARADOX)

The rule in the preceding section bears directly upon the motivation that led diverse writers (including Fournier) to variants of a fractal Universe. They recognized that such universes "exorcise" geometrically the *Blazing Sky Effect*, often (but wrongly) called *Olbers paradox*. Under the assumption that the distribution of celestial bodies is uniform, meaning that D=3 for all scales, the sky is lit near uniformly, during the night and during the day, to the brighness of the solar disc.

This paradox is no longer of interest to physicists, having been eliminated by relativity theory and the theory of the expansion of the Universe, and other arguments. But its demise left a peculiar by-product: numerous commentators invoke their preferred explanation of the Blazing Sky Effect as an excuse for neglecting clustering, and even as an argument for denying its reality. This is a truly odd viewpoint: even if galaxies *need not* be clustered to avoid the Blazing Sky Effect, they *are* clustered, and this characteristic demands careful study. Furthermore, as seen in Chapter 32, the expansion of the Universe is compatible not only with standard homogeneity but also with fractal homogeneity.

The Blazing Sky argument is simplicity itself. When the light emitted by a star is proportional to its surface area, the amount of light reaching an observer at a distance of $R \propto 1/R^2$, but the star's apparent surface is itself $\propto 1/R^2$. Thus, the apparent ratio of light to spherical angle is independent of R. Also, when the distribution of stars in the Universe is uniform, almost any direction in the sky sooner or later intersects some star. Therefore, the sky is uniformly bright, and seems ablaze. (The Moon's disc would form an exceptional *dark* domain, at least, in the absence of atmospheric diffusion.)

On the other hand, the assumption that the

universe is fractal with D < 2 resolves the paradox. In that case, the universe's projection on
the sky is a fractal with the same D, hence a
set of zero area. Even if the stars are given a
nonzero radius, a large proportion of directions go to infinity without encountering any
star. Along these directions, the night sky is
black. When the range in which D < 3 is followed by a range in which D=3, the sky's
background is not strictly black but illuminated extremely faintly.

The Blazing Sky Effect was noticed by
Kepler shortly after Galileo's *Sidereal
Message* had commented favorably on the
notion that the Universe is unbounded. In his
1610 *Conversation with the Sidereal
Messenger*, Kepler rejoined: "You do not hesitate to declare that there are visible over
10,000 stars... If this is true, and if [the stars
have] the same nature as our sun, why do not
these suns collectively outdistance our sun in
brilliance?... But maybe the intervening ether
obscures them? Not in the least... It is quite
clear that...this world of ours does not belong
to an undifferentiated swarm of countless others." (Rosen 1965, pp. 34–35.)

This conclusion remained controversial,
but the argument was not forgotten, witness
the comment by Edmund Halley, in 1720,
that: "Another Argument I have heard urged,
that if the number of Fixt Stars were more
than finite, the whole superficies of their apparent Sphere would be luminous." Later, this
conclusion was discussed by De Chéseaux and
J. H. Lambert, but came to be credited to
Gauss's great friend, Olbers. The term

"Olbers paradox" that became attached to it
is scandalous but symptomatic. Observations
that had been rejected into the "unclassified
residuum" (page 28) become all too often
credited to the first Establishment figure who
decorates them by a classifiable wrapping,
however transient. Historical discussions are
found in Gamow 1954, Munitz 1957, North
1965, Dickson 1968, Wilson 1965, Jaki 1969,
Clayton 1975, and Harrison 1981.

ASIDE ON NEWTONIAN GRAVITATION

The Rev. Bentley kept pestering Newton with
an observation closely related to the Blazing
Sky Effect: if the stars' distribution is homogeneous, the force they exert on one among
them is infinite. One may add that their gravitational potential is infinite. And that any
distribution wherein $M(R) \propto R^D$ for large R
yields an infinite potential unless D < 1. The
modern theory of potentials (Frostman theory) confirms that there is a privileged link
between Newton's gravitation and the value
D=1. The Fournier and Hoyle derivations of
D=1 cannot fail to be related to this link.
◁ Fournier's theme of "the gravitational potential at the surface being always the same"
is central to modern potential theory. ▶

ASIDE ON RELATIVITY THEORY

◁ To paraphrase de Vaucouleurs 1970:
"Relativity theory led us to believe that to be

optically observable, no stationary material ball can have a radius R less than the Schwarzchild limit $R_M = 2GM/c^2$, where c is the velocity of light. In a plot of the mean density ρ and the characteristic radius R of various cosmical systems, $\rho_M = 3c^2/8\pi GR_M^2$ defines an upper limit. The ratio ρ/ρ_M may be called the Schwarzchild filling factor. For most common astronomical bodies (stars) or systems (galaxies), the filling factor is very small, on the order of 10^{-4} to 10^{-6}." The square of the velocity ratio postulated by Fournier is $(300)^{-2} \sim 10^{-5}$, precisely in the range middle of the above. ▶

AN AGGLUTINATED FRACTAL UNIVERSE?

Many authors think one may explain the genesis of stars and other celestial objects by an *ascending* cascade (i.e., the agglutination of greatly dispersed dust particles into increasingly bigger pieces) rather than by a *descending* cascade à la Hoyle (i.e., the fragmentation of very large and diffuse masses into smaller and smaller pieces).

An analogous alternative arises in connection with the cascades postulated in the study of turbulence, Chapter 10. Richardson's cascade descends toward ever smaller eddies, but ascending cascades may also be present; see Chapter 40, under RICHARDSON. Thus it may be hoped that the interrelations between descending and ascending cascades will be clarified soon.

FRACTAL TELESCOPE ARRAYS

To wind up this discussion, nothing can be more appropriate than a comment about the tools used to observe the galaxies. Dyson 1977 suggests that it may be advantageous to replace one piece telescopes by arrays of small telescopes. The diameter of each would be about 0.1 m, equal to the patch size of the smallest optically significant atmospheric disturbance, their centers would form a fractally hierarchical pattern, and they would be linked by Currie interferometers. A rough analysis leads to the conclusion that a suitable value for the dimension would be ⅔. Dyson's conclusion: "A 3-kilometer array of 1024 ten-centimeter telescopes connected by 1023 interferometers is not a practical proposition today. [It is offered] as a theoretical ideal, to show what can be done in principle."

SURVEY OF RANDOM FRACTAL MODELS OF GALAXY CLUSTERS

If one grants the claim that the distribution of galaxies is described usefully by unknowingly fractal models of limited subtlety and versatility, one should not be surprised that knowingly fractal random models provide even more useful descriptions. To begin with, our understanding of Hoyle curdling improves when it is set in its proper context: random fractals (Chapter 23). Of greater significance, I think, are the random models I developed and discuss in Chapters 32 to 35. One reason for dwelling on several models is that improvement in the quality of description is "paid for" by increased complication. A second reason is that each model involves a fractal dust that deserves attention. Let me survey these models here, out of logical order.

Around 1965, my ambition was to implement the relationship $M(R) \propto R^D$ with $D < 3$ with a model in which there is no "center of the universe." I first achieved this goal by the random walk model described in Chapter 32. Then, as an alternative, I developed a trema model, which consists in cutting out from space a collection of mutually independent *randomly placed* tremas of *random* radius, ranging up to an upper cutoff L that may be either finite or infinite.

Since both models had been selected solely on the basis of formal simplicity, it was delightfully surprising to discover they have predictive value. My theoretical correlation functions (Mandelbrot 1975u) agree with the curve-fitted ones reported in Peebles 1980 (see pp. 243-249). ◁ More precisely, my two approaches agree on the 2-point correlation, my random walk yields a good 3-point correlation and a bad 4-point correlation, and my spherical tremas model is very good for all known correlations. ▶

Unfortunately, the appearance of samples generated by either model is quite unrealistic. Using a notion that I developed for this very purpose and describe in Chapter 35, they have unacceptable lacunarity properties. For the trema model this defect is corrected by introducing more elaborate trema shapes. For the

random walk model, I use a less lacunar "subordinator."

Thus, the study of galaxy clusters has greatly stimulated the development of fractal geometry. And today the uses of fractal geometry in the study of galaxy clusters go well beyond the tasks of streamlining and housecleaning accomplished in the present chapter.

CUT DIAMONDS LOOK LIKE STARS

And the distribution of raw diamonds in the Earth's crust resembles the distribution of stars and galaxies in the sky. Consider a large world map on which each diamond mine or diamond rich site—past or present—is represented by a pin. Where examined from far away, these pins' density is extraordinarily uneven. A few are isolated here and there, but most concentrate in a few blessed (or accursed) areas. However, the Earth's surface in these areas is *not* uniformly paved with diamonds. When examined more closely, any of these areas turns out itself to be mostly blank, with scattered subareas of much greater diamond concentration. The process continues over several orders of magnitude.

Is it not irresistible to inject curdling in this context? Indeed, an unknowingly fractal model has been advanced by de Wijs, as seen under NONLACUNAR FRACTALS in Chapter 39. ■

Plate 95 ◻ PROJECTION OF FOURNIER'S MULTIUNIVERSE (DIMENSION D~0.8270)

This plate represents to scale both the projection and the "equatorial" section of a Universe of dimension D=1 described in the text. See also Plate 96.

To paraphrase the caption in Fournier 1907: "A multiuniverse constructed upon a cruciform or octahedral principle is not the plan of the world but is useful in showing that an infinite series of similar successive universes may exist without producing a 'blazing sky.' *The matter in each world sphere is pro-portional to its radius.* This is the condition required for fulfilling the laws of gravitation and radiation. In some directions the sky would appear quite black, although there is an infinite succession of universes. The 'world ratio' in this case is N=7 instead of 10^{22}, as in reality."

In the sense described in Chapter 34, a universe with D=1 and N=10^{22} is of very low lacunarity, but extraordinarily stratified.

■

Plate 96 ¤ A FLAT FOURNIER UNIVERSE WITH D=1

Plate 95, being drawn to exact scale, is not only hard to print and to see, but potentially misleading. Indeed, it is *not* a universe of dimension D=1 but its planar projection, whose dimension is D=log 5/log 7~0.8270<1. Therefore, in order to avoid leaving the wrong impression, we hasten to exhibit a regular Fournier-like planar pattern of dimension D=1. The construction, which involves 1/r=5 instead of 1/r=7, is carried one step further than is possible in Plate 95. ▪▪▪

10 ¤ Geometry of Turbulence; Intermittency

The study of turbulence is one of the oldest, hardest, and most frustrating chapters of physics. Common experience suffices to show that under certain circumstances the flow of a gas or a liquid is smooth, the technical term being "laminar," while under different circumstances it is not smooth at all. But where should we draw a line? Should the term "turbulence" denote all unsmooth flows, including much of meteorology and oceanography? Or is it better to reserve it for a narrow class, and, if so, for which one? Each scholar seems to answer these questions differently.

This disagreement does not matter here, because we focus on unquestionably turbulent flows, whose most conspicuous characteristic resides in the absence of a well-defined scale of length: they all involve coexistent "eddies" of all sizes. This feature can already be recognized in Leonardo's and Hokusai's drawings. It demonstrates that turbulence is necessarily foreign to the spirit of the "old" physics that focused upon phenomena having well-defined scales. But this same reason makes the study of turbulence of direct interest to us.

As some readers know, practically all investigations of turbulence concentrate upon the *analytic* study of fluid flow, and leave the *geometry* aside. I like to think that this lack of balance does not reflect a perceived lack of importance. In fact, many geometric shapes involved in turbulence are easily seen or made visible and cry out for a proper description. But they could not receive the attention they deserve until the development of fractal geometry. Indeed, as I immediately surmised, turbulence involves many fractal facets, of which I describe a few in this and later chapters.

Two disclaimers are necessary. First of all, we leave aside the problem of the onset of turbulence in a laminar fluid. There is strong reason to believe that this onset has fractal aspects of great importance, but they have not been clarified enough to be discussed here. Secondly, such periodic structures as Bénard cells and Kármán streets do not concern us here.

This chapter begins with pleas for a more geometric approach to turbulence and for the use of fractals. These pleas are numerous but each is brief, because they involve suggestions with few hard results as yet.

After that, we focus on the problem of intermittency, which I have investigated actively. My most important conclusion is that the region of dissipation, namely the spatial set on which turbulent dissipation is concentrated, can be modeled by a fractal. Measurements done for different purposes suggest that this region's D lies around 2.5 to 2.6, but probably below 2.66.

Unfortunately, the model cannot be pinpointed accurately, until we determine the topological properties of the region of dissipation. In particular, is it a dust, or a wiggly and branched curve (vortex tube), or a wiggly and layered surface (vortex sheet)? The first conjecture is unlikely, while the second and third suggest models akin to the ramified fractals of Chapter 14. But we are in no position to decide. Progress on the new fractal front does not help the old topological front at all. Our knowledge of the geometry of turbulence remains primitive indeed.

The bulk of this chapter requires no expertise. ◁ But the specialist will observe that part of fractal analysis of turbulence is the geometric counterpart of the analytic analysis of correlations and spectra. The relationship between turbulence and probability theory is an old story. Indeed, G. I. Taylor's earliest work was, after Perrin's Brownian motion, the second major influence on Norbert Wiener's creation of a mathematical theory of stochastic processes. Spectral analysis has long since "paid back" (with accrued interest) what it once borrowed from the study of turbulence and now it is time for the theory of turbulence to take advantage of the development of a sophisticated stochastic geometry. In particular, the Kolmogorov spectrum has a geometric counterpart examined in Chapter 30. ►

CLOUDS, WAKES, JETS, ETC.

A generic problem in the geometry of turbulence concerns the shape of the boundary of the region where some characteristic of the fluid is encountered. Striking examples are the billows upon billows which one finds in the ordinary (water) clouds, as well as in the clouds provoked by volcanic eruptions and in nuclear mushrooms. At this stage of this Essay, it is indeed difficult to escape the impression that, insofar as there is a range of scales wherein a cloud can be said to have a well-defined boundary, cloud boundaries *must* be fractal surfaces. The same remark applies to the patterns of rain squalls seen on radar screens. (For a first confirmation of this hunch, see Chapter 12.)

But I prefer to deal with simpler shapes. Turbulence may be restricted to a portion of an otherwise laminar fluid, say a wake or a jet. In the roughest approximation, each is a rod. If, however, the boundary is examined in detail, it reveals a hierarchy of indentations, whose depth increases with the value of the classic measure of hydrodynamic scale, called Reynolds number. This very visible and complex "local" structure does not evoke a rod as much as a rope with many loosely attached strings floating around. Its typical cross section is not at all circular, but closer in shape to a Koch curve, and even closer to the most rugged among the coastlines with islands investigated in Chapters 5 and 28. In any event, a jet's boundary seems fractal. When vortex rings are present, their topology is of interest, but does not exhaust the structure.

The next comment requires the reader to have a mental picture of a wake, say, the lovely shape of a disabled tanker's oil spill. The "rod" that describes such a wake in the roughest approximation has a great deal of structure: it is not at all a cylinder, since its cross section broadens rapidly away from the ship, and its "axis" is not at all straight but shows meanders whose typical size again increases away from the ship.

Analogous features are found in the turbulence due to the shear between fluids masses rubbing past each other, as shown in Browand 1966 and Brown & Roshko 1974. The resulting coherent structures ("animals") attract wide attention, today. Fractals do not concern their overall form, but I think it is equally clear that the hierarchy of fine features that "ride" on the meanders is strikingly fractal in its structure.

Jupiter's celebrated red eye may also be an example of this sort.

Related but different problems arise when studying the Gulf Stream. It is not a single well-defined sea current but divides into multiple wiggly branches, and these branches themselves subdivide and ramify. An overall specification of its propensity to branch would be useful, and will doubtless involve fractals.

ISOTHERMS, DISPERSION ETC.

Similarly, it is interesting to study the shape of the surfaces of constant temperature or the isosurfaces of any other scalar characteristic of the flow. The isotherms may be delineated by the surface surrounding proliferating plankton that lives only in water at $T>45°$, and fills all the volume available to it. The boundary of such a blob is extremely convoluted; in the specific model in Chapter 30, it is demonstrably fractal.

A broad class of geometric problems occurs when a medium is completely filled by turbulence, but parts are marked by some "passive" or inert characteristic that does not affect the flow. The best example is when turbulence disperses a blob of color. Branches of all kinds shoot off in all directions, endlessly, but existing analyses and standard geometry are of little help in describing the resulting shapes. Plate 55 and Mandelbrot 1976c argue that these shapes must be fractals.

OTHER GEOMETRIC QUESTIONS

CLEAR-AIR TURBULENCE. Some scattered evidence I examined suggests that the set carrying this phenomenon is a fractal.

FLOW PAST A FRACTAL BOUNDARY. This is another typical case where fluid mechanics is bound to involve fractals (Plates 45 and 68).

VORTEX STRETCHING. Fluid motion forces vortices to stretch, and a stretching vortex must fold to accommodate an increasing length in a fixed volume. To the extent that the flow is scaling, I conjecture the vortex tends toward a fractal.

THE TRAJECTORY OF A FLUID PARTICLE. In a crude approximation, inspired by the Ptolemaic model of planetary motion, let our particle be carried up vertically by an overall current of unit velocity, while it is perturbed by a hierarchy of eddies, each of which is a circular motion in a horizontal plane. The resulting functions x(t)−x(0) and y(t)−y(0) are sums of cosine and of sine functions. When the high frequency terms are very weak, the trajectory is continuous and differentiable, hence it is rectifiable and D=1. When, however, the high frequency terms are strong and continue down to 0, the trajectory is a fractal, with D>1. Assuming that eddies are self-similar, said trajectory happens to be identical to a famous counterexample of analysis: the Weierstrass function (Chapters 2, 39, and 41). This leads one to wonder whether or not the transition of all the fluid to being turbulent can be associated with the circumstances under which the trajectory is a fractal.

THE INTERMITTENCY OF TURBULENCE

Turbulence eventually ends in dissipation: due to the fluid's viscosity, the energy of visible motion transforms into heat. Early theories assume that the dissipation is uniform in space. But the hope that "homogeneous turbulence" would be a sensible model was dashed by Landau & Lifshitz 1953–1959, which notes that some regions are marked by very high dissipation, while other regions seem by contrast nearly free of dissipation. This means that the well-known property of wind, that it comes in gusts, is also reflected—in more consistent fashion—on smaller scales.

This phenomenon, *intermittency*, was first studied in Batchelor & Townsend 1949, p. 253. See also Batchelor 1953, Section 8.3, and Monin & Yaglom 1963, 1971, 1975. Intermittency is particularly clear-cut when the Reynolds number is very large, meaning that the outer cutoff of turbulence is large relative to its inner cutoff: in the stars, the ocean, and the atmosphere.

The regions in which dissipation concentrates are conveniently described as *carrying* or *supporting* it.

The fact that this Essay brings together the intermittency of turbulence and the distribution of galaxies is natural and not new. A while ago, physicists (von Weizäcker 1950) attempted to explain the genesis of the galaxies by turbulence. Recognizing that homogeneous turbulence cannot account for stellar intermittency, von Weizäcker sketched some amendments that are in the spirit of the Fournier ("Charlier") model (Chapter 9), hence of the theory presented here. If von Weizsäcker's unifying efforts are taken up again, they may establish a physical link between two kinds of intermittency and the corresponding self-similar fractals.

One goal of such a unifying effect would be to relate the dimension of the distribution of galaxies, which we know to be D~1.23, with the dimensions involved in turbulence, which we noted lies around 2.5 to 2.7.

A DEFINITION OF TURBULENCE

We noted that, odd as it may seem, the same term, *turbulence*, is applied to several different phenomena. This continuing lack of a definition becomes easy to understand if, as I claim and propose to demonstrate, a proper definition requires fractals.

The customary mental image of turbulence is nearly "frozen" in the terms first isolated by Reynolds, about one hundred years ago, for fluid flow in a pipe: when the upstream pressure is weak, the motion is regular and "laminar"; when the pressure is increased sufficiently, everything suddenly becomes irregular. In this prototype case, the support of turbulent dissipation is either "empty," nonexistent, or is the entire tube. In either case there is not only no geometry to study, but also no imperative reason to define turbulence.

In wakes, things become more complicated. There is a boundary between the turbulent zone and the surrounding sea, and one ought to study its geometry. However, this boundary is again so clear that an "objective" criterion to define turbulence is not really necessary.

In fully developed turbulence in a wind tunnel, matters are again simple, the whole appearing turbulent like the Reynolds pipe. Nevertheless, the procedures used to achieve this goal are sometimes curious, if we are to

believe certain stubbornly held stories. It is rumored that wind tunnels when first "blown" are unfit for the study of turbulence. Far from filling up the volume offered to it, turbulence itself seems "turbulent," presenting itself in irregular gusts. Only gradual efforts manage to stabilize the whole thing, after the fashion of the Reynolds pipe. Because of this fact, I am among those who wonder to what extent the nonintermittent "laboratory turbulence" in wind tunnels can be regarded as the same physical phenomenon as the intermittent "natural turbulence" in the atmosphere. Hence we must define the terms.

We approach this task indirectly, starting from an ill-defined concept of what is turbulent and examining the one-dimensional records of the velocity at a point. The motions of the center of gravity of a large airplane illustrate a rough analysis of such records. Every so often, the airplane is shaken about, which shows that certain regions of the atmosphere are strongly dissipative. A smaller airplane acts as a more sensitive probe: it "feels" turbulent gusts that leave the large airplane undisturbed, and it experiences each shock received by the large airplane as a burst of weaker shocks. Thus, when a strongly dissipative piece of the cross section is examined in detail, laminar inserts become apparent. And further smaller inserts are seen when the analysis is refined further.

Each stage demands a redefinition of what is turbulent. The notion of a *turbulent minute of record* becomes meaningful if interpreted as "minute of record that is not completely free of turbulence." On the other hand, the more demanding notion of a *solidly turbulent minute of record* seems devoid of observable significance. Proceeding to successive stages of analysis, turbulence becomes increasingly sharp over an increasingly small fraction of the total record length. The volume of the support of dissipation seems to decrease. Our next task is to model this support.

ROLE OF SELF-SIMILAR FRACTALS

As already said, it is not surprising, in my view, that very few geometric aspects of turbulence have actually been investigated, because the only available techniques have been Euclidean. To escape their limitations, many pre-Euclidean terms are used. For example, papers on intermittency make an uncommonly heavy use of terms such as *spotty* and *lumpy,* and Batchelor & Townsend 1949 envisions "only four possible categories of shapes: blobs, rods, slabs, and ribbons." Some lecturers (but few writers) also use the terms *beans, spaghetti,* and *lettuce,* an imaginative terminology that does not attempt to hide the poverty of the underlying geometry.

By contrast, the investigations I carried out since 1964, and first presented at the 1966 Kyoto Symposium (Mandelbrot 1967k), augment the classical geometric toolbox by the addition of self-similar fractals.

To advocate the use of fractals is a radical new step, but to restrict the fractals of turbulence to be self-similar is orthodox, because the very notion of self-similarity was first conceived to describe turbulence. The pioneer was the Lewis Fry Richardson whom we first encounter in Chapter 5. Richardson 1926 introduced the concept of a hierarchy of eddies linked by a cascade. (See Chapter 40.)

It is also in the context of turbulence that the theory of cascades and of self-similarity achieved its triumphs of prediction between 1941 and 1948. The main contributors were Kolmogorov, Obukhov, Onsager, and von Weizsäcker, but tradition denotes the developments of the period by Kolmogorov's name. However, a subtle change occurred between Richardson and Kolmogorov.

While self-similarity is suggested by the consideration of visually perceived eddies, the Kolmogorov theory is purely *analytic.* Fractals, on the other hand, make it possible to apply the technique of self-similarity to the *geometry* of turbulence.

The fractal approach should be contrasted with the peculiar fact that the blobs, rods, slabs, and ribbons involved in yesterday's four-way choice fail to be self-similar. This may be why Kuo & Corrsin 1972 admit that this choice is "primitive" and that one needs *in-between* patterns.

A number of possible ad hoc changes in the standard patterns come to mind. For example, one might split rods into ropes surrounded with loose strands (remember the analogous situation with wakes or jets) and slice slabs into sheets surrounded with loose layers. Somehow those strands and layers might be made self-similar.

However, an ad hoc injection of self-similarity has never been implemented, and I find it both unpromising and unpalatable. I prefer to follow an entirely different tack, allowing the overall shapes and the details of strand and layer to be generated by the same process. Since the basic self-similar fractals are devoid of privileged direction, our study leaves aside (for now) all the interesting geometric questions that combine turbulence with strong overall motion.

◁ Obukhov 1962 and Kolmogorov 1962 are the first analytic studies of intermittency. In immediate influence, they nearly matched the 1941 papers of the same authors, but they are seriously flawed, and their long run influence promises to be small. See Mandelbrot 1972j, 1974f, 1976o; Kraichnan 1974. ►

INNER AND OUTER CUTOFFS

Due to viscosity, the inner cutoff of turbulence is positive. And wakes, jets, and analogous flows clearly show a finite outer cutoff Ω. But the widespread current belief in the finiteness of of Ω should be subjected to criticism. Richardson 1926 claims that "observation shows that the numerical values [presumed to converge for samples of size about Ω] would depend entirely upon how long a volume was included in the mean. Defant's researches show that no limit is attained within the atmosphere." The meteorologists have discounted, then forgotten, this assertion, far too hastily to my mind. New data in Chapter 11 and the study of lacunarity in Chapter 34 add to my conviction that the matter is not yet closed.

CURDLING AND FRACTALLY HOMOGENEOUS TURBULENCE

In a rough preliminary stage, we may represent the support of turbulence by one of the self-similar fractals which the preceding chapters obtain through curdling. This curdling is a crude "de-randomized" form of the Novikov & Stewart model of Chapter 23. After a finite number m of stages of a curdling cascade, dissipation is distributed uniformly over $N=r^{-mD}$ out of r^{-3m} mth-order nonoverlapping subeddies, whose positions are specified by a generator. After a cascade has continued without end, the limit distribution of dissipation spreads uniformly over a fractal of dimension $D<3$. I propose that the limit be called *fractally* homogeneous turbulence.

G. I. Taylor's homogeneous turbulence is obtained for $D\to3$. The salient fact is that curdling does not exclude $D=3$, but it allows the novel possibility $D<3$.

DIRECT EXPERIMENTAL EVIDENCE THAT INTERMITTENCY SATISFIES D>2

From the viewpoint of linear sections, wide classes of unbounded fractals behave very simply: the section is almost surely empty when $D<2$ and is nonempty with positive probability when D>2. (Chapter 23 proves this result for a simple class of fractals.)

Had the set that supports turbulent dissipation satisfied D<2, the preceding statement should imply that nearly all experimental probes would slip between turbulent regions. The fact that such is not the case suggests that in reality D>2. This inference is extraordinarily strong, because it relies upon an experiment that is repeated constantly, and for which the possible outcomes are reduced to an alternative between "never" and "often."

A tentative topological counterpart $D_T>2$, Mandelbrot 1976o, is tempting, but too special to be recounted here.

GALAXIES & TURBULENCE COMPARED

The inequality D>2 for the set that supports turbulent dissipation, and the opposite inequality D<2 for the distribution of mass in the cosmos, Chapter 9, spring from the closely related effects of the sign of D−2 on the typical section of a fractal and on its typical projection on a plane or the sky. For the phenomenon studied in the present chapter, the section has to be nonempty. In Chapter 9, on the contrary, the Blazing Sky Effect is "exorcised" if the majority of straight lines drawn from the Earth *never* meet a star. This requires the stars' projection on the sky to be of vanishing area.

The contrast between the signs of D−2 in these two problems must have a vital bearing on a constrast between their structures.

(IN)EQUALITIES BETWEEN EXPONENTS (MANDELBROT 1967k, 1976o)

Many useful characteristics of fractally homogeneous turbulence depend solely upon D. This topic is studied in Mandelbrot 1976o, where intermittent turbulence is characterized by a series of conceptually distinct exponents linked by (in)equalities. ◁ The situation is reminiscent of critical point phenomena. ▶

SPECTRUM (IN)EQUALITIES. The (in)equality first stated in Mandelbrot 1967k (which uses the notation $\theta=D-2$), is ordinarily expressed in terms of the spectrum of the turbulent velocity, but is here stated in terms of variance. In fractally homogeneous turbulence, the velocity v at point x satisfies

$$\langle[v(x)-v(x+r)]^2\rangle=|r|^{2/3+B},$$

where B=(3−D)/3.

In Taylor homogeneous turbulence, D=3, and B vanishes, leaving the classic Kolmogorov exponent 2/3, which we meet again in Chapter 30.

Mandelbrot 1976o also shows that the more general model of weighted curdling, as described in Mandelbrot 1974f, involves the inequality B≤(3−D)/3.

THE β MODEL. Frisch, Nelkin & Sulem 1978 grafts a pseudo dynamic vocabulary upon the geometry of fractally homogeneous turbulence, as described in Mandelbrot 1976o. The interpretation has proven helpful, but the mathematical arguments and the conclusions are identical to mine. The term "β-model" given to their interpretation has gained some currency, and is often identified with fractal homogeneity.

THE TOPOLOGY OF TURBULENCE REMAINS AN OPEN ISSUE

The preceding chapters make it abundantly clear that the same value of D can be encountered in sets that differ in terms of topological connectedness. The topological dimension D_T yields a lower bound to the fractal dimension D, but this bound is frequently exceeded by such a wide margin as to be of no use. A shape with a fractal dimension D between 2 and 3 may be either "sheetlike," "linelike," or "dustlike," and can achieve configurations in such variety as to make it hard to coin or find names for them all. For example, even in fractal shapes that are most nearly ropelike, the "strands" can be so heavy that the result is really "more" than ropelike. Similarly, fractal near sheets are "more" than sheetlike. Also, it is possible to mix sheetlike and ropelike features at will. Intuitively, one might have hoped that some closer relationship should exist between fractal dimension and degree of connectedness, but this is a hope mathematicians lost between 1875 and 1925. We turn to a special problem of this kind in Chapter 23, but it may be said that the actual loose relationship between these structures is essentially unexplored territory.

The question of ramification, raised in Chapter 14, is also vital, but its impact on the study of turbulence is as yet unexplored.

KURTOSIS INEQUALITIES. Using a measure of intermittency called kurtosis, the issue of connectedness is tackled in Corrsin 1962, Tennekes 1968, and Saffman 1968. Ostensibly, those models deal with shapes that share the topological dimension of the plane (sheets) or the straight line (rods). However, they test the topology indirectly, through the exponent of a predicted power law relationship between the kurtosis and a Reynolds number. Unfortunately, this attempt fails because the kurtosis exponent is in fact dominated by diverse additional assumptions, and ultimately depends solely on the fractal dimension D of the shape generated by the model. Corrsin 1962 predicts a value of D equal to the topological dimension it postulates, $D_T=2$. The prediction is incorrect, expressing the fact that the data involve fractals, but this model does not. On the other hand, Tennekes 1968 postulates $D_T=1$ but yields the fractional value $D=2.6$, hence does involve an approximate fractal. Nevertheless, the attempted inference from the kurtosis to a combination of intuitive "shape" and topological dimension is unwarranted. ■

11 ¤ Fractal Singularities of Differential Equations

The present chapter concerns a first connection between the fractal geometry of Nature and the mainstream of mathematical physics. The topic is so vital that it deserves a separate chapter. Readers whose interests lie elsewhere should forge ahead.

A SPLIT IN TURBULENCE THEORY

A major defect of the current theoretical study of turbulence is that it separates into at least two disconnected parts. One part includes the successful phenomenology put forth in Kolmogorov 1941 (examined in greater detail in Chapter 30). And the other part includes the differential equations of hydrodynamics, due to Euler for nonviscous fluids, and to Navier (and Stokes) for viscous fluids. These two parts remain unrelated: If "explained" and "understood" mean "reduced to basic equations," the Kolmogorov theory is not yet explained or understood. And Kolmogorov has not helped solve the equations of fluid motion.

My assertion in Chapter 10, that turbulent dissipation is not homogeneous over the whole space, only over a fractal subset, may seem at first sight to make the gap even greater. *But I contended that the opposite is the case.* And there is increasing evidence in my favor.

THE IMPORTANCE OF SINGULARITIES

Let us review the procedure that allows an equation of mathematical physics to be solved successfully. Typically, one draws up a list that combines solutions obtained by solving the equation under special conditions, and solutions guessed on the basis of physical observation. Next, neglecting details of the solutions, one draws a list of elementary "singularities" characteristic of the problem. From then on, more complex instances of the equation can often be solved in the first approximation by identifying the appropriate singularities and stringing them together as required. This is how the student of calculus draws the graph of a rational function. Of course, the standard singularities are standard Euclidean sets: points, curves, and surfaces.

CONJECTURE: THE SINGULARITIES OF FLUID MOTION ARE FRACTAL SETS (MANDELBROT 1976c)

In this perspective, I interpret the difficulties experienced in deriving turbulence from the Euler and Navier-Stokes solutions as implying that *no* standard singularity accounts for what we perceive intuitively to be the characteristic features of turbulence.

I contend instead (Mandelbrot 1976c) that the turbulent solutions of the basic equations involve singularities or "near singularities" of an entirely new kind. The singularities are locally scaling fractal sets, and the near singularities are approximations thereto.

An unspecific motivation for this contention is that, standard sets having proven inadequate, one may as well try the next best known sets. But more specific motivation is available.

NONVISCOUS (EULER) FLUIDS

FIRST SPECIFIC CONJECTURE. Part of my contention is that the singularities of the solutions of the Euler equations are fractal sets.

MOTIVATION. This belief relies on the very old notion that the symmetries and other invariances present in an equation "ought" to be reflected in the equation's solution. (For a self-standing, careful and eloquent description, see Chapter IV of Birkhoff 1960.) Of course, preservation of symmetries is by no means a general principle of Nature, hence one cannot exclude the possibility of "broken symmetry" here. I propose, however, that one try the consequences of symmetry preservation. Since the Euler equations are scale-free, the equations' typical solutions should also be scale-free, and the same should hold of any singularities they may possess. If the failure of past efforts is taken as evidence that the singularities are not standard points or lines or surfaces, they must be fractals.

It may of course happen that a scale is imposed by the boundary's shape and the initial velocities. It is, however, likely that the solutions' local behavior is ruled by a "principle of not feeling the boundary." Hence the solutions should be locally scaleless.

ALEXANDRE CHORIN'S WORK. Chorin 1981 provides strong support for my contention, by applying a vortex method to the analysis of the inertial range in fully developed turbulence. The finding is that the highly stretched vorticity collects itself into a body of decreasing volume, and of dimension $D \sim 2.5$ compatible with the conclusions in Chapter 10. The correction to the Kolmogorov exponents, $B = .17 \pm 0.03$, is compatible with experimental data. The calculations suggest that the solutions of Euler's equations in three dimensions blow up in a finite time.

Unpublished work of Chorin comes even closer to experiment: $2.5 < D < 2.6$.

VISCOUS (NAVIER-STOKES) FLUIDS

SECOND SPECIFIC CONJECTURE. Furthermore, I contended that the singularities of the solutions of the Navier-Stokes equations can only be fractals.

DIMENSION INEQUALITIES. Furthermore, we have the intuitive feeling that the solutions of the Navier-Stokes equations are necessarily smoother, hence less singular, than those of the Euler equations. Hence the further conjecture that the dimension is larger in the Euler than in the Navier-Stokes case. The passage to zero viscosity is doubtless singular.

NEAR SINGULARITIES. A final conjecture in the implementation of my overall contention concerns the peaks of dissipation involved in the notion of intermittency: they are Euler singularities smoothed out by viscosity.

V. SCHEFFER'S WORK. The examination of my conjectures for the viscous case was pioneered by V. Scheffer, recently joined by others in studying in this light a finite or infinite fluid subject to the Navier-Stokes equations with a finite kinetic energy at t=0.

Assuming that singularities are indeed present, Scheffer 1976 shows that they necessarily satisfy the following theorems. First, their projection over the time axis has at most the fractal dimension ½. Second, their projection on the space coordinates is at most a fractal of dimension equal to 1.

It turns out, after the fact, that the first of the above results is a corollary of a remark in an old and famous paper Leray 1934 ends abruptly after a formal inequality of which

Scheffer's first theorem is a corollary, in fact merely a restatement. But is it fair to say "merely"? Restating a result in more elegant terminology is (for sound reasons) rarely viewed as a scientific advance, but I think that the present instance is different. The inequality in Leray's theorem was nearly useless until the Mandelbrot-Scheffer corollary placed it in proper perspective.

The almost routine uses of Hausdorff Besicovitch dimension in recent studies of the Navier-Stokes equations can all be traced back to my conjecture.

SINGULARITIES OF OTHER NONLINEAR EQUATIONS OF PHYSICS

The other phenomena which this Essay claims involve scaling fractals have nothing to do with either Euler or Navier and Stokes. For example, the distribution of galaxies is ruled by the equations of gravitation. But the symmetry preservation argument applies to all scaling equations. As a matter of fact, an obscure remark by Laplace (see the entry SCALING IN LEIBNIZ AND LAPLACE, Chapter 41) can now be construed (with 20/20 hindsight!) as pointing toward the theme of Chapter 9.

More generally, the singularities' fractal character is likely to be traceable to generic features shared by many different equations of mathematical physics. Can it be some very broad kind of nonlinearity? The issue is joined again, in different terms, in Chapter 20. ■

12 ¤ Length–Area–Volume Relations

Chapters 12 and 13 extend the properties of fractal dimension through numerous mini case studies of varying importance and increasing difficulty, and Chapter 14 shows that fractal geometry necessarily involves concepts beyond the fractal dimension.

The present chapter describes, and applies to diverse concrete cases, the fractal counterparts I developed for certain standard results of Euclidean geometry. They can be viewed as parallel to the fractal relations of the form $M(R) \propto R^D$ obtained in Chapters 6, 8, and 9.

STANDARD DIMENSIONAL ANALYSIS

From the facts that the circumferential length of a circle of radius R is equal to $2\pi R$, and the area of the disc bounded by the circle is πR^2, it follows that

$$(\text{length}) = 2\pi^{1/2}(\text{area})^{1/2}.$$

Among squares, the corresponding relation is

$$(\text{length}) = 4(\text{area})^{1/2}.$$

More generally, within each family of standard planar shapes that are geometrically similar and have different linear extents, the ratio $(\text{length})/(\text{area})^{1/2}$ is a number entirely determined by the common shape.

In space (E=3), length, $(\text{area})^{1/2}$, and $(\text{volume})^{1/3}$ provide alternative evaluations of the linear extent of the shape, and the ratio of any two of them is a shape parameter independent of the units of measurement.

The equivalence of different linear extents is very useful in many applications. And its extension when time and mass are added lead to a powerful technique, known to physicists as "dimensional analysis." (Birkhoff 1960 is a recommended exposition of its basic features.)

PARADOXICAL DIMENSIONAL FINDINGS

However, in increasingly numerous instances, the equivalence between alternative linear extents proves distressingly elusive. For example, mammalian brains satisfy

$$(\text{volume})^{1/3} \propto (\text{area})^{1/D},$$

with $D \sim 3$, *far above* the anticipated value of 2. In river drainage basins, Hack 1957 measures length along the main river, and finds

$$(\text{area})^{1/2} \propto (\text{length})^{1/D},$$

with D *definitely above* the anticipated value of 1. Early writers interpret this last result as implying that river basins fail to be self-similar, large ones being elongated and small ones being chubby. Unfortunately, this interpretation conflicts with the evidence.

The present chapter describes how I explain these and related findings in more convincing fashion. My tool is a new, fractal, length-area-volume relation.

FRACTAL LENGTH–AREA RELATION

To pinpoint the argument, consider a collection of geometrically similar islands with fractal coastlines of dimension $D > 1$. The standard ratio $(\text{length})/(\text{area})^{1/2}$ is infinite in this context, but I propose to show it has a useful fractal counterpart. We denote as G-length the coast length measured with a yardstick length of G, and as G-area the island area measured in units of G^2. Knowing the dependence of G-length upon G to be nonstandard, while the dependence of G-area is standard, we form the generalized ratio

$$(\text{G-length})^{1/D}/(\text{G-area})^{1/2}.$$

I claim that this ratio takes the same value for our geometrically similar islands.

As a result, there are two different ways of evaluating the linear extent of each island in units of G: the standard expression $(\text{G-area})^{1/2}$ but also the nonstandard $(\text{G-length})^{1/D}$.

The novel feature is that if G is replaced by a different yardstick length G' the ratio of the alternative linear extents is replaced by

$$(\text{G'-length})^{1/D}/(\text{G'-area})^{1/2},$$

which differs from the original one by a factor of $(\text{G'}/\text{G})^{1/D-1}$.

As for the ratio of linear extents, it varies between one family of mutually similar bounded shapes and another, whether they are fractal or standard. Hence it quantifies one facet of the shapes' form.

Note that the length-area relation may be used to estimate the dimension of a fractal curve that bounds a standard domain.

PROOF OF THE RELATION. The first step is to measure each coastline length with the intrinsic area-dependent yardstick

$$G^* = (\text{G-area})^{1/2}/1000.$$

When we approximate each of our island coastlines by a polygon of side G^*, these polygons are also mutually similar, and their lengths are proportional to the standard linear extents $(G\text{-area})^{1/2}$.

Next replace G^* by the prescribed yardstick G. We know from Chapter 6 that the measured length changes in the ratio $(G/G^*)^{1-D}$. Hence,

$$(G\text{-length}) \propto (G\text{-area})^{1/2}(G/G^*)^{1-D}$$
$$= (G\text{-area})^{1/2-1/2(1-D)}G^{1-D}1000^{D-1}$$
$$= (G\text{-area})^{1/2 D}G^{1-D}1000^{D-1}.$$

Finally, by raising each side to the power $1/D$, we obtain the relation I claimed.

HOW WINDING IS THE MISSOURI RIVER?

The preceding arguments also throw light on the measured river lengths. To define a length for the leading river of a drainage basin, we approximate the river's course by a wiggly self-similar line of dimension $D>1$ going from a point called source to a point called mouth. If all rivers as well as their basins are mutually similar, the fractal length-area argument predicts that

$$(\text{river's G-length})^{1/D} \text{ is proportional to}$$
$$(\text{basin's G-area})^{1/2}.$$

Moreover, standard reasons predict that

$(\text{basin's G-area})^{1/2}$ is proportional to (straight distance from source to mouth).

Combining the two results, we conclude that

$(\text{river's G-length})^{1/D}$ is proportional to (straight distance from source to mouth).

Most remarkably, as already mentioned, Hack 1957 finds empirically that the ratio

$$(\text{river's G-length})/(\text{basin's G-area})^{.6}$$

is indeed common to all rivers. This indirect estimate of $D/2=.6$ yields $D=1.2$, reminiscent of the values inferred from coastline lengths. If one measures the degree of irregularity by D, the degrees of irregularity of local wiggles of the banks and of enormously global bends turn out to be identical!

However, for basins of area $>10^4$ km^2 and correspondingly long rivers, J. E. Mueller observes that the value of D goes down to 1. The two different values of D suggest that if one maps all basins on sheets of paper of the same size, maps of short rivers look about the same as maps of long rivers, but maps of *extremely long* rivers are more nearly straight. It may be that nonstandard self-similarity breaks around an outer cutoff Ω whose value is of the order of 100 km.

CUMULATIVE LENGTH OF A RIVER TREE. The preceding argument also predicts that the cumulative length of all the rivers in a drainage basin should be proportional to that basin's area. I am told this prediction is correct, but I have no reference.

BACK TO GEOMETRY. For the rivers and watersheds relative to the "snowflake sweep" curve of Plates 68 and 69, D~1.2618, somewhat above the observed value. The corresponding dimensions in Plates 70 and 71 are D~1.1291, on the low side.

The Peano curves of Plates 63 and 64 are well off the mark, since D=1.

Note that the identity between the dimensions of the rivers and of the watersheds is not a logical necessity, only a feature of certain specific recursive models. By way of contrast, a river network linked with the arrowhead curve (Plate 141) and described in Mandelbrot 1975m involves rivers of dimension D=1, which is too small, and watersheds of dimension D~1.5849, which is too large.

GEOMETRY OF RAIN AND OF CLOUDS

Pages 1, 10, 11, and 94 mention the possible use of fractals to model clouds. This hunch has now been confirmed by Lovejoy 1982, via the fractal area-perimeter graph in Plate 115. Very few graphs in meteorology involve all the available data over an enormous range of sizes, and are nearly as straight as this one.

The data combine radar observations from tropical Atlantic rain areas (with rainrate above .2 mm/hr), with geostationary satellite infrared observations of cloud areas over the Indian Ocean (= areas where the top of the cloud temperature is below −10°C). The areas range from 1 to over 1,000,000 km^2.

The dimension of the perimeter, fitted over at least six orders of magnitude, is 4/3. The pleasure of providing a physical explanation is left to Dr. Lovejoy.

The largest cloud extended from central Africa to South India, a distance well above the thickness of the atmosphere, to which the outer cutoff L of atmospheric turbulence is all too often assimilated. Richardson's quote on p. 103 may prove prophetic.

THE AREA-VOLUME RELATION. CONDENSATION BY MICRO-DROPLETS

The derivation of the length area relationship generalizes easily to spatial domains bounded by fractal surfaces, and leads to the relation

$$(\text{G-area})^{1/D} \propto (\text{G-volume})^{1/3}.$$

To illustrate this relation, consider the condensation of vapor into liquid. This is a very familiar physical phenomenon, yet its theory is a recent development. To paraphrase Fisher 1967, the following geometric picture was put forward apparently quite independently by J. Frenkel, W. Band, and A. Bijl in the late 1930's. A gas consists of isolated molecules well separated from one another, except for occasional clusters which are bound together more-or-less tightly by the attractive forces. Clusters of different sizes are in mutual statistical equilibrium, associating and disassociating, but even fairly large clusters resembling "droplets" of liquid have a small chance of occurring. For a large enough cluster (which is not too "drawn out," like a piece of seaweed for example!), the surface area is fairly well defined. The surface of a cluster gives it stability. If the temperature now is lowered, it becomes advantageous for clusters to combine to form droplets and for droplets to amalgamate, thereby reducing the total surface area and hence lowering the total energy. If conditions are favorable, the droplets grow rapidly. A macroscopic droplet's presence indicates that condensation has taken place!

Building on this picture, M. E. Fisher proposes that a condensing droplet's area and volume are related by a formula equivalent to $area^{1/D} = volume^{1/3}$. Fisher evaluates D analytically without concern for its geometric meaning, but it is unavoidable that one should now conjecture that the underlying droplet surfaces are fractals of dimension D.

MAMMALIAN BRAIN FOLDS

To illustrate the area-volume relation in the important limit case D=3, and at the same time to buttress the exorcism of Peano shapes presented in Chapter 7, let us interpret a famous problem of comparative anatomy in terms of near-space-filling surfaces.

Mammalian brain volumes vary from 0.3 to 3000 ml, small animals' cortex being relatively or completely smooth, while large animals' cortex tends to be visibly convoluted, irrespective of the animals' positions on the scale of evolution. Zoologists argue that the proportion of white matter (formed by the neuron axons) to gray matter (where neurons terminate) is approximately the same for all mammals, and that in order to maintain this ratio a large brain's cortex must necessarily become folded. Knowing that the extent of folding is of purely geometric origin relieves Man from feeling threatened by Dolphin or Whale: they are bigger than us but need not be more highly evolved.

A quantitative study of such folding is beyond standard geometry but fits beautifully in fractal geometry. The gray matter's volume is roughly equal to its thickness multiplied by the area of the brain's surface membrane, called "pia." If the thickness ϵ were the same in all species, the pia area would be proportional not only to the gray matter volume but also to the white matter volume, hence to the total volume V. Therefore, the area-volume relationship would yield D=3, and the pia would be a surface that comes within ϵ of filling the space.

The empirical area-volume relation is better fitted by $A \propto V^{D/3}$ with $D/3 \sim 0.91$ to 0.93 (Jerison, private communication, based on the data of Elias & Schwartz, Brodman, and others). The most immediate interpretation is that the pia is only partly space filling, with D in the range between 2.79 and 2.73. A more sophisticated argument is sketched when we resume this topic in Chapter 17.

ALVEOLAR AND CELL MEMBRANES

Will a biologist kindly stand up and proclaim
that the preceding section brings no hard result and no unexpected notion? I delight at
hearing this objection because it buttresses
further the argument with which Chapter 7
begins. Despite the fact that a biologist would
run a mile from a Peano surface as adorned
by mathematicians, I claim that the basic idea
is indeed quite familiar to the good theoretical
minds in this field.

Thus, the main novelty of the preceding
sections lies with surfaces of $D < 3$, which (as
we saw) are required for a good fit. Let us
pursue their novel application to biology by
sketching how they help unscramble the detailed structure of several living membranes.

First, a paragraph to summarize Weibel
1979, section 4.3.7. Estimates of the human
lung's alveolar area are conflicting: light microscopy yields 80 m^2, while electron microscopy claims 140 m^2. Does this discrepancy
matter? The fine details to which it is due
play no role with respect to gas exchange, being smoothed by a fluid lining layer (resulting
in an even smaller functional area), but they
are important with respect to solute exchanges. Measurements (triggered by my
Coast of Britain paper) indicate in the first
approximation that over a wide range of
scales the membrane dimension is $D = 2.17$.

Paumgartner & Weibel 1979 examine
*sub*cellular membranes in liver cells. Again,
the sharp past disagreement between different
estimates of area per volume disappear by
postulating that $D = 2.09$ for the outer mitochondrial membrane (which wraps the cell,
and departs only slightly from the smoothness
characteristic of membranes with minimal
area/volume ratio). On the other hand,
$D = 2.53$ for *inner* mitochondrial membranes,
and $D = 1.72$ for the endoplasmic reticulum.

Also let it be noted that many animals'
nasal bone structure is of extraordinary complication, allowing the "skin" that covers this
bone to have a very large area in a small volume. In Deer and Arctic Fox, this membrane
may serve the sense of smell, but (Schmidt-
Nielsen 1981) the goal of an analogous shape
in Camel is to husband scarce water.

MODULAR COMPUTER GEOMETRY

To illustrate the area-volume relationship further, let us tackle a facet of computers. Computers are not natural systems, but this should
not stop us. This and a few other case studies
help demonstrate that, in the final analysis,
fractal methods can serve to analyze any
"system," whether natural or artificial, that
decomposes into "parts" articulated in a self-
similar fashion, and such that the properties
of the parts are less important than the rules
of articulation.

Complex computer circuits are always subdivided into numerous modules. Each contains
a large number C of components and is connected with its environment by a large number T of terminals. Within an error of a few
percent, one finds that $T^{1/D} \propto C^{1/E}$. The way

the exponent is written will be justified in a moment. Within IBM, the above rule is credited to E. Rent; see Landman & Russo 1971.

The earliest raw data suggested $D/E=\frac{2}{3}$, a value that Keyes 1981 extrapolates to huge "circuits" in the nervous system (optic nerve and *corpus callosum*). However, the ratio D/E increases with the circuit's performance. Performance, in turn, reflects the degree of parallelism that is present in the design. In particular, the designs with extreme characteristics lead to extreme values of D. In a shift register, the modules form a chain and $T=2$, independently of C, hence $D=0$. With integral parallelism, each component requiring its own terminal, $T=C$, hence $D=E$.

To account for $D/E=\frac{2}{3}$, R.W. Keyes noted that components are typically arranged within the volume of the modules, while the connections go through their surfaces. To show that this observation demands Rent's rule, it suffices to assume that all the components have roughly the same volume v and surface σ. Since C is the total volume of the module divided by v, $C^{\frac{1}{3}}$ is roughly proportional to the radius of the module. On the other hand, T is the total surface of the module divided by σ, thus $T^{\frac{1}{2}}$ is also roughly proportional to the radius of the module. Rent's rule simply expresses the equivalence of two different measures of the radius in a standard spatial shape. $E=3$ is the Euclidean dimension of the circuit and $D=2$ is the dimension of a standard surface.

Note that the concept of the module is ambiguous and almost indefinite, but Rent's rule

is quite compatible with this characteristic, insofar as any module's submodules are interconnected by their surfaces.

It is just as easy to interpret the extreme cases mentioned above. In a standard linear structure, $E=1$ and the boundary reduces to two points, hence $D=0$. In a standard planar structure, $E=2$ and $D=1$.

However, when the ratio E/D is neither $3/2$, nor $2/1$, nor $1/0$, standard Euclidean geometry does not make it possible to interpret C as an expression of volume and T as an expression of surface. Yet such interpretations are very useful, and in fractal geometry they are easy. In a spatial circuit in contact with the outside by its whole surface, $E=3$, and D is anywhere between 2 and 3. In a plane circuit in contact with the outside by its whole bounding curve, $E=2$ and D is anywhere between 1 and 2. The case of integral parallelism, $D=E$, corresponds to Peano boundaries. Furthermore, if the boundary is utilized incompletely, the "effective boundary" may be any surface with D between 0 and E.   ▰▰

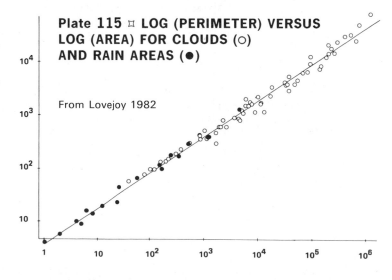

Plate 115 ⌑ LOG (PERIMETER) VERSUS LOG (AREA) FOR CLOUDS (○) AND RAIN AREAS (●)

From Lovejoy 1982

13 ¤ Islands, Clusters, and Percolation; Diameter-Number Relations

This chapter is devoted to fractal σ-curves, that is, to fractals that decompose into an infinity of disjoint fragments, each of them a connected curve. The concrete cases range from the coastlines of islands in an archipelago to an important problem of physics: percolation. The material in the first few sections was new to the 1977 *Fractals*, and the bulk of the chapter's remainder is new.

To begin, let us echo "How Long Is the Coast of Britain" and ask how many islands surround Britain's coast? Surely, their number is both very large and very ill-determined. As increasingly small rock piles become listed as islands, the overall list lengthens, and the total number of islands is practically infinite.

Since earth's relief is finely "corrugated," there is no doubt that, just like a coastline's length, an island's total area is geographically infinite. But the domains surrounded by coastlines have well defined "map areas." And the way in which a total map area is shared among the different islands is an important geographic characteristic. One might even

argue that this "area-number relation" contributes more to geographic form than do the shapes of the individual coastlines. For example, it is difficult to think of the Aegean Sea's shores without also including those of the Greek islands. The issue clearly deserves a quantitative study, and this chapter provides one, by generalizing the Koch curves.

Next, this chapter examines diverse other fragmented shapes obtained by generalizing the familiar fractal-generating processes: either the Koch procedure or curdling. The resulting shapes are called *contact clusters* here, and the diameter-number distribution is shown to be the same for them as for islands.

Special interest attaches to the plane-filling contact clusters, in particular those clusters generated by certain Peano curves, whose teragons do not self-intersect but have carefully controlled points of self-contact. The saga of the taming of Peano monsters is thereby enriched by a new scene!

Last but not least, this chapter includes the first part of a case study of the geometry of percolation, a very important physical phenomenon also studied in Chapter 14.

KORČAK EMPIRICAL LAW, GENERALIZED

List all the islands of a region by decreasing size. The total number of islands of size above a is to be written as Nr(A>a) ◁ patterned after the notation Pr(A>a) of probability theory. ► Here, a is a possible value for an island map's area, and A denotes the area when it is of unknown value.

B and F' being two positive constants, to be called exponent and prefactor, one finds the following striking area-number relation:

$$Nr(A>a)=F'a^{-B}$$

Korčak 1938 (the name is pronounced Kor'chak) comes close to deserving credit for this rule, except that it claims that B = ½, which I found incredible, and which the data showed is unfounded. In fact, B varies between regions and is always >½. Let me now show that the above generalized law is the counterpart of the distribution Chapter 8 obtains for the gap lengths in a Cantor dust.

KOCH CONTINENT AND ISLANDS, AND THEIR DIVERSE DIMENSIONS

To create a Koch counterpart to the Cantor gaps, I let the generator split into disconnected portions. To insure that the limit fractal

remains interpretable in terms of coastlines, the generator includes a connected broken line of $N_C<N$ links, joining the end points of the interval [0,1]. This portion will be called the *coastline generator*, because it determines how an initially straight coastline becomes transformed into a fractal coastline. The remaining $N-N_C$ links form a closed loop that "seeds" new islands and will be called *island generator*. Here is an example:

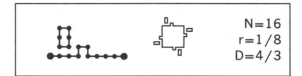

N=16
r=1/8
D=4/3

In later stages, the sub-island always stays to the left of the coastline generator (going from 0 to 1), and of the island generator (going clockwise).

A first novelty is that the limit fractal now involves two distinct dimensions. Lumping all the islands' coastlines together, $D = \log N/\log(1/r)$, but for the coastline of each individual island $D_C = \log N_C/\log(1/r)$, with the inequalities

$$1 \leq D_C < D.$$

The cumulative coastline, not being connected, is not itself a curve but an infinite sum (Σ, sigma) of loops. I propose for it the term *sigma-loop*, shortened into σ-*loop*.

Note that modeling of the observed relation between D and D_C in actual islands requires additional assumptions, unless it can be derived from a theory, as in Chapter 29.

THE DIAMETER-NUMBER RELATION

The proof that the Korčak law holds for last section's islands is simplest when the generator involves a single island, and teragons are self-avoiding. (Recall that the *teragons* are the approximating broken lines). Then the first stage of construction creates 1 island; let its "diameter," defined by $\sqrt{a}$, be λ_0. The second stage creates N islands of diameter $r\lambda_0$, and the mth stage creates N^m islands of diameter $\lambda = r^m\lambda_0$. Altogether, as λ is multiplied by r, $Nr(\Lambda > \lambda)$ is multiplied by N. Hence the distribution of Λ (for all values of λ of the form $r^m\lambda_0$) takes the form

$$Nr(\Lambda > \lambda) = F\lambda^{-D},$$

in which the crucial exponent is the coastline's fractal dimension! As a corollary

$$Nr(A > a) = F'a^{-B}, \text{ with } B = \tfrac{1}{2}D,$$

we have thus derived the Korčak law. For other values of λ or a, one has the staircase curve familiar from the distribution of Cantor gaps' lengths, Chapter 8.

This result is independent of N_c and D_c. It extends to the case when the generator involves two or more islands. We note that the empirical B regarding the whole Earth is of the order of 0.6, very close to one half of D measured from the coastline lengths.

GENERALIZATION TO E>2

In the same construction extended to space, it continues to be true that the E dimensional diameter, defined as $volume^{1/E}$, is ruled by a hyperbolic expression of the form $Nr(volume^{1/E} > \lambda) = F\lambda^{-D}$, wherein the crucial exponent is D.

The exponent D also rules the special case of Cantor dusts for E=1, but there is a major difference. The length outside the Cantor gaps vanishes, while the area outside the "Koch" islands can be, and in general is, positive. We return to this topic in Chapter 15.

FRACTAL DIMENSION MAY BE SOLELY A MEASURE OF FRAGMENTATION

The preceding construction also allows the following generator

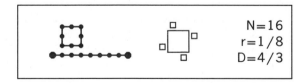

The overall D is unchanged, but the coastline D_c takes the smallest allowable value, $D_c = 1$. In the present model, island coastlines are allowed to be rectifiable! When such is the case, the overall D is not a measure of irregularity, but solely of fragmentation. Instead of the wiggliness of individual curves, D measures the number-area relationship for an infinite family of rectangular islands.

It is still true that, when the length is measured with a yardstick of ϵ, the result tends to infinity as $\epsilon \to 0$, but there is a new reason for this. A yardstick of length ϵ can only measure islands with a diameter of at least ϵ. However, the number of such islands increases as $\epsilon \to 0$, and the measured length behaves like ϵ^{1-D}, exactly as in the absence of islands.

In the general case where $D_C > 1$, the value of D_C measures irregularity alone, while the value of D measures irregularity and fragmentation in combination.

A FRAGMENTED FRACTAL CURVE MAY HAVE TANGENTS EVERYWHERE. By rounding off the islands' corners, one may make every coastline have a tangent at every point, while the areas, hence the overall D, are unaffected. Thus, being a fractal σ-curve and being without tangent are *not* identical properties.

THE INFINITY OF ISLANDS

AN INNOCUOUS DIVERGENCE. As $a \to 0$, $Nr(A > a) = Fa^{-B}$ tends to infinity. Hence, the Korčak law agrees with our initial observation that islands are practically infinite in numbers.

LARGEST ISLAND'S RELATIVE AREA. This last fact is mathematically acceptable because the cumulative *area* of the very small islands is finite and negligible. ◁ All islands of area below ϵ have a total area that behaves like the integral of $a(Ba^{-B-1}) = Ba^{-B}$ from 0 to ϵ. Since $B < 1$, this integral converges, and its

value $B(1-B)^{-1}\epsilon^{1-B}$ tends to 0 with ϵ. ▶

Consequently, the largest island's relative contribution to all the islands' cumulative area tends to a positive limit as the islands increase in numbers. It is *not* asymptotically negligible.

LONGEST COASTLINE'S RELATIVE LENGTH. On the other hand, assuming $D_C = 1$, the coastline lengths have a hyperbolic distribution with the exponent $D > 1$. Hence the cumulative coastline length of small islands is infinite. And, as the construction advances and the number of islands increases, the coastline length of the largest island becomes relatively negligible.

RELATIVELY NEGLIGIBLE SETS. More generally, the inequality $D_C < D$ expresses that the curve drawn using the coastline generator alone is negligible in comparison to the whole coastline. In the same way, a straight line (D=1) is negligible in comparison to a plane (D=2). Just as a point chosen at random in the plane almost never falls on the x-axis, a point chosen at random on the coastline of a "core" island surrounded with sub-islands almost never falls on the core island's coastline.

SEARCH FOR THE INFINITE CONTINENT

In a scaling universe, the distinction between the islands and the continent cannot be based on tradition or "relative size." The only sensible approach is to define the continent as a special island with an *infinite* diameter. Let me now show that the constructions at the beginning of this chapter *practically never* generate a continent. ◁ For those who know probability: the probability of a continent being generated is zero. ▶

In a sensible search for a continent, we must no longer choose the initiator and the generator separately. From now on, the same generator must be made to serve both for interpolation and for extrapolation. The process runs by successive stages, each subdivided into steps. It strongly resembles the extrapolation of the Cantor set in Chapter 8, but deserves to be described even more thoroughly.

The first step upsizes our chosen generator in the ratio of $1/r$. The second step puts a "mark" on *one* of the links of the upsized generator. The third step displaces the upsized generator, to make its marked link coincide with [0,1]. The fourth and last step interpolates the upsized generator's remaining links.

The same process is repeated ad infinitum, its progress and outcome being determined by the sequence of positions of the "marked" links. This sequence can take diverse forms.

The first form requires the coastline generator to include a positive number N_c-2 of "nonextreme" links, defined as belonging to the coastline generator but not ending on either 0 or 1. If the mark is consistently put on a nonextreme link, each stage of extrapolation expands the original bit of coastline, and eventually causes it to be incorporated into a fractal coastline of infinite extent in both directions. This proves that it is indeed *possible* to obtain a continental coastline in this setup.

Secondly, always mark an extreme link of the coastline generator, each possibility being chosen an infinite number of times. Then our bit of coastline again expands without end. If we always choose the same link, the coastline expands in only one direction.

Thirdly, always mark a link that belongs to the island generator. Then the biggest island before extrapolation is made to lie off a bigger island's shore, then off-off a still bigger island's, and so on ad infinitum. No continent is *ever* actually reached.

The next comment involves a bit of "probabilistic common sense," which must be familiar to every reader. We suppose that the marks fall according to the throws of an N-sided die. In order for the extrapolation to generate a continent, it is obviously necessary that all the marks beyond a finite (kth) stage be placed upon one of N_c-2 nonextreme links of the coastline generator. Call them "winning" links. To know one has reached a continent after k stages, one must know that thereafter *every* throw of our die, with *not one* exception, will win. Such luck is not impossible, but it is of vanishing probability.

ISLAND, LAKE AND TREE COMBINATION

The Koch islands being mutually similar, their diameter Λ can be redefined as the distance between any two specified points, best chosen on the coastline. Next, we observe that the derivation of the diameter-number relation makes specific use of the assumption that the generator includes a coastline generator. But the assumption that the generator's remaining links form islands, or are self-avoiding, is never actually used. Thus, the relation

$$Nr(\Lambda > \lambda) = F\lambda^{-D}$$

is of very wide validity. ◁ One can even release the condition the teragons initiated by two intervals must not intersect. ► Let us now show by examples how the configuration of the original $N-N_c$ links can affect the resulting fractal's topology.

COMBINATION OF ISLANDS AND LAKES. Relieve the generator from the requirement of being placed to the left, going clockwise. When it is placed to the right, it forms lakes instead of islands. Alternatively, one may include *both* lakes and islands in the same generator. Either way, the final fractal is a σ-loop whose component loops are nested in each other. For example, consider the generator

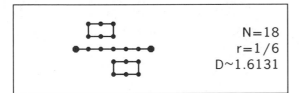

$$N=18$$
$$r=1/6$$
$$D\sim1.6131$$

When initiated by a square, this generator yields the following advanced teragon

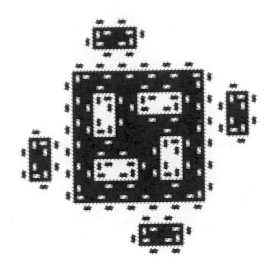

THE ELUSIVE CONTINENT. In the above diagram, the length of the initiator's side injects a nonintrinsic outer cutoff. A more consistent approach is to extrapolate it as we did for islands without lakes. Again, it is almost sure that no continent is ever reached, and that the nesting of islands within lakes within islands continues without bound.

AREA-NUMBER RELATION. In order to define the area of an island (or lake), one may at will take either the total area, or the area of land (or water), within its coastline. The two differ by a fixed numerical factor, hence affect $Nr(A>a)$ through its prefactor F', not its exponent $\frac{1}{2}D$.

COMBINATION OF INTERVALS AND TREES. Now assume that the $N-N_c$ links form either a broken line with two free ends, or a tree. In either case, the fractal splits into an infinite number of disconnected pieces, each of them a curve. This σ-curve is no longer a σ-loop; it is either a σ-tree or a σ-interval.

THE NOTION OF CONTACT CLUSTER

The generator may also combine loops, branches and diverse other topological configurations. If so, the limit fractals' connected portions recall the *clusters* of percolation theory (as seen later in this chapter) and of many other areas of physics. To us, this usage is terribly unfortunate, due to the alternative meaning of *cluster* in the study of dusts (Chapter 9). We need therefore a more specific and cumbersome term. I settled on "contact cluster." Luckily, the term σ-cluster is not ambiguous.

(It may be observed that *contact cluster* has a unique and natural mathematical definition, while the notion of clustering in a dust is diffuse and intuitive, and is at best defined via arguable statistical rules.)

PLANE-FILLING CONTACT CLUSTERS. As D reaches its maximum D=2, the arguments in the preceding section remain valid, but additional comments become necessary. Each individual cluster tends to a limit, which may be a straight line, but in most cases is a fractal curve. On the other hand, all the clusters together form a σ-curve, whose strands fill the plane increasingly tightly. The limit of this σ-curve behaves as in Chapter 7: it is no longer a σ-curve, but a domain of the plane.

THE ELUSIVE INFINITE CLUSTER. *No actually infinite* cluster is involved in the present approach. It is easy to arrange the generator's topology so that any given bounded domain is almost surely surrounded by a contact cluster. This cluster is in turn almost surely surrounded by a larger cluster, etc. There is no upper bound to cluster size. More generally, when a cluster seems infinite because it spans a very large area, the consideration of an even larger area will almost surely show it to be finite.

MASS-NUMBER AND WEIGHTED DIAMETER-NUMBER RELATIONS. THE EXPONENTS $D-D_c$ AND D/D_c.

Now let us reformulate the function $Nr(\Lambda > \lambda)$ in two ways: first by replacing a cluster's diameter λ by its mass μ, then by giving increased weight to large contact clusters.

Here, a cluster's mass is simply the number of links of length b^{-k} in the clusters itself (do *not* count the links *within* a looping cluster!). In effect, Chapters 6 and 12, we create a modified Minkowski sausage (Plate 33), by centering a square of side b^{-k} on each vertex, and adding half a square at each end-point.

The mass of a cluster of diameter Λ being the area of its modified sausage, $M \propto (\Lambda/b^k)^{D_c}(b^k)^2 = \Lambda^{D_c}/(b^k)^{D_c-2}$. Since $D_c < 2$, $M \to 0$ as $k \to \infty$. The mass of all the contact clusters taken together is $\propto (b^k)^{D-2}$; if $D < 2$, it too $\to 0$. And the relative mass of any individual contact cluster is $\propto (b^k)^{D_c-D}$; it tends to 0 at a rate that increases with $D-D_c$.

MASS-NUMBER RELATION. Clearly,

$$Nr(M > \mu) \propto (b^k)^{-D + 2D/D_c} \mu^{-D/D_c}.$$

DISTRIBUTION OF DIAMETER WEIGHTED BY MASS. Observe that $Nr(\Lambda > \lambda)$ counts the number of lines above line λ in a list that starts with the largest contact cluster, continues with the next largest, etc. But we shall momentarily have to attribute to each contact cluster a number of lines equal to its mass. The resulting relation is easily seen to be

$$Wnr(\Lambda > \lambda) \propto \lambda^{-D + D_c}.$$

THE MASS EXPONENT $Q = 2D_c - D$

Denote by $\mathcal{J}$ a fractal of dimension D, constructed recursively with $[0, \Lambda]$ as initiator, and take its total mass to be Λ^D. When $\mathcal{J}$ is a Cantor dust, Chapter 8 shows that the mass in a disc of radius $R < \Lambda$ centered at 0 is $M(R) \propto R^D$. ◁ The quantity $\log[M(R)R^{-D}]$ is a periodic function of $\log_b(\Lambda/R)$, but we shall not dwell on these complications because they vanish when the fractal is modified so that all $r > 0$ are admissible self-similarity ratios. ▶

We know that $M(R) \propto R^D$ also applies to the Koch curve of Chapter 6. Furthermore, this formula extends to the recursive islands and clusters of this chapter, with D replaced by D_c. In all cases, the mass in a disc of radius R centered at 0 takes the form

$$M(R, \Lambda) = R^{D_c} \phi(R/\Lambda),$$

with ϕ a function deducible from the shape of $\mathcal{J}$. In particular,

$$M(R, \Lambda) \propto R^{D_c} \text{ when } R \ll \Lambda,$$
$$\text{and } M(R, \Lambda) \propto \Lambda^{D_c} \text{ when } R \gg \Lambda.$$

Now consider the weighted average of $M(R)$, to be denoted by $\langle M(R) \rangle$, corresponding to the case when Λ is variable with the widely spread-out hyperbolic distribution $Wnr(\Lambda > \lambda) \propto \lambda^{-D + D_c}$. We know that $1 \leq D_c < D \leq 2$. Excluding the combination of $D = 2$ and $D_c = 1$, $0 < D - D_c < D_c$. It follows that

$$\langle M(R) \rangle \propto R^Q \text{ with } Q = 2D_c - D > 0.$$

When the disc's center is a point of $\mathcal{J}$ other than 0, the factor of proportionality changes, but its exponent is unchanged. It also remains unchanged by averaging over all positions of the center in $\mathcal{J}$, and by the replacement of $[0,1]$ by a different initiator. ◁ Usually, an arc of random size Λ is also of random shape. But the above formulas for $M(R, \Lambda)$ apply to $\langle M(R, \Lambda) \rangle$ averaged over all shapes. The final result is unchanged. ▶

REMARK. The preceding derivation does not refer to the clusters' topology: they can be loops, intervals, trees, or anything else.

CONCLUSION. The formula $\langle M(R) \rangle \propto R^Q$ shows that, when Λ is hyperbolically distributed, hence of very wide scatter, one of the essential roles of dimension is taken up by an exponent *other than* D. The most natural exponent is $2D_c - D$, but different weighting function give different Q's.

WARNING: NOT EVERY MASS EXPONENT IS A DIMENSION. The combined quantity Q is important. And, since it is a mass exponent, it is tempting to call it a dimension, but this temptation has no merit. Mixing many clusters with identical D_c but varying Λ leaves D_c unchanged, because dimension is *not* a property of a mixed population of sets, but a property of an individual set. Both D and D_c are fractal dimensions, but Q is not.

More generally, many areas of physics involve relations of the form $\langle M(R) \rangle \propto R^Q$, but such a formula does *not by itself* guarantee that Q is a fractal dimension. And calling Q an *effective dimension*, as some authors propose, is an empty gesture because Q does not possess any of the other properties that characterize D (for example, sums or products of D's have a meaning with no counterpart in the case of Q). Moreover, this empty gesture has proven a source of potential confusion.

NONLUMPED CURDLING CLUSTERS

We now proceed to describe two additional methods for generating contact clusters. One is based on curdling and applies for D<2, while the other is based on Peano curves and applies for D=2. The reader interested in percolation may skip this section and the next.

First, let us replace the Koch construction by the natural generalization of Cantor curdling to the plane. As illustration, consider the following five generators, with the next construction stage drawn underneath

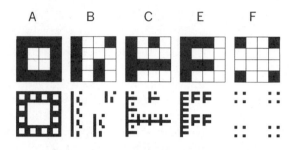

In all these cases, the limit fractal is of zero area and contains no interior point. Its topology can take diverse forms, determined by the generator.

With generator A, the precurd of every stage k is connected, and the limit fractal is a curve, an example of the very important Sierpiński carpet examined in Chapter 14.

With generator F, the precurd splits into disconnected portions, whose maximum linear scale steadily decreases as k→∞. And the limit fractal is a dust, akin to the Fournier model of Chapter 9.

The generators B, C and E are more interesting: in their case, the precurd splits into pieces to be called *preclusters*. Each stage can be said to transform every "old" precluster by making it thinner and wigglier, and to give birth to "new" preclusters. Nevertheless, by deliberate choice of generators, each newborn precluster is entirely contained in a single smallest cell in the lattice prevailing before its birth. By contrast with the "cross lumped clusters" of the next section, the present ones are to be called "nonlumped." It follows that the limit contact clusters have a dimension of the form $\log N_c / \log b$, where N_c is an integer *at most* equal to the number of cells in the generator's largest component. This maximum is attained for generators B and C, for which the contact clusters are, respectively, intervals with $D_c=1$ and fractal trees with $D_c=\log 7/\log 4$. But the fractal based on the generator E does not attain this maximum: in its case, the F-shaped preclusters keep splitting into parts, and the limit, again, is made of straight intervals with $D_c=1$.

Replacing the pseudo-Minkowski sausage by the collection of cells of side b^{-k} intersected by a contact cluster, the diameter-number relation and the other results of the preceding sections extend unchanged.

CROSS LUMPED CURDLING CLUSTERS

Next, let the generator of plane curdling takes either of the following shapes, with the next construction stages drawn to the side

Both cases exhibit massive "cross lumping," meaning that each newborn precluster combines contributions coming from several smallest lattice cells prevailing before its birth.

In the Koch context, an analogous situation prevails when the teragons are allowed to self-contact, resulting in the merger of small cluster teragons. In either case, the analysis is cumbersome, and we cannot dwell on it. But $Nr(\Lambda>\lambda)\propto\lambda^{-D}$ remains a valid relation for small λ.

◁ However, if one attempts to estimate D from this relation, without excluding the large λ's, the estimate is systematically biased and smaller than the true value. ▶

Novel features arise concerning the quantity b^{D_c}: it need not be an integer deducible from the generator by simple inspection, but it may be a fraction. The reason is that every contact cluster combines: (a) an integer number of versions of itself, downsized in the ratio $1/b$, and (b) many downsized versions due to lumping, which involve smaller ratios of the form $r_m=b^{-k(m)}$. The dimension-determining equation $\Sigma r_m^D=1$ of page 56, when rewritten in terms of $x=b^{-D}$, takes the form $\Sigma a_m x^m=1$. Cases where $1/x$ is an integer can only occur as exceptions.

KNOTTED PEANO MONSTERS, TAMED

A plane-filling collection of clusters (D=2) cannot be created by curdling, but I found an alternative approach, using Peano curves beyond those we saw being tamed in Chapter 7. As the reader must recall, Peano curves with self-avoiding teragons create river and watershed trees. But some other Peano curve teragons (for example the teragons in Plate 63, assuming that the corners are not rounded off) are simply chunks of lattice. As the construction proceeds, the open lattice cells separated by such curves "converge" to an every-where dense dust, e.g., to the points for which neither x nor y is a multiple of b^{-k}.

Between these extremes stands a new in-teresting class of Peano curves. Their genera-tors are exemplified by the following, shown together with the next step

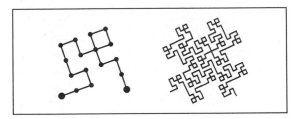

This class of Peano curves is now ready to be tamed. We observe that each point of self-contact "knots off" an open precluster, which may acquire branches and self-contacts, sees chunks of itself "knotted away," and eventu-ally thins down to a highly ramified curve that defines a contact cluster. A cluster's di-ameter Λ, defined as in previous sections of this chapter, is fixed from the moment of birth: roughly equal to the side of the square that "seeded" this cluster. Its distribution is ruled by the familiar relation $Nr(\Lambda > \lambda) \propto \lambda^{-2}$.

Observe in passing that, while Koch con-tact clusters are limits of recursively con-structed curves, the present clusters are limits (in a peculiar sense) of the open components of the *complement* of a curve.

BERNOULLI PERCOLATION CLUSTERS

Whichever method is used to generate fractal contact clusters with $D=E$ and $D_C < D$, they provide a geometric model that had been lack-ing in a very important problem of physics: Bernoulli percolation through lattices. J. M. Hammersley, who posed and first investigated this problem, did not inject Bernoulli's name in this context, but the fractal percolation we encounter in Chapter 23 makes the full term unavoidable here. (It is independently adopted by Smythe & Wiermann 1975.)

LITERATURE. Bernoulli percolation is sur-veyed in Shante & Kirkpatrick 1971, Domb & Green 1972-, especially a chapter by J. W. Essam, Kirkpatrick 1973, deGennes 1976, Stauffer 1979, and Essam 1980.

DEFINITIONS. Percolation involves probabil-istic notions, hence would not be discussed at this stage if we were entirely consistent. But an occasional lack of consistency has its re-wards. The simplest percolation problem for $E=2$ is bond percolation on a square lattice. To illustrate it in homely fashion, imagine we construct a large square lattice with sticks made either of insulating vinyl or of conduct-ing copper. A *Bernoulli lattice* obtains if each stick is selected at random, independently of the other sticks, the probability of choosing a conducting stick being p. Maximal collections of connected copper or vinyl sticks are called copper or vinyl *clusters*. When the lattice in-cludes at least one uninterrupted string of copper sticks, the current can *flow through* from one side of the lattice to the other, and the lattice is said to *percolate*. (In Latin, *per* = through, and *colare* = to flow.) The sticks in uninterrupted electric contact with the top and bottom sides of the lattice form a "percolating cluster," and the sticks actually active in conducting form the percolating cluster's "backbone."

The generalization to other lattices, and to $E > 2$, is immediate.

CRITICAL PROBABILITY. Hammersley's most remarkable finding concerns the special role played by a certain threshold probability: the *critical probability* p_{crit}. This quantity enters in when the Bernoulli lattice's size (measured in numbers of sticks) tends to infinity. One finds that, when $p > p_{crit}$, the probability that there exists a percolating cluster increases with lattice size, and tends to 1. When $p < p_{crit}$, to the contrary, the probability of percolation tends to 0.

Bond percolation on square lattices being such that either copper or vinyl must percolate, $p_{crit} = \frac{1}{2}$.

ANALYTICAL SCALING PROPERTY. The study of percolation long devoted itself to the search for analytic expressions to relate the standard quantities of physics. All these quantities were found to be *scaling*, in the sense that the relations between them are given by power laws. For $p \neq p_{crit}$, scaling extends up to an outer cutoff dependent on $p - p_{crit}$ and denoted by ξ. As $p \rightarrow p_{crit}$, the cutoff satisfies $\xi \rightarrow \infty$. Physicists postulate (see Stauffer 1979, p. 21) that $\langle M(R, \Lambda) \rangle$ follows the rule obtained on p. 123.

THE CLUSTERS' FRACTAL GEOMETRY

THE CLUSTERS' SHAPE. Let $p = p_{crit}$, and let individual sticks decrease in size while the total lattice size remains constant. The clusters become increasingly thin ("all skin and no flesh"), increasingly convoluted, and increasingly rich in branches and detours ("ramified and stringy"). Specifically, Leath 1976, the number of sticks situated outside of the cluster, but next to a stick within the cluster, is roughly proportional to the number of sticks within the cluster.

HYPOTHESIS THAT CLUSTERS ARE FRACTALS. It is natural to conjecture that the property of scaling extends from analytic properties to the clusters' geometry. But this idea could not be implemented in standard geometry, because the clusters are *not* straight lines, Fractal geometry is of course designed to eliminate such difficulties: thus, I conjectured that clusters are representable by fractal σ-curves satisfying $D = 2$ and $1 < D_c < D$. This claim has been accepted, and found to be fruitful. It is elaborated upon in Chapter 36.

◁ To be precise, scaling fractals are taken to represent the clusters that are *not* truncated by the boundary of the original lattice. This excludes the percolating cluster itself. (The term *cluster* has a gift for generating confusion!) To explain the difficulty, start with an extremely large lattice, pick a cluster on it, and a smaller square that is spanned by this cluster. By definition, the intersection of this cluster and the smaller square includes a smaller percolating cluster, but in addition it includes a "residue" that connects with the smaller percolating cluster through links *outside* the square. Note that neglect of this residue creates a downward bias in the estimation of D_c. ►

VERY ROUGH BUT SPECIFIC NONRANDOM FRACTAL MODELS. To be valid, the claim that any given natural phenomenon is fractal must be accompanied by the description of a specific fractal set, to serve as first approximation model, or at least as mental picture. My Koch curve model of coastlines, and the Fournier model of galaxy clusters, demonstrate that rough nonrandom picture may be very useful. Similarly, I expect recursively constructed contact clusters (like those introduced in this chapter) to provide useful fractal models of the ill-known natural phenomena that are customarily modeled by Bernoulli clusters.

However, the Bernoulli clusters themselves are fully known (at least in principle), hence modeling them via explicit recursive fractals is a different task. The Koch contact clusters I studied are not suitable, due to dissymmetry between vinyl and copper, even when there are equal numbers of sticks of both kinds. Next examine the knotted Peano curve clusters. Take an advanced teragon, and cover the cells to the left of the curve with copper, and the other cells with vinyl. The result involves a form of percolation applied to lattice cells (or to their centers, called sites). The problem is symmetric. But it differs from the Bernoulli problem, because the configuration of copper or vinyl cells are *not* the same as in the case of independence: for example, 9 cells forming

a supersquare can all be of copper or vinyl in the Bernoulli case, but not in the knotted Peano curve case. (On the other hand, both models allow groups of 4 cells forming a supersquare to take any of the possible configurations.) This difference has far-reaching consequences: for example, neither copper nor vinyl percolate in the Bernoulli site problem with $p=\frac{1}{2}$, while both percolate in knotted Peano clusters, implying that $\frac{1}{2}$ is a critical probability.

The list of variants of Bernoulli bond percolation is already long, and can easily be lengthened further. And I have already examined many variants of recursively constructed fractal contact clusters. The detailed comparison of these lists is unfortunately complicated, and I shall not dwell on it here.

Let me therefore be satisfied with stating the loose conclusion that significant fractal essentials of the Bernoulli percolation problem seem to be illustrated by nonrandom space-filling σ-clusters defined earlier in this chapter. This model's principal weakness is that it is completely indeterminate beyond what has been said. It can accommodate any observed degree of irregularity and fragmentation. On the matter of topology, see Chapter 14.

MODEL OF CRITICAL CLUSTERS. Specifically, consider the critical clusters, defined as the clusters for $p=p_{crit}$. To represent them, a recursive σ-cluster is extrapolated as indicated in earlier sections of this chapter. Then it is truncated by stopping the interpolation so that the positive inner cutoff is the cell size in the original lattice.

MODELS OF NONCRITICAL CLUSTERS. To extend this geometric picture to noncritical clusters, that is, to clusters for $p \neq p_{crit}$, we seek fractals with a positive inner cutoff and a finite outer cutoff. Analysis calls for the largest copper cluster's extent to be of the order of ξ when $p < p_{crit}$, and to be infinite when $p > p_{crit}$. Either outcome is readily implemented. For example, one can start with the same generator as in the preceding subsection. But, instead of extrapolating it naturally, one initiates it with either of the following shapes

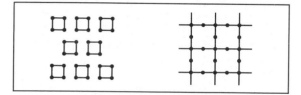

SUBCRITICAL CLUSTERS. The initiator to the left, which is geared towards $p < p_{crit}$, is made of squares of side $\frac{1}{2}\xi$. Now let the chosen generator be positioned *in* through each initiator's left side, and *out* through the other sides. The initiator square will transform into an atypical cluster of length ξ, surrounded by many typical clusters of length $< \xi$.

SUPERCRITICAL CLUSTERS. The initiator to the right, which is geared towards $p > p_{crit}$, is made of those lines of the initial square lattice, whose x or y coordinates are even integers. Four links radiate from each node whose coordinates are even integers; the chosen generator is always positioned to the left. In the special case when the coastline generator involves no loops nor dangling links, the result-

ing picture is a de-randomized and systematized variant of a crude model of clusters based solely on "nodes and links."

Observe that the fractal geometric picture deduces the noncritical clusters from the critical ones, while physicists prefer to consider the critical clusters as limits of the noncritical clusters for $\xi \to \infty$.

CRITICAL BERNOULLI CLUSTERS' D_c

The value of D_c is immediately inferred from either the exponent $D/D_c = E/D_c$ in the formula for $Nr(M > \mu)$, or the exponent $Q = 2D_c - D = 2D_c - E$ in the formula for $\langle M(R) \rangle$. Using the Greek letters τ, δ and η with the meanings customary in this context, we find that $E/D_c = \tau - 1$ and $2D_c - E = 2 - \eta$. Hence,

$$D_c = E/(\tau - 1) = E/(1 + \delta^{-1}),$$
$$\text{and } D_c = 1 + (E - \eta)/2.$$

Due to relations that physicists established between τ, δ and η, the above formulas for D_c are equivalent. Conversely, their equivalence does not reside in physics alone, because it follows from geometry.

Independently of each other, Harrison, Bishop & Quinn 1978, Kirkpatrick 1978, and Stauffer 1979 obtain the same D_c. They start from the properties of clusters for $p > p_{crit}$, hence express their result in terms of different critical exponents (β, γ, ν and σ). These derivations do not involve a specific underlying fractal picture. The dangers inherent in this approach, against which we warned earlier in this chapter, are exemplified by the fact that it misled Stanley 1977 into advancing Q and D_C are equally legitimate dimensions.

For E=2, the numerical value is D_C=1.89. It is compatible with the empirical evidence, as obtained by a procedure familiar in other guises. Pick r, which need *not* be of the form 1/b (b an integer). Then take a big eddy, which is simply a square or cubic lattice of side set to 1. Pave it with subeddies of side r, count the number N of the squares or cubes that intersect the cluster, and evaluate $\log N/\log(1/r)$. Then repeat the process with each nonempty subeddy of side r by forming subsubeddies of side r^2. Continue as far as feasible. The most meaningful results obtain when r is close to 1. Some early simulations gave the biased estimate D^+~1.77 (Mandelbrot 1978h, Halley & Mai 1979), but large simulations (Stauffer 1980) confirm D.

◁ The biased experimental D^+ is very close to Q, hence briefly seemed to confirm the theoretical arguments in Stanley, Birgenau, Reynolds & Nicoll 1976 and Mandelbrot 1978h, which were both in error in claiming that the dimension is Q. The error was brought to my attention by S. Kirkpatrick. A different and even earlier incorrect estimate of D is found in Leath 1976. ▶

THE CYPRESS TREES OF OKEFENOKEE

When a forest that is not "managed" systematically is observed from an airplane, its boundary is reminiscent of an island's coastline. Individual tree patches' outlines are extremely ragged or scalloped, and each large patch is trailed by satellite patches of varying area. My hunch that these shapes may follow the Richardson and/or Korčak laws, is indeed confirmed by an unpublished study of the Okefenokee swamp (Kelly 1951) by H. M. Hastings, R. Monticciolo & D. VunKannon. The patchiness of cypress is great, with D~1.6; the patchiness of broadleaf and mixed broadleaf trees is much less pronounced, with D near 1. My informants comment on the presence of an impressive variety of scales both on personal inspection and on examination of vegetation maps. There is an inner cutoff of about 40 acres, probably a consequence of aerial photography. ■

14 ¤ Ramification and Fractal Lattices

Chapter 6 investigates planar Koch curves that satisfy D<2 and are devoid of double points, hence can be called self-avoiding or nonramified. And Chapter 7 investigates Peano curves, for which everywhere dense double points are unavoidable in the limit. The present chapter takes the next step, and investigates examples of deliberately ramified self-similar shapes: planar curves with 1<D<2, spatial curves with 1<D<3, and surfaces with 2<D<3. In a ramified self-similar curve, the number of double points is infinite.

This chapter's mathematics is old (though known to very few specialists), but my applications to the description of Nature are new.

THE SIERPIŃSKI GASKET AS MONSTER

Sierpiński gasket is the term I propose to denote the shape in Plate 141. An extension to space is shown in Plate 143. The constructions are described in the captions.

Hahn 1956 comments that "A point on a curve is called a branch point if the boundary of any arbitrarily small neighborhood has more than two points in common with the curve... Intuition seems to indicate that it is impossible for a curve to be made up of nothing but... branch points. This intuitive conviction had been refuted [by the] Sierpiński... curve, *all of whose points are branch points*."

THE EIFFEL TOWER: STRONG AND AIRY

Again, Hahn's view is totally without merit, and his uncharacteristic "seems to indicate" is a wise choice of words. My first counter-argument is borrowed from engineering. (As argued before we tackled computers at the end of Chapter 12, there is nothing illogical about including articulated engineering systems in this work concerned with Nature.)

My claim is that (well before Koch, Peano, and Sierpiński) the tower that Gustave Eiffel built in Paris deliberately incorporates the idea of a fractal curve full of branch points.

In a first approximation, the Eiffel Tower is made of four A-shaped structures. Legend has it that Eiffel chose A to express *Amour* for his work. All four A's share the same apex and any two neighbors share an ascender. Also, a straight tower stands on top.

However, the A's and the tower are not made up of solid beams, but of colossal trusses. A truss is a rigid assemblage of interconnected submembers, which one cannot deform without deforming at least one submember. Trusses can be made enormously lighter than cylindrical beams of identical strength. And Eiffel knew that trusses whose "members" are themselves subtrusses are even lighter.

The fact that the key to strength lies in branch points, popularized by Buckminster Fuller, was already known to the sophisticated designers of Gothic cathedrals. The farther we go in applying this principle, the closer we get to a Sierpiński ideal! An infinite extrapolation of the Eiffel Tower design is described in Dyson 1966, p. 646, wherein a former student of Besicovitch seeks strong interplanetary structures of low weight.

CRITICAL PERCOLATION CLUSTERS

Let us now return to nature, or more precisely to an image of nature provided by statistical physics. I think the kin of the Sierpiński gasket is *demanded* by the study of percolation through lattices. Chapter 13, which began our case study of this topic, claims that percolation clusters are fractals. Now I add the further claim that the Sierpiński gasket's branching structure is a promising model of the structure of cluster backbones.

The physicists will mostly judge this model on the fact that it rapidly fulfilled its promise: Gefen, Aharony, Mandelbrot & Kirkpatrick 1981 shows the model allows usual calculations to be carried out *exactly*. But the details are much too technical to be included in this Essay, and the original reasons for my claim remain of interest. It arose from a resemblance I perceived between the gasket and the cluster backbones, as shown in this diagram:

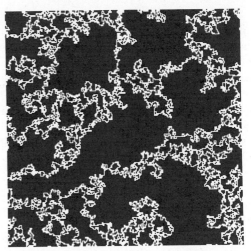

The most conspicuous feature resides in the tremas left vacant by the elimination of dangling bonds (when a cluster was reduced to its backbone), and of clusters contained entirely within the cluster of interest. Second, the fact that the branching is self-similar in a Sierpiński gasket is shown in Chapter 13 to be an eminently desirable property in a geometric model of the percolation cluster. Finally, the dimensions fit to a degree that can hardly be coincidental! S. Kirkpatrick estimates that in the plane D~1.6, astonishingly close to the D of the Sierpiński gasket! And in space, D~2.00, astonishingly close to the D of the fractal skewed web in Plate 143. Furthermore, Gefen, Aharony, Mandelbrot & Kirkpatrick 1981 observes that the identity between the D of the backbone and that of the generalized gasket persists in $\mathbb{R}^4$. An additional argument in favor of the gasket model is mentioned later, as a last application of ramification.

THE TRIADIC SIERPIŃSKI CARPET

Let us now switch from triangular to orthogonal lattices. They allow great versatility in design, yielding *curves* in the plane or in space, or *surfaces* in space. And the curves they yield, despite a superficial resemblance to the Sierpiński gasket, are very different from the fundamental viewpoint of ramification, to which we turn after defining them.

The literal planar extension of Cantor's method of deleting mid-thirds initiates with a square, and is described in the caption on page 142. The fractal obtained by continuing ad infinitum is widely known by the homely term *triadic Sierpiński carpet*. Its dimension is $D=\log 8/\log 3=1.8927$.

NONTRIADIC FRACTAL CARPETS

Given an integer b>3, and writing r=1/b as usual, a "large centered medallion" carpet is obtained by taking as initiator a square, as trema a square of side 1−2r, with the same center, and as generator a thin ring of 4(b−1) squares of side r. The dimensions are $D=\log [4(b-1)]/\log b$. Given an odd integer b>3, a "small centered medallion" carpet is obtained by taking as trema a single subsquare of side r, with the same center as the initiator, and as generator a thick ring of b^3-1 small squares. The dimensions are $D=\log (b^3-1)/\log b$. Thus, any D between 1 and 2 can be approximated arbitrarily closely in a centered carpet.

Noncentered carpets can be defined for b≥2. For example, when b=2 and N=3, a trema made of one subsquare can be positioned in the subsquare on the top right. The corresponding limit set turns out to be the Sierpiński *gasket* built with the triangle forming the bottom left half of the square.

TRIADIC FRACTAL FOAM

The literal spatial extension of the triadic carpet consists in removing a cube's mid 27-th subcube as trema, leaving a shell of 26 subcubes. The resulting fractal is to be called *triadic fractal foam*. Its dimension is $D=\log 26/\log 3=2.9656$.

Here, every trema is entirely enclosed by an uninterrupted boundary split into infinitely many, infinitely thin layers of infinite density. In order to join two points situated in different tremas, it is necessary to cross an infinite number of layers. One is reminded, but this is a topic I do not master thoroughly enough to attempt to account for it here, of the "space-time foam" which characterizes the finest structure of matter according to J. A. Wheeler and G. W. Hawking.

MENGER'S TRIADIC FRACTAL SPONGE

Karl Menger selects a different trema, shaped like a cross with spikes front and back, consisting of N=20 subcubes of side ⅓, connected to one another. Among them, 12 form "rods" or ropes, and the remaining 8 are knots, connectors, or ties. The limit (Plate 145) satisfies $D=\log 20/\log 3=2.7268$. I call it *a sponge*, because both the curd and the whey are connected sets. One can conceive of water flowing between any two points in the whey.

To obtain a mixture of ropes and sheets, let the trema be a triadic cross continued by a single spike in front. By changing the direction of the spike every so often, one may end up with punctured sheets. It may be worth mentioning that I thought of all these shapes before reading Menger, while looking for models of turbulent intermittency.

NONTRIADIC SPONGES AND FOAMS

Given a nontriadic base b>3, generalized Menger sponges are obtained when the trema is the union of three square based cylinders: the axis of each coincides with an axis of the unit cube, its length is 1, and its base has sides parallel to the other axes. The sponge is called "light" when the bases' sides are as large as possible. For E=3, they are of length 1−2/b, leaving as generator a collection of 12b−16 cubes of side r=1/b. Hence the dimension is $D=\log(12b-16)/\log b$. Similarly, a "heavy sponge" is obtained, but only in case b is odd, when the cylinder bases' sides are of length 1/b. For E=3, they leave as generator a collection of b^3-3b+2 cubes of side r=1/b. Now $D=\log(b^3-3b+2)/\log b$.

Fractal foams generalize in analogous fashion. For E=3, "thick wall" foams yield $D=\log(b^3-1)/\log b$, and "thin wall" foams yield $D=\log(6b^2-12b+8)/\log b$. With big holes and D near 2, the foam resembles an overly airy Emmenthaler. With small holes and D near 3, it resembles a different cheese delicacy, Appenzeller.

GAPS' SIZE DISTRIBUTIONS

The sponges' tremas merge together but carpets' and foams' tremas remain as gaps analogous to those of the Cantor dust (Chapter 8). The distribution of their linear scale Λ satisfies

$$Nr(\Lambda > \lambda) \propto F\lambda^{-D},$$

where F is a constant. We know this rule well from the gaps of a Cantor dust, and the islands and clusters of Chapter 13.

THE NOTION OF FRACTAL NET, LATTICE

The *lattices* of standard geometry are formed by parallel lines bounding equal squares or triangles, and analogous regular designs. The same term seems applicable to regular fractals in which any two points can be linked by at least two paths that do not otherwise overlap. When the graph is *not* regular, for example is random, I replace *lattice* by *net*.

However, a closer comparison of standard and fractal lattices reveals considerable differences. The first difference is that the standard lattices are invariant by translation but not by scaling, while for the fractal lattices the contrary is true. A second difference is that any standard lattice, if downsized, converges to the whole plane. Also, several standard lattices in the plane can be interpolated by adding lines halfway between existing parallel lines, and repeating ad infinitum. Again, the result converges towards the whole plane. Similarly, when a standard spatial lattice can be interpolated, its limit is the whole space. Thus, the limit is *not* a lattice. In the fractal context, to the contrary, the limit of an approximate fractal lattice is a fractal lattice.

The term, *ramified fractal lattices* can also be applied to the fractal foams.

THE SECTIONS' FRACTAL DIMENSIONS

A BASIC RULE. In many studies of fractals, it is important to know the dimensions of the linear and planar sections. The basic fact (used in Chapter 10 to show that D>2 for turbulence) concerns the section of a planar fractal shape by an interval "independent of the fractal." One finds that if the section is nonempty, it is "almost sure" that its dimension is D−1.

The corresponding value in space is D−2.

EXCEPTIONS. Unfortunately, this result is hard to illustrate in the case of nonrandom fractals that have axes of symmetry. The intervals that impose themselves upon our consideration are parallel to these axes, hence atypical, and nearly *every* simple section by an interval belongs to the exceptional set wherein the general rule fails to apply.

For example, take the Sierpiński carpet, the triadic Menger sponge and the triadic foam. D−1, which is the almost sure dimension of sections by intervals, is, respectively

$$\log(8/3)/\log 3,$$
$$\log(20/9)/\log 3, \text{ and } \log(26/9)/\log 3,$$

On the other hand, let x be the abscissa of an interval parallel to the y-axis of the Sierpiński carpet. When x, written in counting base 3, ends up by an uninterrupted infinite string of 0's or 2's, the sections are themselves intervals, hence D=1, larger than expected. When x ends up by an uninterrupted infinite string of 1's, to the contrary, the sections are Cantor dusts, hence D=log 2/log 3 is too small. And when x terminates by a periodic pattern of period M, including pM times 1 and (1−p)M times 0 or 2, the sections are of dimension p(log 2/log 3)+(1−p). The expected D prevails for p~.29. ◁ The same holds if the digits of x are random. ► Thus, three dimensions are involved here: the largest, the smallest and the average.

Closely analogous results apply in space.

As to the Sierpiński gasket, the almost sure D is $\log(3/2)/\log 2$, but the D's relative to "natural" cuts range from 1 to 0. For example, a short interval through the midpoint of one of the gasket's sides, if close enough to the perpendicular, intersects the gasket on a *single* point, with D=0.

In part, the variability of these special sections is traceable to the regularity of the original shapes. But in another part, it is inevitable: the most economical section (not necessarily by a straight line) is the basis of the notions of topological dimension and of order of ramification, to which we proceed now.

THE RAMIFIED FRACTALS VIEWED AS CURVES OR SURFACES

As often stated, *curve* is used in this Essay as a synonym of "connected shape of topological dimension $D_T=1$." Actually, this phrase is not fully satisfactory to the mathematicians, and the precise restatements are delicate. Luckily, Chapter 6 could be content with a simple reason why any Koch curve with [0,1] as initiator deserves to be called a curve: like [0,1] itself, it is connected, but becomes disconnected if any point other than 0 or 1 is removed. And a snowflake boundary is like a circle: it is connected, but becomes disconnected if any two points are removed.

Restated more pedantically, as is now necessary, the topological dimension is defined recursively. For the empty set, $D_T=-1$. For any other set S, the value of D_T is 1 higher than the smallest D_T relative to a "cutset" that disconnects S. Finite sets and Cantor dusts satisfy $D_T = 1-1 = 0$, because nothing (the empty set) need be removed to disconnect them. And the following connected sets are all disconnected by the removal of a cutset that satisfies $D_T=0$: circle, [0,1], snowflake boundary, Sierpiński gasket, Sierpiński carpets, Menger sponges. (In the last three cases, it suffices to avoid the special intersections that include intervals.) Hence, all these sets are of dimension $D_T=1$.

By the same token, a fractal foam is a surface, with $D_T=2$.

Here is an alternative proof that $D_T=1$ for the gasket, all carpets, and all sponges with D<2. Since D_T is an integer ≤D, the fact that D<2 means that D_T is either 0 or 1. But the sets in question are connected, hence D_T is *no less* than 1. The only solution is $D_T=1$.

A CURVE'S ORDER OF RAMIFICATION

Topological dimension, and the corresponding notions of dust, curve, and surface, yield only a first level classification. Indeed, two finite sets containing M' and M'' points, respectively, have the same $D_T=0$, but they differ topologically. And Cantor dust differs from *all* finite dusts.

Let us now see how a parallel distinction

based on the number of points in a set ◁ its "cardinality" ► carries on to curves, leading to the topological notion of *order of ramification*, defined by Paul Urysohn and Karl Menger in the early 1920's. This notion is mentioned in few mathematics books other than the pioneers', but is becoming indispensable in physics, hence becoming better known after being tamed than in the wild. It shows that the reasons for discussing first a gasket, then a carpet, go beyond esthetics and the search for completeness.

The order of ramification involves the cutset containing the *smallest* number of points, that must be removed in order to disconnect the set S. And it involves separately the neighborhood of every point P in S.

THE CIRCLE. As background from standard geometry, begin by taking for S a circle of radius 1. A circle $\mathcal{B}$ centered on P cuts S in R=2 points, except if $\mathcal{B}$ has a radius exceeding 2, in which case R=0. The disc bounded by $\mathcal{B}$ is called a neighborhood of P. Thus, any point P lies in arbitrarily small neighborhoods whose boundaries intersect S at R=2 points. This is the best one can do: when $\mathcal{B}$ is the boundary of a general neighborhood of P, not necessarily circular but "not too large," R is at least 2. The terms "not too large" in the preceding sentence are a complication, but are unfortunately unavoidable. R=2 is called the order of ramification of the circle. We note that it is the same at all points of the circle.

THE GASKET. Next, let S be a Sierpiński gasket, constructed via tremas. Here R is no longer the same for every P. Let me show after Sierpiński that, excluding the initiator's vertices, R can be either $3=R_{min}$ or $4=R_{max}$.

The value R=4 applies to the vertices of any finite approximation of S by triangles. A vertex in an approximation of order h≥k is the common vertex P of two triangles of side 2^{-k}. Again, circles of center P and radius 2^{-k}, with h>k, intersect S in 4 points, and bound arbitrarily small neighborhoods of P. And if $\mathcal{B}$ bounds a "sufficiently small" neighborhood of P (in the new sense that the initiator's vertices lie outside $\mathcal{B}$), one can show that $\mathcal{B}$ intersects S in at least 4 points.

The value R=3 applies for every point of S that is the limit of an infinite sequence of triangles, each contained in its predecessor and having vertices distinct from its predecessor's. Circles circumscribed to these triangles intersect S in 3 points, and bound arbitrarily small neighborhoods of P. Also if $\mathcal{B}$ bounds a sufficiently small neighborhood of P (again, the initiator's vertices must lie outside), one can show that $\mathcal{B}$ intersects S at 3 points at least.

THE CARPETS. When S is a Sierpiński carpet, the result is *radically* different. Any neighborhood's boundary, if sufficiently small, intersects S in a nondenumerably infinite cutset, regardless of the parameters N, r, or D.

COMMENT. In this finite versus infinite dichotomy, the gasket does not differ from the standard curves, while the carpets do not differ from the whole plane.

HOMOGENEITY. UNICITY. Denoting by R_{min} and R_{max} the smallest and the largest R attained on a point of S, Urysohn proves that

$R_{max} \geq 2R_{min}-2$. The ramification is called *homogeneous* when the equality $R_{max}=R_{min}$ holds; this is the case when $R \equiv 2$, as in simple closed curves, and when $R \equiv \infty$.

For other lattices with $R_{max} = 2R_{min}-2$, I propose the term *quasi-homogeneous*. One simple and famous example, the Sierpiński gasket, is self-similar. The other nonrandom examples are part of a collection set up by Urysohn 1927, and are *not* self-similar. Thus, the conditions, of being quasi-homogeneous and self-similar, have only one known solution, the Sierpiński gasket. Could this seeming unicity be confirmed rigorously?

STANDARD LATTICES. Here the order of ramification ranges from a minimum of 2 for all points off the lattice sites, to a variable finite maximum attained on the lattice sites: 4 (squares), 6 (triangles or cubes) or 3 (hexagons). However, as a standard lattice of any kind is downsized, it transforms from a curve into a plane domain, and its ramification becomes $R=\infty$.

This last fact is made more obvious by exchanging the infinitely small and the infinitely large, holding to a lattice of fixed cell size, and observing that in order to isolate an increasingly *large* portion of lattice, one must cut points whose number has no finite bound.

FORMAL DEFINITION. ◁ See Menger 1932 and p. 442 of Blumenthal & Menger 1970. ▶

APPLICATIONS OF RAMIFICATION

Let us now face a familiar question. Whatever interest the Sierpiński and Menger shapes, and their kin, may have for the mathematician, is it not obvious that the order of ramification can be of no interest to the student of Nature? The response is as familiar—to us!—as the question. The order of ramification is already meaningful in the "real world" of the finite approximations which obtain when the interpolation leading to a fractal is stopped at some positive inner cutoff, ϵ.

Indeed, given an approximate Sierpiński gasket made of filled triangles of side ϵ, a domain whose linear scale is above ϵ can be disconnected by removing 3 or 4 points, each of which belongs to 2 neighboring gaps' boundaries. This number (3 or 4) does not change as this approximation is refined. Hence, from the viewpoint of ramification, all approximate gaskets are curve-like.

To the contrary, all carpets have the property that the boundaries of any two gaps fail to overlap. To disconnect a finite approximation of such a shape, in which the gaps of diameter $<\epsilon$ are disregarded, it is necessary to remove whole intervals. And these intervals' number increases as $\epsilon \to 0$. Whyburn 1958 shows that all the fractal curves that possess this property are topologically identical ◁ homeomorphic ▶, and are characterized by the fact they contain no part that can be disconnected by the removal of a single point.

Due to the preceding comments, it is not surprising that the finiteness of ramification acquires clearcut implications when fractal geometry is called to determine in detail how much a plane fractal curve partakes of its two

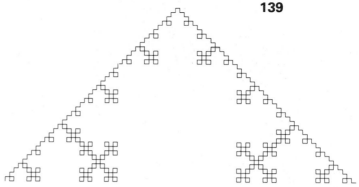

standard limits: the straight line and the whole plane. In general, knowing the fractal dimension does *not* suffice. For example, Gefen, Mandelbrot & Aharony 1980 examines critical phenomena for Ising models on a fractal lattice, and finds that the most important issue ◁ whether the critical temperature is 0 or positive ▶ depends on the finiteness of R.

We are now in a position to give an explanation we had postponed. The reason why a cluster backbone in critical Bernoulli percolation seems better modeled by a gasket than by a carpet lies in this finding reported in Kirkpatrick 197?. Even on extremely large lattices, a critical backbone can be cut by removing an essentially unvarying small number of bonds, of the order of 2. Even allowing for certain biases I could think of, this points out very strongly toward R<∞.

ALTERNATIVE FORM OF RAMIFICATION

Two variants of the Koch snowflake achieve ramification through branches without loops. The first is a plane curve obtained when the initiator is a square and the generator is

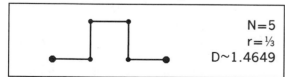

N=5
r=⅓
D∼1.4649

The resulting shape is totally different from the snowflake, as shown overleaf.

The next example is a surface of zero volume, infinite area, and a dimension equal to log 6 / log 2 = 2.58497. The initiator is a regular tetrahedron. On the mid-quarter of each face (= the triangle having as vertices the sides' midpoints), one attaches a tetrahedron reduced in the ratio ½. One repeats the procedure with each face of the resulting regular (skew and nonconvex) 24-hedron, and so on ad infinitum. From the second stage on, the added tetrahedrons self-contact along lines, without self-intersecting. And eventually they swarm all over the initiator. Let each fourth of this shape, growing on a face of the initiator, be called a Koch pyramid.

SECRETS OF THE KOCH PYRAMID

A Koch pyramid is a wondrous shape—plain when seen from above, but with a wealth of hidden chambers to defy the imagination.

Seen from above, it is a tetrahedron whose base is a equilateral triangle, but whose three other faces are right isosceles triangles joined at their 90° vertices. Three Koch pyramids, if put together on the sides of a regular tetrahedron, add to a plain cubic box.

Now lift such a pyramid from the floor of the desert. From a distance, we see its base subdivides into four equal regular triangles. But in place of the middle triangle there is a hole opening up on a "chamber of order 1," shaped like a regular tetrahedron whose fourth vertex coincides with the pyramid's top vertex. Next, as we approach and perceive finer detail, we find that the regular triangles that form the peripheral fourths of the base and the top faces of the chamber of order 1 are not smooth either. Each is broken by a tetrahedral chamber of order 2. Similarly, as we explore the chambers of order 2, each of their triangular walls reveals a chamber of order 3 in its middle portion. And increasingly tiny chambers appear without end.

All the chambers together add up precisely to the Koch pyramid's volume. On the other hand, if the chambers are viewed as including their bases but not their three other faces, they do *not* overlap. Were our pyramid to be dug from a mound, the chamber diggers would have to scoop out all its volume, leaving a mere shell. The curve along which this surface rests on the base's plane, and the chamber "walls," are Sierpiński gaskets.

SPHERICAL TREMAS AND LATTICES

Lieb & Lebowitz 1972 makes an unwitting contribution to fractal geometry, by packing $\mathbb{R}^E$ with balls whose radii are of the form $\rho_k = \rho_0 r^k$, with $r < 1$; the per-unit-volume number of balls of radius ρ_k is of the form $n_k = n_0 \nu^k$, where ν is an integer and is of form $\nu = (1-r)r^{-E}$, which strongly restricts r. Thus,

the exponent of the distribution of gap sizes is

$$D = \log \nu / \log(1/r) = E - \log(1-r)/\log r.$$

First, one centers big spheres of radius ρ_1 on a lattice of side $2\rho_1$. The vertices of a lattice of side $2\rho_2$ that lie outside of the big spheres are numerous enough to serve as centers for the next smaller spheres, and so on. The construction involves these upper bounds on r:

for $E=1$, $r \leq 1/3$;   for $E=2$, $r \leq 1/10$;
for $E=3$, $r \leq 1/27$;   as $E \to \infty$, $r \to 0$.

Packing of $\mathbb{R}^3$ by nonoverlapping balls can proceed more rapidly. For example, on the line, the maximum r is ⅓, corresponding to the triadic dust of Cantor! The existence of Cantor dusts with $r > ⅓$ demonstrates that one-dimensional packing can leave a remainder of arbitrarily low dimension. However, a tighter packing involves richer structure.

PREVIEW OF LACUNARITY

Even after the order of ramification R is added to the dimensions D_T and D, a fractal remains incompletely specified for many purposes. Of special importance is the additional notion of lacunarity that I developed. A very lacunar fractal's gaps are very large, and conversely. The basic definitions could have been described here, but it is more expedient to wait until Chapter 34.   ■

THIS PLATE'S CAPTION
IS FOUND OVERLEAF

Plate 141, OVERLEAF ¤ SIERPIŃSKI ARROWHEAD (BOUNDARY DIMENSION D~1.5849)

In Sierpiński 1915, the initiator is [0,1], and the generator and second teragon are

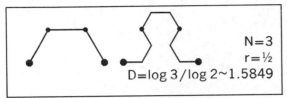

$$N=3$$
$$r=½$$
$$D=\log 3/\log 2 \sim 1.5849$$

This construction's next two stages are

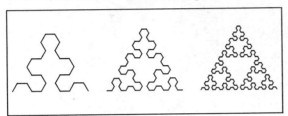

And an advanced stage is shown as the "coastline" of the upper portion of Plate 141 (above the largest solid black triangle).

SELF-CONTACTS. Finite construction stages are free of points of self-contact, as in Chapter 6, but the limit curve *does* self-contact infinitely often.

TILING ARROWHEADS. The arrowhead in Plate 141 (turned sideways, it becomes a tropical fish) is defined as a piece of the Sierpiński curve contained between two suc-cessive returns to a point of self-contact, namely the midpoint of [0,1]. Arrowheads tile the plane, with neighboring tiles being linked together by a nightmarish extrapolation of Velcro. (To mix metaphors, one fish's fins fit exactly those of two other fish). Furthermore, by fusing together four appropriately chosen neighboring tiles, one gets a tile increased in the ratio of 2.

THE SIERPIŃSKI GASKET'S TREMAS. I call Sierpiński's curve a *gasket*, because of an alternative construction that relies upon cutting out "tremas," a method used extensively in Chapters 8 and 31 to 35. The Sierpiński gasket is obtained if the initiator, the generator, and next two stages are these closed sets:

This trema generator includes the above stick generator as a proper subset.

WATERSHED. I first encountered the arrowhead curve without being aware of Sierpiński, while studying a certain watershed in Mandelbrot 1975m. ∎

Plate 143 ¤ A FRACTAL SKEWED WEB
(DIMENSION D=2)

This web obtains recursively, with N=4 and r=½, using a closed tetrahedron as initiator and a collection of tetrahedrons as generator.

Its dimension is D=2. Let us project it along a direction joining the midpoints of either couple of opposite sides. The initiator tetrahedron projects on a square, to be called initial. Each second-generation tetrahedron projects on a subsquare, namely (¼)th of the initial square, etc. Thus, the web projects on the initial square. The subsquares' boundaries overlap. ∎

142

SIERPIŃSKI CARPET. In Sierpiński 1916, the initiator is a filled square, while the generator and the next two steps are

N=8, r=⅓, D~1.8928.

This carpet's area vanishes, while the total perimeter of its holes is infinite.

PLATE 145. THE MENGER SPONGE. The principle of the construction is evident. Continued without end, it leaves a remainder to be called a Menger sponge. I regret having credited it wrongfully in earlier Essays, to Sierpiński. (Reproduced from *Studies in Geometry*, by Leonard M. Blumenthal and Karl Menger, by permission of the publishers, W. H. Freeman and Company, copyright 1970.) The intersections of the sponge with medians or diagonals of the initial cube are triadic Cantor sets.

FUSED ISLANDS. The carpet, as well as the gasket in Plate 143, may also be obtained by yet another generalization of the Koch recursion, wherein self-overlap is allowed, but overlapping portions count only once.

To obtain a gasket, the initiator is a regular triangle, and we take the generator to the left. To obtain a carpet, the initiator is a square, and we take the generator to the right

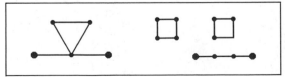

Two phenomena familiar from Chapter 13 are encountered again: each island's coastline is rectifiable and therefore of dimension 1, and the dimension of the gasket or the carpet expresses the degree of fragmentation of land into islands rather than the degree of irregularity of the islands' coastlines.

Otherwise, the result is unfamiliar: in Chapter 13 the sea is connected, which seems to be a proper topological interpretation of nautical openness. It is also open in the set topological sense of not including its boundary. The novelty brought in by the present construction is that it is possible for the Koch islands to "fuse" *asymptotically* into a solid superisland; there is no continent, and the coastlines combine into a lattice.

◁ Topologically, every Sierpiński carpet is a plane universal curve, and the Menger sponge is a spatial universal curve. That is, see Blumenthal & Menger 1970, pp. 433 and 501, these shapes are respectively the most complicated curve in the plane, and the most complicated curve in any higher dimensional space. ► ▨

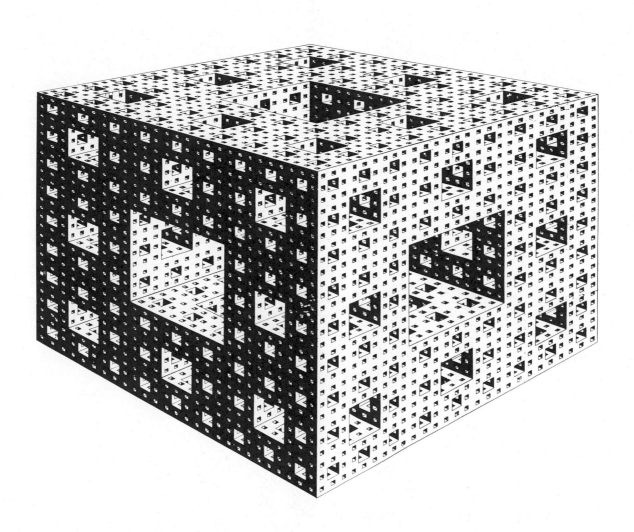

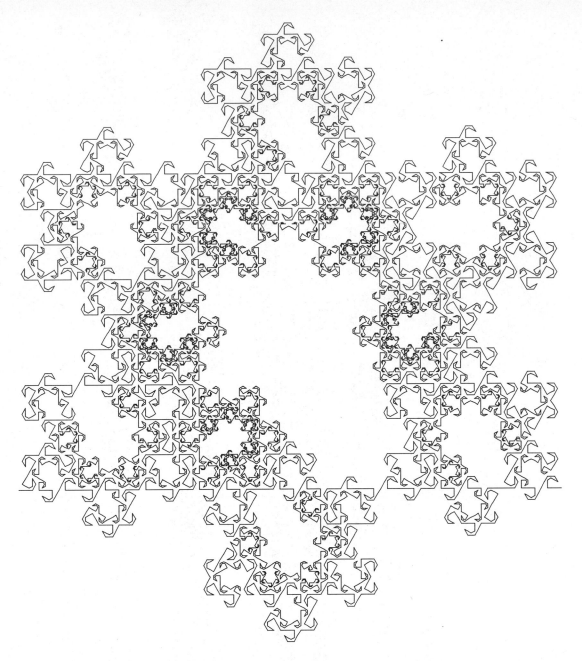

**Plate 146 ¤ SPLIT SNOWFLAKE HALLS
(DIMENSION D~1.8687)**

Long ago and far away, the Great Ruler and
his retinue had sat their power in the splendid
Snowflake Halls. A schism occurs, a war follows, ending in stalemate, and finally Wise

Elders draw a line to divide the Halls between
the contenders from the North and the South.

RIDDLES OF THE MAZE. Who controls the
Great Hall, and how is it reached from outside? Why do some Halls fail to be oriented
toward either of the cardinal points? For
hints, see the Monkeys Tree on Plate 31. ■

15 ¤ Surfaces with Positive Volume, and Flesh

The fractal curves, surfaces, and dusts which the present Part describes and tames for the purposes of science, are only scaling in an asymptotic or otherwise limited sense.

This first chapter centers on surfaces with a positive (nonvanishing!) volume. What a mad combination of contradictory features! Have we not finally come to mathematical monsters without conceivable utility to the natural philosopher? Again, the answer is emphatically to the negative. While believing they were fleeing Nature, two famous pure mathematicians unknowingly prepared the precise tool I need to grasp (among others) the geometry of...flesh.

CANTOR DUSTS OF POSITIVE MEASURE

A preliminary step is to review Cantor's construction of the triadic set $\mathcal{C}$. Its being of zero length (more pedantically, of zero linear measure) follows from the fact that the lengths of the mid third tremas add to

$$1/3 + 2/3^2 \ldots + 2^k/3^{k+1} \ldots = 1.$$

But the fact that $\mathcal{C}$ is totally disconnected, hence of topological dimension $D_T = 0$, is independent of the trema lengths. It comes from the basic fact that each construction stage bisects every interval created in the preceding stage, by removing a trema centered on the "host" interval's midpoint. Denoting the ratio of the trema and host lengths by λ_k, the cumulative length of the intervals that remain after K stages is $\Pi_0^K(1-\lambda_k)$. It decreases as $K \to \infty$ to a limit denoted by P. In Cantor's original construction, $\lambda_k \equiv \frac{2}{3}$, hence P=0. But P>0 whenever $\Sigma_0^\infty \lambda_k < \infty$. In that case, the remainder set $\mathcal{C}_*$ has the positive length 1−P. This set is not self-similar, hence has no similarity dimension, but the Hausdorff Besicovitch definition, Chapter 5, concludes that

D=1. It follows from D>D$_T$ that $\mathcal{C}_*$ is a fractal set. Since D and D$_T$ are both independent of the trema lengths λ_k, their values describe $\mathcal{C}_*$ very superficially.

The construction is even more perspicuous in the plane. Cut out from the unit square a cross of area λ_1, leaving four square tiles. Next cut out from each a cross of relative area λ_2. This cascade generates a dust, D$_T$=0, having the area $\Pi_0^\infty(1-\lambda_k)$. When this area does not vanish, D=2.

In E-dimensional space, one can similarly achieve *a dust with positive volume,* satisfying D$_T$=0 and D=E.

SLOWLY DRIFTING log N/log (1/r)

◁ Although the Cantor dusts with positive length, area or volume have no similarity dimension, it is useful to set $r_k=(1-\lambda_k)/2$, and to investigate the formal dimensions defined as $D_k=\log N/\log(1/r_k)$.

◁ When D$_k$ drifts slowly, it embodies the idea of effective dimension discussed in Chapter 3 when describing a ball of thread. On the line, the dimension D=1 of the limit set $\mathcal{C}_*$ is the limit of $\log 2/\log(1/r_k)$. Furthermore, the conclusion D=1 does not require $\Sigma\lambda_k<\infty$, only the weaker condition $\lambda_k\to 0$. Consequently there are three classes of linear Cantor dusts: (a) 0<D<1 and length=0, (b) D=1 and length=0, and (c) D=1 and length>0.

◁ The counterpart of the above category (c) can occur for Koch curves. It suffices to change the generator at each construction stage and to let its D tend to 2. For example, take $r_k=\frac{1}{2}k$ and adopt for N$_k$, hence for D$_k$, the maximal value discussed in the caption of Plate 53. The limit has a remarkable combination of properties: its fractal dimension D=2 is nonstandard for a curve; but its topological dimension is standard: it is D$_T$=1, and its area is standard: it vanishes.

◁ The same properties coexist in Brownian motion, Chapter 25, but here they are achieved while avoiding double points.

◁ The formal dimension may also drift *away* from D=2. For example, k stages of a plane filling tree construction may be finished off by stages with D<2. The result may be of use in modeling certain river trees that seem plane filling on scales above the inner cutoff η but crisscross finer scale domains less thoroughly. This η would be very big in deserts, and very small in soaked jungles, possibly equal to 0. Such rivers' effective dimension would be D=2 for scales above η, and D<2 for scales below η. ►

CURVES WITH POSITIVE AREA

Our intuition of dusts being imperfect, it is not bothered by dusts of positive length or volume. But *curves* of positive area are truly hard to swallow. Thus, after Lebesgue 1903 and Osgood 1903 showed that swallow them we must, they came to supersede the Peano curve as supreme monsters. After describing an example, I show that the thought is worse than the reality: in the most textual sense, surfaces with positive volume are very close to Man's heart.

The idea is to generalize the midpoint displacement construction of Plate 43. We hold on to bays and promontories, each a triangle that juts through a triangle of marshland, with its base centered on the midpoint of the marshland's base. The new element is that the relative widths λ_k of bays and promontories are no longer constant, but tend to 0 as k increases, in such a way that $\Pi_0^\infty(1-\lambda_k) > 0$. Now, the area covered by marshland fails to tend to 0, hence the limit of the marshland satisfies $D=2$. On the other hand, it is totally different from any standard set of dimension 2. Not only has it no interior points, but it is a curve with $D_T=1$, because any point's neighborhood can be separated from the set's remainder by removing only two points.

The preceding construction follows Osgood 1903, simplifying his fanciful way of making a contrived construction easier to follow. But the usefulness of a discovery must not be judged on the reasons for introducing it.

GEOMETRY OF ARTERIES AND VEINS

To quote from Harvey 1628, "The blood's motion we may be allowed to call circular, in the same way as Aristotle says that the air and the rain emulate the circular motion of the superior bodies... And similarly in the body, through the motion of the blood,... the various parts are nourished, cherished, quickened by the warmer, more perfect, vaporous, spirituous, and alimentive blood; which, on the other hand, owing to its contact with these parts, becomes cooled, coagulated, and so to speak effete."

Harvey led to a view of the circulation of blood which asserts that both an artery and a vein are found within a small distance of nearly every point of the body. (See also *The Merchant of Venice*.) This view excludes the capillaries, but to a first approximation it is best to demand that there should be both an artery and a vein *infinitely near every point*—except of course that points *within* an artery (or a vein) are prevented from being very close to a vein (or an artery).

Stated differently (but this restatement makes the result sound much odder!): every point in nonvascular tissue should lie on the boundary between the two blood networks.

A second design factor is that blood is expensive. Hence the volume of all the arteries and veins must be a small percentage of the body volume, leaving the bulk to tissue.

LEBESGUE-OSGOOD MONSTERS ARE THE VERY SUBSTANCE OF OUR FLESH!

From a Euclidean viewpoint, our criteria involve an exquisite anomaly. A shape must be topologically two-dimensional, because it forms the common boundary of two shapes that are topologically three-dimensional, but it is required to have a volume that not only is nonnegligible compared to the volumes of the shapes it bounds, but is much larger!

A virtue of the fractal approach to anatomy is that it shows the above requirements to be perfectly compatible. A spatial variant of the Osgood construction described in the section before last fulfills all the requirements we impose upon the design of a vascular system.

In this construct, veins and arteries are standard domains, since small balls (the blood cells!) can be drawn entirely within them. On the other hand, vessels occupy only a small percent of the overall volume. Tissue is very different; it contains no piece, however small, that is not crisscrossed by both artery and vein. It is a fractal surface: its topological dimension is 2, and its fractal dimension is 3.

As restated, these properties cease to sound extravagant. No one cares that they first arose in a contrived mathematical flight from common sense. I have shown that they are intuitively unavoidable, that *Lebesgue-Osgood fractal monsters are the very substance of our flesh!*

OF INTUITION, OLD AND NEW

The combination of a lung's pipes and its vasculature also proves to be a very interesting construct, wherein *three* sets—arteries, veins, and bronchioles—have a common boundary. The first example of such a set is due to Brouwer. When introduced in this way, Brouwer's construct agrees perfectly with intuition. But to put it in historical perspective, we must return to our spokesman for the conventional viewpoint, Hans Hahn.

"Intuition seems to indicate that three-country corners occur only at isolated points... Intuition cannot comprehend the Brouwer pattern, although logical analysis requires us to accept it. Once more [we find] that in simple and elementary geometric questions, intuition is a wholly unreliable guide. It is impossible to [let it] serve as the starting point or basis of a mathematical discipline. The space of geometry is... a logical construct...

"[However, if] we become more and more accustomed to dealing with these logical constructs; if they penetrate into the curriculum of the schools; if we, so to speak, learn them at our mother's knee, as we now learn three-dimensional Euclidean geometry—then nobody will think of saying that these geometries are contrary to intuition."

This Essay demonstrates that Hahn is dead wrong. To tame his own examples, I find it necessary to train our *present* intuition to perform new tasks, but it does not suffer any discontinuous change of character. Hahn draws a mistaken diagnosis, and suggests a lethal treatment.

Geometric intuition acknowledged long ago that it needs the assistance of logic, with its strange and tortuous methods. Why should logic keep trying to flee from intuition?

In any event, the typical mathematician's view of what is intuitive is wholly unreliable; it is impossible to permit it to serve as guide in model making; mathematics is too important to be abandoned to fanatic logicians. ■

16 ¤ Trees; Scaling Residues; Nonuniform Fractals

The present chapter discusses filiform fractal trees and other fractals that are almost scaling, that is, are scaling except for a residue that is fractally negligible. It is observed that these fractals are nonuniform, in the sense that D and/or D_T take different values for different parts of these sets. By contrast, a look back shows that all the fractals discussed until now can be characterized as uniform.

THE NOTION OF SCALING RESIDUE SET

STANDARD INTERVALS. The *semi-open* interval]0, 1], including its right but not its left endpoint, is scaling since it is the union of N=2 reduced replicas]0, ½] and]½, 1]. By contrast, the *open* interval]0, 1[fails to be scaling, since in addition to the N=2 reduced scale replicas,]0, ½[and]½, 1[, it includes the midpoint x=½. I propose that this midpoint be called a *scaling residue*. For the calculation of D, and for many other purposes, it is negligible. A physicist would say it is of smaller physical order of magnitude than the whole and the parts.

The preceding example tempts one to view *all* residue terms as pedantic complications that do not affect the consequences of scaling. But in analogous examples relative to fractals, which I call *nonuniform fractals*, the residue may be surprisingly significant. A nonuniform fractal is the sum (or the difference) of parts of varying fractal and topological dimensions. None of these parts can be disregarded completely, even if it is both fractally and topologically negligible. These two viewpoints often clash, with important and interesting effects.

CANTOR DUSTS AND ISOLATED POINTS. ◁ Construct a Cantor dust by dividing [0,1] into b=4 parts, and preserving [0,¼] and [¾,1]. The alternative construction that erases]¼,½[and]½,¾[yields the same dust, plus the residue point x=½. This isolated residue is not a fractal, since both D_T and D equal 0.

In the spatial generalization to $\mathbb{R}^E$, the Cantor dust satisfies $D_T=0$ and $D>0$, while the nonfractal residue set satisfies $D_T=D=E-1$. The residue may well dominate the dust topologically and/or fractally. ►

FRACTAL TREE SKELETONS WHOSE RESIDUE TERMS ARE INTERVALS

Plate 155 shows examples of umbrella trees with infinitely thin stems. They are not capable of life, and their adequacy as models of plants is improved upon in Chapter 17. Yet, tree skeletons are of great interest to many chapters of mathematics. The topologist sees them as identical, because he views any tree as made of infinitely elastic threads, and our trees can be stretched, or pulled back, onto one another. However, these trees differ from each other intuitively, and as fractals.

BRANCH TIPS. A tree is the sum of two parts, branch tips and branches, whose dimensions clash in very interesting fashion. The part easier to study is the set of the branch tips. It is a fractal dust, analogous to many we know well. It is scaling with $N=2$, and a value of r between $1/\sqrt{2}$ and 0. Hence D can range from 2 to 0, though the plate's figures are limited to D between 1 and 2. The interbranch angle takes the same value θ at every fork; it can be varied over a wide range without affecting r and D. Hence the same D allows for a variety of tree shapes.

When $1 < D < 2$, these trees self-overlap when $\theta < \theta_{crit}$, hence self-avoidance narrows the choice of θ. The trees in Plate 155 satisfy $\theta = \theta_{crit}$, but we shall first argue as if they satisfied $\theta = \theta_{crit} + \epsilon$.

TREES. The whole trees also seem self-similar at first blink, because every branch plus the branches it carries is a reduced scale version of the whole. But in fact, the two branch-

es above the main fork do *not* add up to the whole, unless one adds a residue: a trunk. Intuitively, this residue is by no means negligible. As a matter of fact, one tends to give more importance to a tree's trunks and branches than to its branch tips. Intuitively the branches "dominate" the branch tips.

Also, irrespectively of the value of D, the branch tips of a self-avoiding tree form a dust with $D_T = 0$, but the branches form a curve with $D_T = 1$, whether or not their tips are included. Hence, the branches dominate topologically. ◁ Indeed, to disconnect a point P and its neighborhood, one needs to erase either 1 point (if P is a branch tip) or 2 points (if P lies in the interior of a branch) or 3 points (if P is a point of branching). ▶

Now to the fractal viewpoint. The dimension of the branch tips is D, and the dimension of each branch is 1. As to the whole, it is not scaling, but its fractal dimension defined by the Hausdorff Besicovitch formula cannot be less than either D or 1, and it turns out to be the larger of the two. Let us restate the resulting two possibilities separately.

FRACTAL TREES. When $D > 1$, the whole tree's fractal dimension is also D. Even though the branches predominate intuitively and topologically, they are fractally negligible! Since $D > D_T$, the tree is a fractal set in which D measures the abundance of branching. Thus, we encounter yet another facet of fractal dimension, to be added to its roles as measure of irregularity and fragmentation. When, in Chapter 17, we move to nonfiliform trees, we find that a surface that is smooth but involves enough localized sharp "pimples" may become "more" than a standard surface.

SUBFRACTAL TREES. When $0 < D < 1$, to the contrary, the whole tree's linear measure (cumulative length) is finite and positive, so its fractal dimension is necessarily 1. Thus, $D = D_T$, meaning that the tree is *not* a fractal.

In fact, if we choose the units so that the trunk is of length $1 - 2r$, the branches (viewed as open intervals) can be repositioned along the gaps of a linear Cantor dust C that lies on $[0, 1]$ and has the same $N = 2$ and r as the branch tips. And similarly, the branch tips can be repositioned on C. We see that the interval $[0, 1]$ is entirely filled by maps of points on our tree. The only points that fail to be mapped are those which hold the branches together. They form a denumerable residue.

We are reminded of the comment about Plate 83, that the Devil's Staircase curve is peculiar but not fractal. If such shapes' importance increases, they may need a carefully chosen name. For now, *subfractals* will do.

For a last comment, replace the rectilinear branches by fractal curves of dimension $D^* > 1$. When $D < D^*$, the tree's fractal properties are dominated by the branches, and the tree is a fractal of dimension D^*. But when $D > D^*$, the tree is a fractal of dimension D.

NONUNIFORM FRACTALS, ETC.

We are now ready for a new definition. A fractal $\mathcal{J}$ is to be called *uniform* if any set obtained as the intersections of $\mathcal{J}$ with a disc (or ball) centered on $\mathcal{J}$ have identical values of D_T and of $D > D_T$.

We see that Koch curves, Cantor dusts, ramified curves, etc., are uniform fractals. But the preceding section's tree skeletons for $D > 1$ are nonuniform fractals.

As a matter of fact, trees may be called fractal in part: their intersection with a small enough disc centered on a branch is not a fractal but is made of one or a few intervals.

FRACTAL CANOPIES

Thus far, Plate 155 has been viewed as illustrating trees that are barely self-avoiding. But in reality these trees' tips self-contact asymptotically. As a result, the set of branch tips ceases to be a dust with $D_T = 0$, and becomes instead a curve with $D_T = 1$, with no change in its fractal dimension. For this new class of fractal curves, I propose the term *extended fractal canopies*. Observe that their vertical shadow's length increases with D.

The curve that bounds the open region outside the resulting shape is to be called "fractal canopy." Due to the elimination of the "folds" of the extended canopy, the canopy's dimension falls short of D, by an amount that increases with D.

Since light is a vital consideration for trees, branches ending on the folds of the extended fractal canopy can be expected to wither away. A tree designer may either allow some branches to grow, then wither for lack of sunshine, or write a more complicated pro-

gram that instructs these branches never to
grow. I would choose the simpler program.

When D < 1, the merger of a dust of dimension D into a curve is inconceivable. When one
seeks self-contact by diminishing the interbranch angle θ, the goal is not reached until
this angle becomes 0, and the tree collapses
into an interval. Alternatively, if one keeps
the tree's vertical shadow to the fixed length
1, and seeks self-contact by lengthening the
branches, the goal is never reached: the tree
tends to a linear Cantor dust C, plus half lines
hanging down from each point of C.

TREES WITHOUT RESIDUE TERM

Fractal trees are not limited to those constructed in the preceding sections. For example, recall the construction on page 140. Alternatively, take as Koch generator a cross
with branches of length r_t(top), r_b(bottom),
and r_s(sides), such that $r_t^2 + r_b^2 + 2r_s^2 < 1$. In
the resulting fractal tree, every branch, however short, is crowded with subbranches. If
the root point is excluded, such trees are scaling without residue.

HIGH ENERGY PHYSICS: JETS

Feynman 1979 reports that fractal trees made
it possible for him to visualize and model the
"jets" that arise when particles collide head
on at very high energy. The idea is explored
in CERN reports by G. Veneziano.

Plate 155 ▫ FRACTAL UMBRELLA TREES AND FRACTAL CANOPIES

The trees on this plate have infinitely thin
stems, and the same angle θ between the
branches throughout. D ranges from 1 to 2,
and for each D, θ takes the smallest value that
is compatible with self-avoidance.

For D barely above 1 (top left), the result
is whisk-like, then broom-like. As D increases,
the branches open up, and the outline or
"canopy" extends into folds hidden from the
sunshine. One is reminded of the flowers in
several varieties of the species *Brassica
oleracea*: cauliflower (*B. o. botrytis*) and
broccoli (*B. o. italica*). Could it be significant that part of the geometric difference between
cauliflower and broccoli is quantified by a
fractal dimension?

For larger D (bottom left), a Frenchman is
reminded of the fortifications by Vauban. The
values D=2 and $\theta = \pi$ yields a plane-filling
tree. To allow for $\theta > \pi$ (bottom right), one
must again decrease D, all pretense of umbrella being replaced by a contorted pattern
worthy of classical dance sculptures of India.

In one of the best known figures in *On
Growth and Form*, Thompson 1917, the skulls
of different species of fish are mapped onto
each other by continuous and smooth transformations in the spirit of Euclid. The transformations that map the present trees on each
other partake of the same inspiration, but in
very different spirit. ■

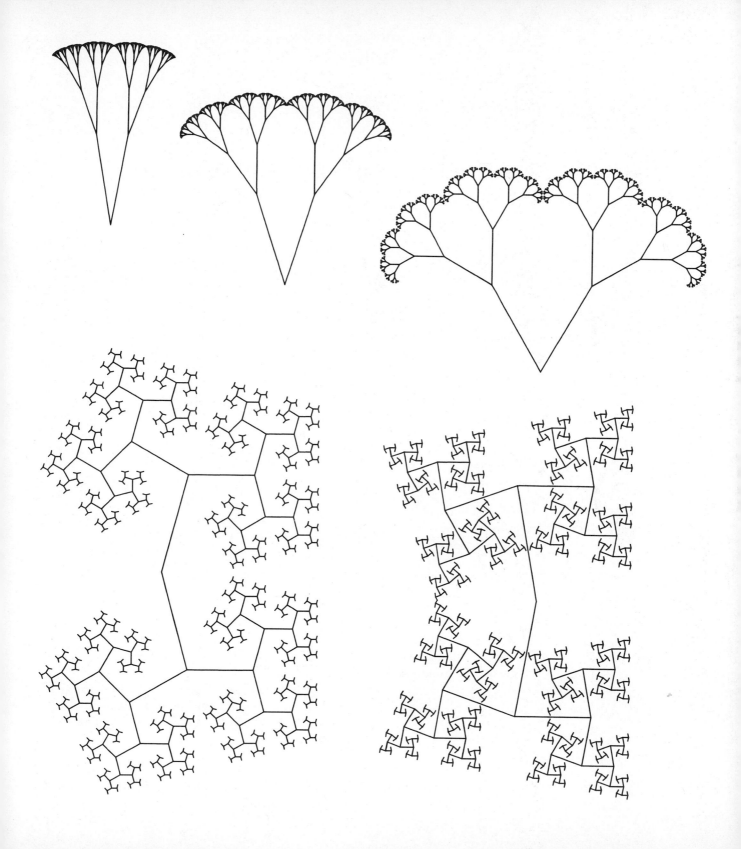

17 ☒ Trees and the Diameter Exponent

The present chapter investigates the geometrically imbedded thick stemmed "trees" involved in lungs, vasculatures, botanical trees, river networks, and the like.

These natural objects are extremely familiar, in fact, no other object illustrates as simply as they do the idea of a shape having a large number of different elements of linear scale. Unfortunately, trees are less simple than they seem. They were not tackled earlier because of a complication encountered in the preceding chapter: trees cannot be self-similar. The best one can hope is that self-similarity holds for the branch tips, as will be assumed in this chapter. In addition to the tips' fractal dimension D, trees involve a parameter to be called the diameter exponent, Δ. When the tree is self-similar with a residue, as in Chapter 16, Δ coincides with the D of the branch tips. Otherwise, Δ and D are separate characteristics, and we deal with an instance of the phenomenon biologists call "allometry." We encounter examples of both $\Delta = D$ and $\Delta < D$.

THE DIAMETER EXPONENT Δ

Leonardo da Vinci claims in his *Notebooks,* note No. 394, that "All the branches of a tree at every stage of its height when put together are equal in thickness to the trunk (below them)." The formal expression is that a botanical tree's branch diameters before and after a bifurcation, d, d_1, and d_2,

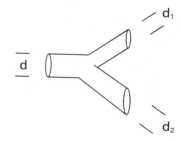

satisfy the relation

$$d^\Delta = d_1{}^\Delta + d_2{}^\Delta,$$

the exponent being $\Delta = 2$. The implication is this: if branches' thickness is taken into account, botanical trees are *not* self-similar with near space-filling bark. Indeed, self-similarity requires $\Delta=D$, and near space-filling requires D to be near $E=3$.

In other words, whenever the above relation is satisfied, Δ is a new parameter to be added to D; it will be called *diameter exponent*. It has been considered by many people, often unaware of each other, witness the references in Thompson 1917–1942–1961. This chapter shows that for bronchi, $\Delta \sim 3$. For arteries, $\Delta \sim 2.7$. Botanical trees are close to Leonardo's $\Delta=2$. And $\Delta=2$ for the rivers' widths. This chapter also explores a few physical, physiological, and geometric consequences of the value of Δ.

◁ PARADIMENSION. The 1977 *Fractals* call Δ a *paradimension* (from $\pi\alpha\rho\alpha$ = besides), but I no longer advocate this term. The awkward role of Δ—sometimes a dimension and other times not—is shared by the exponent in Besicovitch & Taylor 1954; See Chapter 39. ►

THE LUNG'S BRONCHIAL TREE

As a first example, the subdivision of the lung's air pipes is for all practical purposes self-similar, with $\Delta=D$, and $D \sim E=3$.

The inner shape of the lung is not well known, hence it would be instructive to insert an actual photograph at this point (examples are found in Weibel 1963 and Comroe 1966). However, this Essay's policy (this may be the only occasion for regretting it) is to keep to simulations. Therefore, a brief verbal descrip-

tion must suffice. After the lung's air is replaced by noncured plastic, then the plastic cured and the tissue dissolved, one is left with an extremely heavily branched tree that fills the outline of the lung with a degree of tightness, uniformity, and visual impenetrability that botanical trees never achieve. Between the first two bifurcations, which are beyond our concern, and the last three, which lead to alveoli (discussed in Chapter 12), there are 15 successive bifurcations of striking regularity.

From the data in Weibel 1963, the pipe intervals are in a first approximation similar to each other, and $\Delta \sim 3$. The airflow is a concrete quantity divided between bifurcating branches, and since airflow equals pipe cross-sectional area times air velocity, we see that the velocity varies like $d^{\Delta-2}$: air slows down as it moves toward thinner bronchi.

The precise value $\Delta=3$ is important. A first interpretation involves an argument by Murray 1927, thus presented by Thompson 1942, p. 954, or 1961, p. 129: "[T]he increasing surface of the branches soon means increased friction, and a slower pace of the [fluid] traveling through; and therefore the branches must be more capacious than at first appears. It becomes a question not of capacity but of resistance; and in general terms the answer is that the ratio of resistance to cross section shall be equal in every part of the system, before and after bifurcation, as a condition of least possible resistance in the whole system; the total cross section of the branches, therefore, must be greater than that of the trunk in proportion to the increased resistance. An approximate result, familiar to students of hydrodynamics [for a modern treatment, see Hersfield & Cummings 1967, Wilson 1967], is that the resistance is a minimum, and the condition an optimum," when the branching ratio is $2^{1/3} \sim 1.26$ throughout.

Hence $\Delta = 3$ is the best value that either a goal-oriented design or selective evolution could strive to achieve. Of course, Murray's optimality criterion is purely *local*, and the designer can never be sure whether locally optimal pieces can be made to fit together.

PACKING 3-SPACE WITH BRONCHI

My alternative fractal argument for $\Delta = 3$ is very different: it invokes the effect of *nonwillful* geometric constraints upon the lung's prenatal growth and upon its pipes' fully grown shape. An obvious advantage is that here the branching ratio of $2^{1/\Delta} \sim 2^{1/3}$ need not be part of the genetic code (as should be the case in the Murray approach).

The basic datum is that a lung's prenatal growth starts with a bud, which grows into a pipe, which forms two buds, each of which behaves as above. Furthermore, this growth is self-similar (with the trunk constituting a residue!). To account for self-similarity, we need *not* argue that it is best, only that it is simplest: the growth-governing program is shortest when each step repeats the previous one on smaller scale, or on the same scale after

the previous stage had grown. If so, the outcome of growth is determined fully by the branches' width/length ratio and the diameter exponent. And one needs in addition a rule that indicates when growth is to stop.

Now, depending on the value of Δ (the width/length ratio being held fixed), growth according to these rules achieves one of three outcomes: (a) after a finite number of stages, branches run out of space in which to grow; (b) branches never fill more than a part of the available space; *or* (c) they find the available space to be precisely what they need. When one wishes the limit to be a space-filling tree, no detailed instructions need be incorporated into the growth program, because competition for space leaves little room for indeterminacy. A two-dimensional reduction of the process is illustrated on Plates 164 and 165, where we see that, as the branches' width/length ratio decreases toward 0, the plane-filling branching ratio increases toward $2^{1/2}$, yielding $\Delta = E = 2$. Similarly, the space-filling branching ratio corresponding to infinitely thin branches is $2^{1/3}$, yielding $\Delta = E = 3$.

Since $\Delta = 3$ corresponds to the limit of infinitely thin pipes, it cannot be actually implemented. What a pity, since a tree made of infinitely thin bifurcations continuing to zero has a space-filling "skin." This last property could have been given a teleological interpretation to rival Murray's: it would be best from the viewpoint of allowing the largest possible surface for the purpose of chemical exchanges between air and blood.

But actual pipes are not infinitely thin, so

the best one can achieve is a value of D and Δ a bit below 3, quite compatible with the empirical evidence. This involves the same degree of imperfection at all branch points—but this property is obtained as a side consequence of self-similarity with a residue, and need not be set up as a goal.

DIMENSION. The branches add up to a standard set: topologically and fractally of dimension E. When each branch's skin is smooth, the whole skin is of dimension Δ.

ALVEOLAR INNER CUTOFF

As usual, the interpolation to increasingly thin bronchi is interrupted by a cutoff. The cutoff is gradual after the 15th bifurcation, and I find it to be of excellent geometric design.

A basic remark is that, while infinite self-similar bifurcation would eventually fill all the available space, it proceeds slowly, so that the lung's first 15 bifurcation stages fill only a small proportion of the lung's box. To fill the remaining space in few stages, the pipes must be made markedly larger than suggested by self-similar extrapolation. Indeed, Weibel 1963, pp. 123–124, can be interpreted as indicating that, in stages beyond the 15th, pipe width ceases to decrease (Δ is no longer defined). And that pipe lengths are longer than suggested by similarity, the ultimate multiplier being 2. Since Plate 165 suggests that self-similar branches enter about half way into the nearest available gap, a multiplier

equal to 2 is eminently sensible, and suggests again that much of the program for the lung's design is imposed by the properties of space and need not be separately encoded.

MORE ABOUT VASCULAR GEOMETRY

Let us now return to the high point of Chapter 15, where I proclaim that *Lebesgue-Osgood fractal monsters are the very substance of our flesh*. Granted that a branching domain $\mathcal{A}$ (arteries) has a volume of about 3% of the volume of a domain $\mathcal{B}$ (body), but is supposed to come infinitely close to every point of $\mathcal{B}$, I argue that the branches of $\mathcal{B}$ must thin out more rapidly than in self-similar trees. Now that we have established that in some cases the rate of thinning is measurable by Δ, we can inquire whether or not Δ is defined for arteries.

Not only Δ is indeed defined in a wide subrange of the 8 to 30 bifurcations one observes between the heart and the capillaries, but the fact has been known for nearly a century. Indeed, Thoma 1901 and Groat 1948 summarized their experimental findings by asserting that Δ=2.7. Their estimate is remarkably well confirmed by Suwa & Takahashi 1971.

BOTANICAL TREES

After playing with objects to which the term *tree* is applied figuratively, we return to the trees that botanists study. The "normal" values the analysis will suggest are D=3 and Δ=2. They are hardly universal, however: given the astounding diversity of botanical shape, specific deviations may be more interesting than the "norm." A consequence of Δ=2 is that, seen next to the near self-similar branches of lung casts, plant branches are extremely sparse; one cannot see through a lung cast, but one can see through a leafless tree.

The reason behind the fact that D and Δ take up the integer Euclidean dimensions of solids and surfaces, is that, in the words of D'Arcy Thompson, "a tree is governed by the simple physical rules which determine relative changes in volume and area." In more specific terms, Hallé, Oldeman & Tomlinson 1978, "The problem of energy interchange in trees can be simplified by considering the tree as a system in which as large [an area] as possible must be irrigated with the minimum production of volume while at the same time guaranteeing the evacuation of absorbed energy." Since volumes and areas are incommensurable within the framework of Euclid, the geometric problem of the architecture of trees is intrinsically a fractal problem. When D and/or Δ cease to be integers, the problem's fractal character is even more obvious.

BOTANICAL TREES' D AND Δ

THE VALUE D=3. The reader knows well that the largest possible leaf area is implemented by a space filling surface—as approximated by those bushes whose leaves or needles come very close to every point within a certain outline (except perhaps for a dead core we overlook). A very small 3−D suffices to allow sunlight and wind to enter.

UMBRELLAS. However, diverse additional constraints imposed upon tree architecture may prevent D=3 from being implementable. The only standard alternative is a standard surface of dimension D=2, for example the surface of a spherical "umbrella" hiding a core empty of leaves but crisscrossed by branches. This is why Horn 1971, which limits itself to standard geometry, allows for either D=3 or D=2. However, there is no clear advantage to D=2; in fact, in order to terminate on a spherical umbrella, the branches have to follow very peculiar rules.

On the other hand, the freedom of design of the "tree architect" is immensely increased by fractals. First of all, the repeatedly scalloped surfaces of many large trees can be represented by scaling fractals of dimension D between 2 and 3, and can be distinguished by the value of D. Broccoli and cauliflower also come to mind, but they raise a different issue, to which we turn in a moment. And one can conceive of sparse climbing plants of dimension below 2 (and conjecture that bonzai trees well contrived to be "harmonious" are also fractal with D<3.)

THE VALUE Δ=2. The Leonardo da Vinci quote at the beginning of this chapter is invalid for lungs (Δ=3) and for arteries (Δ=2.7). But plant anatomy differs from human anatomy. The value $\Delta=2$ rests on the mental image of trees as bundles of nonbranching vessels of fixed diameter, connecting roots to the leaves and occupying a fixed proportion of each branch's cross section. Zimmermann tells us that this image is called the "pipe model" by Japanese workers.

MEASUREMENT OF Δ. The empirical evidence turns out to be astonishingly scant and indirect. Murray 1927, quoted in Thompson 1917, finds empirically that branch weight is proportional to (branch diameter)^M, with M $\sim$2.5, but I would say his M is larger than that. And he claims that $M=\Delta$, but my own analysis yields $M=2+\Delta/D$. For D=3, Leonardo's value of $\Delta=2$ would correspond to M$\sim$2.66, while M$\sim$2.5 would yield $\Delta=1.5$. Recently, the data concerning 3 "McMahon's trees" used in writing McMahon & Kronauer 1976 were kindly communicated to me by Prof. McMahon, and they have been analyzed. Denoting d_1/d by x, and d_2/d by y, we sought a value of Δ such that $X=x^\Delta$ and $Y=y^\Delta$ fall along the line $X+Y=1$. Unfortunately, the empirical scatter is extremely large for every Δ, and the estimate of Δ is necessarily unreliable. Again, the value $\Delta=2$ is not disproved, but a slightly smaller Δ is suggested. The safe conclusion at present is that $\Delta=2$ is a reasonable rough value, but that tree architecture is on the conservative side, with daughter branches thinner than strictly necessary.

COROLLARIES OF D=3 AND $\Delta=2$. A first corollary is that a branch's leaf area is proportional both to the volume of the branch's outline, and to the cross-sectional area of the branch. This inference is indeed empirically correct, having been made by Huber in 1928.

Another corollary is that the ratio (tree height)^3/(trunk diameter)^2 is constant for each species, and that it is equal to the ratio (linear scale of a branch's drainage volume)^3/(branch's diameter)^2. One may also expect this ratio to vary comparatively little between species. Observe that the force the wind exerts on a bare (respectively, leaf-carrying) tree is roughly proportional to the branch (respectively, branch and leaf) area, and proportional to (height)^3 in this model. And the trunk's counterresistance is proportional to (diameter)^2. This suggests that the ratio of these quantities is a factor of safety.

In an umbrella shaped tree with $\Delta=2$ and D=2, the ratio (height)^2/(trunk diameter)^2 is constant, so is more generally the ratio (height)^D/(trunk diameter)$^\Delta$.

DIGRESSION ON HINDLEG BONES. The relation between height and diameter that is characteristic of botanical trees with D=3 and $\Delta=2$ also applies to animal skeletons, with d the diameter of the main supporting bone.

GREENHILL'S ELASTIC SCALING

While pulmonary and vascular trees are supported from the outside, most plants support themselves. Greenhill (quoted in Thompson 1961) injects at this point the notion of elastic as opposed to geometric similarity. The idea of static elastic similarity is that a tree must limit its overall height to a fixed percent of the critical buckling height of a uniform cylinder of the same base diameter loaded under its own weight. This requirement yields precisely the same results as fractals with D=3 and Δ=2. Thus, a "pipe model" tree with space-filling leaves will not buckle.

McMahon & Kronauer 1976 elaborate on Greenhill's idea: they inject *dynamic* elastic similarity, and again obtain the same result.

PLANTS WITH D=Δ<3

In some plants, wood is not specialized to bear weight and carry sap, but also serves to store nutrients. If so, and even when the vasculature obeys the "pipe model," the value of Δ=2 need not apply.

An example wherein the branch tips form a nonstandard "umbrella" with D<3, and Δ=D, is illustrated (in plane reduction showing D−1 and Δ−1) in Plate 163. One observes that the geometric cauliflower shape has empty occlusions..., just like the botanical cauliflowers. Is this a mere coincidence? Characteristics preordained by geometry need not burden the genetic code.

MORE ABOUT THE BRAIN'S GEOMETRY

When discussing the brain's surface in Chapter 12, we did not consider the network of axons that join different parts together. In the case of the cerebellum, the axons join the surface to the outside, and one deals with a gray matter surface that envelops a white matter tree. I revised the argument of Chapter 12 to take account of this tree, and found that the resulting corrective terms in the volume-area relation yield an improved fit to the data. But the story is too long to be told here.

NEURON BRANCHING. The Purkinje cells in mammalian cerebellum are practically flat, and their dendrites form a plane-filling maze. From mammals to pigeon, alligator, frog, and fish, the degree of filling decreases (Llinas 1969). It would be nice if this corresponded to a decrease in D, but the notion that neurons are fractals remains conjectural.

THE RALL LAW. Rall 1959 observes that neuronal trees which preserve the quantity d^{Δ} with Δ=1.5 are electrically equivalent to cylinders, hence especially convenient to study. Further detail is provided by Jack et al. 1975.

HOW WIDE IS THE MISSOURI RIVER?

Now let us turn to rivers. Despite its conceptual importance, my "Peano" model of Chapter 7 can only be a first approximation. In particular it implies that river widths vanish, while in fact they are of positive width.

An important empirical question is whether or not the rivers' bifurcations have the same diameter exponent Δ throughout. If Δ is indeed defined, the next question is whether 2−Δ is =0 or >0. No direct test is known to me, but the discharge through a river, Q, is preserved in bifurcation, hence could stand in

for d$^\Delta$. Maddock (see Leopold 1962) finds that d$\sim$Q$^{1/2}$, hence $\Delta = 2$. Furthermore, a river's depth is proportional to Q$^{0.4}$, and its velocity is proportional to Q$^{0.1}$. The exponents duly add to $0.5 + 0.4 + 0.1 = 1$.

G. Lacey observed in the 1930's that $\Delta = 2$ also holds for stabilized irrigation channels in India, which pose a well-defined problem of hydraulics. One may therefore hope for a fluid mechanics explanation playing the role that Murray's argument plays for the lung.

$\Delta = 2$ has an interesting implication: if rivers are drawn on a map as ribbons of correct relative width, guessing a map's scale from the shape of the river tree is impossible. (This is also impossible for river meanders, but that is a totally different story.)

Those who believe that Leonardo knew everything will read the value $\Delta = 2$ in the continuation of the quote with which this chapter begins: "All the branches of a water (stream) at every stage of its course, if they are of equal rapidity, are equal to the body of the main stream." ▬

Plate 163 ⌗ FLATTENED FRACTAL MODELS OF PLANT FLOWERS

Select one of the umbrella trees of Plate 155, with $\theta < \pi$, and replace each stick by an isosceles triangle of which said stick is a side, the angles at this stick's ends being $\frac{1}{2}\theta$ (root end) and $\pi - \theta$. Since θ is the smallest value that avoids self-overlap of the tree, the triangular thickened stems do not overlap either, and they fill in the umbrella's "insides." To make the figures more transparent, the triangles in one of them are trimmed slightly on one side.

Observe that the branches thin out rapidly as D approaches either 1 or 2, that is, as the spatial D approaches 2 or 3. Do actually observed D's correspond to the thickest possible branches? ▬

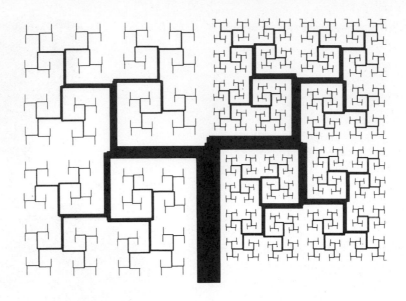

Plates **164** and **165** ◻ **PLANE-FILLING RECURSIVE BRONCHI**

PLATE 165. In Koch recursion, every straight interval in a finite approximation is eventually broken up into shorter pieces. In many applications, it is useful to generalize this procedure by allowing certain intervals to be "infertile," so that in later stages they remain untouched.

Here, this generalized procedure is used to grow a "tree." One starts with a trunk having barren sides and a fertile "bud." The bud generates two "branches," on which again only two terminal "buds" are fertile. And so on ad infinitum. The growth is asymmetric to insure that the tree fills a roughly rectangular portion of the plane with no gap and no overlap. However, asymptotic self-contact is not avoided, and indeed every point on the "bark" line can also be obtained as a limit branch tip.

The "subtrees" constructed starting with the main leaders are similar to the whole tree in two different similarity ratios, r_1 and r_2. The whole tree is not self-similar because in addition to the subtrees it includes a trunk. On the other hand, the set of asymptotic branch tips is self-similar. From the legend of Plates 56-57, the similarity dimension is the D that satisfies the equation $r_1^D + r_2^D = 1$. In the top Figure of Plate 165, the tips are nearly plane-filling and 2−D is small; in the bottom Figure of Plate 165, D is much below 2.

Incidentally, the diameter/length ratio having been set, the codimension 3−D of a full spatial picture is smaller than the codimension 2−D of this planar reduction.

PLATE 164. This composite Figure results from a Koch tree construction in which the generator is changed at each stage, so that the ratio of width to length decreases to 0. On the left side of the composite Figure, this ratio decreases even faster than on the right side. The result is that the branch tips are no longer self-similar. However, the tips can achieve the dimension D=2. This is a new way of achieving the same goal as in Chapter 15. ■

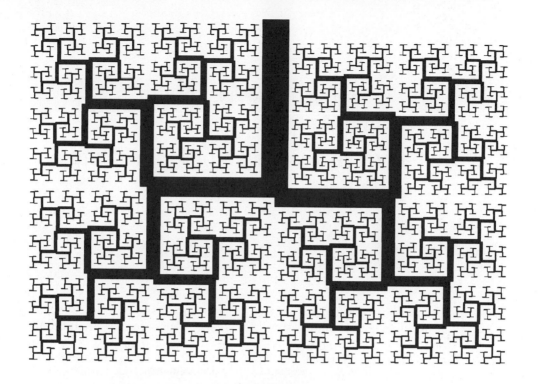

18 ¤ Self-Inverse Fractals, Apollonian Nets, and Soap

The bulk of this Essay is devoted to fractals that are either fully invariant under similitudes or, at least, "nearly" self-similar. As a result, the reader may have formed the impression that the notion of fractal is wedded to self-similarity. Such is emphatically *not* the case, but fractal geometry must begin by dealing with the fractal counterparts of straight lines... call them "linear fractals."

Chapters 18 and 19 take the next step. They sketch the properties of fractals that are, respectively, the smallest sets to be invariant under geometric inversion, and the boundaries of the largest bounded sets to be invariant under a form of squaring.

Both families differ fundamentally from the self-similar fractals. Appropriate linear transformations leave scaling fractals invariant, but in order to generate them, one must specify a generator and diverse other rules. On the other hand, the fact that a fractal is "generated" by a nonlinear transformation, often suffices to determine, hence generate, its shape. Furthermore, many nonlinear fractals are bounded, i.e., have a built-in finite outer cutoff $\Omega < \infty$. Those who find $\Omega = \infty$ objectionable ought to be enchanted by its demise.

The first self-inverse fractals were introduced in the 1880's by Henri Poincaré and Felix Klein, not long after the discovery by Weierstrass of a continuous but not differentiable function, roughly at the same time as the Cantor sets, and well before the Peano and Koch curves and their scaling kin. The irony is that scaling fractals found a durable niche as material for well-known counterexamples and mathematical games, while self-inverse fractals became a special topic of the theory of automorphic functions. This theory was neglected for a while, then revived in a very abstract form. One reason why the self-inverse fractals were half-forgotten is that their actual shape has remained unexplored until the present chapter, wherein an effective new construction is exhibited.

The chapter's last section tackles a problem of physics, whose star happens to be the simplest self-inverse fractal.

BIOLOGICAL FORM AND "SIMPLICITY"

As will be seen, many nonlinear fractals "look organic," hence the present aside concerned with biology. Biological form being often very complicated, it may seem that the programs that encode this form must be very lengthy. When the complication seems to serve no purpose (as is often the case in fairly simple creatures), the fact that the generating programs were not rubbed off to leave room for useful instructions is paradoxical.

However, the complications in question are often highly repetitive in their structure. We may recall from the end of Chapter 6 that a Koch curve must *not* be viewed as either irregular or complicated, because its generating rule is systematic and simple. The key is that the rule is applied again and again, in successive loops. Chapter 17 extends this thought to the pre-coding of the lung's structure.

In Chapters 18 and 19 we go much further and find that some fractals generated using nonlinear rules recall either insects or cephalopods, while others recall plants. The paradox vanishes, leaving an incredibly hard task of actual implementation.

STANDARD GEOMETRIC INVERSION

After the line, the next simplest shape in Euclid is the circle. And the property of being a circle is not only preserved under similitude, but also under inversion. Many scholars have never heard of inversion since their early teens, hence the basic facts bear being restated. Given a circle C of origin O and radius R, inversion with respect to C transforms the point P into P' such that P and P' lie on the same half line from O, and the lengths |OP| and |OP'| satisfy $|OP\|OP'| = R^2$. Circles containing O invert into straight lines not containing O, and conversely (see below). Circles not containing O invert into circles (third figure below). Circles orthogonal to C, and straight lines passing through O, are invariant under inversion in C (fourth figure).

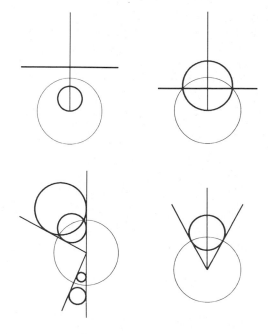

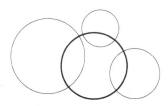

Now consider jointly the three circles C_1, C_2, and C_3. Ordinarily, for example when the open bounded discs surrounded by the C_m are nonoverlapping, there exists a circle Γ orthogonal to every C_m, see above. When Γ exists, it is jointly self-inverse with respect to the C_m.

The preceding bland results nearly exhaust what standard geometry has to say about self-inverse sets. Other self-inverse sets are fractals, and most are anything but bland.

GENERATOR. SELF-INVERSE SETS. As usual, we begin with a *generator*, which is in the present case made up of any number M of circles C_m. The transformations made of a succession of inversions with respect to these circles form what algebraists call the group generated by these inversions; call it G. The formal term for "self-inverse set" is "a set invariant under the operations of the group G."

SEEDS AND CLANS. Take any set S (call it *a seed*), and add to it the transforms of S by all the operations of G. The result, to be called here the *clan* of S, is self-inverse. But it need not deserve attention. For example, if S is the extended plane $\mathbb{R}^*$ (the plane $\mathbb{R}$ plus the point at infinity), the clan of S is identical to $\mathbb{R}^* = S$.

CHAOTIC INVERSION GROUPS. Furthermore, given a group G based upon inversions, it may happen that the clan of every domain S covers the whole plane. If so, the self-inverse set must be the whole plane. For reasons that transpire in Chapter 20, I propose that such groups be called *chaotic*. The nonchaotic groups are due to Poincaré, but are called Kleinian: Poincaré had credited some other work of Klein's to L. Fuchs, Klein protested, Poincaré promised to label his next great discovery after Klein—and he did!

Keeping to nonchaotic groups, we discuss three self-inverse sets singled out by Poincaré, then a fourth set of uncertain history, and a fifth set whose importance I discovered.

HYPERBOLIC TESSELLATION OR TILING

Few of Maurits Escher's admirers know that this celebrated draftsman's inspiration often came straight from "unknown" mathematicians and physicists (Coxeter 1979). In many cases, Escher added decorations to self-inverse tessellations known to Poincaré and illustrated extensively in Fricke & Klein 1897.

These sets, to be denoted by $\mathcal{T}$, are obtained by merging the clans of the circles C_m themselves.

◁ G being assumed nonchaotic, the complement of the merged clans of the C_m is a collection of circular polygons called "open tiles." Any open tile (or its closure) can be transformed into any other open (closed) tile by a sequence of inversions belonging to G. In other words, the clan of any closed tile is $\mathbb{R}^*$. More important, the clan of any open tile is the complement of $\mathcal{I}$. And $\mathcal{I}$ is, so to speak, the "grout line" of these tiles. $\mathbb{R}^*$ is self-inverse. $\mathcal{I}$ and the complement of $\mathcal{I}$ are self-inverse and involve a "hyperbolic tiling" or "tessellation" of $\mathbb{R}^*$. (The root is the Latin *tessera* = a square, from the Greek τεσσαρες = four, but tiles can have any number of corners greater than 2.) In Escher's drawings, each tile bears a fanciful picture. ▶

AN INVERSION GROUP'S LIMIT SET

The most interesting self-inverse set is the smallest one. It is called the limit set, and denoted by $\mathcal{L}$, because it is also the set of limit points of the transforms of any initial point under operations of the group G. It belongs to the clan of any seed S. To make a technical point clearer: it is the set of those limit points that cannot also be attained by a finite number of inversions. Intuitively, it is the region where infinitesimal children concentrate.

$\mathcal{L}$ may reduce to a point or a circle, but in general it is a fragmented and/or irregular fractal set.

◁ $\mathcal{L}$ stands out in a tessellation, as the "set of infinitesimally small tiles." It plays, with respect to the finite parts of the tessellation, the role the branch tips (Chapter 16)

play with respect to the branches. But the situation is simpler here: like $\mathcal{L}$, the tesselation $\mathcal{I}$ is self-inverse *without* residue. ▶

APOLLONIAN NETS AND GASKETS

A set $\mathcal{L}$ is to be called *Apollonian* if it is made of an infinity of circles plus their limit points. In this case, its being fractal is solely the result of fragmentation. This case was understood (though in diffuse fashion) at an early point of the history of the subject.

First we construct a basic example, then show it is self-inverse. Apollonius of Perga was a Greek mathematician of the Alexandrine school circa 200 B.C. and close follower of Euclid, who discovered an algorithm to draw the five circles tangent to three given circles. When the given circles are mutually tangent, the number of Apollonian circles is two. As will be seen momentarily, there is no loss of generality in assuming that two of the given circles are exterior to each other but contained within the third, as follows:

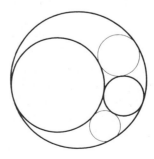

These three circles define two circular triangles with angles of 0°. And the two Apollonian circles are the largest circles inscribed in these triangles, as follows:

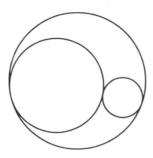

The Apollonian construction concludes with five circles, three given and two Apollonian, which together define six circular triangles. Repeating the same procedure, we draw the largest inscribed circle in each triangle. Infinite further repetition is called *Apollonian packing.* To the resulting infinite collection of circles one adds its limit points, and one obtains a set I call *Apollonian net.* A portion of net within a circular triangle, as exemplified to the right, is to be called *Apollonian gasket.*

If one of the first generation Apollonian circles is exchanged for either of the inner given circles, the limit set is unchanged. ◁ If said Apollonian circle is made to replace the outer given circle, the construction starts with three given circles exterior to each other, and one of the first stage Apollonian circles is the smallest circle *circumscribed* to the three given circles. After this atypical stage, the construction proceeds as above, proving that our figures involve no loss of generalities ▶.

LEIBNIZ PACKING. Apollonian packing recalls a construction I call *Leibniz packing* of a circle, because Leibniz described it in a letter to de Brosses: "Imagine a circle; inscribe within it three other circles congruent to each other and of maximum radius; proceed similarly within each of these circles and within each interval between them, and imagine that the process continues to infinity...."

APOLLONIAN NETS ARE SELF-INVERSE

Let us now return to the starting point of the construction of Apollonian net: three circles tangent to each other. Add *either one* of the corresponding Apollonian circles, and call the resulting 4 circles Γ circles. Here they are shown by bold curves.

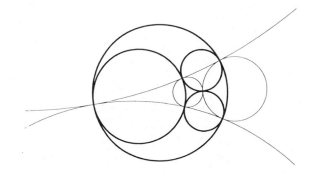

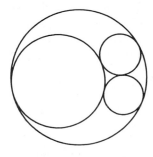

There are 4 combinations of the Γ circles 3 by 3, to be called triplets, and to each corresponds a circle orthogonal to each circle in the triplet. We take these new circles as our generator, and we label them as C_1, C_2, C_3, and C_4, (the diagram below shows them as thin curves). And the Γ circle orthogonal to C_i, C_j, and C_k will be labeled as Γ_{ijk}.

Having set these tedious labels, here is the payoff: Simple inspection shows that the smallest (closed) self-inverse set with respect to the 4 generating circles C_m is the Apollonian net constructed on the 4 circles Γ. Curiously, this observation is nowhere explicit in the literature, but it must be widely known.

A more careful inspection shows that each circle in the net transforms into one of the Γ circles through a *unique* sequence of inversions with respect to the C circles. In this way, the circles in the Apollonian net can be sorted out into 4 clans; the clan descending from Γ_{ijk} will be denoted as $\mathcal{G}\,\Gamma_{ijk}$.

NET KNITTING WITH A SINGLE THREAD

The Apollonian gasket and the Sierpiński gasket of Plate 141 share an imporant feature: the complement of the Sierpiński gasket is a union of triangles, a σ-triangle, and the complement of an Apollonian net or gasket is a union of discs, a σ-disc.

But we also know that the Sierpiński gasket admits of an alternative Koch construction, in which finite approximations are teragons (broken lines) without self-contact, and double points do not come in until one goes to the limit. This shows that the Sierpiński gasket can be drawn without ever lifting the pen; the line will go twice over certain points but will never go twice over any interval of line.

To change metaphors, the Sierpiński gasket can be knitted with a single loop of thread!

The same is true of the Apollonian net.

NON-SELF-SIMILAR CASCADES, AND THE EVALUATION OF THE DIMENSION

The circular triangles of Apollonian packing are *not* similar to each other, hence the Apollonian cascade is not self-similar, and the Apollonian net is not a scaling set. One must resort to the Hausdorff Besicovitch definition of D (as exponent used to define measure), which applies to every set, but the derivation of D proves surprisingly difficult. Thus far (Boyd 1973a,b), the best one can say is that

$$1.300197 < D < 1.314534,$$

but Boyd's latest (unpublished) numerical experiments yield $D \sim 1.3058$.

In any event, since D is a fraction while $D_T = 1$, the Apollonian gasket and net are fractal curves. In the present context, D is a measure of fragmentation. When, for example, the discs of radius smaller than ϵ are "cut off," the remaining interstices have a perimeter proportional to ϵ^{1-D} and a surface proportional to ϵ^{2-D}.

$\mathscr{L}$ IN NON-FUCHSIAN POINCARÉ CHAINS

Inversions with respect to less special configuration of the generating circles C_m, lead to self-inverse fractals that are less simple than any Apollonian net. A workable construction of mine, to be presented momentarily, characterizes $\mathscr{L}$ suitably in most cases. It is a great improvement over the previous method, due to Poincaré and Klein, which is cumbersome and converges slowly.

But the older method remains important, so let us go through it in a special case. Let the C_m form a configuration one may call *Poincaré chain*, namely a collection of M circles C_m numbered cyclically, so that C_m is tangent to C_{m-1} and to C_{m+1} (modulo M), and intersects no other circle in the chain. In that case, $\mathscr{L}$ is a *curve* that separates the plane into an inside and an outside. (As homage to Camille Jordan, who first saw that it is not obvious that the plane can thus be subdivided by a single loop, such loops are called Jordan curves.)

When all the C_m are orthogonal to the same circle Γ, $\mathscr{L}$ is identical to Γ. This case, called Fuchsian, is excluded in this chapter.

POINCARÉ'S CONSTRUCTION OF $\mathscr{L}$. The customary construction of $\mathscr{L}$ and my alternative will be fully described in the case of the following special chain with M=4:

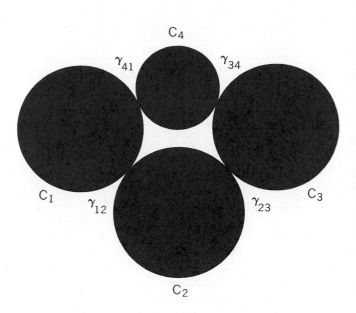

To obtain $\mathcal{L}$, Poincaré and Fricke & Klein 1897 replace the original chain, in stages, by chains made of an increasing number of increasingly small links. The first stage replaces every link C_i by the inverses in C_i of the links C_m other than C_i, thus creating $M(M-1) = 12$ smaller links. They are shown in the facing column, superimposed on a (gray) photographic negative of the original links. And each stage takes the chain with which it started and inverts it in each of the original C_m. Here several stages are shown in black, each being superposed on the preceding one, shown in white on gray background. Ultimately, the chain thins out to its thread, which is $\mathcal{L}$.

Unfortunately, some links remain of substantial size after large numbers of stages, and even fairly advanced approximate chains give a poor idea of of $\mathcal{L}$. This difficulty is exemplified in horrid fashion in Plate 179.

THE NOTION OF FRACTAL OSCULATION

My alternative construction of $\mathcal{L}$ involves a new fractal notion of osculation that extends an obvious facet of the Apollonian case.

STANDARD OSCULATION. This notion is linked to the concept of curvature. To the first order, a standard curve near a regular point P is approximated by the tangent straight line. To the second order, it is approximated by the circle, called *osculating*, that has the same tangent and the same curvature.

To index the circles tangent to the curve at P, a convenient parameter, u, is the inverse of the (arbitrarily oriented) distance from P to the circle's center. Write the index of the osculating circle as u_0. If $u < u_0$, a small portion of curve centered at P lies entirely on one side of the tangent circle, while if $u > u_0$ it lies entirely on the other side.

This u_0 is what physicists call a *critical value* and mathematicians call a *cut*. And $|u_0|$ defines the local "curvature."

GLOBAL FRACTAL OSCULATION. For the Apollonian net, the definition of osculation through the curvature is meaningless. However, at every point of the net where two packing circles are tangent to each other, they obviously "embrace" the rest of $\mathcal{L}$ between them. It is tempting to call *both* of them *osculating*.

To extend this notion to a non-Apollonian sets $\mathcal{L}$, we take a point where $\mathcal{L}$ has a tangent, and start with the definition of ordinary osculation based on criticality (= cut). The novelty is that, as u varies from $-\infty$ to $+\infty$,

the single critical u_0 is replaced by two distinct values, u' and u'' > u', defined as follows: For all $u < u'$, $\mathcal{L}$ lies entirely to one side of our circle, while for all $u < u''$, $\mathcal{L}$ lies entirely to the other side, and for $u' < u < u''$, parts of $\mathcal{L}$ are found on both sides of the circle. I suggest that the circles of parameters u' and u'' *both* be called *fractally osculating*.

Any circle bounds two open discs (one includes the circle's center, and the other includes the point at infinity). The open discs bounded by the osculating circles and lying outside $\mathcal{L}$ will be called osculating discs.

It may happen that one or two osculating circles degenerate to a point.

LOCAL VERSUS GLOBAL NOTIONS. Returning to standard osculation, we observe that it is a local concept, since its definition is independent of the curve's shape away from P. In other words, the curve, its tangent, and its osculating circle may intersect at any number of points in addition to P. By contrast, the preceding definition of fractal osculation is global, but this distinction is not vital. Fractal osculation may be redefined locally, with a corresponding split of "curvature" into 2 numbers. However, in the application at hand, global and local osculations coincide.

OSCULATING TRIANGLES. ◁ Global fractal osculation has a counterpart in a familiar context. To define the interior of our old friend the Koch snowflake curve K as a sigma-triangle (σ-triangle), it suffices that the triangles laid at each new stage of Plate 42 be lengthened as much as is feasible without intersecting the snowflake curve. ▶

σ-DISCS THAT OSCULATE $\mathcal{L}$

Osculating discs and σ-discs are the key of my new construction of $\mathcal{L}$, which is free from the drawbacks listed on p. 173. This construction is illustrated here for the first time (though it was previewed in 1980, in *The 1981 Springer Mathematical Calendar*!). The key is to take the inverses, not of the C_m themselves, but of some of circles Γ_{ijk}, which (as defined on page 171) are orthogonal to triplets C_i, C_j, and C_k. Again, we assume that the Γ_{ijk} are not all identical to a single Γ.

RESTRICTION TO M=4. The assumption M=4 insures that, for every triplet i,j,k, either one or the other of the two open discs bounded by Γ_{ijk}—namely, either its inside or its outside—contains none of the points γ_{mn} which we define on page 173. We shall denote this γ-free disc by Δ_{ijk}.

My construction of $\mathcal{L}$ is rooted in the following observations: every γ-free Δ_{ijk} osculates $\mathcal{L}$; so do their inverses and repeated inverses in the circles C_m; and the clans built using the Δ_{ijk} as seeds cover the whole plane except for the curve $\mathcal{L}$.

Plate 177 uses the same Poincaré chain as already used on page 173, but is drawn on larger scale. As is true in most cases, the first stage outlines $\mathcal{L}$ quite accurately. Later stages add detail very "efficiently," and after few stages the mind can interpolate the curve $\mathcal{L}$ without the temptation of error present in the Poincaré approach.

GENERALIZATIONS

CHAINS WITH FIVE OR MORE LINKS. When the number of original links in a Poincaré chain is M>4, my new construction of $\mathcal{L}$ involves an additional step: it begins by sorting the Γ circles into 2 bins. Some Γ circles are such that *each* of the open discs bounded by Γ contains at least one point γ_{mn}; as a result, Δ_{ijk} is *not* defined. Such Γ circles intersect $\mathcal{L}$ instead of osculating it. But they are not needed to construct $\mathcal{L}$.

The remaining circles Γ_{ijk} define osculating discs Δ_{ijk} that fall into two classes. Adding up the clans of the Δ_{ijk} in the first class, one represents the interior of $\mathcal{L}$, and adding up the clans of the Δ_{ijk} in the second class, one represents the exterior of $\mathcal{L}$.

The same is true in many (but not all) cases when the C_m fail to form a Poincaré chain.

OVERLAPPING AND/OR DISASSEMBLED CHAINS. When C_m and C_n have two intersection points γ'_{mn} and γ''_{mn}, these points jointly replace γ. When C_m and C_n are disjoint, γ is replaced by the two mutually inverse points γ'_{mn} and γ''_{mn}. The criterion for identifying Δ_{ijk} becomes cumbersome to state, but the basic idea is unchanged.

RAMIFIED SELF-INVERSE FRACTALS. $\mathcal{L}$ may borrow features from both a crumpled loop (Jordan curve), and an Apollonian net, yielding a fractally ramified curve akin to those examined in Chapter 14, but often much more baroque in appearance, as in Plate C7.

SELF-INVERSE DUSTS. It may also happen that $\mathcal{L}$ is a fractal dust.

THE APOLLONIAN MODEL OF SMECTICS

This section outlines the part that Apollonian packing and fractal dimension play in the description of a category of "liquid crystals." In doing so, we cast a glance toward one of the most active areas of physics, the theory of *critical points*. An example is the "point" on a temperature-pressure diagram that describes the physical conditions under which solid, liquid, and gaseous phases can coexist at equilibrium in a single physical system. The analytic characteristics of a physical system in the neighborhood of a critical point are scaling, therefore governed by power laws, and specified by critical exponents (Chapter 36). Many of them turn out to be fractal dimensions; the first example is encountered here.

Since liquid crystals are little known, we describe them by paraphrasing Bragg 1934. These beautiful and mysterious substances are liquid in their mobility and crystalline in their optical behavior. Their molecules are relatively complicated structures, lengthy and chainlike. Some liquid crystal phases are called *smectic*, from the Greek σμηγμα signifying soap, because they constitute a model of a soaplike organic system. A smectic liquid crystal is made of molecules that are arranged side by side like corn in a field, the thickness of the layer being the molecules' length. The resulting layers or sheets are very flexible and very strong and tend to straighten out when bent and then released. At low temperatures, they pile regularly, like the leaves of a book, and form a solid crystal. When temperatures rise, however, the sheets become able to slide easily on each other. Each layer constitutes a two-dimensional liquid.

Of special interest is the focal conics structure. A block of liquid crystal separates into two sets of pyramids, half of which have their bases on one of two opposite faces and vertices on the other. Within each pyramid, liquid crystal layers fold to form very pointed cones. All the cones have the same peak and are approximately perpendicular to the plane. As a result, their bases are discs bounded by circles. Their minimum radius ϵ is the thickness of the liquid crystal's layers. Within a spatial domain such as a square-based pyramid, the discs that constitute the bases of the cones are distributed over the pyramid's base. To obtain an equilibrium distribution, one begins by placing in the base a disc of maximum radius. Then another disc with as large a radius as possible is placed within each of the four remaining pieces, and so on and so forth. If it were possible to proceed without end, we would achieve exact Apollonian packing.

The physical properties of of this model of soap depend upon the surface and perimeter of the sum of interstices. The link is affected through the fractal dimension D of a kind of photographic "negative," the gasket that the molecules of soap fail to penetrate. Details of the physics are in Bidaux, Boccara, Sarma, Sèze, de Gennes & Parodi 1973. ■■

PLATE 177 ☐ A SELF-INVERSE FRACTAL (MANDELBROT CONSTRUCTION)

This Plate illustrates page 175.

TOP FIGURE. In Poincaré chains with $M=4$, at least one of the discs Δ_{ijk} is always unbounded, call it Δ_{123}, and it intersects the disc Δ_{341}. (Here, Δ_{341} is also unbounded, but in other cases it is not.) The union of Δ_{123} and Δ_{341}, shown in gray, provides a first approximation of the outside of $\mathcal{L}$. It is analogous to the approximation of the outside of Koch's $\mathcal{K}$ by the regular convex hexagon in Plate 43.

The discs Δ_{234} and Δ_{412} intersect, and their union, shown in black, provides a first approximation of the inside of $\mathcal{L}$. It is analogous to the approximation of the inside of $\mathcal{K}$ by the two triangles that form the regular star hexagon in Plate 43.

MIDDLE FIGURE. A second approximation of the outside of $\mathcal{L}$ is achieved by adding to Δ_{123} and Δ_{341} their inverses in C_4 and C_2, respectively. The result, shown in gray, is analogous to the second approximation of the outside of $\mathcal{K}$ in Plate 43.

The corresponding second approximation of the inside of $\mathcal{L}$ is achieved by adding to Δ_{234} and Δ_{412} their inverses in C_1 and C_3, respectively. The result, shown in black, is analogous to the second approximation of the inside of $\mathcal{K}$ in Plate 43.

BOTTOM FIGURE. The outside of $\mathcal{L}$, shown in gray, is the union of the clans of Δ_{123} and Δ_{341}. And the inside of $\mathcal{L}$, shown in black, is the union of the clans of Δ_{234} and Δ_{412}. The fine structure of the inside of $\mathcal{L}$ is seen in the bottom Plate 179, using a different Poincaré chain. Together, the black and gray open regions cover the whole plane, minus $\mathcal{L}$. ■

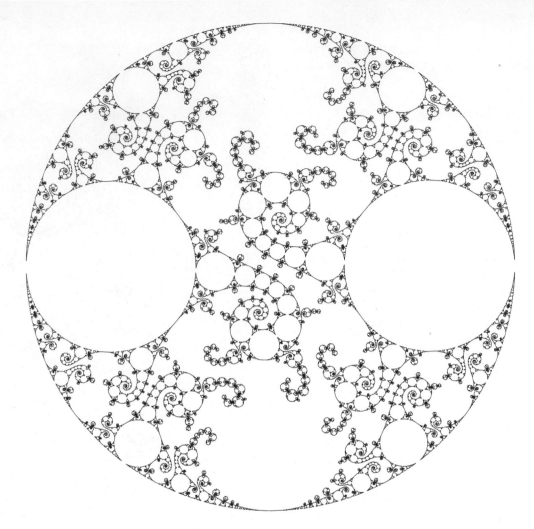

Plate 178 ⊔ SELF-HOMOGRAPHIC FRACTAL, NEAR THE PEANO LIMIT

To the mathematician, the main interest of groups based upon inversions resides in their relation with certain groups of homographies. An homography (also called Möbius, or fractional linear transformation) maps the z-plane by $z \rightarrow (az+b)/(cz+d)$, where $ad-bc=1$. The most general homography can be written as the product of an inversion, a symmetry with respect to a line (which is a degenerate inversion), and a rotation. This is why, in the absence of rotation, the study of homographies learns much from the study of groups based on inversions. But it is obvious that allowing the rotations brings in new riches.

Here is an example of limit set $\mathcal{L}$ for a group of homographies. David Mumford devised it (in the course of investigations inspired by the new results reported in this chapter), and kindly allowed its publication here. This shape is almost plane-filling, and shows uncanny analogies and differences with the almost plane-filling shape in Plate 191.

The fact that the limit set of a group of homographies is a fractal has been proven under wide conditions by T. Akaza, A. F. Beardon, R. Bowen, S. J. Patterson, and D. Sullivan. See Sullivan 1979. ■

Plate 179 ◻ A CELEBRATED SELF-INVERSE FRACTAL, CORRECTED (MANDELBROT CONSTRUCTION)

The top left reproduces Figure 156 of Fricke & Klein 1897, which claims (in my terminology) to represent the self-inverse fractal whose generator is made of the 5 circles that bound the blackened central region. This Figure has been reproduced very widely.

The outline of the black shape on the top right shows the actual shape of this fractal, as given by my osculating σ-disc construction. The discrepancy is horrid. Fricke knew that $\mathcal{L}$ incorporates circles, and he instructed his draftsman to include them. But otherwise Fricke did not know what sort of very irregular shape he should expect.

The actual $\mathcal{L}$ includes the boundary $\mathcal{L}^*$ of the shape drawn on the bottom right using my algorithm. This $\mathcal{L}^*$ is the self-inverse fractal corresponding to the four among the generating circles that form a Poincaré chain. Transforms of $\mathcal{L}^*$ by other inversions are clearly seen to belong to $\mathcal{L}$. Mandelbrot 1982i elaborates upon this plate. ■

19 ¤ Cantor and Fatou Dusts; Self-Squared Dragons

This chapter takes up two very simple families of nonlinear transformations (mappings) and investigates certain fractal sets which these transformations leave invariant, and for which they can serve as generators.

First, a broken line transformation of the real line deepens our understanding of an old acquaintance, the Cantor dust. These remarks could have been squeezed into Chapter 8, but they are better appreciated at this point.

In particular, they help appreciate the effect of the real and complex quadratic transforms, of the form $x \to f^*(x) = x^2 - \mu$, where x and μ are real numbers, or $z \to f^*(z) = z^2 - \mu$, where $z = x + iy$ and μ are complex numbers.

The elementary case $\mu = 0$ is geometrically dull, but other values of μ involve extraordinary fractal riches, many of them first revealed in Mandelbrot 1980n.

The invariant shapes in question are best obtained as a by-product of the study of iteration, that is, of the repeated application of one of the above transformations. The initial values will be denoted by x_0 or z_0, and the k times iterated transforms by f^* will be denoted by x_k or z_k.

Iteration was studied in three rough stages. The first, concerned with complex z, was dominated by Pierre Fatou (1878-1929) and by Gaston Julia (1893-1978). Their publications are masterpieces of classic complex analysis, greatly admired by the mathematicians, but exceedingly difficult to build upon. In my work, of which this chapter is a very concise sketch, some of their basic findings are made intuitive by combining analysis with physics and detailed drawing. And innumerable new facts emerge.

The resulting revival makes the properties of iteration essential to the theory of fractals. The fact that the Fatou-Julia findings did *not* develop to become the *source* of this theory suggests that even classical analysis the needs intuition to develop, and can be helped by the computer.

The intermediate stage includes P. J. Myrberg's studies of iterates of real quadratic mappings of $\mathbb{R}$ (e.g., Myrberg 1962), Stein & Ulam 1964, and Brolin 1965.

The current stage largely ignores the past,

and concentrates on self-mappings of [0,1], as surveyed in Gurel & Rössler 1979, Helleman 1980, Collet & Eckman 1980, Feigenbaum 1981, and Hofstadter 1981. This chapter's last section concerns the exponent δ due to Grossmann & Thomae 1977 and Feigenbaum 1978: the existence of δ is proven to follow from a more perspicuous (fractal) property of iteration in the complex plane.

THE CANTOR DUST CAN BE GENERATED BY A NONLINEAR TRANSFORMATION

We know from Chapter 8 that the triadic Cantor dust C is invariant by similitudes whose ratio is of the form 3^{-k}. This self-similarity is a vital property, but it *does not* suffice to specify C. In sharp contrast, C is *entirely* determined as the largest bounded set that is invariant under the following nonlinear "inverted V" transformation:

$$x \to f(x) = \{ \tfrac{1}{2} - |x - \tfrac{1}{2}| \} / r, \text{ with } r = \tfrac{1}{3}.$$

More precisely, we apply this self-mapping of the real axis repeatedly, with x_0 spread out over the x-axis, and the final values reduce to the point $x = -\infty$, plus the Cantor dust C. The fixed points $x = 0$ and $x = \tfrac{3}{4}$ belong to C.

SKETCH OF A PROOF OF THE INVARIANCE OF C. Since $f(x) = 3x$ when $x < 0$, the iterates of all the points $x_0 < 0$ converge to $-\infty$ directly, that is, without ceasing to satisfy $x_n < 0$. For the points $x_0 > 1$, direct convergence is preceded by one preliminary step, since $x_k < 0$ for all

k≥1. For the points in the gap $\tfrac{1}{3} < x_0 < \tfrac{2}{3}$, there are 2 preliminary steps, since $x_1 > 0$ but $x_k < 0$ for all k≥2. For the points in the gaps $1/9 < x_0 < 2/9$ or $7/9 < x_0 < 8/9$, there are 3 preliminary steps. More generally, if an interval is bounded by a gap that is sent to $-\infty$ after k preliminary steps, this interval's (open) mid third will proceed directly to $-\infty$ after the (k+1)st step. But *all* the points of C are found to fail to converge to $-\infty$.

FINITENESS OF THE OUTER CUTOFF

To extend these results to the general Cantor dust with $N = 2$ and r between 0 and $\tfrac{1}{2}$, it suffices to plug in the desired r in $f(x) = \{ \tfrac{1}{2} - |x - \tfrac{1}{2}| \} / r$. To obtain any other Cantor dust, the graph of $f(x)$ must be an appropriate zigzag curve.

However, no comparable method is available for the Cantor dust extrapolated to the whole real axis. This is a special case of a very general feature: Typically, a nonlinear $f(x)$ carries *within itself* a finite outer cutoff Ω. To the contrary, as we know well, all linear transformations (similarities and affinities) are characterized by $\Omega = \infty$, and a finite Ω (if one is required) must be imposed artificially.

ANATOMY OF THE CANTOR DUST

We know from Chapter 7 that C is a very "thin" set, yet the behavior of the iterates of $f(x)$ leads to a better understanding of fine distinctions between its points.

Everyone must be tempted, at first acquaintance, to believe that $\mathcal{C}$ reduces to the end points of the open gaps. But this is very far from being the case, because $\mathcal{C}$ includes by definition all the limits of sequences of gap end points.

This fact is *not* reputed intuitive. With many fellow students, I would have agreed if our battered acquaintance Hans Hahn had listed these limit points among the concepts whose existence must be imposed by cold logic. But the present discussion yields *intuitive* proof that these limit points have strong and diverse personalities.

For example, the point $x=\frac{3}{4}$, which $f(x)$ leaves unchanged, lies neither within any mid third interval, nor on its boundary. Points of the form $x=(\frac{1}{4})/3^k$ have iterates that converge to $x=\frac{3}{4}$. In addition, there is an infinity of limit cycles, each made up of a finite number of points. And $\mathcal{C}$ also contains points whose transforms run endlessly around $\mathcal{C}$.

THE SQUARING GENERATOR

The inverted V generating function $f(x)$ used in the preceding sections was chosen to yield a familiar result. But it makes the Cantor dust seem contrived. Now we replace it by

$$x \to f(x) = \lambda x(1-x),$$

whose unexpected wealth of properties was first noted in Fatou 1906. Changing the origin

and the scale of the x, and writing $\mu = \lambda^2/4 - \lambda/2$, this function can be written as

$$x \to f^*(x) = x^2 - \mu.$$

Convenience is served by using sometimes $f(x)$, and sometimes $f^*(x)$.

It is nice to call $f(x)$ or $f^*(x)$ the *squaring generator*. Squaring is, of course, an algebraic operation, but it is given a geometric interpretation here, so that the sets it leaves invariant can be called *self-squared*. Strict squaring replaces the point of abscissa x by the point of abscissa x^2. Thus, the self-squared points on the line reduce to $x=\infty$, $x=0$, and $x=1$. The addition of $-\mu$ may seem totally innocuous, but in fact it introduces totally unexpected possibilities we now consider.

FATOU'S REAL SELF-SQUARED DUSTS

Having yielded a familiar end product, the Cantor dust, the V transformation makes an extraordinary but never widely known discovery of Pierre Fatou easier to state. Fatou 1906 assumes that λ is real and satisfies $\lambda > 4$, and he investigates the largest of the bounded sets on $\mathbb{R}$, that are left invariant under $f(x)$. This is a close relative to the Cantor dust, which I call *real Fatou dust*. It requires no further explanation, and is illustrated in Plate 192.

In the complex plane, the largest bounded self-squared set, for the above λ's, remains the real Fatou dust.

SELF-SQUARED JULIA CURVES IN THE PLANE (MANDELBROT 1980n)

The simplest self-squared curve is obtained for $\mu=0$: it is the circle $|z|=1$. By the transformation $z \to z^2$, a belt wound *once* around the circle stretches into a belt wound *twice*, the "buckle" at $z=1$ remaining fixed. The corresponding largest bounded self-squared domain is the disc $|z| \leq 1$.

However, introducing a real $\mu \neq 0$ (Plates 186 and 187), then a complex μ (Plates 190 and 191), opens Pandora's boxes of possibilities, the *Julia fractal curves*. They satisfy the eye no less than they satisfy the mind.

THE SEPARATOR S. The topology of the largest bounded self-squared set depends on where μ lies with respect to a ramified curve S, which I discovered and now call *separator*. It is the connected boundary of the black shape in bottom Plate 188 ; it is a "limit lemniscate," namely the limit for $n \to \infty$ of the algebraic curves called lemniscates, defined by $|f_n^*(0)|=R$ for some large R. See Plate 189 for the structure of S.

THE ATOMS. The open domain within S splits into an infinity of maximal connected sets I now propose to call "atoms." Two atoms' boundaries either fail to overlap, or have in common one point, to be called "bond," that belongs to S.

TOPOLOGICAL DIMENSION. When μ lies outside S, the largest bounded self-squared set is a (Fatou) dust. When μ lies within S, or is a bond, the largest such set is a domain bounded by a self-squared curve. At least some μ on

S yield a tree-like curve.

SELF-SQUARED FRACTALS. These dusts and curves being fractal when $\mu \neq 0$ is rumored to have been proven fully in some further cases by Dennis Sullivan, and I harbor no doubt it will be proven in all cases.

The shape of a self-squared dust or curve varies continuously with μ, hence D is bound to be a smooth function of μ.

RAMIFICATION. When λ lies in one of the open empty discs of top Plate 189, the self-squared curve is a closed simple curve (not ramified, a loop), as in Plates 186 and 187.

When λ lies on the circles $|\lambda|=1$ or $|\lambda-2|=1$, or in the surrounding open connected region, the self-squared curve is a ramified net, with tremas bounded by fractal loops, like the dragons in Plate 191.

When λ lies in the very important island molecules, which will soon prove to be *regions of nonconfluence to 1*, the self-squared curve is either a σ-loop, or a σ-dragon, as in bottom Plate 190. The σ introduces no new loop.

μ-ATOMS AND μ-MOLECULES

To dissect the parameter map further is easier when the parameter is μ. A μ-atom may be heart-shaped, in which case it is the "seed" to which an infinity of oval-shaped atoms bind either directly or through intermediate atoms. Mutually bound atoms, plus their bonds, form a "molecule." A seed's cusp is never a bond.

To each atom is attached an integer w, its "period." When μ lies in an atom of period w, the iterates $f_n^*(z)$ converge to ∞ or to a stable limit cycle containing w points. Within an atom of period w, $|f_w^{*\prime}(z_\mu)| < 1$, where z_μ is any point of the limit cycle corresponding to μ. On the atom's boundary, $|f_w^{*\prime}(z_\mu)| = 1$, with $f_w^{*\prime}(z_\mu) = 1$ characterizing a cusp or a "root." Each atom contains a point to be called "nucleus," satisfying $f_w^{*\prime}(z_\mu) = 0$ and $f_w^*(0) = 0$.

The nuclei on the real axis were introduced by Myrberg (see Myrberg 1962), and rediscovered in Metropolis, Stein & Stein 1973. The corresponding maps are often called "superstable" (Collet & Eckman 1980).

Viewed as algebraic equation in μ, $f_w^*(0) = 0$ is of order 2^{w-1}. Hence, there could be at most 2^{w-1} atoms of period w, but there are fewer, except for w=1. For w=2, $f_2^*(0) = 0$ has 2 roots, but one of them is already the nucleus of an "old" atom of period 1. More generally, all the roots of $f_m^*(0) = 0$ are also roots of $f_{km}^*(0) = 0$ where k is an integer > 1. Next, observe that each rational boundary point on the boundary of an atom of period w, defined as satisfying $f_w^{*\prime}(z_\mu) = \exp(2\pi i m/n)$, where m/n is an irreducible rational number < 1, carries a "receptor bond" ready to connect to an atom of period nw. As a result, some new atoms bind to existing receptor bonds. But not all new atoms are thereby exhausted, and the remaining ones have no choice but to seed new molecules. The molecules are therefore infinite in number.

When μ varies continuously in a molecule, each outbound traversal of a bond leads to bifurcation: w is multiplied by n. Example: increasing a real-valued μ leads to Myrberg's period doubling. The inverse of bifurcation, which Mandelbrot 1980n investigates and calls *confluence*, must stop at the period of the molecule's seed. The continent molecule is the region of confluence to c=1, and each island molecule is a region of confluence to c>1. The dragon's or sub-dragon's shape is ruled by the values of $f_w^{*\prime}(z_\mu)$ and w/c.

THE SEPARATOR IS A FRACTAL CURVE; FEIGENBAUM'S δ AS A COROLLARY

I conjecture ◁ via a "renormalization" argument ► that atoms increasingly removed from their molecule's seed come increasingly close to being *identical* in shape.

A corollary is that the boundary of each molecule is locally self-similar. Since it is not smooth on small scales, it is a fractal curve.

This local self-similarity generalizes a fact concerning Myrberg bifurcation, due to Grossmann & Thomae and to Feigenbaum. The widths of increasingly small sprouts' intercepts by the real axis of λ or μ, converge to a geometrically decreasing sequence, of ratio $\delta = 4.66920...$ (Collet & Eckman 1980). In its original form, the existence of δ seems a technical analytic result. Now it proves to be an aspect of a broader property of fractal scaling.

Each bifurcation into m>2 introduces an additional basic ratio.   ■

Plate 185 ¤ SELF-SQUARED FRACTAL CURVES FOR REAL λ

The shapes in Plates 185 to 192 are presented here for the first time, except for a few that are reproduced from Mandelbrot 1980n.

The left side of this plate represents the maximal bounded self-squared domains for λ = 1, 1.5, 2.0, 2.5 and 3.0. The central black shape spans the segment [0,1].

λ=1: SCALLOP SHELL.

λ=3: SAN MARCO DRAGON CURVE. This is a mathematician's wild extrapolation of the skyline of the Basilica in Venice, together with its reflection in a flooded Piazza; I nicknamed it the *San Marco dragon*.

The right side of this plate is relative to λ=3.3260680. This is the nuclear λ (as defined on p. 184) corresponding to w=2. The corresponding self-squared shape is turned by 90° to make it fit in. ■

CAPTION CONTINUED FROM P. 188

TOP PLATE 188. This is part of the inverse of the λ-map with respect to $\lambda=1$. Examining on the λ-map the sprouts whose roots are of the form $\lambda=\exp(2\pi i/n)$, one gains the impression that "corresponding points" lie on circles. The present plate provides confirmation. Other perceived circles are confirmed by different inversions.

ISLAND MOLECULES. Many of the "spots" around the maps are genuine "island molecules," first reported in Mandelbrot 1980n. They are shaped like the whole μ map, except for a nonlinear distortion.

SEPARATOR, SPINES AND TREES. The boundary of the filled-in black domain in the λ- or μ map is a connected curve I discovered and call *separator* S. The set within S decomposes into open *atoms* (see text). When the atom's period is w, let us define its *spine* as the curve where $f_w^*(z_\mu)$ is real.

The spines lying on the real axis are known in the theory of self-mapping as $[0,1]$, and their closure is known to be $[-2,4]$.

I discovered more generally that the closure of the other atom spines decomposes into a collection of trees, each rooted on a receptor bond. The list of orders of ramification at different points of such a tree is made up of 1 for the branch tips, plus the orders of bifurcation leading to the tree's root. Furthermore, when the tree is rooted on an island atom, one must add the orders of bifurcation leading from $|\lambda-2|\leq1$ or $|\lambda|\leq1$ to this atom.

BOTTOM LEFT PLATE 189. This is a detailed λ map near $\lambda=2-\exp(-2\pi i/3)$. The set within S is the limit of domains of the form $|f_n(\frac{1}{2})|<R$, whose boundaries are algebraic curves called lemniscates. A few such domains are shown here in superposition. For large n, these domains seem disconnected, and so does the λ map, but in fact they connect outside the grid used in the computation.

BOTTOM RIGHT PLATE 189. This is a detailed λ map near $\lambda=2-\exp(-2\pi i/100)$. This hundred-fold branching tree shares striking features with the z map in Plate 191. ■

Plate 187 ¤
COMPOSITE OF SELF-SQUARED FRACTAL CURVES FOR REAL λ

This draped "sculpture" was made within a computer's memory, by a process that amounts to whittling away all points in an initial cube, whose iterates by $z\rightarrow\lambda z(1-z)$ converge to infinity. The parameter λ is a real number ranging from 1 to 4. The λ axis runs vertically along the sculpture's side. And x and y form the complex number $z=x+iy$.

Each horizontal section is a maximal bounded self-squared shape of parameter μ.

For the special value $\lambda=2$, this section's boundary is a circle: the drape's "belt."

For all other values of λ, the self-squared shape's boundaries are fractal curves, including those shown in Plate 185. One perceives striking "pleats" whose position varies continuously with λ; they are pressed *in* below the belt, and pressed *out* above the belt.

Of special interest are the blobs on the wall holding the drape. This sculpture cannot possibly do justice to the complication of the top of the drape. A) For every value of λ, the drape includes, as "backbone," a fractal tree formed by the iterated pre-images of the x-interval $[0,1]$. For all small, and some high values of $\lambda<3$, this tree's branches are completely "covered by flesh." For other high values of λ, however, there is no flesh. The branches along either $x=\frac{1}{2}$ or $y=0$ are visible here, but the graphic process unavoidably misses the rest. B) Certain horizontal stripes of the wall behind the drape are entirely covered with tiny "hills" or "corrugations," but only a few of the largest ones can be seen. These stripes and hills concern the "island molecules" (Plates 188 and 189) intersected by the real axis. Observations A) and B) generalize the Myrberg-Feigenbaum theory. ■

Plates 188 and 189 ¤ THE SEPARATORS OF $z \to \lambda z(1-z)$ AND OF $z \to z^2 - \mu$

BOTTOM PLATE 188. μ-MAP. The μ in the closed black area (bounded by a fractal curve) are such that the iterates of $z_0 = 0$ under $z \to z^2 - \mu$ fail to converge to ∞. The large cusp is $\mu = -\frac{1}{4}$, and the right-most point is $\mu = 2$.

TOP PLATE 189. λ-MAP. The λ in the closed black area, plus the empty disc, satisfy $\text{Re}\lambda > 1$ and are such that the iterates of $z_0 = \frac{1}{2}$ under $z \to \lambda z(1-z)$ fail to converge to ∞. The full λ map is symmetric with respect to the line $\text{Re}\lambda = 1$.

THE DISC $|\lambda - 2| \leq 1$, AND THE DISC $|\lambda| \leq 1$ LESS $\lambda = 0$. The λ in these domains are such that the iterates of $z_0 = \frac{1}{2}$ converge to a bounded limit point.

CORONA AND SPROUTS. The λ-map outside the empty discs forms a "corona." It splits into "sprouts," whose "roots" are "receptor bonds" defined as the points of the form $\lambda = \exp(2\pi i m/n)$ or $\lambda = 2 - \exp(2\pi i m/n)$, with m/n an irreducible rational number <1.

CAPTION
CONTINUES
ON P. 186

Publisher's Note (Fall 1985). The filament structure of the μ map (now known as the *Mandelbrot set*) was not visible on the rendering that was used in earlier printings. We have therefore substituted a new variant, due to J. Milnor and I. Jungreis, which thickens the filaments so they can be seen.

$\lambda'' $

θ θ

$\lambda = \lambda' + i\lambda''$

-1   0   1   2   3   4 λ'

λ plane

Plates 190 and 191 ¤ SELF-SQUARED DRAGONS; APPROACH TO THE "PEANO" LIMIT

Each self-squared curve is attractive in its own way. And the most attractive ones to me are the "dragons" shown in the present figures and in Plate C5.

DRACONIC MOLTING. To watch a dragon in the process of self-squaring would be a fascinating sight! A monstrous "molting" detaches the skins of a dragon's belly and back from their innumerable folds. Then, it stretches each skin to twice its length, which of course remains infinite all along! Next, it folds each skin around the back as well as the belly. And finally, it re-attaches all the folds neatly in their new positions.

FRACTAL HERALDRY. The self-squared dragons must not be confused with the self-similar one of Harter & Heightway, Plates 66 and 67. The reader may find it amusing to detail the similarities and the many differences.

CAPTION CONTINUES ON P. 192

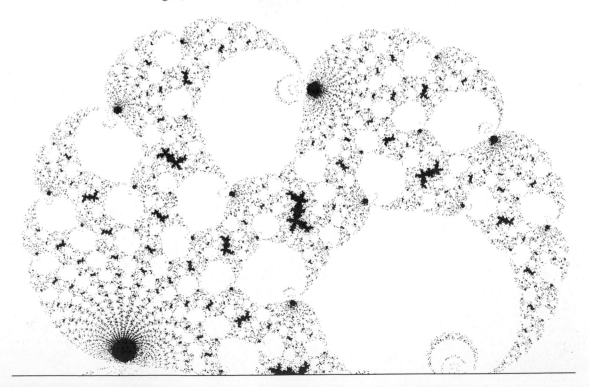

191

CAPTION CONTINUED FROM P. 190

SUCCESSIVE BIFURCATIONS. The best self-squared dragons obtain where λ lies in a sprout of Plate 189 that corresponds to $\theta/2\pi = m/n$, with small integers n and m. Given the bifurcation order n, the number of dragon heads or tails (or whatever these domains should be called) around each articulation point is n. A second bifurcation of order m'/n' splits each of these domains into n' "sausage links," and thins them down.

Dragons with a nice heft, neither obese nor skinny, obtain when λ lies within a sprout, at some distance away from the root. Dragons with a nice twist obtain when λ lies near one of the 2 subsprouts corresponding to an order of bifurcation of 4 to 10: one subsprout yields a leftward, the other a rightward, twist.

RIGHT TOP OF PLATE 190. "STARVED DRAGON." A dragon subjected to infinitely many bifurcations loses all flesh and collapses into a skeletal branched curve.

The topological dimension of the set that fails to go to ∞ is 0 for the Fatou dusts, 1 for starved dragons, and 2 for other dragons.

BOTTOM OF PLATE 190. σ-DRAGON. This shape is connected; its λ lies in the large "offshore island" in bottom right Plate 189.

PLATE 191. THE SINGULAR LIMIT $\lambda=1$. PEANO DRAGONS. Let λ lie in an island off-shore of the bond at $\theta=2\pi/n$. As $n\to\infty$, $\theta\to0$, hence λ tends to 1. The corresponding dragon must necessarily converge to the scallop shape at the base of the drape in Plate 187. But a qualitative difference separates $n=\infty$ from n large but finite.

As $n\to\infty$, the dragon's arms grow in number, the skin crumples, and the skin's dimension increases. The whole really attempts to converge to a "hermit-dragon" that would fill the shell of a $\lambda=1$ scallop to the brim, i.e., to the dimension D=2. A self-squared Peano curve? Yes, but we know from Chapter 7 that Peano curves are not curves: as it attains D=2, our dragon curve dies as a curve to become a plane domain. ∎

Plate 192 ⊏ REAL SELF-SQUARED FATOU DUSTS ON [0,1]

Fatou 1906 is a masterpiece of an odd literary genre: the *Comptes Rendus Notes* of the Paris Academy of Sciences. In many cases, the purpose is to reveal little, but to squirrel evidence that the author had thought of everything.

Among other marvelous remarks best understood after long self-study, Fatou 1906 points out the following. When λ is real and either $\lambda>4$ or $\lambda<-2$, the largest bounded set that the transformation $x\to f(x) = \lambda x(1-x)$ leaves invariant is a dust contained in [0,1]. This plate illustrates this dust's shape for $\lambda>4$. Along the vertical coordinate, $-4/\lambda$ varies from -1 to 0. The black intervals mark the end points of the tremas of order 1 to 5. The end points x_1 and x_2 of the mid trema are solutions of the equation $\lambda x(1-x)=1$; they draw a parabola. Second-order tremas end at the points $x_{1,2}$, $x_{1,2}$, $x_{2,1}$, and $x_{2,2}$, such that $\lambda x_{m,n}(1-x_{m,n}) = x_m$, etc.

The remarkable relation between Cantor-like dusts and one of the most elementary among all functions deserves to be known beyond the circle of specialists. ∎

20 ¤ Fractal Attractors
and Fractal ("Chaotic") Evolutions

This chapter seeks to acquaint the reader with a theory that evolved independently of fractals, but is being penetrated by them. Its most common name is "theory of strange attractors and of chaotic (or stochastic) evolution," but reasons for giving it the new name in the title will, I hope, emerge in this chapter.

Its involvement with fractals would suffice to justify mentioning this theory in this Essay, and I see reasons for devoting a full chapter to it. A practical reason is that little special exposition is required, because several major themes can be presented by merely reinterpreting the results of Chapters 18 and 19.

Secondly, several features of the fractal geometry of nature become clarified when contrasted with the theory of fractal attractors. Indeed, my work is concerned primarily with shapes in the real space one can see, at least through the microscope, while the theory of attractors is ultimately concerned with the temporal evolution in time of points situated in an invisible, abstract, representative space.

This contrast is especially striking in the context of turbulence: turbulent intermittency was the first major problem I attacked (starting in 1964) using early forms of fractal techniques, and (quite independently) the theory of strange attractors took off for earnest with the study of turbulence in Ruelle & Takens 1971. Thus far, the two approaches have not met, but they are bound to meet soon.

Those interested in the sociology of science will savor the fact that, while my case studies that linked the mathematical monsters to *real* physical shapes encountered resistance, the *abstract* attractors' being monstrous shapes was accepted with equanimity.

A third reason for mentioning fractal attractors is suggested by the fact that the corresponding evolutions look "chaotic" or "stochastic." As seen in Chapters 21 and 22, many scholars question the use of randomness in science; now the hope has arisen that it will be justified via fractal attractors.

Finally, those who have accepted many chapters ago (or one or two Essays ago) my contention that many facets of nature can only be described with the help of certain sets previously reputed pathological may be impatient to move from "how" to "why." Expository accounts have demonstrated in several cases that it is not difficult to sugar-coat the geometric frames in previous chapters, making them more immediately palatable. But a taste for fractals is one I wanted the reader to acquire, however bitter it may first seem to most grown scientists. Furthermore, pseudo-explanation via sugar-coating is never compelling, in my opinion, as explained in Chapter 42. Therefore, explanation was downplayed, except when a compelling one is available, as in Chapter 11. In addition, I suspect that many further genuine explanations will come forth when fractal attractors become a foundation of the fractal geometry of visible natural shapes.

Since the transforms that have attractors are *non*linear, the visible fractals are likely *not* to be self-similar. This is fine: there was a paradox in my use of the fractal counterpart of the straight line to handle phenomena ruled by *non*linear equations. The scaling fractals that account well for a natural phenomenon would be local approximations to nonlinear fractals.

THE NOTION OF ATTRACTOR

The present chapter centers around a long neglected observation due to Henri Poincaré: The "orbits" of nonlinear dynamical systems may be "attracted" to odd sets that I identify as nonlinear fractals.

Let us first examine the simplest attractor:

a point. The "orbit" followed by the motion of a small ball put inside a funnel begins with wiggles that depend on its initial position and velocity, but converges eventually to the funnel's tip; if the ball is bigger than the funnel aperture, it comes to rest at the tip. The tip is a stable equilibrium point, or stable fixed point, for the ball. In a nice alternative descriptive terminology (which one must be careful not to interpret in anthropocentric terms), the funnel's tip is called an *attractor point*.

A physical system may also have a stable attracting circle or ellipse. For example, it is believed (and fervently hoped, though no one will live long enough to care) that the solar system is stable, meaning that Earth's orbit, if perturbed, would eventually be "attracted back" into its present rut.

More generally, a dynamical system is customarily defined as follows: Its state at time t is a point $\sigma(t)$ on the line, in the plane, or in some higher dimensional Euclidean "phase space" $\mathbb{R}^E$, and its evolution between the times t and $t + \Delta t$ is determined by rules in which the value of t does not enter explicitly. Each point in phase space can be taken as the initial state $\sigma(0)$ at $t=0$, and it is followed by an orbit defined by the $\sigma(t)$ for all $t > 0$.

The major distinction between such systems concerns the geometric distribution of $\sigma(t)$ for large t's. A dynamical system is said to have an *attractor* if there exists a proper subset $\mathcal{A}$ of the phase space $\mathbb{R}^E$, such that for almost all starting points $\sigma(0)$, and t large enough, $\sigma(t)$ is close to some point of $\mathcal{A}$.

THE NOTION OF REPELLER

On the other hand, a ball can be poised in unstable equilibrium on a pencil's point. When the initial position is near this equilibrium, the ball seems to be pushed away, before converging to stable equilibrium elsewhere.

The set of all unstable equilibrium states, plus their limit points, is called *repeller*.

In many cases, the repellers and attractors exchange roles by turning the equations around. When the force is gravitation, it suffices to invert the direction of gravity. For example, consider a largely horizontal sheet with a dip in both directions. When a ball is positioned on the sheet's upper side and gravity points down, let A denote the attractor dip and R the repeller dip. When the ball is repositioned on the sheet's lower side and gravity points up, A and R exchange roles. Such exchanges play a central role in this chapter.

FRACTAL ATTRACTORS. "CHAOS"

Much of textbook mechanics concerns dynamical systems whose attractors are points, near-circles, or other shapes from Euclid. But these are rare exceptions, and the behavior of most dynamic systems is incomparably more complicated: their attractors or repellers tend to be fractals. The next few sections describe examples where time is discrete, with $\Delta t = 1$.

AN ATTRACTOR THAT IS A DUST. THE α OF FEIGENBAUM. The simplest example is obtained through squaring (Chapter 19). As prelude, consider yet another representation of the Cantor dust $\mathcal{C}$ with $N=2$ and $r<\frac{1}{2}$, spanning $[-r/(1-r), r/(1-r)]$. This $\mathcal{C}$ is the limit of $\mathcal{C}_n$, defined as the set of points of the form $\pm r \pm r^2 \pm \ldots \pm r^n$. As $n \to n+1$, each point of $\mathcal{C}_n$ bifurcates into 2, and $\mathcal{C}$ is the outcome of an infinity of bifurcations.

Interpreting P. Grassberger (preprint), the attractor $\mathcal{A}_\lambda$ of $x \to \lambda x(1-x)$ for real λ is analogous to $\mathcal{C}_n$, but with 2 distinct ratios of similarity, one of which is Feigenbaum's $1/\alpha \sim .3995\ldots$ (Feigenbaum 1981). After an infinity of bifurcations, this attractor is a fractal dust $\mathcal{A}$ with D $\sim .538$.

"CHAOS". No point of $\mathcal{A}$ is visited twice in finite time, Many authors describe evolutions on fractal attractors as "chaotic."

◁ SELF-AFFINE TREES. Juxtaposing the $\mathcal{A}_\lambda$ in the (x,λ) plane, one obtains a tree. Since $\delta \sim 4.6692 \neq \alpha$, this tree is asymptotically self-affine with a residue. ▶

COMMENT. The theory should ideally focus upon intrinsically interesting and realistic (but simple) dynamical systems, whose attractors are fully understood fractals. The strange attractors literature—though extremely important—is far from this ideal: its fractals are usually incompletely understood, few are intrinsically compelling, and most fail to be solutions to well-motivated problems.

I was therefore led to devise "dynamical systems" that amount to seeking new questions to obtain old and pleasant answers. That is, I contrived problems so that their solutions are familiar fractals. Somewhat surprisingly, these systems are of interest.

SELF-INVERSE ATTRACTORS

Chapter 18 describes the $\mathcal{L}$ sets of Poincaré chains as being both the smallest self-inverse sets and limit sets. To restate this last property: given an arbitrary starting point P_0, every point of $\mathcal{L}$ is approached arbitrarily closely by transforms of P_0 by sequences of inversions. Now suppose that this sequence of inversions is selected by a separate process, independent from the present and past positions of P. Under wide conditions, the resulting sequences of P's can always be expected, and is often actually shown, to be attracted by $\mathcal{L}$. In this fashion, the enormous literature concerning the groups based upon inversions is interpreted in terms of dynamical systems.

"TIME" REVERSAL

My search for further systems with interesting fractal attractors moved on to the trove of known systems with geometrically standard attractors but interesting repellers. To invert the roles of these two sets, thus making time run backward, is possible as long as the operations of the dynamical systems have inverses (orbits never join or cross) so that knowledge of $\sigma(t)$ determines all $\sigma(t')$ for $t' < t$. However, the specific systems in which we want to reverse time, are different. Their orbits are like rivers: the path is uniquely determined in the downhill direction, but in the uphill direction each fork involves a special decision.

For example, let us try and invert the V-transformation f(x) that gives the Cantor dust in Chapter 19. Two different inverse functions are defined for $x > 1.5$, and one may agree to transform all $x > 1.5$ into $x = \frac{1}{2}$. Similarly, $x \rightarrow \lambda x(1-x)$ has two possible inverses. In either case, a meaningful inversion requires choosing between two functions. In other examples, the number of possibilities is even larger. Again, we want them to be selected by a separate process. These thoughts point to generalized dynamical systems, to be introduced and described in the following section.

DECOMPOSABLE DYNAMIC SYSTEMS (MANDELBROT 1980n)

We demand that one of the coordinates of the state $\sigma(t)$—call it *determining index*, and denote it by $\sigma^\dagger(t)$—evolves independently of the state of the other $E-1$ coordinates—call it $\sigma^*(t)$—while the transformation from $\sigma^*(t)$ to $\sigma^*(t+1)$ is determined by both $\sigma^*(t)$ and $\sigma^\dagger(t)$. In the examples I studied most, the transformation $\sigma^*(t) \rightarrow \sigma^*(t+1)$ is chosen in a finite collection of G different possibilities $\mathcal{I}_g$, which may be selected according to the value of some integer-valued function $g(t) = \gamma[\sigma^\dagger(t)]$. Thus, I studied dynamics in the product of the σ^*-space by a finite index set.

In fact, in the examples that motivate this generalization, the sequence g(t) either is random or behaves as if it were. This Essay does not tackle randomness until the next chapter, but I doubt this is a serious difficulty. More serious is the fact that dynamical systems are the very model of fully deterministic behavior, hence are forbidden to accommodate randomness! However, one can inject its effects without actually postulating it, by taking for g(t) the value of a sufficiently mixing ergodic process. For example, one can take an irrational number β, and make g(t) the integer part of $\sigma^\dagger(t) = \beta^t \sigma^\dagger(0)$. The necessary statements, being easy in principle but cumbersome, will not be written here.

THE ROLE OF "STRANGE" ATTRACTORS

Students of "strange" attractors advance the following two-part argument: A) Granted that dynamic systems with standard attractors cannot explain turbulence, perhaps it can be explained by *topologically* "stranger" attractors. (This recalls my independent argument, Chapter 11, that when a differential equation has no standard singularities, one ought to try fractal singularities.) B) The attractors of absurdly simple systems, such as $z \rightarrow \lambda z(1-z)$ for real λ and z in [0,1], are strange and in many ways characteristic of more complex and more realistic systems. Therefore, there can be no doubt that topologically strange attractors are the rule.

THE TERMS "FRACTAL" VS. "STRANGE'

EVERY KNOWN "STRANGE" ATTRACTOR IS A FRACTAL. D has been evaluated for many "strange" attractors. In all cases, $D > D_T$.

Hence, these attractors are fractal sets. For many strange attractor fractals, D is not a measure of irregularity but of the way smooth curves or surfaces pile upon each other—a variant of fragmentation (Chapter 13).

A famous attractor, called solenoid, was introduced in two stages by S. Smale. The original definition was purely topological, leaving D undefined, but a revision was made metric (Smale 1977, p. 57). For this revision, D was evaluated in Mandelbrot 1978b which injected D into the study of strange attractors. For the Saltzman-Lorenz attractor with $v=40$, $\sigma=16$, and b=4, the value D=2.06 was obtained independently by M. G. Velarde and Ya. G. Sinai, (private conversations). This D is above 2, but not by much, meaning that this attractor is definitely not a standard surface, but that it is not far from being one. Mori & Fujisaka 1980 confirms my D for the Smale attractor and the D for the Saltzman-Lorenz attractor. For the Hénon mapping. with a=1.4 and b=0.3, they find D=1.26. Many other articles to the same effect are on the way.

CONVERSE. Whether or not all fractal attractors are strange is a matter of semantics. Increasing numbers of authors agree with me that *for most purposes an attractor is strange when it is a fractal*. This is a healthy attitude, if "strange" is taken to be a synonym to "monstrous," "pathological," and other epithets once applied to individual fractals.

But "strange" is sometimes given a technical sense, ◁ one so exclusive that the Saltzman-Lorenz attractor is not "strange," but "strange-strange." ► In this light, an attractor's "strangeness" involves nonstandard *topological* properties, with nonstandard *fractal* properties coming along as an "overhead." A closed curve without double points is not "strange" in this sense, however crumpled it may be; hence, many fractal attractors I examined are not strange.

With this definition of "strange," the argument in the preceding section ceases to be compelling. But it becomes compelling again if strangeness is modified from being a topological to being a fractal notion. Thus, I think that those who define "strange" as "fractal" deserve to win. Since indeed they are winning, there is little reason to preserve a term whose motivation vanished when I showed that fractals are no stranger than coastlines or mountains. Anyhow, I cannot conceal a personal dislike for the term "strange." ■

Plates 198 and 199 ¤
ATTRACTION TO FRACTALS

These two shapes illustrate long orbits of successive positions of two decomposable dynamical systems. The *Pharaoh's Breastplate* in Plate 199 is self-inverse (Chapter 18), being based upon 4 inversions selected to insure that the limit set $\mathcal{L}$ is a collection of circles. The *San Marco dragon* in Plate 198 is self-squared (Chapter 19), being based upon the two inverses of $x \to 3x(1-x)$.

The determining index is chosen among 4, respectively 2, possibilities, using a pseudorandom algorithm repeated 64,000 times. The first few positions are not plotted.

Regions in the neighborhoods of cusps and self-intersections are very slow to fill. ■

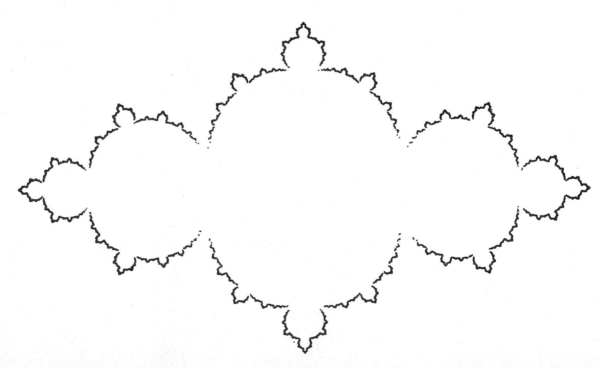

21 ¤ Chance as a Tool in Model Making

Although the basic fractal themes involve exclusively deterministic constructions, the full meaning and practical relevance of these themes are not apparent until one tackles random fractals. And conversely, the study of fractals seems, at least to this writer, to increase one's understanding of randomness.

A first reason to inject chance is familiar to every scientist, yet deserves comment in this chapter, among less generally familiar remarks of a general nature. The following chapter opens new vistas and shows that chance is also needed for reasons specific to the study of fractals.

⟨X⟩ DENOTES AN EXPECTATION; THE ABBREVIATION FOR PROBABILITY IS Pr

Each discipline seems to denote the expectation of the random variable X differently. The physicists' notation ⟨X⟩ is adopted in this Es-say, because it has the virtue of including its own portative parentheses.

Given a function B(t), and its $\Delta B(t) = B(t+\Delta t) - B(t)$, I call $\langle \Delta B(t) \rangle$ the *delta mean*, and $\langle [\Delta B(t) - \langle \Delta B(t) \rangle]^2 \rangle$ the *delta variance*.

RANDOM MODELS' STANDARD ROLE

Let us go back to the question "How long is the coast of Britain?" Much as it reminds us of real maps, the Koch curve has major defects which we encounter almost unchanged in early models of every other phenomenon studied in this Essay. Its parts are identical to each other, and the self-similarity ratio r must be part of a scale of the form b^{-k}, where b is an integer, namely, ⅓, (⅓)^2, and so on.

One might improve the model by invoking more complicated deterministic algorithms. However, this approach would be not only tedious, but doomed to failure, because each coastline is molded throughout the ages by multiple influences that are not recorded and cannot be reconstituted in any detail. The goal of achieving a full description is hopeless, and should not even be entertained.

In physics, for example in the theory of Brownian motion, the key out of this difficulty lies in statistics. In geomorphology, statistics is even harder to avoid. Indeed, while the laws of mechanics affect molecular motion directly, they affect geomorphological patterns through many ill-explored intermediates. Hence, even more than the physicist, the geomorphologist is compelled to forsake a precise description of reality and to use statistics. In other fields which we shall explore, the current knowledge of local interactions lies somewhere between physics and geomorphology.

SEARCH FOR THE RIGHT AMOUNT OF CHANCE IRREGULARITY

Can chance bring about the strong degree of irregularity encountered, say, in coastlines? Not only does it, but in many cases it goes *beyond* the desired goal. In other words, the power of chance is widely underestimated. The physicists' concept of randomness is shaped by theories in which chance is essential at the microscopic level, while at the macroscopic level it is insignificant. Quite to the contrary, in the case of the scaling random fractals that concern us, the importance of chance remains constant on all levels, including the macroscopic one.

A PRAGMATIC USE OF CHANCE

The relationship between statistical unpredictability and determinism raises fascinating questions, but this Essay has little to say about them. It makes the expression "at random" revert to its intuitive connotation at the time when medieval English borrowed it from French. The phrase *"un cheval à randon"* is reputed to have been unconcerned with either mathematical axiomatics or equine psychology, and merely denoted irregular motion the horseman could not *predict*.

Thus, while chance evokes all kinds of quasi-metaphysical anxieties, this Essay is determined to be little concerned with whether or not, in Einstein's words, "the Lord plays with dice." The theory of probability is the only mathematical tool available to help map the unknown and the uncontrollable. It is fortunate that this tool, while tricky, is extraordinarily powerful and convenient.

FROM RECURSIVITY TO RANDOMNESS

Furthermore, probability theory can be introduced to fit smoothly within the recursive methods that predominate in this Essay. In other words, this Essay's second half follows the first half without discontinuity. We shall continue to concentrate on cases where both the mathematical definition and the graphics algorithm can be written in the form of a "processor program" with an internal loop, and each run around the loop adds fresh detail to what has been drawn on previous runs.

The familiar loop that generates the triadic Koch curve reduces to this processor program. But other nonrandom fractals involve in addition a "control program," which we must now emphasize, and whose functions evolve interestingly but progressively toward increased generality. In a first step, the caption of Plate 46 observes that certain Koch generators can be used either in straight (S) or flipped (F) variants, hence their processor needs a controller to tell it before each loop whether to use S or F. In general, different control sequences yield different fractals. Hence, for each choice of M and of the corresponding D, the fractal loop of Plate 46 is not really *one* curve but an infinite (denumerable) *family* of curves, one for each control sequence. The controller may either read his sequence from a tape, or *interpret* a compact instruction of the form "alternate S and F," or "let the k-th stage use S (or F) whenever the k-th decimal of π is even (or odd)."

RANDOMNESS/PSEUDORANDOMNESS

Many random fractals involve precisely the same pattern: an interpreting controller followed by a processor. This fact is often hidden (sometimes to make things look harder), but is clearcut in the desirable cases whose definition is explicitly recursive.

The very simplest controller is called "sequence of throws of a fair coin," but I have never used one. In today's computer en-

vironment, the controller is a "random number generator." Its input, called *seed*, is an integer with a prescribed number M of binary digits. (M is determined by the equipment; when fewer than M digits are typed in, the front is filled in with zeros.) The controller's output is a sequence of 0 and 1. In simulations of a Bernoulli game, each digit stands for the result of the toss of a fair coin. And a game of 1,000 coin tosses is really a sequence of 1,000 individual pseudo-random digits.

But one can also imagine that there exists somewhere a big book of 2^{1000} pages, in which each possible outcome of 1,000 coin tosses is recorded on a separate page. Thus, any game of 1,000 tosses can be specified by selecting a page in this book. The parameter of chance is simply the page number, i.e., the seed.

More generally, the controller's output is often sliced into chunks of A integers. Then, by adding a decimal point in front, each chunk is made into a fraction U, and this fraction is called "random variable uniformly distributed between 0 and 1."

The output of a practical random set generator is not a single function or shape, but a virtual "grand portfolio" of 2^A pages, each devoted to a single shape. Again, the page numbers are the seeds.

The botanical analogy implies of course that the seeds are all of the same species and variety. One allows for "defective seeds" that produce very atypical plants, but one expects the overwhelming majority of plants to differ in detail but be the same on essentials.

The random number generator is the hinge of any simulation. Upstream are operations that involve in every case the same interface between number theory and probability theory, and are independent of the program's goals. They are exemplary of deterministic transformations that mimic randomness as described by the theory of probability. Downstream lie steps which vary according to the simulation's objective.

The move from this practical environment to full-fledged recursive probability is a natural one. The main change is that fractions with a finite number of digits are replaced by real numbers. The seeds become the mysterious "elementary events" which mathematical probabilists denote by the letter ω. ◁ To "interpret" ω into an infinite sequence of real control variables, Paley & Wiener 1934 suggests converse Cantor diagonalization. ▶

EMPTY INVOCATION OF CHANCE VERSUS ACTUAL DESCRIPTION

The preceding section argues that the theory of chance is not *really* difficult. Unfortunately, it is not really easy. One is tempted to say that, to achieve a model of coastlines free of the defects of the Koch curve but preserving its assets, it suffices to deform the different portions of the curve and to modify their sizes, all at random, then string them together in random order.

Such an *invocation* of chance is allowable in preliminary investigations, and our early chapters indulge in it freely. It is not sinful, unless it is hidden from the reader or is not recognized by the writer. And in some cases it can be implemented. In other cases, merely to invoke chance is an empty gesture. Indeed, rules that generate acceptable random curves are very hard to describe, because geometric sets are imbedded in a space. By merely varying at random the shapes, the sizes, and the order of a coastline's parts, one tends to be left with pieces that will not fit together.

NONCONSTRAINED AND SELF-CONSTRAINED CHANCE

Thus we hit immediately upon an informal distinction of great practical impact. Sometimes our controller followed by a processor may go through their loops without having to inspect the earlier loops' effects, because there is no fear of a resulting mismatch. One can say such models involve a *nonconstrained* form of chance. Otherwise, late stages of the construction are constrained by the outcome of earlier stages, and/or chance is *strongly self-constrained by the geometry of space.*

To exemplify the contrast, the 2n-sided polygons on a lattice, including the self-intersecting ones, raise an easy problem of combinatorics. And one can generate such a polygon by nonconstrained chance. But coastlines must *not* self-intersect, and counting the numbers of polygonal approximations of coastlines is a problem of strongly self-constrained chance, that continues to elude the best minds.

Since the problems involving self-constrained chance are hard ones, they are avoided except in Chapter 36.

HYPERBOLIC RANDOM VARIABLES

A nonuniform random variable X is simply a monotone nondecreasing function $x=F^{-1}(u)$. The inverse function $U=F(x)$ is called the probability $Pr(X<x)$. (Discontinuities in $F(x)$ or $F^{-1}(u)$ require careful wording.)

The expression $Nr(U>u) \propto u^{-D}$ stars in Chapters 6, 13, and 14. Its probabilistic counterpart, $Pr(U>u) \propto u^{-D}$, is called hyperbolic distribution and stars in many of this Essay's remaining chapters. The property that $Pr(U>0)=\infty$ is curious but must *not* provoke panic. It turns out to be just as desirable and manageable as $Nr(U>0)=\infty$ was in Chapter 13. It will have to be handled carefully, but the technicalities can and will be avoided.

A RANDOM SET'S TYPICAL D AND D_T

When a set is random, the notions of dimension demand elaboration. In our "grand portfolio" that brings together a population of random sets, each page is a set, hence has values of D and D_T attached to it, as in earlier chapters. These values vary between samples (= pages), but in all the cases we encounter their distribution is simple.

There is a batch of aberrant samples ("defective seeds") for which D takes all kinds of values, but this batch has a vanishing overall probability. All other samples are characterized by some common D called "almost sure value."

I believe the same holds for D_T and hope that the topic will draw the attention of the mathematicians.

The almost sure values are in every way "typical" of the population. For example, the expected value of D is identical to the almost sure value.

On the other hand, one should avoid even thinking of the dimension of the "average set." For example, assuming the reader has a mental picture of a symmetric random walk, let us try to define the average walk. If it is a process whose positions are the averages of all the walks in a population, then the average does not walk but sits: it never leaves its initial position, hence D=0, ◁ while for almost every walk, Chapter 25 implies that D=2. ▶ The only average set that is "safe" for purposes of handling dimensions is the set characterized by the average D; this definition is safe because it is circular.

Any method applicable to nonrandom fractals can serve to evaluate D. But recall a warning made in Chapter 13: when the portion of a fractal set contained within a ball of radius R centered on the set tends to have a measure ("mass") satisfying $M(R) \propto R^Q$, the exponent Q need not be a dimension.   ■

22 ⋈ Conditional Stationarity and Cosmographic Principles

The preceding chapter's retelling of the usual reasons for favoring randomness does not distinguish between the standard and the fractal models. In the former context, randomization brings considerable improvements, but nonrandom models remain acceptable for many purposes. Let us now show that in the fractal context, randomness is necessary for a model to be really acceptable.

TRANSLATION INVARIANCE, SYMMETRY

The argument involves the old philosophical notion of *symmetry*. It is *not* understood here as in "bilateral" symmetry with respect to a line, but in a combination of the original meaning of $\sigma\upsilon\mu\mu\epsilon\tau\rho\iota\alpha$ in Greek, as "resulting from the commensuration of the various constituent parts with the whole" (Weyl 1952), and of the physicists' current use, which makes symmetry a synonym of invariance.

The nonrandom fractals' essential failing is that they are not symmetric enough. A first failing, stated in the vocabularies of different sciences, is that it is inconceivable for a nonrandom fractal to be translationally invariant, or stationary, and that it cannot satisfy the cosmological principle.

Second, a nonrandom fractal cannot be uniformly scaling, in the sense that it only allows for a discrete scale of similarity ratios of the form r^k.

The problem of galaxy clusters is so important that the present discussion will center around it, making this chapter the second stage of this Essay's contribution to astronomy.

THE COSMOLOGIC PRINCIPLE

The postulate that time and our position on Earth are neither special nor central, that the laws of Nature must everywhere and always be the same, is called *cosmologic principle*.

This assertion, formalized by Einstein and E.A. Milne (North 1965, p. 157), is discussed at length in Bondi 1952.

STRONG COSMOGRAPHIC PRINCIPLE

A brutal application of the cosmologic principle demands that the distribution of matter follow *precisely* the same laws regardless of the system of reference (origin and axes) used to examine it. In other words, the distribution must be translationally invariant.

One must be careful in selecting a term to denote this corollary. Since it does not deal with theory (λογος), but with description (γραφη), and since we shall momentarily propose series of a weakened versions, it is best to speak of the *strong cosmographic principle.*

The underlying idea can already be read into the doctrine of "learned ignorance" of Nicholas of Cusa (1401–1464): "Wherever one is, one thinks is the center;" "The world has its center everywhere, and thus nowhere, and its circumference is nowhere."

COSMOGRAPHIC PRINCIPLE

However, the distribution of matter is *not* strictly homogeneous.

The most obvious weakening of the principle is to introduce chance, in the standard framework described in the preceding chapter. The resulting assertion is called *statistical stationarity* by probabilists, but for the sake of consistency we shall call it *uniform statistical cosmographic principle*: The distribution of matter follows the same *statistical* laws regardless of the system of references.

A QUANDARY

The application of the above principle to galactic clustering poses hard problems. The Fournier universe of Chapter 9 is of course grossly nonhomogeneous, but one may have hoped to be able to randomize it in order to satisfy the uniform statistical cosmographic principle. To preserve the model's spirit, however, the randomization must preserve the property that the approximate density $M(R)R^{-3}$ in a sphere of radius R tends toward 0 when R tends toward infinity. Unfortunately, this last feature and the uniform statistical cosmographic principle are incompatible.

It is tempting to attach less weight to mere data than to a general principle, and to conclude that hierarchical clustering must end at a finite upper cutoff, so that all fluctuations are local in extent, and the overall density of matter is nonzero, after all.

To implement this idea, one may for example take infinitely many Fournier universes, and scatter them around in statistically uniform fashion. A variant proposed by R. M. Soneira is discussed in Peebles 1980.

CONDITIONAL STATIONARITY

However, I believe that the uniform statistical cosmographic principle goes beyond what is reasonable and desirable, and that it should be replaced by a weaker form, to be called *conditional,* which does not refer to *all* observers, only to *material* ones. Astronomers should find this weaker form acceptable, and might have studied it long ago had they known it had the slightest substantive interest. And indeed it does: the conditional form implies no assumption concerning the global density, and it allows $M(R) \propto R^{D-3}$.

To restate my point less assertively, it is either difficult or impossible to reconcile the strong cosmographic principle with the notion that the actual galaxies' distribution is extremely far from uniform. On the one hand, if the global density of matter δ in the Universe vanishes, the strong cosmographic principle must be wrong. On the other hand, if δ is small but positive, the strong cosmographic principle holds asymptotically, but for the scales in which we are interested it is of no use. One may like to keep it in the background, if it is reassuring. One may prefer to avoid it as being potentially misleading. Finally, one may settle on replacing it with a statement that is meaningful for all scales, and is independent of whether $\delta=0$ or $\delta>0$. This last approach amounts to subdividing the strong cosmographic principle into two parts.

THE CONDITIONAL
COSMOGRAPHIC PRINCIPLE

CONDITIONAL DISTRIBUTION. When the frame of reference satisfies the condition that its origin is itself a material point, the probability distribution of mass is called *conditional*.

PRIMARY COSMOGRAPHIC ASSUMPTION. The conditional distribution of mass is the same for all conditioned frames of reference. In particular, the mass $M(R)$ contained within a ball of radius R is a random variable independent of the frame of reference.

The statement of the conditional cosmographic principle involves precisely the same words regardless of whether $\delta>0$ or $\delta=0$. This is esthetically pleasing and has the philosophical advantage of satisfying the spirit of contemporary physics. By subdividing the strong cosmographic principle into two parts, we can highlight a statement that concerns everything that is observable, and we downgrade a statement that constitutes an act of faith or a working hypothesis.

THE AUXILIARY ASSUMPTION OF
POSITIVE OVERALL MATTER DENSITY

AUXILIARY COSMOGRAPHIC ASSUMPTION. The quantities

$$\lim_{R \to \infty} M(R)R^{-3} \text{ and } \lim_{R \to \infty} \langle M(R) \rangle R^{-3}$$

exist, are almost certainly equal, and are positive and finite.

THE STANDARD CASE WHERE $\delta>0$

The statistical laws of distribution of matter can be stated in different ways. One can use the absolute probability distribution, which is relative to an arbitrary frame of reference. Alternatively, one can use the conditional probability distribution relative to a frame centered on a material point. In case the above auxiliary assumption is verified, the conditional probability distribution derives from the absolute distribution by the usual Bayes rule. And the absolute probability derives from the conditional probability by taking the average relative to origins that are uniformly distributed over space.

◁ The uniform distribution of origins integrated over the whole space results in an infinite mass. The nonconditional distribution may be re-normalized to add up to 1, if, and only if, the global density be positive. See Mandelbrot 1967b. ▶

THE NONSTANDARD CASE WHERE δ=0

Suppose to the contrary that the auxiliary assumption is *not* valid, more precisely, that $\lim_{R \to \infty} M(R)R^{-3}$ vanishes. If so, the absolute probability distribution merely states that a ball with a finite radius R chosen at random is almost certain to be empty. Hence, one who could peek around from a point selected at random would almost surely see nothing. However, Man is only interested in the probability distribution of mass in the case of the actual Universe, where it is known that mass does *not* vanish in Man's neighborhood. *After* an event has occurred, its absolute probability of occurrence is of limited interest.

The very fact that the nonconditional distribution automatically disregards such cases implies it is grossly inadequate when δ=0.

Not only is it compatible with mass carried by any fractal satisfying D < 3, but tells absolutely nothing beyond δ=0.

The conditional probability distribution, on the contrary, distinguishes among fractals having different fractal dimensions, among fractals that are or are not scaling, and among other alternative assumptions.

NONSTANDARD "NEGLIGIBLE EVENTS"

The nonstandard case δ=0 faces the physicist with an almost sure event which *can* be disregarded, and an event of zero probability which not only *cannot* be disregarded but must be analyzed into finer subevents.

This contrast is precisely inverse of the one to which one is accustomed. The average number of heads in an increasing sequence of tosses of a fair coin may fail to converge to ½, but the cases of nonconvergence are of zero probability and *therefore* devoid of interest. When a statistical mechanics conclusion (such as the principle of the increase of entropy) holds almost surely, the opposite conclusion has a vanishing probability and is *therefore* negligible. Clearly, the *therefore* in the preceding two sentences yields the precise opposite of what I propose in cosmography.

AVOIDANCE OF STRATIFICATION

A second form of symmetry concerns scaling. When the reduction ratios of the parts of a nonrandom fractal all equal r, the admissible scaling ratios are of the form r^k. When the parts' reduction ratios are r_1, r_2..., the admissible overall ratios are less restricted, but still cannot be chosen freely.

In other words, nonrandom fractals embody a strong hierarchical structure or, as I prefer to say, are *strongly stratified*. Some stratified models look good to the physicists, because they are very manageable computationally. Nevertheless, this characteristic is philosophically unpalatable, and in the case of galaxies there is no direct evidence of the clusters' reality. This is why the call is heard, notably in de Vaucouleurs 1970, for "the extension of Charlier's work to quasi-continuous models of density fluctuations that would replace the original oversimplified discrete hierarchical model."

This desire cannot be fulfilled by a nonrandom fractal, but random ones can fulfill it, as I shall show.

NONSTRATIFIED CONDITIONALLY COSMOGRAPHIC FRACTAL WORLDS

As previously indicated, astronomers are unlikely to object a priori to the idea of conditioning, and this idea would be commonplace, were it acknowledged as having consequences worthy of attention. I propose to prove that it is indeed an authentic generalization, and not merely a formal refinement with this goal in mind, Chapters 32 to 35 describe explicit constructions with the following properties:

- They induce a zero global density.
- They satisfy the conditional statistical cosmographic principle.
- They fail to satisfy any other form of the cosmographic principle.
- They are scaling with respect to every r.
- They are not stratified by design, but instead induce an apparent hierarchical structure as a corollary of a dimension < 2.
- Finally, they fit the quantitative data.

All these properties but the last are satisfied by every one of my models. As to the quantitative fit, it improves from Chapter 32 to Chapter 35. Thus, it suffices to order my models naturally by increasing complication, to achieve increasingly perfect fit to the best analyses of the data.

PREVIEW

Having hailed the splendid vistas opened by thoroughly random fractals, we cannot rush to contemplate those models, because they exhibit mathematical complications that are best postponed. Chapters 23 to 30 keep to comparatively familiar probabilistic ground. ■

23 ¤ Random Curds:
Contact Clusters and Fractal Percolation

This group of chapters shows that diverse devices of almost ridiculous simplicity lead to effective random fractals. Chapter 23 randomizes curdling, a procedure used to rough out a Cantor model of noise (Chapter 8), a spatial Cantor dust model of galaxies (Chapter 9), one of turbulent intermittency (Chapter 10), etc. Chapter 24 is primarily meant to introduce my squigs, a new randomized form of Koch curve. Chapter 25 concerns Brownian motion and Chapter 26 defines other "random midpoint displacement" fractals.

The term "stratified" in the title of this group of chapters expresses that in all these case studies we deal with fractals constructed by a superposition of layers (= *strata* in Latin), each involving finer detail. In many cases, the strata are hierarchical. Without saying so, all the earlier chapters deal exclusively with stratified fractals. But later chapters establish that random fractals need not be stratified.

The fractals in this chapter involve a grid or lattice, made of intervals, squares, or cubes, each divided into b^E subintervals, subsquares, or subcubes; b is the lattice base.

RANDOMIZED LINEAR DUSTS

The simplest random dust on the line, which may improve upon the Cantorian model of errors of Chapter 8, starts as the simplest form of Cantor curdling: with a lattice of intervals of base b and an integer $N < b$. But, instead of a specific generator, one is given the list of all possible Cantor generators, that is, of all the distinct rows of N full and $b-N$ empty boxes. Each time, one chooses one of these generators at random, with equal probabilities.

Any point P of the curd is defined by a sequence of imbedded "precurd" intervals of lengths $R_k = b^{-k}$. If the total initial mass is 1, each precurd contains the same mass R_k^D. The mass in the interval of length $2R_k$, centered as P, is R_k^D multiplied by a random variable lying between 1 and 2, independent of k.

Observe that D is bound to the sequence $\log(b-1)/\log b$, $\log(b-2)/\log b$, ... This restriction is often inconvenient. More important, the above definition of curdling is awkward to implement on the computer and to manipulate analytically. Since the main virtue of curdling resides in its simplicity, an alternative definition given in the next sections should be preferred. To distinguish this section's definition, let us call it *constrained*. (Mandelbrot 1974f calls it *microcanonical*.)

CURDLED RANDOM LINEAR DUSTS

A better definition of curdling, found in Mandelbrot 1974f, which calls it *canonical*, is obtainable by a sequence of binary random choices, each of them ruled by mere coin tossing. By throws of a coin, the first stage of a cascade decides the later fate of each of b subintervals. When the coin falls on heads, an event of probability $p < 1$, the subinterval "survives" as part of a precurd; otherwise, it dies off. After each stage, the isolated points left between two dead subintervals of any length are erased. They are only a small nuisance, but their plane or spatial counterparts, (isolated lines, etc.), would introduce spurious connections in the set. The expected number of surviving subintervals is $<N> = pb = p/r$.

Then the process resumes with each subinterval, independently of all others.

BIRTH PROCESS FORMALISM. Calling the subintervals "children," and the whole cascade a "family," shows that the distribution of the number of children is ruled by the well-known birth and death process (Harris 1963).

The fundamental result is the existence of a critical value for $\langle N \rangle$: this fact was discovered by Irénée Bienaymé in 1845 (see Heyde & Seneta 1977), and deserves to be called the *Bienaymé Effect*.

The value $\langle N \rangle = 1$ is *critical* in the sense that the number N(m) of offspring present after the mth generation is ruled by the following alternative. When $\langle N \rangle \leq 1$, it is almost certain that the family line eventually dies out, meaning in the present interpretation that the cascade yields an empty set. When $\langle N \rangle > 1$, to the contrary, the family line of each curd has a nonzero probability of extending to an infinite number of generations. In this case, random curdling yields a random dust on the line.

MEANING OF SIMILARITY DIMENSION. The ratio $\log N(m)/\log(1/r)$ being random here, similarity dimension requires fresh thinking. The almost sure relation,

$$\lim_{m \to \infty} \log N(m)/\log(1/r^m)$$
$$= \log \langle N \rangle / \log(1/r),$$

suggests a generalized similarity dimension

$$D^* = \log \langle N \rangle / \log(1/r) = E - \log p / \log r.$$

With this D*, the condition for the exist-ence of a nonempty limit set, $\langle N \rangle > 1$, takes a very sensible form: $D^* > 0$. When $D^* > 0$, one has $D = D^*$. Formally applied when $\langle N \rangle \leq 1$, this formula would yield $D \leq 0$, but in fact the empty set is always of dimension $D = 0$.

IMBEDDED CURDS OF DECREASING D

Let us construct a series of random curds of decreasing dimension D, each imbedded in the preceding one.

A preliminary step is *in*dependent of D: it attaches to each eddy of any order a random number U between 0 and 1. We know (Chapter 21) that all these numbers taken together are equivalent to a single number that measures the contribution of chance. Next, D is selected, and the last written for-mula uses it to yield a probability threshold p. Finally, curdling involves the following "fractal decimation process." Whenever $U > p$, the eddy "dies off" as whey, taking along all its subeddies. When $U \leq p$, the eddy survives, to curdle again.

This method makes it possible to follow all the characteristics of curd, whey, and all oth-er sets of interest as functions of a continu-ously varying dimension. It suffices to hold all the random numbers U fixed, while p decreas-es from 1 to 0, and D decreases from 3 to 0.

Given the curds Q_1 and Q_2 corresponding to the probabilities p_1 and $p_2 < p_1$ and having the dimensions D_1 and $D_2 < D_1$, the transfor-mation from Q_1 to $Q_2 \subset Q_1$ can be called "relative fractal decimation" of relative prob-ability p_2/p_1 and relative dimension $D_2 - D_1$. To perform relative decimation directly, one seeks the eddies of side $1/b$ that belong to Q_1, and one lets them live on with the new probability p_2/p_1. Then one proceeds like-wise with the surviving eddies of side $1/b^2$ etc. If the sequence $Q_1, Q_2,..., Q_g$ is obtained by successive decimations, the relative proba-bilities multiply, and the relative dimensions add...until their sum falls below 0, and Q be-comes empty.

HOYLE CURDLING OF GALAXIES

Constrained curdling has a spatial counterpart that can serve to implement the Hoyle cur-dling model of galaxy distribution, Plates 218 and 219.

NOVIKOV-STEWART TURBULENT DISSIPATION INVOLVES CURDLING

Spatial random curdling also arises unwitting-ly in a very early model of the intermittency of turbulence. Novikov & Stewart 1964 as-sume that the spatial distribution of dissipa-tion is generated by a cascade; each stage takes the precurd of the preceding stage and curdles it further into N pieces smaller in the ratio r. See Plates 220 through 223.

This is a very crude model, even cruder than the model that Berger & Mandelbrot 1963 give for certain excess noises (Chapter 8 and 31). It attracted little favorable attention, and failed to be pursued and developed. But the scorn directed toward it turns out to be unwarranted. My investigations reveal that many features of refined but complicated models are already present in curdling.

CHEESE. The image incorporated in the term *curdling*, and in the term *whey* to denote its complement, should not be taken literally, but the formation of real cheese may result from biochemical instability in the same way Novikov & Stewart curdling is presumed to result from hydrodynamical instability. However, I have no data to tell whether or not any edible cheese is also a fractal cheese.

CONSEQUENCES OF RANDOM CURDS' BEING "IN-BETWEEN" SHAPES

The standard shapes in space for which $D < 3$ (points, lines, and surfaces) are known to have a vanishing volume. The same is true for random curds.

The area of the precurds also behaves very simply. When $D > 2$, it tends to infinity. When $D < 2$, it tends to zero. When $D = 2$, curdling leaves it essentially constant.

Similarly, when $m \to \infty$, the cumulative length of the edges of the precurds tends to infinity when $D > 1$, and to zero when $D < 1$.

These volume and area properties confirm that curds with a fractal dimension satisfying $2 < D < 3$ lie somewhere between an ordinary surface and a volume.

◁ PROOFS. They are simplest when cur-

dling is constrained. The volume of the mth precurd is $L^3 r^{3m} N^m = L^3 (r^{3-D})^m$, which $\to 0$ with the inner scale $\eta = r^m$. For the area, the case $D < 2$ is settled on the basis of an upper bound. The area of the mth order precurd at most equals the sum of the areas of the contributing eddies, because the latter sum also includes subeddy sides that neutralize each other by being common to adjacent curds. The area of each mth order eddy being $6L^2 r^{2m}$, the total area is at most $6L^2 r^{2m} N^m = 6L^2 (r^{2-D})^m$. When $D < 2$, the upper bound tends to 0 with $m \to \infty$, which proves our assertion. In the case $D > 2$, a lower bound is obtained by noting that the surface of the union of mth order eddies contained in the mth order precurd includes at least one square of side r^m and area r^{2m} that is contained in said (m−1)th order precurd and cannot possibly be erased. Hence the total area is at least $L^2 r^{2m} N^{m-1} = (L^2/N)(r^{2-D})^m$, which $\to \infty$ with m. Finally, when $D = 2$, both bounds are finite and positive. ▶

THE D'S OF FRACTALS' SECTIONS: RULE THAT THE CODIMENSIONS ADD

Our next topic is mentioned in several earlier chapters. Now we are ready to tackle it explicitly and fully in a special case.

As background, recall that it is a standard property of Euclidean plane geometry that, if a shape's dimension D satisfies $D \geq 1$, its section by a line, if nonempty, is "typically" of dimension D−1. For example, a nonempty linear section of a square (D=2) is an interval, of dimension 1=D−1. And the linear section of a line (D=1) is a point, of dimension 0=1−1, except when the two lines coincide.

More generally, the standard geometric rules concerning the behavior of dimension under intersection are summarized as follows: If the sum of the codimensions C=E−D is smaller than E, this sum is the codimension of the typical intersection; otherwise, the intersection is typically empty. (The reader is encouraged to check this claim for diverse configurations of planes and lines in space.)

It is fortunate that this rule extends to fractal dimensions. Thanks to it, many arguments about fractals are far simpler than one may have feared. The numerous exceptions must, however, be kept in mind. In particular we saw in Chapter 14 that when a nonrandom fractal $\mathcal{F}$ is cut by a specially positioned line or plane, the section's dimension cannot always be deduced from the dimension of $\mathcal{F}$. But random fractals are simpler from this viewpoint.

THE D'S OF RANDOM CURDS' SECTIONS

To prove the basic rule in the case of fractal curds, consider the traces (squares and intervals) that the eddies and subeddies of the curdling cascade leave upon either a face or an edge of the original eddy of side L. Each cascade stage replaces a piece of precurd by pieces whose number is determined by a birth and death process. Denote by $N_1(m)$ the number of mth generation offspring aligned along an edge of the original eddy. Classical results, already used earlier in this chapter, show that $N_1(m)$ is ruled by the following alternative. When $\langle N_1 \rangle = Nr^2 \leq 1$, that is, $D \leq 2$, it is almost certain that the family eventually dies out, meaning that the edge eventually becomes empty, hence of zero dimension. When $\langle N_1 \rangle > 1$, that is, $D > 2$, the family line of each edge has, to the contrary, a nonzero probability of extending to an infinite number of generations. And the similarity dimension is D−2, due to the almost sure relation

$$\lim_{m \to \infty} \log N_1(m)/\log(1/r^m)$$
$$= \log \langle N_1 \rangle / \log(1/r) = D-2.$$

Two-dimensional eddy traces obey the same argument, after replacement of N_1 by a random N_2 such that $\langle N_2 \rangle = Nr$. When $\langle N_2 \rangle \leq 1$, that is, $D \leq 1$, each eddy face eventually becomes empty. When $\langle N_2 \rangle > 1$, that is, $D > 1$, the similarity dimension is D−1, due to the almost sure relation

$$\lim_{m \to \infty} \log N_2(m)/\log(1/r^m)$$
$$= \log \langle N_2 \rangle / \log(1/r) = D-1.$$

Constrained curdling yields identical conclusions.

As a further confirmation that fractal dimension behaves under intersection in the same way as Euclidean dimension, the intersection of several curdled fractals of respective dimensions D_m, carried by the same grid, satisfies $E-D = \Sigma(E-D_m)$.

THE CURDS' TOPOLOGY: CLUSTERS

Although this disclaimer may become tiresome, the basic inequalities $D < 2$ for galaxies (Chapter 9) and $D > 2$ for turbulence (Chapter 10) are not topological but fractal.

In nonrandom curdling, for $E \geq 2$, Chapters 13 and 14, the designer also controls the topology. Connected plane curds include the Sierpiński carpets ($D > D_T = 1$), and connected spatial curds include the sponges ($D > D_T = 1$) and foams ($D > D_T = 2$). Other curds are σ-clusters or dusts. Thus, when $E = 3$ and $D > 2$, which is the case of interest in the study of turbulence, a nonrandom cascade can yield either $D_T = 0$ (dust) or $D_T = 1$ (curves or σ-curves) or $D_T = 2$ (surfaces or σ-surfaces). When $E = 3$ and $D < 2$, which is the case of interest in astronomy, D_T can be 0 or 1.

A random curdling cascade, to the contrary, amounts to a statistically mixed generator that almost surely imposes a certain determined topology (end of Chapter 21). By its very crudeness, curdling is so simple that it is essential to examine its predictions on this account. The present knowledge combines proven facts with inferences from circumstantial evidence.

CRITICAL DIMENSIONS. The curd's D_T changes discontinuously as D crosses certain critical thresholds, to be denoted by D_{crit}, $D_{2crit}, \ldots, D_{(E-1)crit}$. In other words, mixed curds that split into portions with different values of D_T are almost never encountered.

The most important threshold is D_{crit}. It is, at the same time, an upper bound for the D's such that the curd is almost surely a dust, and a lower bound for the D's such that the curd almost surely separates into an infinite collection of disjoint pieces, each a connected set. For reasons explained in Chapter 13, these pieces are called contact clusters.

The next threshold, D_{2crit}, separates the D's where the curd is a σ-curve from those where it is a σ-surface, etc. If and when the whey's topology becomes of interest, it too may lead to new critical thresholds.

CLUSTERS' DIMENSION. When $D > D_{crit}$, the contact clusters have a fractal dimension $D_c < D$. As D decreases from E to D_{crit}, D_c decreases from E to $D_{cmin} > 1$, then crashes to 0.

SIZE NUMBER DISTRIBUTIONS. $\Pr(\Lambda > \lambda)$, $\Pr(A > a)$, etc. obtain by replacing Nr by Pr in the formulas in Chapter 13.

BOUNDS ON D_{crit} AND D_{2crit}. Obviously, $D_{crit} \geq 1$ and $D_{2crit} \geq 2$. And it is proven in the next section that D_{crit} has an upper bound less than E, showing that the above definitions have actual content.

In addition, tighter lower bounds apply regardless of b. It is shown momentarily that a sufficient condition for $D_T = 0$ is $D < \frac{1}{2}(E+1)$. Hence $D_{crit} > \frac{1}{2}(E+1) > 1$. And a sufficient condition for D_T to be either 0 or 1 is $D < \frac{1}{2}E + 1$. Hence, $D_{2crit} > \frac{1}{2}E + 1 > 2$.

For $E = 3$, we find $D < \frac{1}{2}(E+1) = 2$, which is satisfied (with room to spare) by the Fourier-Hoyle value $D = 1$, and by the empirical galaxy value of $D \sim 1.23$. Thus, a random curd with either D is a dust, as we want it to be.

The condition $D < \frac{1}{2}E + 1$ yields $D < 2.5$ when $E = 3$. This threshold value also happens to be the estimated dimension of the carrier of turbulent intermittency. Past experience with sufficient conditions obtained by crude means suggests that they are rarely optimal. So, it would follow that the curdling model carrier of turbulence is *less than* sheetlike.

DERIVATION OF THE LOWER BOUNDS. Their background resides in the fact stressed in Chapter 13, that curd contact clusters arise where the content of neighboring cells becomes lumped together. Consider therefore the intersection of the curd with a plane perpendicular to an axis, with a coordinate of the form $\alpha b^{-\beta}$, where α and β are integers. We know that, if $D > 1$, this intersection has a positive probability of being nonempty. However, lumping demands an overlap between the partial contributions to the intersection coming from opposite sides of a side of length $b^{-\beta}$. If nonempty, these contributions are statistically independent, hence their overlap is formally of dimension $D^* = E - 1 - 2(E - D) = 2D - E - 1$.

When $D^* < 0$, i.e., when $D < \frac{1}{2}(E + 1)$, the contributions fail to overlap. Hence, the curd cannot possibly contain a continuous curve crossing our plane, and $D_T < 1$.

When $D^* < 1$, i.e. when $D < \frac{1}{2}E + 1$, the overlap, if there is one, cannot contain a curve. Hence, the curd cannot contain a continuous surface crossing our plane, and $D_T < 2$.

When $D^* < F$, with $F > 1$, i.e., when $D < \frac{1}{2}(E + 1 + F)$, the same argument excludes an hypersurface of dimension $D_T = F$.

Granted these results, the remainder of the proof of the above inequalities is straightforward: when the curd contains a curve (or sur-face), any point P on this curve (or surface) is contained in a box of side of the form $b^{-\beta}$, which the curve (or surface) intersects at some point (or curve). One ascertains that it is almost sure no such point (or curve) exists when $D < \frac{1}{2}(E + 1)$ (or $D < \frac{1}{2}E + 1$).

PERCOLATING FRACTAL CLUSTERS

The discussion of topology is best continued using percolation vocabulary. According to the definition in Chapter 13, a shape drawn on a square or a cube is said to percolate if it includes a connected curve joining opposite sides of the square or cube. Percolation is ordinarily tackled in the Bernoulli context discussed in Chapters 13 and 14. But the same problem arises in the context of random fractal. Here we tackle it for random curds.

The basic fact is that, when a shape is a σ-cluster, it percolates if and only if one of its contact clusters percolates. When the contact clusters are fractals and their lengths follow a scaleless hyperbolic distribution, the probability of percolation is independent of the square's side, and does not degenerate to either 0 or 1. In Bernoulli percolation, the "when" in the preceding sentence is satisfied under the narrow condition $p = p_{crit}$. In percolation through fractal curds, the condition broadens to $D > D_{crit}$. This is a considerable difference. Nevertheless, to understand Bernoulli percolation helps us understand curds' percolation, and vice versa.

AN UPPER BOUND ON D_{crit}. Let me argue

that, if $b \geq 3$, D_{crit} satisfies $b^{D_{crit}} > b^E + \frac{1}{2}b^{E-1}$. More precisely, when N is fixed (constrained curdling), this condition makes percolation almost certain. In nonconstrained curdling, this condition insures that failure to percolate has a positive but small probability.

First of all, consider the case of nonrandom N. Under the stronger condition $b^E - N \geq \frac{1}{2}b^{E-1} - 1$, there is no way that any given face between two precurd cells can fail to survive. Even if the worst happens, and all the nonsurviving subeddies crowd along said face, these eddies are so insufficient in numbers that it is sure (*not almost*, but absolutely) that no path becomes disconnected. Under the weaker condition $b^E - N \geq \frac{1}{2}b^{E-1}$, the same result is not absolutely, but almost, certain. The resulting curd is made of sheets surrounding separate gaps filled with whey. Two points of the whey can be linked only when they are in the same gap. The topology is almost surely that of Sierpiński carpet, or of foam; Chapter 14.

With the same condition applied to unconstrained curdling, failure to percolate is no longer an impossibility, but an unlikely event.

Let us examine numerical examples for $E = 2$. When $b = 3$, the weaker and more useful of the above conditions become $N > 7.5$, which has only one solution $N = 8$ (its value for the Sierpiński carpet)! As $b \to \infty$, the above upper bound on D_{crit} gets increasingly closer to 2.

LOWER BOUND TO D_{crit}. When $b \gg 1$, $D_{crit} > E + \log_b p_{crit}$, where p_{crit} is the critical probability in Bernoulli percolation. The background to this bound is that the first stage of random fractal curdling amounts to building a Bernoulli floor with a tile having the probability b^{D-E} of being conducting. If this probability is less than p_{crit}, a floor's being conductive is an event of small probability. And, if it does occur, it is likely to be due to a single string of conductive tiles. The second stage of random fractal curdling builds a Bernoulli floor with the same probability b^{D-E} on each conductive first stage tile. This step is very likely to destroy the percolating link.

As $b \to \infty$, the new bound tends to E, and in its domain of validity ($b \gg 1$), it exceeds the bound $\frac{1}{2}(E+1)$. Thus, $D_{crit} \to E$. ∎

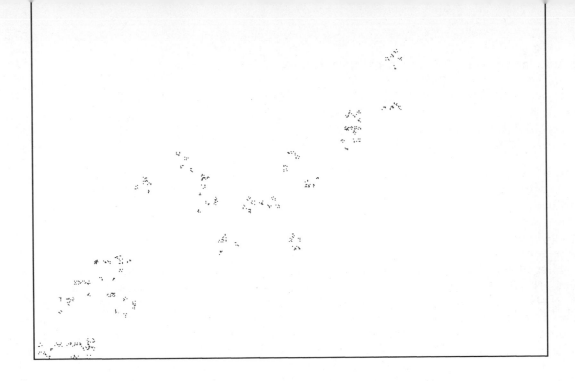

Plates 218 and 219 ◻ IMPLEMENTATION OF HOYLE'S MODEL (DIMENSION D=1) USING RANDOM CURDLING IN A GRID

In Hoyle's model (Chapter 9), a very low density gas cloud collapses repeatedly to form clusters of galaxies, then galaxies, and so on. Hoyle's description, however, is extremely schematic, and actual geometric implementation requires specific assumptions. Those plates show a plane projection of the simplest implementation.

PLATE 219. An initiator cube of side 1 is subdivided into $5^3=125$ subcubes of side 5^{-1}, and so on successively into 125^k subcubes of the kth order, each of side 5^{-k}. In the kth cascade stage, the matter contained in a (k−1)th order subcube collapses into a set of 5 subcubes of the kth order, to be called k-precurd. Hoyle curdling always reduces the dimension from D=3 down to D=1.

In this plate, the first three stages are illustrated in superposition, using increasingly dark shades of gray to represent increasing gas density. Compared to Hoyle 1975, p. 286, this plate may seem crude. But it is carefully

drawn to scale, because questions relative to dimension demand accuracy.

Because we present a plane projection of a curd, it is not rare that two contributing cubes should project on the same square. In the limit, however, the projections of two points almost never coincide. The dust is so sparse as to leave space essentially transparent.

PLATE 218. Here, the fourth stage of curdling (with a different seed) is represented alone. There is little evidence of the underlying grid, which is fortunate, because there is no evidence of such grids in nature (Chapter 27). The top part of the eddy, which is cut by the edge of the page, is empty in this instance.

◁ CONTROL OF LACUNARITY. The notion of lacunarity, presented in Chapter 34, applies directly to random curdling on the line and to Hoyle curdling. If Hoyle's N=5 is replaced by Fournier's "real" value of $N=10^{22}$ (Plate 95), a random curd's lacunarity becomes very small indeed. ► ■

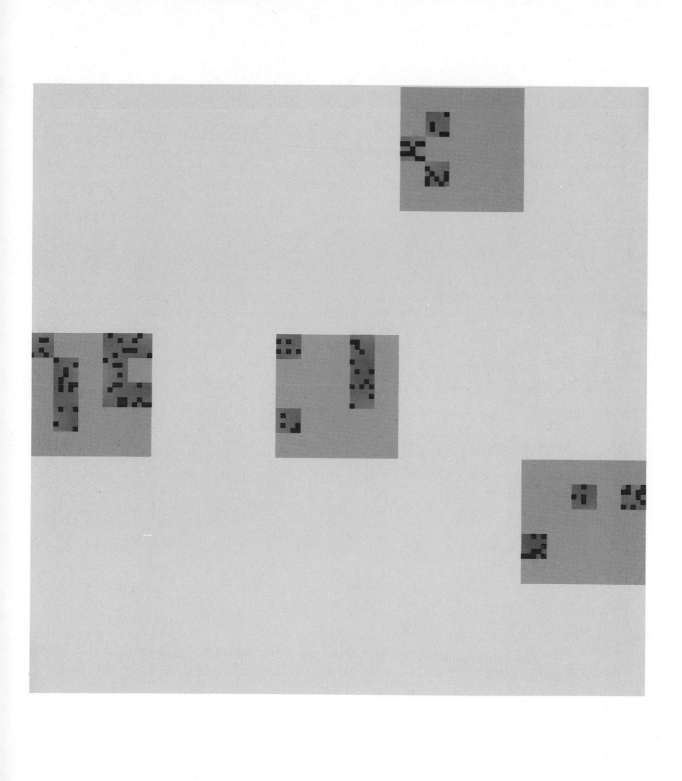

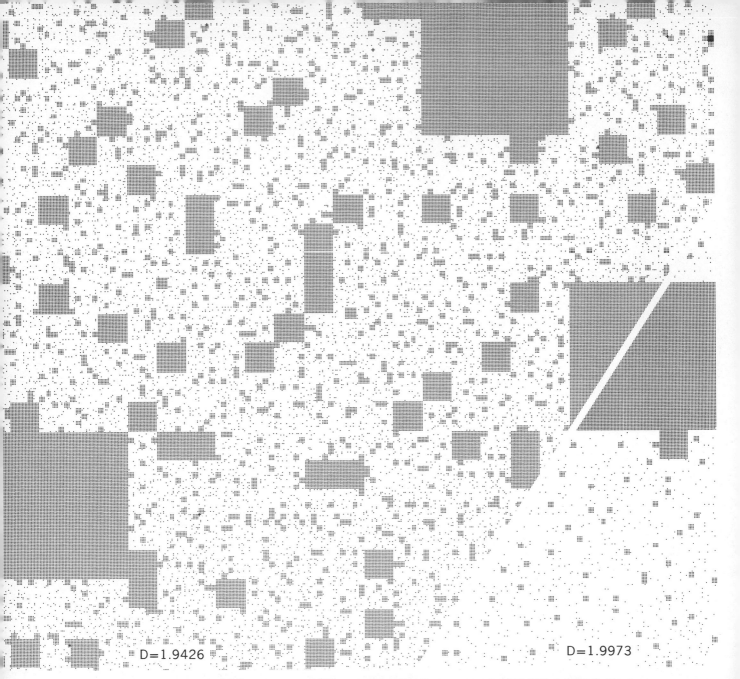

D=1.9426

D=1.9973

**Plates 220 through 223 ¤ NOVIKOV-STEWART RANDOM CURDS IN A
PLANE GRID (DIMENSIONS D=1.5936 TO D=1.9973) FOLLOWED BY PERCOLATION**

The Novikov-Stewart cascade provides a useful general idea of how turbulent dissipation in a fluid curdles into a small relative volume. Conceptually, it is very similar to the Hoyle cascade illustrated in the preceding plates, but the values of the fractal dimension D are very different. For galaxies, D~1, while in turbulence D>2, and D~2.5 to 2.6 is a good guess. The present plates illustrate several different values of dimension, for the sake of a general understanding of

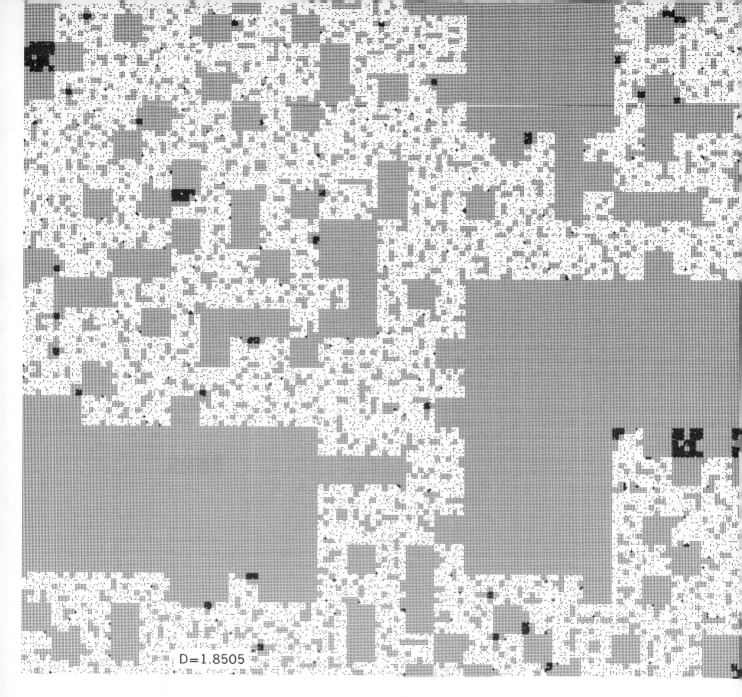

D=1.8505

the process of curdling. Throughout, r=1/5, and

N=5×24, N=5×22, N=5×19, N=5×16, and N=5×13,

respectively. Hence the dimensions take the values

D=1+log 24/log 5=2.9973, D=2.9426, D=2.8505, D=2.7227, and D=2.5936.

The whey is represented in gray, while the curd is drawn in either black or white. The white portion is a percolating contact cluster, namely, the connected portions touching both the upper

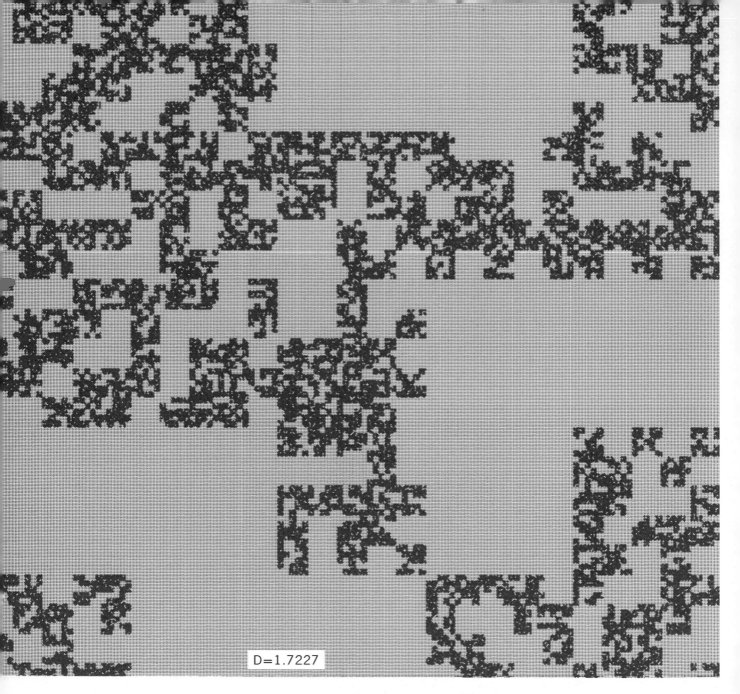

D=1.7227

and the lower sides of the graph. The black portion combines all the other contact clusters.

Because turbulence satisfies D>2, these curds are essentially opaque, and (contrary to Hoyle curds) these plates illustrate their plane cross sections, whose dimensions are

D=1.9973, D=1.9426, D=1.8505, D=1.7227, and D=1.5936.

In Plate 220, the lower-right corner illustrates D~1.9973, a case barren of interesting detail, and the remainder illustrates D~1.9426.

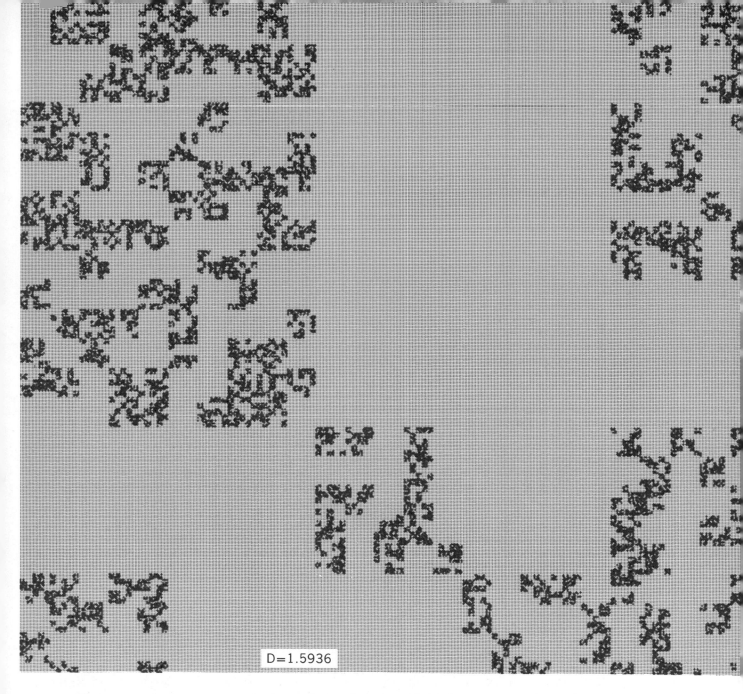

D=1.5936

The generating program and the seed are the same throughout, and one can follow the progressive disappearance of the grays. One began by stacking at random the 25 subeddies of each eddy. Then for successive integer values of $5^D = N$, the top 25−N subeddies in the stack were "grayed away."

For the two smaller dimensions, there is no percolation. For N=19, there is a bit of black and much of white. A few seeds percolate already for N=18. But the numbers of stages in this illustration is to small for reliable estimation of D_{crit}. ∎

24 ¤ Random Chains and Squigs

The preceding chapter shows that curdling can be randomized without disturbing the underlying spatial grid of base b. In random curdling, the "stuff" present in a lattice cell at stage k remains within it forever, while its distribution becomes less and less uniform. This process is very simple, because the evolution of each cell is independent from what happens in other cells. However, the resulting fractals' topology must be allowed to be determined by chance and the properties of space.

The present chapter shows how curdling can be constrained to force the resulting fractal to have specified connectedness properties. For example, a "self-avoiding" curve is needed when the goal is to model a coastline, or river's course. A different example arises in the totally different field of polymer science: an immensely long molecule floating in a good solvent wanders around but is obviously prevented from occupying the same portion of space more than once.

In the recursive methods that insure that the set created by curdling is connected and self-avoiding, the initiator continues to be a plane domain, say a square, and the generator continues to be a collection of smaller domains contained in the initiator. In Chapter 23, the only condition on these smaller domains is that they must not overlap, except that common vertices or sides are *permissible*. In the present chapter, to the contrary, the presence of common vertices or sides is *imposed*.

Common vertices, which are examined first, involve "random chains" that yield a direct generalization of certain Koch or Peano curves.

Common sides turn out to yield a much more attractive and interesting family of fractals, introduced in Mandelbrot 1978r, 1979c. Some are self-avoiding and nonbranching "simple curves", others are loops or trees; and the process extends to surfaces. From now on, I propose to call these new shapes *squigs*.

The main reason for preferring them to random chains is that their being *less* versatile seems to reflect a basic property of space.

Linear squigs are rough models of linear polymers and river courses, looped squigs model coastlines, and tree squigs model river trees.

RANDOM CHAINS AND CHAIN CURVES

The white domains in Plate 43 can be viewed as forming a chain of triangles joined by vertices. The next construction stage replaces every triangle by a substring entirely contained within it and yields a chain made of smaller triangles joined by single points. This sequence of imbedded chains converges to the Koch curve. (The procedure recalls the Poincaré chains of Chapter 18.)

Many other Koch curves can be constructed in this fashion, for example the Sierpiński gasket, Plate 141, whose chain is the shape obtained after removal of the central triangular tremas.

This method of construction is readily randomized, for example a triangle can be replaced either by two triangles with $r=1/\sqrt{3}$, as in Plate 43, or by three triangles with $r=\frac{1}{3}$.

SIMPLEST SQUIGS (MANDELBROT 1978r)

The simplest "squig curve" is a random fractal curve designed in Mandelbrot 1978r, 1979c, and studied further in Peyrière 1978, 1979, 1981. It is a model of a river's course, patterned after the well-known pictures in geology or geography that show the successive stages of a river that burrows into its valley, defining its course with increasing precision.

Before the kth stage of burrowing, the river flows within a "pre-squig" valley made of cells in a regular triangular lattice of side 2^{-k}. Of course, no lattice cell can be visited more than once, and each link in the valley must be in contact with 2 neighbors, through a shared side, while the third side is "locked."

The k-th stage of burrowing replaces this pre-squig by a finer one, drawn on an interpolated lattice of side 2^{-k-1}. Clearly, the pre-squig of order (k+1) necessarily incorporates one half of every side shared between two neighboring links of order k. And a strong converse holds, namely: the position of the shared (unlocked) halves determines the pre-squig of order (k+1) without ambiguity.

SYMMETRICALLY RANDOM SQUIGS. Pick the half side to be locked at random, the alternatives having equal probabilities. The number of links of order k+1 within a link of order k is 1 with a probability of $\frac{1}{4}$, and 3 with a probability of $\frac{3}{4}$. The average number is 2.5.

The valley narrows down at every stage, and it converges asymptotically to a fractal curve. Naturally, I conjectured that the limit is of dimension $D=\log 2.5/\log 2=1.3219$. The proof (which is delicate) is provided in Peyrière 1978.

ASYMMETRICALLY RANDOM SQUIGS. After a side has been split into 2 halves, let $p\neq\frac{1}{2}$ be the probability that the subvalley crosses the "half to the left." One can define this notion with respect to either an observer looking downstream, or an observer standing at the center of the triangle being subdivided. In the first case, $D = \log[3-p^2-(1-p^2)]/\log 2$, which ranges from 1 to $\log 2.5/\log 2$. And in the second case, $D = \log[3-2p(1-p)]/\log 2$, which ranges from $\log 2.5/\log 2$ to log 3/log 2. Altogether, all the D's for 1 to log 3/log 2 are attainable.

ALTERNATIVE LATTICES AND SQUIGS

Alternative squig curves are obtained by using different interpolated lattices. The generalization is straightforward whenever knowing the intervals where a pre-squig of order (k+1) crosses the sides between the cells of order k suffices to identify the pre-squig of order (k+1). An example is the rectangular lattice wherein the ratio of the long to the short sides is of the form $\sqrt{b}$, and cells interpolate into b cells placed across.

But such is *not* the case for triangular lattices whose cells interpolate into $b^2 \geq 9$ triangles, or for square lattices whose cells interpolate into $b^2 \geq 4$ squares. In either case, the interpolation of the pre-squigs requires additional steps.

When b=3 in the case of triangles, or b=2 in the case of squares, one very natural extra step suffices. Consider indeed the 4 "rays" that radiate from a square's center and divide it into 4, or the 6 rays that help divide a triangle into 9. As soon as one of these rays is locked, the subvalley becomes fully determined. In my definition of the squigs, the ray to be locked is chosen at random, with equal probabilities. The D~1.3347 for the triangles split into 9, and D~1.2886 for the squares split into 4. Recalling that the simplest squigs yield D~1.3219, we see that a squig's D is near universal: in the neighborhood of 4/3.

When a triangle is divided into b^2 parts, with b>3, or a square is divided into b^2 parts, with b>2, further decisions are needed to specify the subvalley, and the construction becomes increasingly arbitrary. In the spirit of the next section's discussion, the merits of the squig construction becomes lost.

CHAIN AND SQUIG CURVES, COMPARED

Let us stop to recall that, when a fractal curve is obtained by either the chain method of Cesàro or the original method of Koch, the error committed by truncating the process is very *nonuniform* along the curve. The fact that certain points are attained with infinite precision after a finite number of stages may be advantageous. For example, it helped in Koch's search for the simplest curve devoid of tangent at all points. But the essential meaning of the notion of curve becomes far clearer when the curve is the limit of a strip of *uniform* width. My squig curves satisfy this desideratum.

Another element of comparison involves the number of arbitrary decisions each approach demands from its "designer." The Koch approach to nonrandom or random fractals is very powerful (in particular, achieves any D one may wish, by a simple curve), but it involves on the part of the designer a large number of specific choices for which there is no independent motivation. The base b is especially nonintrinsic.

Science having long suffered from Euclid's

barrenness in models for the unsmooth patterns of nature, the fact that fractals release us from unquestionable inappropriateness was reason to rejoice. But at the present stage of the theory, we must sober up and do with fewer arbitrary decisions.

In this light, the fact that the squig construction is very much constrained by the geometry of the plane (meaning that it is less versatile than the chain model) is a virtue.

THE DIMENSION D~4/3

In particular, the squigs' dimension $D \sim 4/3$ must be kept in mind. The fact that this value is also encountered in Chapter 25 (Plate 243) and Chapter 36 cannot be coincidental, and may eventually lead to basic insights about the geometric structure of the plane.

BRANCHING SQUIG CURVES

Let us return to the construction of a river's course. After a triangular interval of a valley has been replaced by a bit of subvalley made of either 1 or 3 subtriangles, imagine that the remaining 3 or 1 subtriangles drain into the new subvalley. Their pattern of drainage is fully determined. The points where the subrivers cross the divides between triangles are selected by the same system as for the main river. The resulting construction converges to a tree that fills a triangle at random, as seen in the facing column.

TWO LIGHTNING CASE STUDIES

It is interesting and possibly significant that a model as crude as my linear squig curves should suffice to account—albeit only roughly—for rivers' observed dimensions.

And it also yields the dimension of the usual model of highly diluted linear polymers, the self-avoiding random walk on a lattice (SARW) (Chapter 36).

The reason why the constraints due to the geometry of the plane are far easier to manage for the squig curves than for SARW clearly resides in that squigs are constructed by interpolation.

SQUIG SURFACES

They are defined on a cube subdivided into b^3 subcubes, I identified appropriate "locking" procedures to determine uniquely a kind of comforter of constant but decreasing thickness. The algorithm is unfortunately too lengthy to be given here.   ▬

**Plate 228 ⌑ RANDOM KOCH
COASTLINE (DIMENSION D=1.6131)**

In many instances, a Koch curve with prescribed D and no self-contact can be achieved
in several different ways by using the same
overall grid, and the same initiator. Suppose
in addition that at least two different generators can fit within the same overall outline.
Then it is easy to randomize the construction
by selecting among said generators by chance.
For example, one can alternate between the
following generators

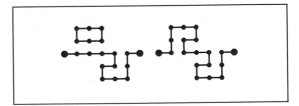

The result is shown above.

The overall form of a random Koch island
constructed in this fashion is very dependent
on the initial shape. In particular, all the initial symmetries remain visible throughout. For
this reason, and other reasons described in
Chapter 24, random shuffling of the parts of a
Koch curve is a method of limited scope. ■

Plate 229 ⊠ RANDOM
PEANO CURVE (DIMENSION D=2)

The following generator, acting on the initiator [0,1], yields a way of sweeping a triangle

N=4
r=½
D=2

The generator's position depends on the parity of the teragon interval. On odd-numbered in-

tervals, the above (straight) generator is positioned to the right. On even numbered intervals, its flipped form (Plate 68) is positioned to the left. The method of randomization used here consists in selecting these focal points at random. In this instance, the distribution is symmetric with respect to the midpoint. Each subtriangle is later subdivided into four, independently of its neighbors, ad infinitum.

To make the teragon easier to follow, each contributing interval is replaced by two, the added end point being the center of this interval's shelter. ▄▄

229

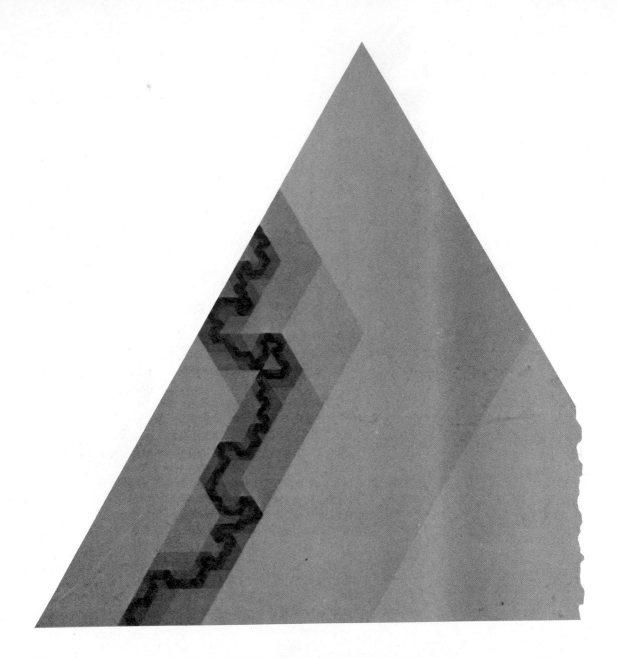

Plate 230 ¤ TRIANGLE & SQUIG CURVE

The simplest squig construction is illustrated here by a superposition of several diagrams, each shade of gray being viewed as continuing under those of darker hue. The illustration begins with the triangle drawn in light gray, and stops with a curve in black. The scale is larger for stages 6 to 10 than for stages 0 to 5. The steps are described in the text.

Plate 231 ⌑ **A HEXA-SQUIG COASTLINE**

This plate strings six squigs together into a self-avoiding loop. The dimension is very close to D=4/3. This value also occurs in numerous other instances of self-avoidance, for ex- ample in the boundary of the Brownian hull of Plate 243, whose resemblance to a hexa-squig is particularly worthy of notice. ■

25 ¤ Brownian Motion and Brown Fractals

The position of the present chapter in this Essay is the child of compromise. It would belong more obviously in the following Part, but some of it is a prerequisite to Chapter 26.

THE ROLES OF BROWNIAN MOTION

As seen in Chapter 2, Jean Perrin had the brilliant idea of comparing physical Brownian motion with continuous nondifferentiable curves. He thus inspired the young Norbert Wiener, around 1920, to define and study a mathematical implementation often called *Wiener process*. Much later, it became known that the same process had been considered in detail, though without rigor, in Bachelier 1900 (Chapters 37 and 39).

Oddly, given its extraordinary importance elsewhere, Brownian motion itself finds no new application in this Essay. On occasion, it helps rough out a problem, but even in those cases the next stage of investigation must supersede it by a different process. However, one can go surprisingly far in many cases modifying Brownian motion, while making sure the modifications remain scaling.

For this and other reasons, other random fractals cannot be appreciated without a thorough understanding of the concrete properties of this prototype. However, the millions of pages devoted to this topic either slight or neglect the issues to be tackled in this chapter. If the reader finds the going becomes rough, he should—as usual—forge ahead to the next section or the next chapter.

BROWN FRACTALS: FUNCTION & TRAIL

Unfortunately, the term *Brownian motion* is ambiguous. It can, first of all, designate the *graph* of B(t) as *function* of t. When B(t) is the ordinate of a point in the plane, the graph is a plane curve like in Plate 241. When B(t) is a point in E-space, the graph is a curve in a (1+E)-space (the time coordinate being added to the E coordinates of B). In many instances, however, one is only interested in the curve in E-space, which a motion leaves behind as its *trail*. When the trail bends at uniformly spaced instants of time, the function and the trail deduce from each other. However, in a continuous Brownian motion, the two aspects are not equivalent, and to designate both by the same term is confusing.

When ambiguity threatens, I use either *Brown function* or *Brown trail*. The same ambiguity exists for Koch curves, but it is more apparent here because of the term "motion."

In addition, the variable of the Brown functions of Chapters 28 to 30 is multidimensional. For example, one of the models of Earth's relief in Chapter 28 assumes that the altitude is a Brown function of latitude and longitude. Therefore, further specification of the terminology is often required. When necessary, we speak of Brown line-to-line, or line-to-space, or space-to-line, or line-to-E-space, etc., functions or trails.

BROWN "FIELDS." A "random field" is not a randomized (algebraic) field, but a fashionable synonym (e.g., Adler 1981) for "random function of several variables." This term cannot be justified, and ought to be banished before it becomes entrenched. It seems an incompetent translation from the Russian, ◁ as *automodel* (whose spread I stopped in time) was an incompetent translation from the Russian word for self-similar. ▶

PLANAR BROWN TRAIL, CONSTRUCTED AS RANDOM PEANO CURVE WITH N=2

The Brown trail casts fresh light on the Peano curves, of which it turns out to be a randomized variant. This construction was not identi-

fied as such by a haphazard group of scholars that I polled, nor is it mentioned as such in a haphazard pile of books on the subject that I scanned. Anyhow, mathematicians shun this approach, because its basic ingredient (a hierarchy of strata with increasingly fine detail, controlled by a dyadic time grid) is not intrinsic to the construction's outcome. Hence, this approach is called artificial by mathematicians, but for this very same reason it fits beautifully in this Essay.

The procedure starts with any Peano curve with $N=2$ and $r=1/\sqrt{2}$. The trick is to release various constraints in successive steps.

The intermediate fractals, "Peano-Brown hybrids," deserve to be studied in their own right on more suitable occasions.

TRANSVERSAL MIDPOINT DISPLACEMENT. In the Plates 64 to 67, the $(k+1)$st stage transforms the kth teragon by displacing each side's midpoint transversally by $|\Delta M| = \sqrt{2}^{-k-1}$, to the left or the right, according to specific rules, e.g., the parity of k.

Now let a Peano curve's displacements over a time span $\Delta t = t^{-k}$, and over its two halves $\Delta_1 t$ and $\Delta_2 t$, be denoted by ΔP, $\Delta_1 P$, and $\Delta_2 P$. We have the Pythagorean identity

$$|\Delta P|^2 = |\Delta_1 P|^2 + |\Delta_2 P|^2.$$

ISOTROPIC DISPLACEMENT DIRECTIONS. In a first departure from any Peano curve's rules, we randomize the displacement *directions*. One approach is to go left or right with equal probabilities, leading to a "random flip-flop curve." A different approach consists in throwing a point at random (with uniform density) on a circle graduated in degrees, and reading off an angle. This procedure defines the displacements as being *isotropic*.

Either form of randomization preserves the Pythagorean identity: the isotropic motion's increments over dyadic subintervals of a dyadic interval are geometrically orthogonal.

RANDOM DISPLACEMENT LENGTHS. Our second departure from the nonrandom rules is to allow the displacement *length* to be random: from now on, 2^{-k-1} will not be the square of a nonrandom $|\Delta M|$, but the mean square of a random $|\Delta M|$. The resulting displacements ΔP^* satisfy

$$\langle |\Delta_1 P^*|^2 \rangle = \langle |\Delta_2 P^*|^2 \rangle = \tfrac{1}{4}\langle |\Delta P^*|^2 \rangle + \langle |\Delta M|^2 \rangle$$

$$\langle |\Delta_1 P^*|^2 + |\Delta_2 P^*|^2 \rangle = \tfrac{1}{2}\langle |\Delta P^*|^2 \rangle + 2^{-k}$$

RANDOM INITIATOR. The next step is to take the initiator itself to be random of mean square length equal to 1. It follows necessarily that $\langle |\Delta P^*|^2 \rangle = 2^{-k-1}$, and we have the mean Pythagorean identity

$$\langle |\Delta_1 P^*|^2 + |\Delta_2 P^*|^2 - |\Delta P^*|^2 \rangle = 0$$

In other words, *geometrically* orthogonal sides are replaced by sides that probabilists call *statistically* orthogonal, or *uncorrelated*.

INDEPENDENT INCREMENTS. The midpoint displacements are made statistically independent, both within and between the stages.

GAUSSIAN INCREMENTS. The randomized Peano curve becomes the Brown trail B(t)

when the midpoints' displacements are made to follow an isotropic Gaussian distribution. ◁ In the plane, this variable's square modulus is exponentially distributed. Hence a direct construction picks U uniformly on [0,1] and draws $|\Delta M| = [-2 \log_e U]^{1/2}$. ▶

GENERALIZATION TO SPACE. The final construction remains meaningful when E > 2.

THE DIMENSION D=2. The mean Pythagorean identity is a generalized definition of similarity dimension. It is suitable for the Brown trail, because the Hausdorff Besicovitch dimension is also equal to 2. Its suitability in case the midpoint displacement is not Gaussian remains to be studied.

BROWN FRACTAL NETS (LATTICES)

MULTIPLE POINTS. Even if randomization stops after the first stage described in the last section, it results in the utter destruction of the exquisite long and short range orders that make the Peano curves avoid self-intersection. The randomized teragons self-intersect after few steps, and the limit trail almost surely self-intersects ceaselessly.

BROWN GAPS. It is widely known that a Brown trail extrapolated for all t's from −∞ to +∞ covers the plane densely. This property will be rederived momentarily. However, a trail drawn during a unit time span has its own most peculiar geometry—which I do not recall seeing described anywhere.

In apparent compensation for points that are covered repeatedly when $t \in [0,1]$, B(t) leaves other points uncovered. The uncovered points form an open set that splits into an exterior set containing the point at infinity, and an infinite number of disjoint *Brown gaps*. The exterior set and each gap are bounded by fractal curves which are subsets of the trail. The Brown trail is therefore a fractal net. Examples are shown in Plates 242 and 243.

◁ Chapter 14 describes nets of dimension D, for which the number of gaps with area U exceeding u is $Nr(U>u) \propto u^{-D/E}$. In a random context with D=E=2, a formal extension is $P(u) = Pr(U>u) \propto u^{-1}$. However, this would not do, because $\int_0^\epsilon P(U>u)du$ must converge. Hence I conjecture that $Pr(U>u) \propto u^{-1}L(u)$, where L(u) is a slowly varying function that decreases fast enough for the above integral to converge. Because of the need for a nonconstant L(u), the dimension D=2 is not achievable in a self-similar ramified net, just as Chapter 15 shows D=2 is not achievable in a self-similar simple curve. ▶

THE BROWN NET'S AREA VANISHES. Despite the value of its dimension, D=2, a Brown net has a vanishing area. The same must be true of the Peano-Brown hybrids.

THE UNBOUNDED TRAIL IS DENSE IN THE PLANE. This property hinges on the fact, to be established in a later section concerned with zerosets, that the unbounded trail "recurs" infinitely often into any prescribed plane domain $\mathcal{D}$, such as a disc. By making $\mathcal{D}$ arbitrarily small and centered on any point P, we see that the unbounded trail comes infinitely often arbitrarily close to every point in the plane.

However, as we shall also see when we examine the zerosets, the probability that an individual trail hits a prescribed point *exactly* is zero, hence a prescribed point is almost surely *not hit* by the unbounded trail.

The portion of an unbounded trail within a domain $\mathcal{D}$ can be mentally approximated by a denumerable infinity of independent bounded nets suitably thrown upon $\mathcal{D}$. The result recalls a denumerable infinity of points thrown at random upon [0,1], independently of one another. As is well-known, the resulting set is everywhere dense, but its length vanishes.

DEPENDENCE OF MASS ON RADIUS

Scaling by $\sqrt{t}$ is characteristic of most aspects of Brownian motion. For example, the distance it covers in time t, measured as the crow flies, is a random multiple of $\sqrt{t}$. Also, the total time spent in a circle of radius R around B(0)=0 is a random multiple of R^2.

Weighting the different pieces of a Brown trail by "masses" proportional to the time it takes to run through them, one finds that, in the plane or in the space (E≥2), the total mass in a circle of radius R is $M(R) \propto R^2$.

Formally, this relationship is precisely the same as in the case of the Koch curve examined in Chapter 6 and the Cantor dust examined in Chapter 8. It is a fortiori the same as in the classical cases of an interval, disc, or sphere of uniform density.

THE BROWN TRAIL IS "CREASELESS," HAS STATIONARY INCREMENTS

As the result of what may be described as a windfall, randomizing the Peano curve achieves more than has been bargained for. As a preliminary comment, observe that the nonrandom Koch and Peano curves exhibit permanent "creases" at the time instants of the form N^{-k}. For example, if we break one-third of the snowflake boundary into quarters, the angle between quarters 1 and 2 differs from the angle between quarters 2 and 3. Hence the left half *cannot* be mistaken for the mid half.

But the Brown trail is "creaseless." Given an interval corresponding to the time span t, one cannot tell this span's position along the time axis. Probabilists say that Brown trail has "stationary increments."

This property is noteworthy because a) it is the foundation stone of the alternative grid-free definition described later in this chapter, and b) it has no counterpart among the analogous randomized forms of simple fractal curves or surfaces.

THE BROWN TRAIL IS SELF-SIMILAR

A corollary of creaselessness is a strong form of statistical self-similarity. Setting $B(0)=0$ and picking two positive numbers h and h', a chapter of probability called theory of weak convergence shows that the functions $h^{-\frac{1}{2}}B(ht)$ and $h'^{-\frac{1}{2}}B(h't)$ are statistically identical. Also, setting $T<\infty$ and $h<1$, and varying t from 0 to T, we find that $h^{-\frac{1}{2}}B(ht)$ is a rescaled form of a portion of B(t). This portion's being statistically identical to the whole is a form of self-similarity.

Self-similarity as applied to random sets is less demanding than the notion introduced in Chapter 5, since the parts need no longer be precisely similar to the whole. It suffices that the parts and the whole reduced by similarity should have identical distributions.

Observe that the Koch curves require similarity ratios of the form $r=b^{-k}$, where b is the base, a positive integer, but any r is acceptable for Brown trail. This feature is valuable.

THE BROWN ZEROSET IS SELF-SIMILAR

Of special importance to the study of Brown functions are the sets of constancy, or *isosets*, of its coordinate functions X(t) and Y(t). For example, the zeroset is defined as those instants t for which $X(t)=0$.

The isosets are self-similar, and the obvious fact that they are extremely sparse is confirmed by their having the fractal dimension $D=\frac{1}{2}$. They are a special case of the Lévy dusts to be investigated in Chapter 32.

BROWN ZEROSETS' GAP DISTRIBUTION. The lengths of a Brown zeroset's gaps satisfy $Pr(U>u)=u^{-D}$ with $D=\frac{1}{2}$. This is the counterpart of the relation $Nr(U>u)=u^{-D}$ we know to be applicable to Cantor gaps. However, Nr is replaced by Pr, and the stairs are eliminated due to randomization.

THE BROWN FUNCTION IS SELF-AFFINE

By contrast, the graphs of X(t) and Y(t), and of the vector function B(t), are *not* self-similar, merely *self-affine*. That is, the curve from t=0 to t=4 can be paved by M=4 portions obtained if the space coordinate(s) continue to be reduced in the ratio r=½, while the time coordinate is reduced in the *different* ratio $r^2=1/M$. Hence, similarity dimension is not defined for the graphs of either X(t), Y(t), or B(t).

Furthermore, affine spaces are such that distances along t and X or Y cannot be compared to each other, hence discs cannot be defined. As a result, the formula $M(R) \propto R^D$ has no counterpart that could serve to define D for the Brown functions.

On the other hand, the Hausdorff Besicovitch definition does extend to them. This example agrees with the assertion in Chapters 5 and 6 that the Hausdorff Besicovitch dimension is the most general way of catching the intuitive content of fractal dimension (and the most unwieldy!). The value of D is 3/2 for X(t), and 2 for B(t).

◁ ROUGH PROOF. During a time span Δt, max X(t)−min X(t) is of the order of √Δt. Covering this subgraph of X(t) by squares of side Δt requires on the order of 1/√Δt squares. Therefore, covering the graph from t=0 to t=1 requires on the order of $(\Delta t)^{-3/2}$ squares. This number being $(\Delta t)^{-D}$ (Chapter 5), it follows heuristically that D=3/2. ▶

THE SECTIONS' FRACTAL DIMENSIONS

The zeroset of the Brown line-to-line function is a horizontal section of a Brown function X(t). Applying again a rule stated in Chapter 23, the zeroset's dimension is expected to be 3/2−1=1/2, as we know is the case. Other applications of this rule are also of extraordinary heuristic value, as we now proceed to show. This rule suffers exceptions, however, especially for fractals that are not isotropic. For example, the section of the Brown line-to-line function by a vertical line is simply a point.

Similarly, a linear section of a Brown line-to-plane trail should have the dimension 2−1=1, and such is indeed the case.

More generally stated, the standard rule is this: excluding special configurations, the codimensions E−D add under intersection. Hence, the codimension of the intersection of k planar Brown trails is k.0=0. In particular, a Brown trail's self-intersections are expected to, and do, form a set of dimension 2. (However, just like the Brown trail itself, the trail's multiple points *fail* to fill the plane.)

The rule of addition of codimensions can be used to argue that (as asserted earlier) Brownian motion almost surely does not return to its point of departure B(0)=0, but almost surely returns infinitely often to the neighborhood of O. To add generality to these arguments, and make them usable again without change in Chapter 27, the dimension of the Brown zeroset will be written as H.

The time instants where B(t) returns to 0

are those when $X(t)=0$ and $Y(t)=0$ simultaneously. Hence, they belong to the intersection of the zerosets of $X(t)$ and $Y(t)$, which are independent sets. The intersection's codimension is $1-2H$, with $H=\frac{1}{2}$, hence their dimension is $D=0$. Hence, the strong hint (but a full proof is more involved!) that $B(t)$ almost surely fails to return to $B(0)=0$.

On the other hand, consider the set of instants when $B(t)$ returns to the horizontal square of side 2ϵ centered on O. This is approximately the intersection of the sets where t is within the distance of $\epsilon^{1/H}$ from a point in the zeroset of $X(t)$, resp., of $Y(t)$. For each of these sets, the mass in the time span $[0,t]$ is $\propto \epsilon^{1/H} t^{1-H}$, and the probability of this span's containing the instant t is $\propto \epsilon^{1/H} t^{-H}$. Hence, the probability of t being contained in these sets' intersection is $\propto \epsilon^{2/H} t^{-2H}$. Since $H=\frac{1}{2}$, we have $\int^{\infty} t^{-2H} dt = \infty$; hence a theorem due to Borel and Cantelli concludes that the number of returns to the square around O is almost surely infinite. But one may call it "barely" infinite. As a result, the gaps in bounded Brownian nets become filled slowly and with seeming reluctance.

DOWNSIZED LATTICE RANDOM WALKS

One can also generate Brownian motion through a random walk on a lattice. We mention this approach here, but diverse complications postpone a discussion to Chapter 36.

A point $P(t)=\{X(t), Y(t)\}$ in $\mathbb{R}^2$ performs a lattice random walk if, at successive instants of time separated by the interval Δt, it moves by steps of fixed length $|\Delta P|$ in randomly selected directions restricted to a lattice.

When the lattice is made of the points in the plane whose coordinates are integers, the quantities $(X+Y)/\sqrt{2}$ and $(X-Y)/\sqrt{2}$ both change by ± 1 at every step. Each is said to perform a random walk on the line; an example is shown as Plate 241. On rough scale, that is, when Δt is small and $\Delta P = \sqrt{\Delta t}$, the walk is indistinguishable from a Brownian motion.

GRID-FREE DIRECT DEFINITIONS OF B(t)

The preceding definitions of Brownian motion begin with either a time grid or with time and space lattices, but these "props" are absent from the final result. And it is possible to characterize the final result without them.

The direct characterization in Bachelier 1900 postulates that, over an arbitrary succession of equal time increments Δt, the displacement vectors $\Delta B(t)$ are independent, isotropic, and random, with a Gaussian probability distribution. Thus,

$$\langle \Delta B(t) \rangle = 0 \text{ and } \langle [\Delta B(t)]^2 \rangle = |\Delta t|.$$

Hence the root mean square of ΔB is $\sqrt{|\Delta t|}$. This definition is independent of the coordinate system, but the projection of $\Delta B(t)$ on any axis is a Gaussian scalar random variable, with zero mean and a variance equal to $\frac{1}{2}|\Delta t|$.

The definition favored by mathematicians goes further and dispenses with the division of time into equal steps. It requires isotropy for the motions between any pair of instants t and $t_0 > t$. It requires independence of future motion with respect to the past position. Finally, it requires the vector from $B(t)$ to $B(t_0)$, divided by $\sqrt{|t_0 - t|}$, to have the reduced Gaussian probability density for all t and t_0.

DRIFT AND THE CROSSOVER TO D=1

The motion of a colloid particle in a uniformly flowing river, or of an electron in a conducting copper wire, can be represented as $B(t) + \delta t$. This function's trail is indistinguishable from that of $B(t)$ when $t << 1/\delta^2$, and from that of δt when $t >> 1/\delta^2$. Thus, the trail's dimension crosses over from D=2 to D=1 for $t_c \propto 1/\delta^2$ and $r_c \propto 1/\delta$. ◁ In the terminology of critical phenomena, δ is the distance from a critical point, and the exponents in the formulas for t_c and r_c are critical exponents. ▶

ALTERNATIVE RANDOM PEANO CURVES

◁ The randomizing of Peano curves through midpoint displacement benefits from exceptional circumstances. Analogous constructions starting with a Peano curve for which N>2 are much more complicated. Also, a closer parallelism with nonrandom scaling is achieved if the midpoint's displacement follows a Gaussian distribution of root mean square equal to $\frac{1}{2}|\Delta B|$, implying that r_1 and r_2 are Gaussian and independent with the more familiar relation $\langle r_1^2 + r_2^2 - 1 \rangle = 0$. The resulting process is very interesting. But it is not Brownian motion. It is not creaseless. ▶

DIMENSION OF PARTICLE PATHS IN QUANTUM MECHANICS

This discussion can close by mentioning a new fractal wrinkle to the presentation of quantum mechanics. Feynman & Hibbs 1965 notes that the typical path of a quantum mechanical particle is continuous and nondifferentiable, and many authors observe similarities between Brownian and quantum-mechanical motions (see, for example, Nelson 1966 and references herein). Inspired by these parallels and by my early Essays, Abbot & Wise 1980 shows that the observed path of a particle in quantum mechanics is a fractal curve with D=2. The analogy is interesting, at least pedagogically. ▰

The longest running (and least demanding!) of all games of chance started around 1700, when the Bernoulli family was ruling over probability theory. When an eternally fair coin comes up heads, Henry wins a penny; when it comes up tails, Thomas wins. (They used to be called Peter and Paul, but I never remembered which one bets on heads.)

Some time ago, William Feller came by to observe this game, and he reported Henry's cumulative wins on the upper Figure of this plate, which is from Feller 1950. (Reproduced from *An Introduction to Probability Theory and Its Applications, Volume I*, by William Feller, by kind permission of the publishers, J. Wiley and Sons, copyright 1950.)

The middle and bottom Figures represent Henry's cumulative winnings during a longer game, using data at intervals of 200 tosses.

When increasingly long sets of data are reported on increasingly fine graph paper, one obtains asymptotically a sample of values of a Brown line-to-line function.

Feller has confided in a lecture that these Figures are "atypical," and were selected in preference to several others that looked too wild to be believable. Be that as it may, seemingly endless contemplation of these Figures played a decisive part in elaborating two theories incorporated into this Essay.

WHOLE GRAPH. Mandelbrot 1963e observes that the whole graph's shape is reminiscent of a mountain's silhouette or of a vertical section of Earth's relief. Through several generalizations, this observation led to the successive models described in Chapter 28.

GRAPH'S ZEROSET. The graph's zeroset is the set of moments when Henry's and Thomas' fortunes come back to what they were when we started reporting them. By construction, the time *intervals* between the zeros are mutually independent. However, it is clear that the *positions* of the zeros are far from independent. They are *very distinctly clustered*. For example, when the second curve is examined in the same detail as the first curve, almost every zero is replaced by a whole cluster of points. When dealing with mathematical Brownian motion, one can subdivide these clusters in a hierarchical manner, ad infinitum.

When asked to help model the distribution of telephone errors, I was fortunate to think of Feller's diagram. Although such errors were known to be grouped in bursts (this being the gist of the practical problem being raised), I suggested that the intervals between the errors might be independent. A detailed empirical study did confirm this conjecture and led to the models discussed in Chapters 8 and 31.

◁ The Brownian zeroset constitutes the simplest Lévy dust, namely, a random Cantor dust of dimension D=½. Any other D between 0 and 1 may likewise be obtained through the zeros of other random functions. Through this model it is possible to define the fractal dimension of a telephone channel. Actual D's depend on the precise characteristics of the underlying physical process. ► ■

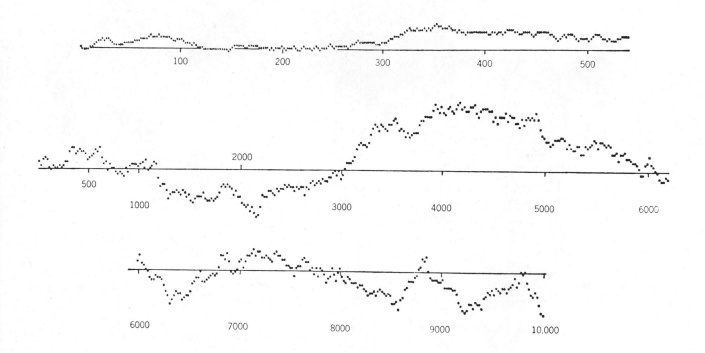

Plate 242 and 243 ☐ BROWN HULLS/ISLANDS; SELF-AVOIDING BROWNIAN MOTION

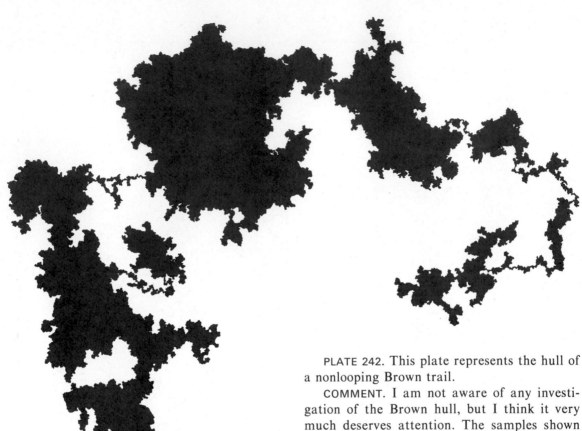

BROWN LOOP. By this term, I denote a trail that is covered in a finite time Δt, by a planar Brownian motion that returns to its point of departure. This is a random Peano curve whose initiator is of zero length.

PLATE 243. BROWN HULL. Being (almost certainly) bounded, a Brown loop separates the plane into two parts: an *exterior* which can be reached from a distant point without intersecting the loop, and an interior which I propose to call *Brown hull* or *Brown island*.

PLATE 242. This plate represents the hull of a nonlooping Brown trail.

COMMENT. I am not aware of any investigation of the Brown hull, but I think it very much deserves attention. The samples shown to the right involve 200,000 Brownian steps, each drawn on a raster of $(1,200)^2$.

By construction, Brown hulls corresponding to different values of Δt are statistically identical, except for scale. And there is every reason (short of actual proof) to believe that the fine details of the hull's boundary are asymptotically self-similar. The boundary cannot be strictly scaling, because a loop cannot be subdivided into pieces having the same structure, but small subpieces become increasingly close to scaling.

SELF-AVOIDING BROWNIAN MOTION. For reasons detailed in Chapter 36, when we examine the self-avoiding random walk, I propose for the Brown hull's boundary the term *self-avoiding Brownian motion*.

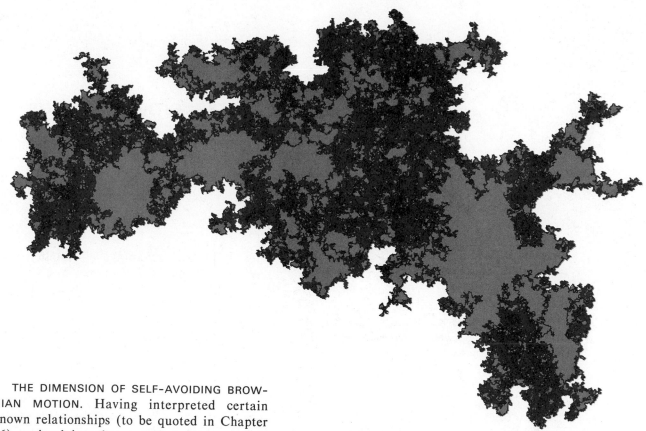

THE DIMENSION OF SELF-AVOIDING BROW-
NIAN MOTION. Having interpreted certain
known relationships (to be quoted in Chapter
36) as implying that a self-avoiding random
walk is of dimension 4/3, I conjecture that
the same is true of self-avoiding Brownian
motion.

An empirical test of this conjecture provides an excellent opportunity to test also the
length-area relation of Chapter 12. The plate
is covered by increasingly tight square lattices, and we count the numbers of squares of
side G intersected by a) the hull, standing for
G-area, and b) its boundary, standing for
G-length. Graphs relating G-length to G-area,
using doubly logarithmic coordinates, were
found to be remarkably straight, with a slope
indistinguishable from $D/2 = (4/3)/2 = 2/3$.

The resemblance between the curves in
Plates 243 and 231, and their dimensions, is
worth stressing.

NOTE. In Plate 243, the maximal open domains that B(t) does not visit are seen in
gray. They can be viewed as tremas bounded
by fractals, hence the loop is a net in the
sense of Chapter 14.

◁ The question arises, of whether the
loop is a gasket or a carpet from the viewpoint
of the order of ramification. I conjectured
that the latter is the case, meaning that
Brown nets satisfy the Whyburn property, as
described on p. 133. This conjecture has been
confirmed in Kakutani & Tongling
(unpublished). It follows that the Brown trail
is a universal curve in the sense defined on
page 144. ► ■

26 ¤ Random Midpoint Displacement Curves

This chapter's logical thread starts back in the middle of Chapter 25, after the section where Brownian motion is generated by randomizing a Peano curve.

Recall that the kth teragon of a Brownian B(t) is linear between successive instants of the form $h2^{-k}$. And that the (k+1)st teragon is obtained by displacing at random the midpoints of the kth teragon's sides. The same words apply to the teragons $X_k(t)$ and $Y_k(t)$ of the coordinate processes $X(t)$ and $Y(t)$ of B(t).

The midpoint displacement procedure being completely successful for D=2, one can hardly wait to adapt it to the original snowflake and to other Koch curves with N=2 and then to use it to construct surfaces. This is what we now proceed to do.

The same general approach has been taken by numerous authors of computer generated films and graphics who attempted to duplicate and improve the graphics in the 1977 *Fractals*, and in addition sought a more direct and less costly procedure. These authors failed to recognize that the method of random midpoint displacement yields a result substantially different from the goal they were seeking. It has the advantage of simplicity, but also many undesirable features.

SPATIALLY UNCONSTRAINED RANDOM KOCH CURVES WITH TIME GRID

Recall that one can construct the Koch snowflake curve in the base N=2, using a generator made of two intervals of length $1/\sqrt{3}$. In this case, and more generally whenever the generator is made up of two intervals of length $2^{-1/D}$, with D<2, the construction tells whether to displace the midpoint of the kth teragon's sides to the left or to the right. The displacement is always orthogonal to the side and its length squared is given by

$$2^{-2(k+1)/D} - 2^{-2(k/D+1)}.$$

The randomization of this construction proceeds as the transformation of a Peano curve into a Brownian motion. The displacement's direction is made random and isotropic, independently of anything that came before, the displacement length's distribution is made

Gaussian, and the above formula is made to apply to the *mean* square displacement. Nothing is done to prevent self-intersection, and the limit fractal curve is rife with self-intersections. We denote it as $B^*_H(t)$, using the notation $H = 1/D$, which will be justified momentarily.

As a result, the relation between the displacement ΔB^*_H over the time space 2^{-k} and the two interpolated displacements $\Delta_1 B^*_H$ and $\Delta_2 B^*_H$ now takes the form

$$\langle |\Delta_1 B^*_H|^D + |\Delta_2 B^*_H|^D - |\Delta B^*_H|^D \rangle = 0,$$

with an arbitrarily prescribed $D < 2$.

A corollary is that when the time interval $[t', t'']$ is dyadic, that is, if $t' = h2^{-k}$ and $t'' = (h+1)2^{-k}$, we have

$$\langle |\Delta B^*_H|^2 \rangle = \Delta t^{2/D} = |\Delta t|^{2H}.$$

We selected H as parameter because it is the exponent of the root mean square displacement.

It can also be shown that, if $B^*_H(0) = 0$, the function $B^*_H(t)$ is statistically self-similar with respect to reduction ratios of the form 2^{-k}. This is a desirable generalization of what we know for $D = 2$.

NONSTATIONARY INCREMENTS

We must not rejoice too hard, however. Except in the Peano-Brown case $D = 2$, when it reduces to $B(t)$, $B^*_H(t)$ is *not* statistically self-similar with respect to reduction ratios other than 2^{-k}.

A more serious problem develops whenever the interval $[t', t'']$ is nondyadic though of the same length $\Delta t = 2^{-k}$, for example, if it is the interval from $t' = (h-0.5)2^{-k}$ to $t'' = (h+0.5)2^{-k}$. Over such intervals, the increment ΔB^*_H has a *different* and smaller variance, dependent on k. A lower bound to this variance is $2^{1-2H}\Delta t^{2H}$. Moreover, if one knows Δt but not t, the distribution of the corresponding ΔB^*_H is not Gaussian, but is a random mixture of different Gaussian.

As a result, the creases that characterize the dyadic points of the approximating teragon remain forever. With D barely below 2, hence H barely above ½, the creases are slight. However, with H nearly 1 (Chapter 28 shows that modeling of Earth's relief involves $H \sim .8$ to .9), the creases are very important and can be seen on the sample functions. The only way to avoid them is to give up the recursive midpoint displacement scheme, as we do in the next section and in Chapter 27.

RANDOMLY POSITIONED STRATA

◁ To trace the reason for the nonstationarity of the midpoint displacement curves and surfaces, consider the coordinate function X(t) of a curve $B^*_H(t)$. Each stage contributes a broken line function $\Delta_k X(t) = X_k(t) - X_{k-1}(t)$ whose zeroset a) is periodic of period 2^{-k}, and b) includes the zeroset of $\Delta_{k-1} X(t)$. Thus, each contribution can be said to be in synchrony with all the following ones.

◁ The fact that the zerosets are periodic and synchronous ("hierarchical") prevents the increments from being stationary. Conversely, one may seek stationarity by destroying these features.

◁ One approach is to construct the broken-line function $\Delta B_k^\dagger(t)$ as follows. Select a Poisson sequence of time instants $t_n^{(k)}$, with an average number of points per unit time equal to 2^k, then let the $\Delta B_k^\dagger(t_n^{(k)})$ be independent and identically distributed random values, and finally interpolate linearly between the $t_n^{(k)}$. The infinite sum $B_H^\dagger(t)$ of such contributions is a stationary random function, pioneered in the Ph.D. dissertation of the hydrologist O. Ditlevsen (1969). See Mejía, Rodriguez-Iturbe & Dawdy 1972 and Mandelbrot 1972w.

◁ Looking back, we see that this generalization no longer requires the average number of zeros per unit time to be 2^k. It may be of the form b^k, with b any real base >1.

◁ The admissible reduction ratios of the corresponding fractal are given by the discrete sequence $r=b^{-k}$. As $b \to 1$, this sequence becomes increasingly tight, and asymptotically it becomes, in effect, as good as continuous. Thus, $B_H^\dagger(t)$ becomes increasingly acceptable to those who seek stationarity and a wide choice of scaling ratios. But in the process $B_H^\dagger(t)$ loses its specificity. The argument in Mandelbrot 1972w implies that $B_H^\dagger(t)$ converges to the random function $B_H(t)$ studied in next chapter. ▪

Plate 246 ⬚ THE COMPUTER "BUG" AS ARTIST, OPUS 1

This plate can be credited in part to faulty computer programming. The "bug" was promptly identified and corrected (but only after its output had be recorded, of course!), and the final outcome was Plates 306 to 309.

The change that had been wrought by a single tiny bug in a critical place had gone well beyond anything we had expected.

It is clear that a very strict order had been designed into the correct plates. Here, this order is hidden, and no other order is apparent.

The fact that, at least at first blush, this plate could pass for High Art, cannot be an accident. My thoughts on this account are sketched in Mandelbrot 1981l, and are to be presented fully in the near future. ▪

IX ¤¤ FRACTIONAL BROWN FRACTALS

27 ¤ River Discharges; Scaling Nets and Noises

Moving on to the fractional Brown fractals marks a major turning point of this Essay. Until now, we have kept to fractals that involve grids of time and/or space, with resulting restrictions on a fractal's invariance properties, i.e., on the admissible translations and similarities that map this fractal upon itself.

Such restrictions contradict the second reason for randomizing fractals, as expounded in Chapter 22. Moreover, in most cases of interest they have no physical reality. Chapters 27 to 35, to the contrary, move on to fractals whose translational and scaling invariances are both unrestricted.

This chapter investigates a generalized Brownian motion, to be denoted $B_H(t)$, which Mandelbrot & VanNess 1968 calls fractional Brownian motion (fBm for short). The motivation resides in annual river discharges, but scaling nets and scaling ("$1/f$") noises are also mentioned. And Chapters 28 to 30 investigate related surfaces.

THE IMPORTANCE OF BEING GAUSSIAN

A first feature shared by Chapters 27 to 30, is that they all involve Gaussian processes exclusively. To statisticians, being Gaussian is something extraordinarily special, but I have long ceased to share this view. (See my comments in Chapter 42 on this account.) Nevertheless, Gaussian processes remain a benchmark, and demand to be investigated with great care before one steps beyond them.

NONRECURSIVE DEFINITIONS

Chapters 27 to 30 also share a feature that is not present anywhere else in this Essay.

All the other chapters' constructions, whether random or not, proceed recursively, by adding increasing detail to less detailed shapes obtained earlier in the construction. The resulting fractals' properties are derived from the generating rules.

Now, to the contrary, we begin by declaring certain properties to be desirable, and only after that do we seek generating rules that fulfill our desires. Unfortunately, while the desirable properties are easy to state and *look* simple, the implementing rules are not recursive, in fact are rather disagreeable.

If so, why should we insist upon these properties? The answer is that they include self-similarity and creaselessness, that is, stationarity, which lie at the very heart of science, and also of the theory of fractals.

The relative cost of the "axiomatic" approach in this chapter is especially apparent when its outcome is paralleled by a fractal obtained recursively. For example, anyone investigating a concrete case that calls for a plane fractal curve of dimension D between 1 and 2 may hesitate between a midpoint displacement process from Chapter 26 and a process to be described in this chapter. The former is *not* creaseless, which is a drawback the latter avoids. And the sequence of discrete stages that makes recursive constructions so attractive is in most cases reflected in strata that are meaningless and undesirable.

JOSEPH AND NOAH EFFECTS

The claim made in Chapter 1, that many unsmooth patterns of Nature have long attracted Man's attention, is in many cases difficult to document precisely. But the Bible offers two marvelous exceptions:

...were all the fountains of the great deep broken up, and the windows of heaven were opened. And the rain was upon the earth forty days and forty nights. Genesis, 6: 11–12.

...there came seven years of great plenty throughout the land of Egypt. And there shall arise after them seven years of famine. Genesis, 41: 29–30.

It is hard not to view the story of Noah as a parable about the unevenness of Middle Eastern precipitation, and the story of Joseph as a parable about the tendency of wet and of dry years to cluster into wet periods and droughts. In lectures on *New Forms of Chance in the Sciences* (not published, but sketched in part in Mandelbrot & Wallis 1968 and Mandelbrot 1973f), I pinned upon these stories the terms *Noah Effect* and *Joseph Effect*.

As controllable data confirm, the Biblical "seven and seven" is a poetic oversimplification of reality, and (not so obvious) any appearance of periodicity in actual Nile records is an illusion. On the other hand, it is a well-established fact that successive yearly discharges and flood levels of the Nile and many other rivers are extraordinarily persistent.

This persistence is as fascinating to diverse scholars as it is vital to those involved in the design of dams. For a long time, however, it remained beyond the scope of measurement, hence of analysis. Like every field taking its first step into statistics, hydrology first assumed that every river's successive discharges are independent, identically distributed Gaussian variables, a white Gaussian noise. The traditional second step assumed Markov dependence. Both models, however, are grossly unrealistic. A breakthrough occurred with Mandelbrot 1965h, based upon empirical results in Hurst 1951, 1955. (Hurst's story is told in Chapter 40.)

HURST PHENOMENON. H EXPONENT

Denote by $X^*(t)$ a river's cumulated discharge between the beginning of year 0 and the end of year t. Adjust by subtracting the *sample* average discharge between the years 0 and d, and define R(d) as the difference between the maximum and the minimum of the adjusted $X^*(t)$ as t ranges from 0 to d. After the fact, R(d) is the capacity one should have attributed to a reservoir to insure ideal performance over the d years in question. A reservoir performs ideally if it ends as full as it begins, never empties and never overruns, and produces a uniform outflow. This ideal is obviously unattainable, but R(d) is the basis of a method of reservoir design, due to Rippl, which was to be used for the Aswan High Dam. Hurst realized that one can use R(d) as a tool to investigate the actual behavior of river discharge records. For reasons of convenience, he divided R(d) by a scaling factor S(d) and examined the dependence of R(d)/S(d) upon d.

Under the assumption that the annual discharges follow a white Gaussian noise, the factor S is not significant, and a known theorem shows that the cumulated discharge $X^*(t)$ is approximately a line-to-line Brown function B(t). Hence R(d) is proportional to the root mean square of $X^*(d)$, which is $\propto \sqrt{d}$. This argument yields $R/S \propto \sqrt{d}$ (Feller 1951). The same result holds if the yearly discharges are dependent but Markovian ◁ with a finite variance ▶, or if their dependence takes any of the forms described in elementary books of probability or statistics.

However, the evidence led Hurst to the sharply different and totally unexpected conclusion that $R/S \propto d^H$, with H nearly always *above* ½. The annual discharges of the Nile, being furthest from independent, show H=0.9. For the rivers Saint Lawrence, Colorado, and Loire, H is between 0.9 and ½. The Rhine is an exceptional river, with no Joseph legend and no Hurst phenomenon, and for it H=½ within experimental error. Diverse data are collected in Mandelbrot & Wallis 1969b.

HURST NOISE AS A SCALING NOISE

When a fluctuation or noise X(t) is such that $R/S \propto d^H$, I propose that X(t) be called a *Hurst noise*. Mandelbrot 1975w shows that one must have $0 \le H \le 1$.

Challenged by H. A. Thomas Jr. to account for the Hurst phenomenon, I conjectured it is a symptom of scaling. To define a scaling noise in intuitive fashion, let us recall that any natural fluctuation can be processed to be heard—as implied by the term *noise*. Tape it, and listen to it through a speaker that reproduces faithfully between, say, 40 Hz to

14,000 Hz. Then play the same tape faster or slower than normal. In general, one expects the character of what is heard to change considerably. A violin, for example, no longer sounds like a violin. And a whale's song, if played fast enough, changes from inaudible to audible. There is a special class of sounds, however, that behave quite differently. After the tape speed is changed, it suffices to adjust the volume to make the speaker output "sound the same" as before. I propose that such sounds or noises be called *scaling*.

White Gaussian noise remains the same dull hum under these transformations, hence it is scaling. But other scaling noises can be made available for model making.

FRACTIONAL DELTA VARIANCE

Chapter 21 defines a random function's delta variance as the variance of the function's increment during the time increment Δt. The ordinary Brown function's delta variance is $|\Delta t|$ (Chapter 25). To account for Hurst's $R(d)/S(d) \propto d^H$, with any desired H, Mandelbrot 1965h observes that it would *suffice* that the cumulative process X* be Gaussian with a vanishing delta expectation and a delta variance equal to $|\Delta t|^{2H}$. These conditions determine a unique scaling Gaussian random process. And, the exponent 2H being a fraction, this unique process is entitled to be termed (reduced) *fractional Brown line-to-line function*. For detail and illustrations, see Mandelbrot & Van Ness 1968, Mandelbrot &

Wallis 1968, 1969abc.

Moving from a line-to-line to a line-to-plane $B_H(t)$, an alternative definition by way of desiderata is this: Among the curves of dimension $D = 1/H$ parametrized by time, the trail of $B_H(t)$ is the only one whose increments are Gaussian, stationary with respect to *any* translation, hence "creaseless," and scaling with respect to *any* ratio $r > 0$.

The value $H = \frac{1}{2}$, hence $D = 2$, yields the ordinary Brownian motion, which we know is a process *without persistence* (independent increments). The remaining fBm's fall into two sharply distinct subfamilies. The values $\frac{1}{2} < H < 1$ correspond to *persistent* fBm, whose trails are curves of dimension $D = 1/H$ lying between 1 and 2. The values $0 < H < \frac{1}{2}$ correspond to *antipersistent* fBm.

FRACTIONAL INTEGRODIFFERENTIATION

Having pinpointed a desirable delta variance, it remains to implement it. If one starts with Brownian motion, one must inject persistence. A standard method is to integrate, but it injects more persistence than is needed. By luck, there is a way of achieving only a fraction of the standard effects of integration. When $0 < H < \frac{1}{2}$, the same applies to differentiation. The idea hides in one of the many "classical but obscure" corners of mathematics. It harks back to Leibniz (Chapter 41), and was implemented by Riemann, Liouville and H. Weyl.

As background, recall from calculus that, m being an integer > 0, one transforms the function $x^{1/2}$ into $x^{1/2-m}$ by m repeated differentiations, and into $x^{1/2+m}$ by m repeated integrations (followed in each case by multiplication by a constant). The Riemann-Liouville-Weyl algorithm generalizes this transformation to the case where m is not an integer. And fractional integrodifferentiation of order $1/D-1/2$ applied to Brownian motion yields fBm. Thus, the usual Brownian formula, displacement $\propto \sqrt{\text{time}}$, is replaced by the generalization displacement $\propto (\text{time})^{1/D}$, with $1/D \neq 1/2$. Our goal is reached!

The relevant formulas are given in Mandelbrot & VanNess 1968, and (honest) approximations are described in Mandelbrot and Wallis 1969c and Mandelbrot 1972f.

◁ Here is yet another complication and potential pitfall. The Riemann-Liouville-Weyl algorithm involves a convolution, hence it is tempting to implement it through fast Fourier techniques (fFt). This approach yields a periodic function, hence a function adjusted to have no systematic trend. In investigations of standard time series, detrending hardly matters, because dependence is limited to the short term. In the case of fBm, on the contrary, detrending does matter, to an extent that increases with $|H-1/2|$, and may be very significant. This effect is illustrated, in an expanded context, by comparing diverse pictures of mountains in the next chapter. Plates 264 and 265, being obtained by fFt, show no overall trend, hence mimic mountain tops, while Plate 268, being obtained without shortcuts, shows a clearcut overall trend.

◁ Given the favorable economics of fFt, it is often best to use them anyhow, but one must take a period much longer than the desired sample size, and allow wastage that increases as $H \to 1$. ▶

H > ½: LONG (= INFINITE) TERM PERSISTENCE & NONPERIODIC CYCLES

In the case $H > 1/2$, the vital property of the function $B_H(t)$ is that its increments' persistence takes a very special form: it extends *forever*. Therefore, the link between fBm and the Hurst phenomenon suggests that the persistence encountered in river discharge records is not limited to short time spans (like the term in office of Pharaoh's ministers), but extends over centuries (some are wet, others are dry) and even millennia. The strength of persistence is measured by the parameter H.

Persistence manifests itself very clearly on graphs of increments of $B_H(t)$, and of the yearly river discharges that these increments model. Nearly every sample looks like a "random noise" superposed upon a background that performs several cycles, whichever the sample's duration. However, these cycles are *not* periodic, that is, *cannot* be extrapolated as the sample lengthens. In addition, one often sees an underlying trend that need not continue in the extrapolate.

The interest of these observations is expanded by the fact that analogous behavior is often observed in economics, where economists like to decompose *any* set of data into a trend, a few cycles, and noise. The decomposition purports to help understand the underlying mechanism, but the example of fBm demonstrates that the trend and the cycles may be due to a noise that signifies nothing.

◁ INTERPOLATION. When the ordinary Brown B(t) is known at the instants t_1, t_2,..., not necessarily equidistant ones, the expected values of B(t) between these instants are obtained by linear interpolation. In particular, the interpolate on $[t_j, t_{j+1}]$ depends *solely* on the values of B_H at the instants t_j and t_{j+1}. Quite to the contrary, in all cases $H \neq \frac{1}{2}$, the interpolate of $B_H(t)$ is *nonlinear*, and it depends on all the t_m and all the $B_H(t_m)$. As $t_m - t_j$ increases, the influence of $B_H(t_m)$ decreases, but slowly. Therefore, the interpolation of B_H can be described as being global. The random midpoint displacement curves investigated in Chapter 26 behave very differently, since their interpolates are linear over certain time intervals. This is the crux of the difference between these two processes. ▶

THE FUNCTION'S & THE ZEROSET'S D

The persistence in the increments is synonymous with a graph of $B_H(t)$ being less irregular *at all scales* than the ordinary Brown graph B(t). This is expressed by its dimension being $2-H$. Its zeroset's dimension is $1-H$.

H > ½: FRACTIONAL BROWN TRAILS

When we move on to two-dimensional vector-valued $B_H(t)$, we seek motions whose direction tends to persist at all scales. Persistence includes an appropriately intense tendency, but not an obligation, to avoid self-intersection. Since we also want to preserve self-similarity in the present Essay, we assume that the coordinate functions $X_H(t)$ and $Y_H(t)$ are two fractional Brown line-to-line functions of time, statistically independent with the same parameter H. In this way, one obtains a fractional Brown line-to-plane trail. (Plate 255).

Its fractal dimension is $D = 1/H$; it is at least $1/1 = 1$, as must be the case for a curve, and at most $1/(\frac{1}{2}) = 2$. This last result suggests that the trail of $B_H(t)$ fills the plane less "densely" than the ordinary Brown trail. To confirm this suggestion, we examine bounded and unbounded trails separately.

The effect of H on bounded trails is one of degree. For $H > \frac{1}{2}$ just as for $H = \frac{1}{2}$, a bounded Brown trail is a fractal net pierced by an infinite number of gaps. Strong heuristic considerations suggest that the gaps' areas satisfy $Pr(U > u) \propto u^{-D/E} = u^{1/2H}$.

Furthermore, I investigated empirically the boundaries of bounded trails of varying D, looking for departure from the value of 4/3 which plate 242 reports is observed in the Brownian case. No clearcut departure was found!

On the other hand, the *unbounded* trails are affected by H qualitatively. When a trail starts at O at time 0, its expected number of returns to a small box around O was found to be infinite for the Brown prototype, but it is finite when $H > \frac{1}{2}$. ◁ The reason is that the integral $\int_1^\infty t^{-2H}dt$, derived in the last but one section of Chapter 25, diverges when $H=\frac{1}{2}$, but converges when $H>\frac{1}{2}$. ► When a finite number of fractal nets are superposed upon a box, it becomes covered in less lacunar fashion, but dense covering is almost surely not achieved. The number of superposed lattices is small when H is near 1 and grows to infinity for $H=\frac{1}{2}$.

H < ½. ANTIPERSISTENT
FRACTIONAL BROWNIAN MOTIONS

The fractional Brownian motions with $0<H<\frac{1}{2}$ yield *antipersistent* functions and trails. To be antipersistent is to tend to turn back constantly toward the point one came from, hence to diffuse *more slowly* than the Brown counterparts.

The formula $D=1/H$ is valid only if $E>1/H$. When $E<1/H$, (in particular, in the case of the plane, $E=2$), the fractal dimension attains its greatest conceivable value, $D=E$. We are reminded that the highest possible dimension for a Brown trail is $D=2$, and that this maximum can only be implemented when $E\geq2$. When squeezed into a real line with $E=1$, a Brown trail must accommodate itself to $D=1$. When $H=\frac{1}{3}$, the fBm trail barely fills the ordinary 3-space.

Returning to the plane, $E=2$, dimensional analysis shows that the unbounded trail with $H<\frac{1}{2}$ almost surely visits any prescribed point infinitely often. Thus, contrary to B(t), which fails to measure up to what is expected from

D=2, and fills the plane densely but not completely, any excess of $1/H$ over 2 achieves complete filling. To prove that $B_H(t)$ almost surely returns infinitely often to its point of departure, recall from Chapter 25 that the dimension of the instants of return is $1-2H$, hence is positive when $H<\frac{1}{2}$. The argument extends to points other than 0. Thus, the intersection of an unbounded fractional Brown trail for $H<\frac{1}{2}$ with a box of side 1 is of unit area.

A bounded trail is a net with gaps, but has a positive area (shades of Chapter 15!).

FRACTIONAL BROWNIAN MODEL
OF RIVER DISCHARGE, "MOTIVATED"

Again, the initial motivation for introducing B_H had resided in this geometer's personal experience of which mathematical and graphical tricks are likely to work. I am prepared to argue that a lack of serious motivation in a model that fits and works well is much preferable to a lack of fit in a model that seems well motivated, but scientists are greedy for both. Unfortunately, present "explanations" are contrived, in my opinion, and carry less conviction than the fact to be explained.

To understand why successive yearly discharges of rivers are interdependent, one begins by taking into account the water which natural reservoirs carry over from one season to the next. However, natural storage yields short-term smoothing of the records, and can at best introduce short-term persistence. From the long-term viewpoint, the graph of the cumulative discharge continues in "effect" (as defined in Chapter 3) to be of dimension equal to $3/2$.

To go further, many writers are more prepared than I am to invoke a whole hierarchy of processes, each with its own different scale. In the simplest case, the contributions are additive. The first component takes account of natural reservoirs, the second takes account of microclimatic changes, the third of climatic changes, and so forth.

Unfortunately, an infinite range of persistence demands an infinite number of components, and the model ends up with infinitely many parameters. It remains necessary to explain why the sum of various contributions is scaling.

At one point of the discussion, a function (the correlation) is written as an infinite sum of exponentials. I spent endless hours pointing out that showing this sum to be hyperbolic is no easier than explaining why the original curve is hyperbolic, and arguing that an invocation of possible causes can only be if magical (not scientific) value, as long as it remains empty. What a pleasure it was, therefore, to discover that I had been working alongside James Clerk Maxwell; see the entry SCALING: DURABLE ANCIENT PANACEAS in Chapter 41.

Of course, the practicing hydrological engineer can impose on every process a finite outer cutoff of the order of magnitude of the horizon of the longest engineering project.

OTHER SCALING NOISES. 1/f NOISES

FORMAL DEFINITION. A noise X(t) is to be called *scaling* if X itself or its integral or derivative (repeated, if need arises) is *self-affine*. That is, if X(t) is statistically identical to its transform by contraction in time followed by a corresponding change in intensity. Thus, there must exist an exponent $\alpha > 0$ such that for every $h > 0$, X(t) is statistically identical to $h^{-\alpha}X(ht)$. More generally, and especially in case t is discrete, X(t) is to be called asymptotically scaling if there exists a slowly varying function L(h) such that $h^{-\alpha}L^{-1}(h)X(ht)$ tends to a limit as $h \to \infty$.

This definition requires that one check every mathematical characteristic of X(t) and $h^{-\alpha}X(ht)$. Thus, scaling can never be *proved* in empirical science, and in most instances the scaling property is inferred from a single test that is only concerned with one facet of sameness, for example the distribution of gap lengths (Chapter 8) or Hurst's R/S.

◁ The most widely used test of scaling is based on spectra. A noise is spectrally scaling if its measured spectral density at the frequency f is of the form $1/f^{\beta}$ with β a positive exponent. When β is close enough to 1 to justify $1/f^{\beta}$ being abbreviated into $1/f$, one deals with a "$1/f$ noise." ▶

Many scaling noises have remarkable implications in their fields, and their ubiquitous nature is a remarkable generic fact.   ▰

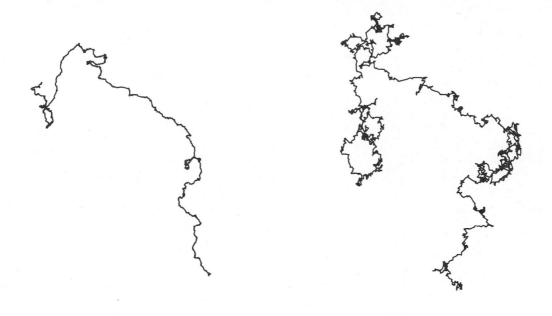

Plate 255 ¤ **FRACTIONAL BROWN TRAILS (DIMENSIONS D~1.1111, D~1.4285)**

The Figure on the left constitutes an example of a statistically self-similar fractal curve with D=1/0.9000~1.1111. Its coordinate functions are independent fractional Brown functions of exponent H=0.9000, which accounts for the Joseph Effect for the Nile. The fact that H is close to 1 does not suffice to prevent self-intersections, but greatly discourages them by forcing the curve's "trend" to persist in any direction upon which it has embarked. Thinking of complicated curves as the super-impositions of large, medium, and small con-

volutions, it may be said that in the case of high persistence and dimension close to 1, small convolutions are barely visible.

The Figure to the right uses the same computer program with D~1/0.7000~1.4285. The pseudo-random seed is not changed, hence the overall shape is recognizable. But the increase in the value of D increases the relative importance of the small convolutions, and to a lesser extent, of the medium ones. Previously invisible details become very apparent. ■■

28 ◻ Relief and Coastlines

This chapter, whose prime exhibits reside in thoroughly artificial pictures that mimic maps and photographs of mountains and islands, proposes to show that mountains like the Alps are usefully modeled in a first approximation by appropriately selected fractal surfaces ruled by Brownian chance. And we encounter, at last, a sensible model of the natural patterns with which this Essay begins, but which have so far eluded us: coastlines.

The point of departure is the notion that mountains' surfaces are scaling shapes. Is this a new idea? Certainly not! It had failed to be stated and explored scientifically, but it is a literary commonplace. For an example to add to the quote that opens Chapter 2, we read on p. 88 of Edward Whymper's *Scrambles Amongst the Alps in 1860-1869* that "It is worthy of remark that...fragments of...rock... often present the characteristic forms of the cliffs from which they have been broken... Why should it not be so if the mountain's mass is more or less homogeneous? The same causes which produce the small forms fashion the large ones: the same influences are at work—the same frost and rain give shape to the mass as well as to its parts."

One need not take Whymper's poetic view literally to agree that it is worthwhile to explore its consequences. In this chapter, I do so within the most manageable mathematical environment I can think of: Brownian and fractional Brownian surfaces.

Even with my first simulations of fractional Brownian mountains (Plates 70 and 71), "to see is to believe." As the quality of the graphics began to improve, so did the quality of belief. But eventually discrepancies between the model and our experience became very clear, and a new model had to be introduced, as is seen in the following chapter.

BROWN RELIEF ON A FLAT EARTH (MANDELBROT 1975w)

We approach the relief by way of the vertical sections. As already indicated in Chapter 4 and Plate 241, one of the sources of this Essay was a feeling reported in Mandelbrot 1963e that a scalar random walk is a rough first approximation of a mountain's cross section.

Hence, I searched for a random surface whose vertical sections are Brown line-to-line functions. The tool box of the builder of statistical models contained no such surface, but a somewhat obscure candidate turned up for adoption.

It is the Brown plane-to-line function of a point, B(P), as defined in Lévy 1948. In order to become familiar with it on short acquaintance and to apply it concretely, there is no substitute for a careful examination of the actual simulation in Plate 264. The Brown imaginary landscape is of fractal dimension $D=5/2$, and it is definitely rougher than most of Earth's relief.

Therefore, it is a crude model, begging to be returned to the bench. But is it not a beautiful long jump forward!

WARNING. DO NOT BE CONFUSED BY THE BROWNIAN SHEET. The proliferation of variants of Brownian motion is endless, and terminology is casual. The Brown plane-to-line function used here must not be confused with the Brownian sheet. The latter is an entirely different process that vanishes along the coordinate axes and is strongly isotropic. See Adler 1981, especially the illustrations found on pp. 185 and 186.

A BROWN RELIEF'S COASTLINES

Let us stop and check for progress in the study of ocean coastlines, defined as zerosets: points located at ocean level, inclusive of points situated on offshore islands. The Brown coastline included in Plate 270 was the first example I encountered of a curve that (a) is devoid of self-intersections, (b) is practically devoid of self-contacts, (c) has a fractal dimension clearly greater than 1, and (d) is isotropic. A more recent variant is included in Plate 267.

More precisely, the dimension is $3/2$. This value being higher than most of Richardson's values from Plate 33, a Brown coastline is of limited applicability. It does recall northern Canada, Indonesia, perhaps western Scotland and the Aegean, and is applicable to many other examples, but certainly not to all. Because of the Richardson data, it would, anyhow, be foolish to expect any single D to apply universally.

GENERATING A BROWN RELIEF (MANDELBROT 1975c)

It is a pity that the simple Brown relief of dimension $D=5/2$ and coastlines of dimension $D=3/2$ do not suffice, because they would be easy to account for. Indeed, the Brown function is an excellent approximation to the "Poisson" relief that is created by superimposing independent rectilinear faults. A horizontal plateau is broken along a straight line chosen at random. Then the difference between the levels on the two sides of the resulting cliff is also chosen at random: for example ± 1 with equal probabilities, or Gaussian. Then we start all over again, and follow the kth stage by division by $\sqrt{k}$ (thus making each individual cliff become negligible in size, compared to the cumulative sum of the other cliffs).

The result obtained by continuing ad infinitum generalizes the usual Poisson process in time. With no need for mathematical or physical details, we can see that the argument seizes at least one aspect of tectonic evolution.

Because of the simplicity of this mechanism, it would be comforting to believe that in some early and especially "normal" state of affairs, Earth had a Brownian relief with $D=5/2$ throughout. But this topic must be withheld for a later section.

GLOBAL EFFECTS IN BROWN RELIEF

Lévy found that the Brown space-to-line function has a property that surprises at first blush and has very direct practical implications. Loosely stated, this property asserts that the different parts of a Brown relief are *far* from being statistically independent. Thus, in order to imbed the Brown line-to-line function in a Brown plane-to-line function, it is necessary to give up one aspect that until now had been the characteristic virtue of Brownian chance: independence of the parts.

Consider two points located, respectively, east and west of a meridian section of the relief. Along the meridian, relief is a Brown line-to-line function, hence "slopes" at different points are independent. Furthermore, one may expect our meridian to act as screen, in such a way that knowledge of the relief at the eastern point does not affect the relief's distribution at the western point. ◁ If such were the case, the relief would be Markovian. ▶ In fact, the west *does affect* the east, meaning that the generative process involves inevitably a strong overall dependence.

This dependence implies that a Brown surface is much harder to construct than a Brown line-to-line function. The random midpoint displacement process of Chapter 25, whose failure to extend to the fractional Brown line-to-line function is documented in Chapters 26 and 27, also fails to extend to the *ordinary* Brown plane-to-line function. That is, one *cannot* proceed by first pinning this function down on a rough grid, and then filling in its values within each cell, independently of the other cells. It is also impossible to construct it layer by layer: first for $x=0$, then for $x=\epsilon$ without regard for its values for $x<0$, then for $x=2\epsilon$, without regard for the values for $x<\epsilon$, etc.

More generally, every algorithm that promises an easy step-by-step generalization of the Brown line-to-line function to "multidimensional time" inevitably turns out to lead to a function that differs systematically from what was intended.

As mentioned in this chapter's last section, the simulations in which I had a part rephrase the unmanageable theoretical definitions in ways that involve successive approximations with known error terms. But I cannot vouch for those who, stimulated by reading my earlier Essays, have joined us in this game.

BROWN RELIEF ON A SPHERE

Next, let the base surface of Earth's relief be a sphere. Fortunately, the corresponding Brown sphere-to-line function $B_O(P)$ has also been provided by my mentor; see Lévy 1959. It is easy to describe, it is fun, and it may even be significant. But we shall see that it is not realistic either, because it too predicts coastlines with $D=3/2$, a serious drawback.

The simplest definition of $B_O(P)$ uses noise theory terms, which we cannot stop to define, but which are familiar to many readers. One lays on the sphere a blanket of white Gaussian noise, and $B_O(P)$ is the integral of this white noise over the hemisphere centered at P.

Within angular distances less than $60°$, $B_O(P)$ looks very much like a Brown plane-to-line function. Globally, however, it does not.

For example, $B_O(P)$ has the striking property that, when P and P' are antipodal points on the sphere, the sum $B_O(P) + B_O(P')$ is independent of P and P'. Indeed, this sum is simply the integral taken over the whole sphere of the white noise used to build $B_O(P)$.

Thus, a big hill at the point P corresponds to every big hole at the antipodal point P'. Such a distribution has a center of gravity distinct from the center of the base surface, and it could hardly be in a stable equilibrium. But we need not worry: it is saved from static instability—hence from early dismissal as a model—thanks to the theory of isostasy. This theory claims that Earth's near-solid crust is very thin at the ocean's deepest points and very thick below the highest mountains, in such a way that a sphere concentric to Earth and drawn a bit below the Ocean's deepest point nearly bisects the crust. After it is agreed that a mountain's visible crest must always be considered in conjunction with its invisible root under the reference sphere, the constancy of $B_O(P)+B_O(P')$ does not cease to surprise, but does not necessarily imply gross static unbalance.

BROWN PANGAEA AND PANTHALASSIA

How well does the above variant of Brown relief fit the evidence? On the basis of today's continents and oceans, D is wrong, hence the fit is poor.

On the other hand, plate tectonics (the theory of continental split and drift) suggests that the test of adequacy be carried out on the primeval Earth as it appeared 200 million years ago. The evidence being flimsier, the test is less certain to fail in this case. Wegener told us, and his account has become accepted (for example, see Wilson 1972), that once upon a time the continents were linked within a *Pangaea,* while the oceans formed a super-ocean, *Panthalassia.*

Like Pangaea, the relief in Plate 269 is a blob of land, dented here and there by broad sinuses. But this first-glance resemblance is misleading. It tends to over-emphasize the very large-scale detail due to the combination of the geometry of the sphere and the fact that on the sphere the Brownian rules of dependence involve a strong positive correlation for angles below 60°, and a strong negative correlation between antipodal points. Under a more attentive second glance focused on less global features, the fit deteriorates; for angles below 30° (say), a Brown coastline on the sphere becomes indistinguishable from a Brown coastline on the plane. All the defects of the latter float back to the surface.

A fractal flake in which the altitude function is the same as in the above Pangaea, but with a scale of the order of magnitude of half the radius looks like one of the irregular moons of the outer planets. In contrast to Plates 10 and 11, it is not accompanied by flotsam or jetsam, hence its D is a measure of irregularity alone and not of fragmentation.

FRACTIONAL BROWN RELIEF ON A FLAT EARTH (MANDELBROT 1975w)

The trouble with either of the above two Brownian models of the relief is that $D=3/2$ is too high for coastlines. As a consequence, our search for a more widely applicable model acquires an unexpected flavor. Long ago, Chapters 5 and 6 determined that $D>1$, and we started looking for ways to force D to rise *above* 1. Now we must squeeze D *below* $3/2$. To obtain less unsmooth coasts, we must have a less unsmooth relief and less unsmooth vertical sections.

Fortunately, the preceding chapter prepared us well. To achieve a model of vertical sections, I replaced the Brown line-to-line function by its fractional variant. Random plane-to-line functions $B_H(P)$ possessing such sections do indeed exist. The D of their surfaces is $3-H$ (Adler 1981), and the D of their level lines and vertical sections is $2-H$.

Therefore, there is no longer any difficulty in modeling and simulating any dimension that the empirical data may require.

DETERMINATION OF D. Richardson's data (Chapter 5) makes us expect coastline dimensions to be "typically" around 1.2, and relief dimensions to be around 2.2. We can therefore go a long way with $H=0.8$—a value that justifies Plate 265. However, other values are needed to account for specific areas of Earth. Values of $D{\sim}2.05$ or so account for relief dominated by very slowly varying components. When this component is a big slope, the relief is an inclined uneven table and the coastline differs from a straight line by no more than mild irregularities. Near a summit, the relief is an uneven cone and the coastline a mildly irregular oval.

Reliefs with a D near to 3 are also potentially useful but hard to illustrate in rewarding fashion. It suffices to observe that in Plate 270, the coastline with D near 3 is reminiscent of a flooded alluvial plain. Therefore, all values of H will find a place in the tool box of the builder of statistical models.

COSMOGRAPHIC PRINCIPLES

The cosmographic principles of Chapter 21 can be rephrased in terms of relief. The strong cosmographic principle combines the probabilistic notions of stationarity and isotropy. Hence the relief $Z(x,y)$ on the flat Earth may be said to be strongly cosmographic, if the rules generating relief are the same in every frame of reference in which the origin (x_0,y_0,z_0) satisfies $z_0=0$ and the z-axis is vertical. In particular, said rules must be left invariant by change in x_0 and y_0 and by rotation of the horizontal axes. My Brown relief on a flat Earth, and its fractional version, both *fail* to satisfy this principle.

But they satisfy a "conditional" version, in which the origin is conditioned to satisfy $z_0 = B(x_0,y_0)$ so that it lies upon Earth's surface.

Attempts to fit the relief by a stationary process have been made. They cover the $z=0$ plane with a regular lattice, and take altitudes within distinct lattice cells to be independent random variables. Such models cannot account for any of the scaling laws examined throughout this chapter.

Brown relief on a spherical Earth fulfills the cosmographic principle in its strong form, which deals usefully with large portions of the Earth, the strong form is the more useful one. A fortiori, the conditional form holds, and it is preferable when dealing with local effects.

THE HORIZON

For an observer sited at a finite distance above Earth's surface, the horizon is made up of the nonhidden points of greatest apparent height, along every direction of the compass.

When the relief is a perturbation upon a spherical Earth, the horizon is obviously at a finite distance from the observer.

When the relief is a Brownian or fractional Brownian perturbation upon a flat horizontal plane, the horizon's existence is not obvious: each mountain might be backed at a distance by a higher mountain, and so forth ad infinitum. In fact, a mountain located at the distance R from an observer has a relative height of the order of R^H, so the tangent of its apparent height in degrees above the horizontal plane is about R^{H-1}, and tends to 0 as $R \to \infty$. Hence, the horizon is again defined.

To gain further insight, divide the distance from observer to the horizon by its average. On a flat Earth, this function is statistically independent of the observer's height. On a round Earth, to the contrary, the horizon tends to a circle as the observer grows taller. Also, a flat Earth's horizon lies *above* a plane passing through the observer, independent of the observer's height. But a round Earth horizon falls *below* such a plane if the observer is tall enough. In summary, the observed properties of the horizon confirm that Earth is round. The opposite conclusion would have been devastating.

FRACTIONAL BROWNIAN MODEL OF EARTH'S RELIEF, "MOTIVATED"

As usual, one wonders why models selected on their virtues of simplicity prove so attractively applicable. I have suggestions, but cannot claim they are convincing (Chapter 42).

◁ First of all, one can construct $B_H(P)$ as was done for $B(P)$, by superimposition of rectilinear faults (Mandelbrot 1975f). However, the faults' profile must no longer be a sharp cliff; its slope must increase as one approaches the fault. Sadly, the appropriate profile is contrived, so this is not a good approach.

◁ It seems preferable to begin with a Brownian model, and then to try and decrease the dimension as Chapter 27 did for rivers. Exclusively local smoothing transforms a surface whose area is infinite into a surface whose area is finite. On the other hand, it leaves large features completely unaffected. Therefore, local smoothing replaces an object having the same well-defined dimension on all scales by an object that exhibits a global effective dimension of 5/2, and a local effective dimension of 2.

◁ More generally, K distinct smoothings having different fundamental scales end up with K+1 zones of distinct dimensions connected by transition zones. However, the whole may become indistinguishable from a fractal of intermediate dimension. In other words, a superposition of phenomena with well-defined scales may mimic scaling.

◁ On the other hand, a scaling phenomenon is often spontaneously analyzed by the mind into a hierarchy in which each level has a scale. For example, the galaxy clusters of Chapter 9 need not be real, as will be shown in Chapters 32 to 35. Therefore, one must not hasten to follow Descartes's recommendation and begin to subdivide every difficulty into parts. While our mind spontaneously analyzes

geomorphological configurations into superpositions of features having sharply distinct scales, these features need not be real.

◁ Fortunately, Earth's relief has an intrinsic finite outer cutoff, because its base surface is round. Therefore, it is safe to assume that the various planings undergone throughout geological history involve spatial scales that stop at the order of magnitude of the continents. The realistic assumption that H varies from place to place allows this planing to vary in relative intensity. ▶

BROKEN STONES, AIRPORT STRIPS, AND TRIBOLOGY

As mentioned long ago, in Chapter 1, I coined *fractal* from the Latin *fractus*, which describes the appearance of a broken stone: irregular and fragmented. Etymology cannot force an actual stone's surface to be fractal, but it is surely not a standard surface, and it should be a fractal if it is scaling.

The argument for scaling is that stone is made of grains stuck together into domains organized hierarchically, bigger domains sticking less strongly together than their smaller components. The energy generated when a stone is hit would dissipate itself easiest by separating big domains, but there is no reason to expect such separation to be allowable geometrically, therefore the break is likely to combine portions belonging to interdomain walls of diverse hierarchical levels.

The science of wear and of friction styles

itself *tribology*, from the Greek $\tau\rho\iota\beta\omega$ = to rub, to grind. The evidence in Sayles & Thomas 1978 (after correction of a flawed analysis; see Berry & Hannay 1978) supports the belief that *fractional* Brown surfaces provide first approximation representations for airport strips, and for many natural rough surfaces. The empirical values of D (deduced from a plot of 7−2D in Sayles & Thomas, Figure 1) range from 2 to 3.

SPATIAL DISTRIBUTION OF OIL AND OTHER NATURAL RESOURCES

Now that my "principle" that the relief is scaling has been tested in various ways, let us examine a corollary. As shown in Chapter 38, we may expect every quantity associated with the relief to follow a hyperbolic probability distribution ("Zipf law", "Pareto law"). Such is indeed often the case. As a matter of fact, my study of coastlines (Chapter 5), which suggested that the relief is scaling, had been preceded by Mandelbrot 1962n, which found the distributions related to oil and other natural resources to be hyperbolic. This finding disagrees with the dominant opinion, that the quantities in question are lognormally distributed. The difference is extremely significant, the reserves being much higher under the hyperbolic than under the lognormal law. My conclusion did not get much hearing in 1962, but I shall try again.

Minerals are discussed again in Chapter 39, in the entry on NONLACUNAR FRACTALS.

SHORTCUTS: PERIODIC SURFACES AND MIDPOINT DISPLACEMENT SURFACES

Since my Brown or fractional Brown reliefs are based on involved algorithms, approximations or shortcuts are needed. Thus, Plates 268, 270 and 271 involve a Poisson approximation to our Gaussian process. And Plates 264 to 267, and C5 to C13 replace a nonperiodic function of x and y by a periodic function computed by fast Fourier methods, then "cropped" to keep to a central portion unaffected by periodicity.

In addition, I used midpoint displacement, as in Chapter 26, to generate fractal surfaces to be denoted by $B^*_H(x,y)$. Such a surface is most easily implemented using as initiator an equilateral triangle $\mathcal{J}$. The values of $B^*_H(x,y)$ being prescribed at the vertices of $\mathcal{J}$, the first stage interpolates this function separately on the 3 midpoints of the sides of $\mathcal{J}$, using the same process as for the coordinate functions of $B^*_H(t)$. The next stage interpolates at 9 second-order midpoints. And so on.

The outcome is more realistic, to be sure, than any nonfractal surface, or most nonrandom fractal surfaces. But is it stationary? $\Delta B^*_H = B^*_H(x,y) - B^*_H(x+\Delta x, y+\Delta y)$ should depend only on the distance between the points (x,y) and $(x+\Delta x, y+\Delta y)$. In fact, the present ΔB^*_H depend explicitly on x, y, Δx, and Δy. Thus, B^*_H is *not* stationary, even if $H = \frac{1}{2}$.

I have also examined and compared a dozen shortcuts that *are* stationary, and some day I hope to publish the comparison.

Plates 264 and 265 ¤ BROWN LAKE LANDSCAPES, ORDINARY AND FRACTIONAL (DIMENSIONS D~2.1 TO D=5/2, PROCEEDING CLOCKWISE)

The figure on top of Plate 265 is an example of fractional Brown relief of dimension fairly close to 2, which is my model of Earth's landscape. The other Figures extrapolate this model to higher D's, ending on top of Plate 264, with an ordinary plane-to-line Brown relief. The latter has as defining characteristic that every vertical cut is an ordinary Brown line-to-line function, as in Plate 241. A Brownian relief is a poor model of Earth because it is conspicuously too irregular in

its detail. The poor fit is quantified by the fact that its surface dimension D=5/2 and its coastline dimension D=3/2 are too large.

For each landscape the attitude is computed for latitudes and longitudes forming a square grid. The computer is programmed to simulate lighting from a source located 60° over the left, while the viewer is located 25° over the base level. For further details, see the captions of the color illustrations. ▄▄

SURFACE D=8/3 ¤ COASTLINE D=5/3

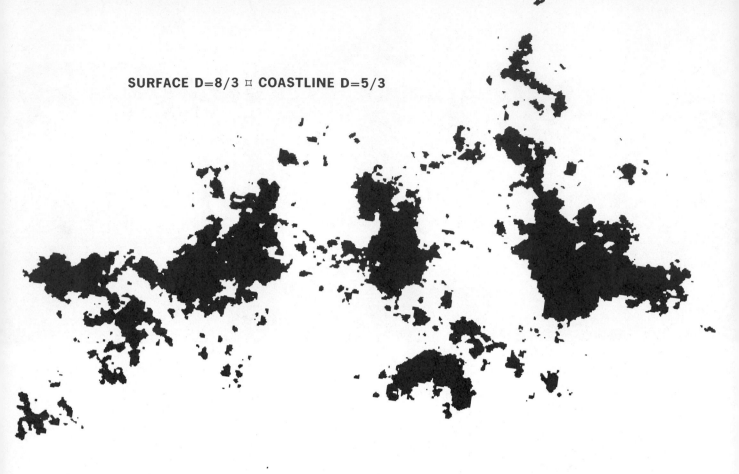

Plates 266 and 267 ¤ BROWN COASTLINES, AND ISLAND "STRINGS"

These plates are primarily meant to underline an important, newly discovered effect. When the relief D reaches and exceeds 2.5, there is a strong and increasing tendency for the ocean to split into roundish separate "seas." These seas intercommunicate, nevertheless each has a sharp individuality. On the other hand, the islands seem to come in "strings." The same effect is also visible (but not quite so clearly) in the ridges present in all the "landscapes": Plates 264, 265, and 271.

This lack of isotropy in the samples is entirely compatible with the fact that the generating mechanism is isotropic.

These plates are equivalent (except for the seed) to planar sections of the flakes in Plates 10 and 11 (which are explained at the end of Chapter 29). Here, as in Plates 10 and 11, we use a trimmed version of one period of a periodic variant of the desired process. This diminishes the overall shape's dependence upon D. The actual Brownian coastlines' overall shapes differ more than shown on these plates.

An effect related to the present strings is discussed in Chapters 34 and 35. ■■

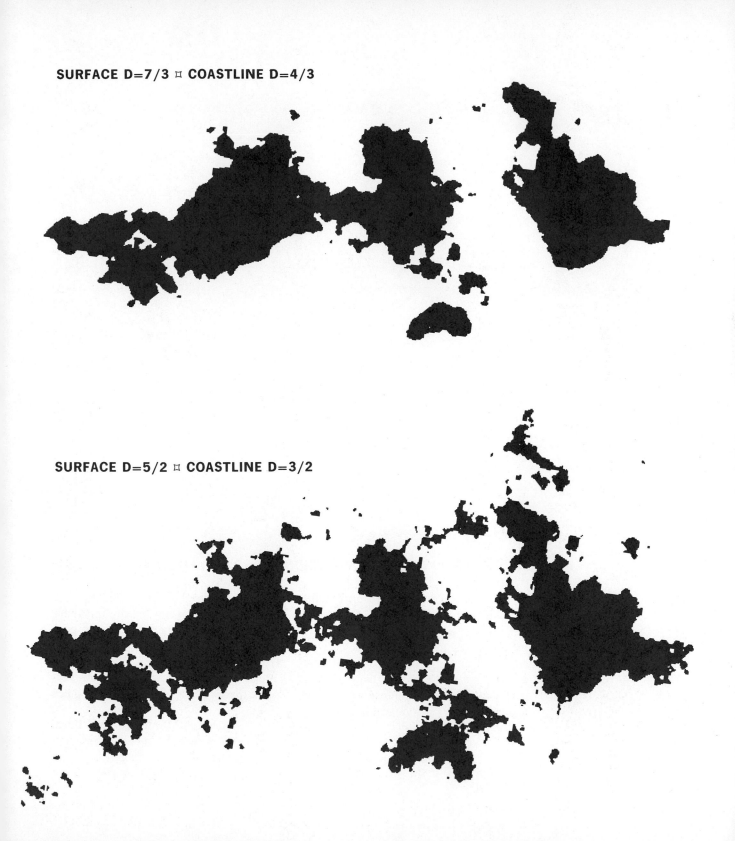

SURFACE D=7/3 ¤ COASTLINE D=4/3

SURFACE D=5/2 ¤ COASTLINE D=3/2

Plate 268 ◻ **CONTOUR LINES IN FRACTIONAL BROWN LANDSCAPES**

Both the figures in this plate combine two or three contour lines (the bold lines being coastlines) for fractional Brown functions. The figures involve different dimensions but the same program and seed: the top figure uses D~1.3333, and the bottom figure uses D~1.1667. By inspection, both dimensions are credible from the viewpoint of geography, but one is on the high and the other on the low side.

These curves seem much less "rugged" than those in Plate 267 having the same D.

The reason is that in the earlier plates each section exhibits a very strong maximum; there is little systematic slope there. Here, by contrast, we see the *side* of a huge mountain, with a strong overall slope. This plate is close in its "generic" appearance to a blown-up version of some particularly rugged small piece of Plate 267.

By comparing these different contour lines, we become better aware of the wide margin left for the interplay between irregularity and fragmentation even after D is fixed. ■

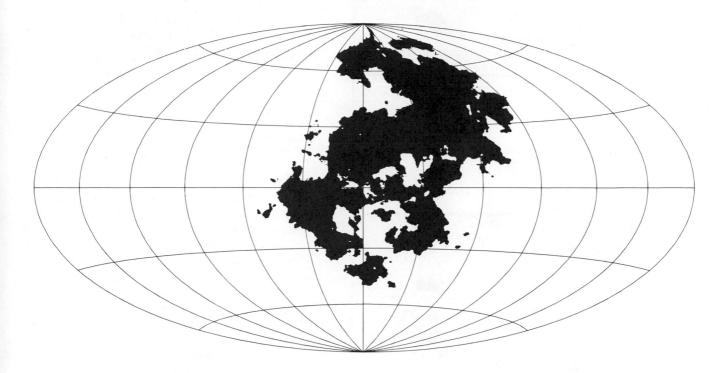

Plates 269 and C9 (top) ⌑ **BROWNIAN PANGAEA (COASTLINE DIMENSION D=3/2)**

The "distant planet" in Plate C9 represents a fictitious fractal Pangaea seen from far away in space. Its relief was generated by implementing on the computer (to the best of my knowledge, for the first time) a random surface due to Paul Lévy: a Brown function from the points of a sphere (the latitude and the longitude) to scalars (the altitude). Sea level was adjusted so that three-quarters of the total area is underwater. The coastline was obtained by interpolation.

This plate shows the same Pangaea on a Hammer map—a projection favored by students of Wegener's theory of continental drift.

How closely does this model Pangaea resemble the "real" one? The specific local detail is not expected to be right, only the degrees of wiggliness, both local and global. The resemblance is imperfect, as expected. Indeed this model Pangaea's coastline satisfies D=3/2, while the imaginative drawings in books of geology attribute to the real Pangaea the same D as observed for today's continents, D~1.2. If new evidence turns out to be compatible with D=3/2, one could account for the geometry of Pangaea with the help of rather elementary tectonic assumptions.

FRACTALS IN NON-EUCLIDEAN SPACE. In Riemann's non-Euclidean geometry, the role of the plane is played by the sphere. Thus, the non-Euclidean geometries go half way: they study Euclidean shapes in a non-Euclidean substratum. The bulk of this Essay also goes half way, since it studies non-Euclidean shapes in a Euclidean substratum. The present Pangaea unites both departures: it is an example of non-Euclidean shape in a non-Euclidean substratum. ■

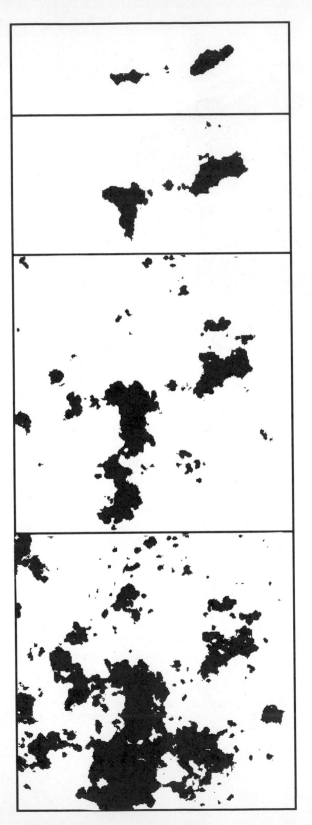

My claim that appropriately selected fractional Brown functions are reasonable models of Earth's relief was originally founded upon these four model coastlines. They are, like Plate 269, a sentimental carry-over from my 1975 French Essay, except that the black areas were filled in more carefully, thus extracting more detail from the original.

When D is near 1, top Figure, the coastline is too straight to be realistic.

On the other hand, the coastline corresponding to D=1.3000, second Figure from the top, strongly reminds us of the real Atlas. We see unmistakable echos of Africa (big island to the left), of South America (big island to the left, as seen in mirror image), and of Greenland (big island to the right, after the top of the page is turned from twelve o'clock to nine o'clock). Finally, if the page is turned to three o'clock, both islands together simulate a slightly undernourished New Zealand, together with a double Bounty Island.

When D rises to 3/2, third Figure from the top, the Atlas guessing game is harder to play.

When D increases again, closer to 2, bottom Figure, the geographic game becomes even more difficult, or at least more specialized (Minnesota? Finland?). Eventually it becomes impossible.

Other seeds yield the same result. However, the same tests based on finer graphics favor D~1.2000. ■■

Plate 271 □ THE FIRST KNOWN EXAMPLES OF FRACTIONAL BROWN ISLANDS (DIMENSION D=2.3000)

Including this plate may involve sentimental overkill, because it does not say anything that is not better expressed by other plates. But these views of an island with varying sea level were featured in Mandelbrot 1975w and in my 1975 Essay, and I am fond of them. They were part of a more complete sequence of fractional Brown islands of varying D and varying sea levels, the first such islands to be drawn anywhere. (In 1976, we made a film of this special island emerging from the sea; in 1981, the film looks ridiculously primitive, but it may acquire antiquarian value.)

Constantly, I lapse into wondering during which trip I actually saw the bottom vista, with its small islands scattered like seeds at the tip of a narrow peninsula.

The original illustration had been photographed from a cathode ray tube that lacked sharpness; the data have therefore been reprocessed. Here (as opposed to Plates 264 and 265, and C9 through C15), no deliberate simulation of side-lighting is required. As luck will have it, the ancient graphic process creates the impression that the sea shimmers toward the horizon.

The reader will observe that, compared with the most recent landscapes, this plate involves a surprisingly high dimension. The reason is that early graphic techniques were incapable of representing small details, hence the early landscapes' dimension seemed smaller than the D that had been fed into the generating programs. To compensate, we increased D beyond the range suggested by the bulk of the evidence. As graphics improved, however, the bias became conspicuous, hence counter-productive. Today, we are at the point where the D's suggested by Richardson's data yield perfectly acceptable landscapes. ■■

29 ¤ The Areas of Islands, Lakes, and Cups

We further explore my Brownian model of the relief, as advanced in the preceding chapter. The consequences concerning island areas prove acceptable, but the consequences concerning lakes and cups are *not* acceptable. To correct this discrepancy, an improved model is put forward.

PROJECTIVE ISLAND AREAS

As pointed out in Chapter 13, the variability of the projective areas A of ocean islands is an obvious characteristic of maps, often more striking than the shape of coastlines. We report that Korčak 1938 gives the distribution of A as hyperbolic: $Pr(A>a)=Fa^{-B}$. (We are now in a position to replace Fr by Pr.) Finally, we show that this empirical result holds when the coastline is self-similar. We are now in a position to add that it is a fortiori sufficient to assume that the relief is self-similar.

There can be no doubt that the relationship 2B=D extends from the nonrandom Koch coastlines examined in Chapter 13 fractional Brown zerosets. But the argument is still part-

ly heuristic as of now. The distribution corresponding to the fractional Brown relief with H=0.800 comes really very close to the empirical data regarding all of Earth.

The dimension D_C of each fractional Brown island taken by itself is not known yet.

PROJECTIVE LAKE AREAS

The areas of lakes also are claimed to follow the hyperbolic distribution, hence one might be tempted to dismiss lakes as involving no new element. At second thought, however, the definitions of lakes and ocean islands are by no means symmetric.

A special analysis sketched in this chapter clarifies many issues concerning two lake surrogates, "deadvalleys" and "cups." And it makes us face the fact that river and watershed trees are asymmetric in Nature, but in none of my Brown models. Hence it leads to a suggested improvement of the latter.

But the distribution of lake areas remains mysterious. Perhaps its being hyperbolic is merely due to the "robustness" of the hyperbolic distribution under diverse forms of torture (Mandelbrot 1963e, and Chapter 38). For example, the product of a hyperbolic random multiplicand and a largely arbitrary multiplier is itself hyperbolic. The multiplicand may be due to a primeval state in which the relief and everything about it is hyperbolic. And the multiplier may be due to the thousand geological and tectonic factors that affect lake shapes. But this "explanation" is really nothing more than hand waving.

THE NOTION OF DEADVALLEY

The concept symmetrical to an ocean island is an area, enclosed by a continent, whose altitude is below the level of the ocean. We shall denote such areas by the self-explanatory mixed term *deadvalleys*. Some contain water—ordinarily at a level below that of the ocean, e.g., the areas centered upon the Dead Sea (filled to −1280 ft.), the Caspian Sea (−92 ft.), and the Salton Sea (−235 ft.). Other deadvalleys are dry, like Death Valley (bottoming at −282 ft.) or the Qattara Depression (−436 ft.). There is also the borderline case of the Lowlands.

Information concerning the projective areas within deadvalleys' contour lines at the ocean level is not available to me. But inspection of maps suggests that deadvalleys are fewer in number than islands. In the context of the model that assumes Earth to be flat except for an added Brown plane-to-line relief, this asymmetry is to be expected. The fact that the distributions of islands and deadvalleys have the same exponent means that the 10th largest island or lake areas are in about the same ratio to the 20th largest island or lake areas. But Korčak's law also involves a "prefactor" F that sets the *absolute* value of the 10th largest island or lake area. A comparative inspection of the various plates clearly shows that in the case of a continent surrounded by an ocean (and conversely) the prefactor is greater for islands than for deadvalleys (and conversely). And within the Brown sphere-to-line model, the lesser area (Pangaea) is more cut up in pieces than the greater one (Panthalassia).

However, the preceding argument tells us nothing about lakes: save for rare and irrelevant exceptions (such as areas near the seashore filled by salt water seepage), deadvalleys and lakes are distinct notions. The altitude of a lake's bottom need not satisfy $z < 0$, and the altitude of its surface need not be $z = 0$. Further complications: most lakes fill to just above the brim, which is a saddle point, but this rule suffers exceptions (e.g., Great Salt Lake and the lakes that cover the deadvalley bottoms listed in the preceding section).

THE NOTION OF CUP

Now we examine a second lake surrogate, to be denoted by the neutral geometric term *cup*.

To define this notion, think of an impermeable landscape, in which every dip is filled exactly to the brim. In order to move out of a dip, a drop of water has to move up, then down. But a drop *added* upon this landscape can conceivably escape along a path that never goes up, but proceeds either horizontally or down. Each dip has a positive area, hence the number of dips is either finite or infinite but denumerable. It is safe to assume that the different outlets have different altitudes. At the precise altitude of an outlet, the relief's contour line is made of a certain number of self-avoiding loops, plus a loop having a point of self-contact. At slightly higher altitudes, this self-contact vanishes. And at slightly lower altitudes, the loop divides into 2 loops nested within each other.

Once filled, the dips according to the above contruction will be called cups.

THE DEVIL'S TERRACES

Now assume that the relief is Brownian with $0 < H < 1$. Because of self-similarity, individual cups' areas are doubtless hyperbolically distributed. When D is not much above 2, the exponent of the distribution of areas is doubtless close to 1.

More specifically, I conjecture that a drop of water falling at random is almost certain to fall within a cup. If this conjecture is correct, the cups' surfaces are a wild extrapolate of the terraced fields in southeast Asia. I call them *Devil's terraces*. The points which fail to fall within cups form the cups' cumulative coastline, and add up to a ramified net, a random form of the Sierpiński gasket. If I am

wrong and the cups' cumulative coastline is in fact of positive rather than zero area (Chapter 15), my fallback conjecture is that there is a cup arbitrarily close to every point that does not lie in a cup.

ERODED BROWN MODEL: MIXTURE OF RIDGES AND FLAT PLAINS

One is now irresistibly drawn to modify my Brownian models by imagining that every cup of a Brown mainland B_H is filled with dirt and made into a flat plain. We need not illustrate the resulting function B^*_H graphically, because, in the interesting cases when D is not much above 2, filling the small cups makes little visible difference.

To obtain the dirt with which it will fill the cups, erosion must wear off the hills; but we shall see that (if D is not much above 2) one does not need an overwhelming quantity of dirt, hence it is useful to assume that the hills' shape is little changed. The fact that erosion wears off the saddle points by which cups empty cannot be handled here.

From this Essay's viewpoint, a major virtue of the proposed modification is that if ocean level is chosen appropriately, the eroded Brown relief on a flat Earth continues to be scaling. What about the effect of such an erosion upon dimension? There is evidence that the dimension of B^*_H lies between 2 and the dimension $3-H$ of B_H.

Let us now argue that the relative amount of dirt needed to fill in all the cups is not large when $D = 2 + \epsilon$. Mainland's volume is of the order of magnitude of (mainland projection's typical length)$^{2+H}$ $\propto$ (mainland area)$^{1+H/2}$, and a cup's volume relative to mainland's is (cup's relative area)$^{1+H/2}$. Since relative area is hyperbolically distributed with an exponent near 1 and Σ(relative area) $= 1$, it follows that Σ(relative area)$^{1+H/2}$ is fairly small. The exceptions concern cases where the largest cup is extremely large; such cups need not be filled, as in the case of Great Salt Lake.

RIVERS AND WATERSHEDS

In a first approximation that plays a central role in Chapter 7, I suggest that rivers and watersheds form conjugate plane-filling trees. Actually, this characterization may only apply to maps; as soon as altitude is introduced, the beautiful symmetry between the river and watershed trees is destroyed. Indeed, neglecting lakes, the points on a watershed tree are *always* either local maxima (hills) or saddle points (passes), while the points on a river tree are *never* either local minima or saddle points. The fact that Brownian and fractional Brownian models *do* have local minima implies they *do not* have river trees. This is a fresh strike against my Brownian models.

After the cups are filled, there are no rivers as such, only branching strings of (infinitely shallow) lakes, reminiscent of cacti with disc-shaped branches. The watersheds form a tree; I believe it is a branching curve

with $D < 2$, but it may be a curve of positive area, hence of dimension $D = 2$. Diverse further variants impose themselves, but are better reserved for a more suitable occasion.

PROPERTIES OF THE CUPS

To put in perspective the claims made in an earlier section, we first examine the one-dimensional reduction, namely a line-to-line fractional Brown function $B_H(x)$. Here, an island is merely an interval $[x', x'']$ wherein $B_H(x) > 0$ when $x' < x < x''$, while $B_H(x') = B_H(x'') = 0$. Denote by $x = x_0$ the point where B reaches its maximum (cases where there are several maxima x_0 are of zero probability), and define $B^*_H(x)$ as follows:

for x in $[x', x_0]$, $B^*_H(x) = \max_{x' \leq u \leq x} B_H(x)$
for x in $[x_0, x'']$, $B^*_H(x) = \max_{x \leq u \leq x''} B_H(x)$.

It is clear that $z \geq B^*_H(x)$, is the necessary and sufficient condition for a droplet starting at the point (x, z) to find its way to the ocean along a nonascending path. Droplets that satisfy $B_H(x) < z < B^*_H(x)$ remain trapped forever, and $z = B^*_H(x)$ is the water level attained when all the cups have been filled. This function B^* is simply a Lévy Devil's staircase (Plates 286 and 287), going up from x' to x_0, followed by a staircase going down from x_0 to x''. It is continuous but not differentiable and varies over a set of length zero. Any drop of water added near mainland's highest point will rejoin the ocean through flat regions alternating with "white water" regions.

The droplets that cannot escape fill the domain $B_H(x) < z \leq B^*_H(x)$. This domain is disconnected, since it contains no point for which $B^*_H = B_H$, and its connected portions are the mainland's cups. A cup's length is the distance between consecutive zeros of $B^*_H - B_H$. Its distribution is hyperbolic because of scaling; its exponent is known to be ½ when $H = ½$, and I am convinced it is always H. The longest cup's length, divided by $|t' - t''|$, is largest when H is close to 0, and is smallest when H is close to 1.

Now we return to a Brownian mainland $B_H(x,y)$ on a flat Earth, the function $B^*_H(x,y)$ is again defined by the condition that a water droplet that starts at a height $z > B^*_H(x,y)$ can escape to the ocean following a nonascending path that keeps above mainland. As before, the spatial domain in which $B_H(x,y) < z \leq B^*_H(x,y)$ decomposes into connected open domains that define the cups.

Now compare these cups to those of a very thin slice of mainland, retained by parallel walls at $y=0$ and $y=\epsilon$. We apply to them the preceding notations $B_H(x)$ and $B^*_H(x)$. The definition of $B^*_H(x)$ restricts water escape to paths lying between the above walls, while the definition of $B^*_H(x,0)$ allows a much wider choice of escape paths. It follows that $B^*_H(x,0) < B^*_H(x)$ for almost every x. Hence the function $B^*_H(x,0)$, and any other vertical cut of $B^*_H(x,y)$, are much more interesting than $B^*_H(x)$. They are devilishly terraced singular function with (an infinity of) peaked local maxima and flat local minima. If my

strongest conjecture is valid, the latter cover almost every point of mainland.

Since the cup areas' sum is at most equal to mainland's area, the cups can be ranked by decreasing area, hence are denumerable. A consequence is that the coastline of B_H that corresponds to a random value of z_0 is almost surely without double point.

The cumulative boundary of all the cups can therefore be obtained as follows. Take a denumerable set of values z_m—which will almost surely fail to include a value for which the coastline involves a loop. Censor the coastlines by erasing the deadvalley coastlines from all $z_0 = z_m$. Take the union of the censored coastlines, and add its limit points.

For any $M > 2$, the generalization to Brown function of M-dimensional $x = \{x_1 \ldots x_M\}$ is straightforward. Given $B_H(x)$, the argument already used for $M=2$ shows that the difference between B^*_H and B_H decreases as M increases. In the limit case where $M = \infty$ and B_H is a Brown function in Hilbert space, it follows from classical results of Paul Lévy that $B^*_H - B_H \equiv 0$. Does this identity hold for all $M > M_{crit}$ with $M_{crit} < \infty$? ▶   ■

Plates C1, C3, and C16 ◻ THREE GREAT ARTISTS OF THE PAST ILLUSTRATE NATURE, AND THEREBY BRING THE READER TO THE THRESHOLD OF FRACTALS

This signature is a book-within-the-book and is dedicated to the proposition that if "to see is to believe," then to see in color may lead to an even higher intensity of belief, however awkward our first efforts in this medium. Of course, the reader is *supposed* to open this book on page 1, *not* on page C1, nevertheless the captions in this signature are somewhat independent of the rest.

The Fractal Geometry of Nature was first set forth by this author. This geometry combines the mathematics and the science necessary to tackle a certain broad and widespread class of natural shapes.

Many of these shapes are very familiar, but the problems they raise had been rarely mentioned by writers of the past. On the other hand, Plates C1, C3, and C16 are ready examples of old works of art that exemplify the issues tackled by fractal geometry.

PLATE C1. THE FRONTISPIECE OF A BIBLE MORALISÉE. The period of Western European history centered at 1200, while stagnant in science and philosophy, was exuberantly active in engineering. In the age that built the Gothic cathedrals, to be a master mason was a very high calling. Thus, the "Bibles Moralisées illustrées" of that time ("comic strip" Bibles) often represent the Lord holding mason's dividers (Friedman 1974).

Plate C1 is an example. It is the frontispiece of a famous Bible Moralisée, written between 1220 and 1250, in the Eastern Champagne dialect of French. It now resides in the Austrian National Library in Vienna (codex 2554), and is reproduced with the Library's kind permission. The legend reads:

ICI CRIE DEX CIEL ET TERRE
SOLEIL ET LUNE ET TOZ ELEMENZ.

(HERE CREATES GOD SKY AND EARTH
SUN AND MOON AND ALL ELEMENTS.)

We perceive three different kinds of form in this newly created world: circles, waves, and "wiggles." The studies of circles and waves benefited from colossal investments of effort by man, and they form the very foundation of science. In comparison, "wiggles" have been left almost totally untouched.

The goal of the present Essay is to face the challenge of building a Natural Geometry of certain "wiggles," to be called "fractals."

A most attractive feature of this plate is that it begs the scientist to "take the measure of the universe." To apply dividers to circles and waves had long proven an easy task. But what if we apply dividers to the wiggles on this plate,...or to coastlines on Earth? The result is unexpected; it is discussed in Chapter 5, and later chapters explore its consequences, and thereby guide the reader along a path one may describe as science-filling.

FROM *THE FRACTAL GEOMETRY OF NATURE*, BY BENOIT B. MANDELBROT
Published by W. H. Freeman and Company. Copyright © 1982 by Benoit B. Mandelbrot

PLATE C3. *THE DELUGE* BY LEONARDO DA VINCI. (From the Windsor Castle Collections. Reproduced by gracious permission of Her Majesty the Queen.)

This is one of many drawings in which Leonardo represented water flow as the superposition of eddies of many diverse sizes. Awareness of this eddy structure entered science belatedly, becoming partly formalized by Lewis F. Richardson in the 1920's into the "scaling" view of the nature of turbulence. However, this view promptly drifted into a search for formulas, losing all geometric flavor, and also (this may not be a coincidence!) proving of limited effectiveness.

The theory expounded in this book allows a return of geometry into the study of turbulence, and shows that many other fields of science are very analogous geometrically and can be handled by related techniques.

LEGEND CONTINUES ON PAGE C16

Plate C5 ¤ SELF-SQUARED FRACTAL DRAGON

Flamboyant this design may be, but its black background must be viewed as an example of extreme minimal art. Indeed, the formula

$$\{z : \lim_{n \to \infty} |f_n(z)| = \infty, \text{ where } f(z) = \lambda z(1-z)\},$$

is all that is needed to duplicate the background in question with complete accuracy. Let me explain this formula: having chosen the complex number λ determining the "generating function" $f(z)$, we construct $f_2(z) = f(f(z))$, then $f_3 = f(f_2(z))$, that is, $f_3(z) = f(f(f(z)))$, and so on ad infinitum.

The complex number λ yielding this plate is $\sim 1.64 + .96i$. Clearly, it was not hit by random fire. The dragon's shape is very sensitive to λ, but a special theory I developed (and sketch in Chapter 19) allows one to choose λ so as to obtain the dragon one wishes among many very varied possibilities.

THE "STONES." As to the design that stands out from the black background, it is made of 25 kinds of "stones," each defined by

$$\{z : \lim_{n \to \infty} f_{25n}(z) = z_g\},$$

where the 25 complex numbers z_g are roots of the equation $f_{25}(z) = z$, and in addition satisfy $|(d/dz)f_{25}(z)| < 1$.

Looking carefully, one sees 5 different reds, 5 different blues, etc. This coloring scheme is chosen because 25 values of z_g fall into 5 "genera," each made up of 5 "species." We attach a color to each genus, and a hue or intensity to each species. For example, all 5 species of gold are strung along the dragon's golden main body, and they come together at this body's wasp's waists.

A PREVIOUSLY HIDDEN FACE OF CLASSICAL MATHEMATICS. The formula for $f(z)$ is so short, and looks so uninteresting (because it comes from an elementary chapter of calculus), that little was expected from it. Thus, previewing this kind of design on the computer screen provoked surprise as well as a deep esthetic shock.

Classical mathematical analysis (which is the most advanced form of calculus) had played a joke on all those who either loved or hated it. It is now revealed that analysis has two very different faces. The face it had been showing us for centuries, and which became its pride (or its curse), was unremittingly austere. But I show that analysis also has a hidden face that is often strikingly attractive and playful.

Respect and admiration for the Great Masters of austere analysis make one hasten to say that the extreme complication of the outline of this black velvet was *not* a surprise to the handful of mathematicians (of whom I had the good fortune of being one) aware of "ancient" (mostly circa 1920) works by Pierre Fatou and Gaston Julia. But such shapes' complication had contributed to enhancing the starkness of analysis, and nothing had made us expect that so many witnesses would perceive this complication as beautiful.

ALGORITHMS THAT INCLUDE A LOOP. Fatou's and Julia's discoveries confirm, in effect, that a very complex artifact can be made with a very simple tool (think of it as a sculptor's chisel), as long as the tool can be applied repeatedly. Here, the tool is the function $f(z)$ from which one generates the functions $f_n(z)$.

Therefore, one does not deal here with an operation that is performed once, then stops when completed, but with an operation that is performed, then repeated, etc. Such iterated functions are examples of treadmills or loops, each turn of which can deal with a fresh task.

The simplest loop programs are linear, which means that they add detail that merely echoes the overall shape on a smaller scale. The resulting shapes are called *self-similar*.

In this instance, to the contrary, the detail becomes deformed as it becomes smaller, because the function $f(z)$ is not linear. This function being quadratic, the boundary of the velvet background is denoted in Chapter 19 by the term, *self-squared*. ∎

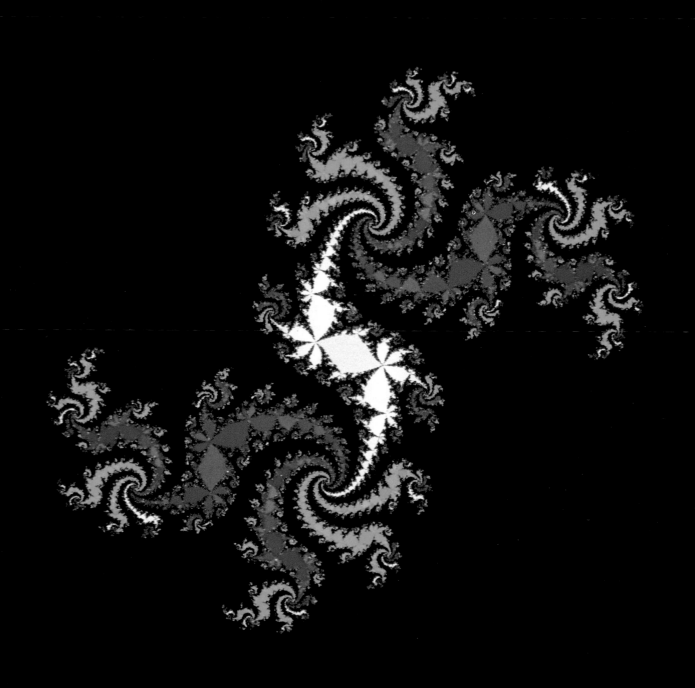

This hanging is patched of six different kinds of transparent cloth. A multitude of open discs (that is, of interiors of circles) are cut from cloth of 6 different colors, and sewn upon a transparent scrim, either singly or in superposition. Most of these discs are too far away or too small to be seen.

This shape is a more intricate variant of one discussed in Chapter 18. Its construction begins by selecting a generator, which in this instance is a collection of 4 circles and 4 straight lines, arranged as follows

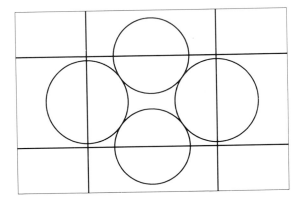

For many reasons explained in Chapter 18, a great deal of interest is attached to the shape $\mathcal{L}$ that is the smallest shape to remain completely unchanged if one performs a symmetry with respect to any of the generating straight lines, or an inversion with respect to any of the generating circles.

In theory, the difference between the notions of line and circle is not basic here; indeed, if the above lines and circles are subjected to geometric inversion with respect to a point that lies on none of them, they transform into 8 circles. Therefore, instead of calling $\mathcal{L}$ "self-inverse and self-symmetric," it suffices to call it "self-inverse."

But the fact that this figure involves 4 symmetries, across lines that form a rectangle, is advantageous, and was built-in to insure that the present set $\mathcal{L}$ is periodic. The first period is bounded by our rectangle, and the others are obtained by translation along either axis.

The problem of determining the structure of $\mathcal{L}$ is an old and famous one, to which I give the workable solution illustrated here. This new solution shows that $\mathcal{L}$ is made up of the points where disc-shaped cloth patches are in contact along the circles that bound them. Points *within* a disc never count as part of $\mathcal{L}$ even when they are on the boundary of a different disc of the same or a different color.

Now to the explanation of how these disc-shaped patches are selected. Starting with the generating shape, one draws 6 circles, call them Γ-circles, each of which is orthogonal to 3 of the 8 generating shapes. There are many other circles orthogonal to 3 of the generator's 8 shapes, but only the present 6 are needed as Γ-circles.

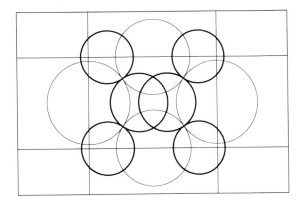

Each Γ-circle bounds a disc associated with a different color of cloth, then the same color is also used in every disc obtained by transforming one of the Γ-discs by inversion in the 4 circles, or by symmetry in the 6 lines in the generator. The discs in the central "medallion" overlap with each other, but neither overlaps with any of its inverses. The corner discs, to the contrary, overlap with certain of their inverses. ■

Plate C9 ◻ PLANETRISE OVER LABELGRAPH HILL
(SOUVENIR FROM A SPACE MISSION THAT NEVER WAS)

Plates C9 to C15 may look "realistic." And, in their own way, some are works of art. However, these plates *are not photographs* and were not *intended* to be artistic. Furthermore, they are *not* examples of the popular fake landscapes one can obtain by processing actual landscapes, in the same way as one synthesizes a chemical by transforming other chemicals. The present plates are *exactly* as *artificial* as Plates C5 and C7. They are the fractal equivalent of the "complete" synthesis of hemoglobin from the component atoms and (a great deal of) time and energy.

Plate C9 combines the implementations of two of my theories of the surfaces of planets, first advanced in Mandelbrot 1975w on the basis of Plates 270 and 271, and explored in Chapters 28 and 29 of this Essay. Various features of the present plate *fail* to fit reality, but the chapters in question show how some of these defects can be improved.

A planet on which water concentrates in oceans and snow (e.g., in polar caps), while the sky remains completely cloudless, is—to put it mildly—a rough approximation. Color is added after the fact to the best of our present abilities, and the color selection is completely independent of my theories. A first stage algorithm showed altitude using the same colors as *The Times Atlas*. Then it became clear that a slight refinement in the coloring scheme would yield considerably better results, without requiring a multiplicity of separate decisions.

This art cannot claim to be as minimal as that in Plates C5 and C7, because the definitions of the two "planets" cannot reduce to a single line without undue artificiality.

A second reason this art cannot be called minimal is that implementing the shadows involves great ingenuity; one would need tomes to explain every detail. In addition, the algorithm is very much influenced by the available tools, hence to duplicate this work one would have to use exactly the same computer equipment.

Since an earlier version of this "Planetrise" appeared on the back jacket, and other fractal landscapes appeared in the Plates of the 1977 *Fractals*, they have been honored by innumerable imitations. The low relative quality of the imitations is further proof of the nonminimality of this art.

Nevertheless, the main feature of either planet can be characterized uniquely by a very small number of very basic properties of continuity and invariance, to be explored in the following captions.

DEDICATION. Labelgraph Hill is named in memory of "lblgraph," an independent-minded and often very ill-mannered heap of graphics programs that originated in work by Alex Hurwitz and Jack Wright of IBM Los Angeles. It graced the T. J. Watson Research Center from 1974 to 1981, responded when treated with consideration, and (with its lively successor, "yogi") made it possible to illustrate my Essays. R.I.P. ■

Plate C11 ◻ GAUSSIAN HILLS THAT NEVER WERE

The name of Carl Friedrich Gauss (1777-1855) appears in nearly every chapter of mathematics and of physics, making him the first (*princeps*) among the mathematicians (including the physicists) of his time. But these imaginary hills being called *Gaussian* is motivated by a probability distribution for which Gauss receives undeserved credit. It is the distribution whose graph is the famous "bell-shaped curve" or "Galton ogive." On Plates C9 to C15, this distribution rules the difference in altitude between any two prescribed points on the map, at least after a suitable transformation.

Many scholars resort to the Gaussian probability distribution in their disquisitions, without feeling that this choice has to be justified. Either it is the only distribution they know intimately and trust, or they believe it accounts for the distribution of every random quantity in Nature, from conscripts' heights to astronomers' errors of measurement.

Actually, this last belief is quite without foundation. This Essay includes many examples that show the world to be full of grossly non-Gaussian phenomena. Therefore, the resort to the Gaussian distribution requires a different and less controversial justification. To me, the only sound justifications are based on the fact that the Gaussian is the only distribution that possesses certain properties of scale invariance, yet leads to continuously varying reliefs. The conclusion is that the simplest possible reliefs are ruled by a "Brown function," or at least by a variant thereof which I called "fractional Brown function."

The only parameter that these desiderata leave indeterminate, so that it remains to be selected on independent grounds, is called *fractal dimension* of the relief, and is denoted by D.

When D attains its minimum value of D=2, the relief is extremely smooth. As D increases, the relief becomes increasingly "corrugated," and begins to resemble high Earth mountains. Eventually, it becomes too corrugated to be mountain-like, and ultimately it becomes near space-filling.

A Brown function's defining characteristic is that every vertical cut is an ordinary Brown line-to-line function.

For every landscape other than the distant planet in Plate C9, the attitude is computed for latitudes and longitudes forming a square grid. Then a semblance of roundness is injected by rolling this relief's flat base surface around a cylinder whose axis runs from left to right. The computer is programmed to simulate lighting from a source located 60° over the left.

Oddly enough, several observers, after commenting briefly that a characterization of relief based solely on invariance and continuity criteria is ingenious and effective, proceed to criticize this approach at length, because its criteria are too abstract and fail to be deduced from explicit "models" or generating mechanisms, either before or after the fact.

I am reluctant to reply (heavy-handedly) by criticizing the concrete "mainstream" theories of relief for failure to come forth with fake landscapes anywhere close in realism to those due to my "abstract" theories. It seems better to point out that many among the finest theories of science did start with exquisite combinations of pistons, strings, and pulleys, only to end (several generations later) with bare-bones invariance principles. From this viewpoint, the work that led to the present illustrations, and other case studies in this Essay, start at the finish line. Is this sufficient reason for unhappiness? ■

The bottoms of all the Gaussian landscapes in this Essay, including those in Chapter 28, are flattened to form an arbitrarily set reference level. This procedure was first used to generate islands. And in mountain landscapes, it was originally meant to help the eye distinguish between different surfaces.

Let me elaborate. When preparing my 1975 Essay, we did not want to waste any data and we plotted all we had, but the result was distressing: Our eyes found it surprisingly hard to discriminate between landscapes we knew to be characterized by significantly different values of D. Then the desire to represent island coastlines together with the relief led us to introduce a flat reference surface into the same picture, and suddenly the differences in D became extremely conspicuous. We should have remembered that, in order to assess motion, one needs a standard to be called rest. The same is true of roughness.

Now we find that, when the same procedure was applied to valleys as well as to mountains, it also had a second effect, an unplanned but most fortunate one. Creating the flats (reminiscent of lakes or banks of snow or alluvia) hides the valley bottoms, hence forces us to concentrate on high mountains, where the model proves powerful beyond expectation. Had we looked too soon at the whole relief, we would have been sorely disappointed, because in the Gaussian models the valley bottoms are as "unsmooth" as the mountain tops, while real valleys are *much* smoother. At present, there is no way I like for accounting for this difference.

But there are ways of "fixing" the Gaussian model of mountains to account better for the valleys. The simplest fix assumes that the sole differences between the various portions of the relief concern vertical scale, the value of D being the same throughout. To justify this assumption, let us reduce the vertical scale of the Gaussian Sierras in Plate C11. Amazingly, they turn into rolling terrain! Conversely, consider almost any near-flat surface, like that of an airport strip, and magnify its asperities. In a first approximation, the result turns out to be very often like the Gaussian Hills of Plate C11, with a dimension that depends upon detailed circumstances. There is no reason that I know for thinking that this result fails to apply to valley bottoms. Hence, one cannot help being curious of the consequences of assuming that the D valid for the mountain tops also applies in a first approximation to the valley bottoms.

A more specific idea is to restrict scaling to apply in small domains, with the same dimension throughout, while the vertical scale increases with the altitude above the valley's bottom. To achieve this goal in the top of this plate, and in the Labelgraph Hill of Plate C9, the altitudes above either lake level or valley bottom are raised to the third power.

When, to the contrary, the vertical scale is made to *decrease* with the altitude above the bottom (by raising the altitude to a power below 1), one obtains the mesa and canyon at the bottom of the present plate.

The trick may be crude, but it is astonishingly effective. ■

Plate C15 ¤ FRACTAL ISLANDS THAT
NEVER WERE, SEEN FROM THE ZENITH

The algorithm used in the bottom of Plate C9, and in Plates C11 to C15, is based on numerical Fourier methods, hence yields a periodic smooth surface, whereas a fractal surface is by definition extremely rough. One can imagine, however, that we inspect our mountains using light whose wavelength is the width of the cells in the grid. Under such light, all the finer details remain totally invisible.

In order to obtain islands we center the relief around a maximum, and omit to plot the altitudes below a certain reference level taken as 0.

The top archipelago corresponds to an ordinary Brown relief. This is a poor model of Earth, because it is clearly too irregular in its detail. The fit is poor because a surface fractal dimension of $D=5/2$ and a coastline dimension of $D=3/2$ are too large.

In the bottom archipelago, the ordinary Brownian function is replaced by a *persistent* fractional Brownian function of dimension $D=2.200$, and the coastline takes the sensible dimension $D=1.200$. The clearcut ridges in the Figure are entirely compatible with the fact that it was generated by an isotropic mechanism.

The resemblance with Hawaii is better than deserved, because there is no reason why the model should be valid for volcanic archipelagoes.

The coastlines' perceived form is much influenced by how tightly they fill the picture. This facet of form is *not* totally determined by D: because Plates C11 and C15 relate to a region near a minimum or a maximum, the reference level plays a central role. ∎

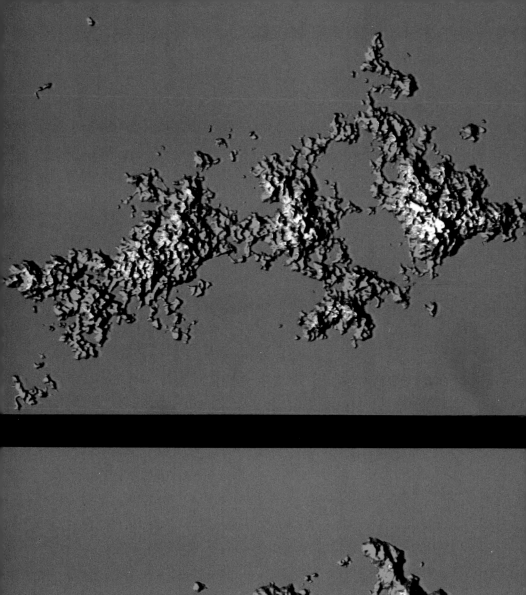
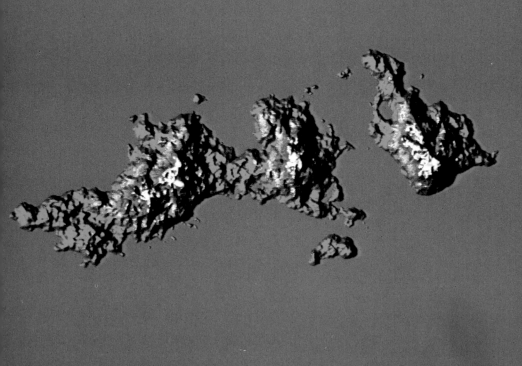

LEGEND FROM PAGE C3, CONT'D

PLATE C16. THE *GREAT WAVE* BY HOKUSAI. Katsushika Hokusai (1760-1849) was a painter and engraver of extraordinary power and versatility, a giant by any standard. He was fascinated by eddies and whorls of every kind, as exemplified by one engraving that reached such fame that a stamp-size reproduction will suffice.

THE NOTION OF FRACTAL. I put together certain geometric shapes whose form is very irregular and very fragmented, and coined the term *fractal* to denote them. Fractals are characterized by the coexistence of distinctive features of every conceivable linear size, ranging between zero and a maximum that allows for two cases. When a fractal is bounded, the maximal feature size is of the order of magnitude of the fractal's overall size. When a portion of an unbounded fractal is drawn within a box of side Ω, the picture has a maximal feature size of the order of Ω. Examples of mathematically constructed fractals are found in Plates C5 to C15.

Fractals star in two distinct stories, separated in time by nearly a century, between which they underwent a total role reversal.

In the first stage, some fractals (not those illustrated in this signature) were deliberately designed from 1875 to 1925 to eat away at the foundations of the prevailing mathematics. *Everyone* viewed these sets as "monsters."

While the rest of mathematics was regarded as a potentially promising hunting ground for physicists in need of new tools, *everyone* agreed that the monsters could safely be assumed to be totally irrelevant to the description of Nature. Hardly any variant of these monsters was created for fifty years.

The role reversal started as I began to find in my research work that one of these monsters after another could serve as the central conceptual tool to answer some old question that Man had been asking about the shape of his world. This led to the emergence of many new examples and to the formulation of fractal geometry, in my Essays on this topic.

THE ROLE OF GRAPHICS. Computer graphics played a central role in the *acceptance* of fractal geometry, but a peripheral role in its genesis. That is, granted the fascination that fractals now hold for the computer practitioners, one is tempted to credit the emergence of the new geometry to the availability of this new tool. Actually, I formulated the theory of fractals when computer graphics was in its infancy. However, I let its development be biased toward topics that lend themselves to intuition-building illustrations.

CLASSICAL PICTORIAL COMPOSITION. Now examine Plates C1 and C3 again. Here, as in almost any other classically "composed" picture, it is strikingly easy to identify at least one "feature" for nearly every scale between the total picture size and an inner cutoff below which details become invisible. Thus, the property of scaling that characterizes fractals is not only present in Nature, but in some of Man's most carefully crafted creations.

ACKNOWLEDGMENT

The computer programs used to draw the color plates were developed by Richard F. Voss (Plates C9 to C15) and V. Alan Norton (Plates C5 and C7). ■

30 ¤ Isothermal Surfaces
of Homogeneous Turbulence

The present chapter culminates in an explanation of Plates 10 and 11. The text is primarily devoted to fractional Brown functions of 3 variables with an antipersistent exponent $H < \frac{1}{2}$. The case $H = \frac{1}{3}$ receives special emphasis, with $H = \frac{1}{2}$ serving again as point of departure.

TURBULENT SCALARS' ISOSURFACES

When a fluid is turbulent, the isothermal surface where the temperature is exactly 45°F is topologically a collection of spheres. However, it is intuitively obvious that this surface is by far more irregular than a sphere or the boundary of any solid described in Euclid.

It reminds us of Perrin's quote in Chapter 2 that describes the form of a colloid flake obtained by salting a soap solution. The resemblance may extend beyond mere geometric analogy. It may be that a flake fills the zone in which the soap concentration exceeds some threshold, and that in addition this concentration acts as an inert marker of very mature turbulence.

Anyhow, the analogy with colloid flakes suggests that the isothermal surfaces are approximate fractals. We wish to know whether or not this is the case, and if it is, to evaluate the fractal dimension. To do so, we need to know the distribution of temperature changes in a fluid. Corrsin 1959d, among others, reduces this question to a classical one, which Kolmogorov and others had faced in the 1940s. In part, these early authors had triumphed to an extraordinary extent; in part, they had failed. A review of these classical results is inserted here for the sake of the nonspecialist.

BURGERS DELTA VARIANCE

The delta variance of X is defined in Chapter 21 as the variance of an increment of X. J.M. Burgers assumed that the delta variance of velocity between two given points P and $P_0 = P + \Delta P$ is proportional to $|\Delta P|$. This crude but simple postulate defines *Burgers turbulence.*

A precise mathematical model of a Burgers function is the Poisson function which results from an infinite collection of steps with directions, locations, and intensities given by three infinite sequences of mutually independent random variables. This description should ring a bell. Except for the addition of the variable z to x and y, and the replacement of the altitude (which is one-dimensional) by a velocity (which is three-dimensional), a Gaussian Burgers function served in my ordinary Brownian model of Earth's surface described in Chapter 28.

KOLMOGOROV DELTA VARIANCE

As a model of turbulence, the Burgers delta variance suffers from deadly defects, the worst being that it is incorrect from the viewpoint of standard dimensional analysis. A correct dimensional argument, advanced by Kolmogorov and simultaneously by Obukhov, Onsager, and von Weiszäcker, shows that only two possibilities exist for the delta variance. Either it is universal, that is, the same regardless of the conditions of experiment, or it is an unholy mess. To be universal, the delta variance must be proportional to $|\Delta P|^{2/3}$. Derivations are found in many books; the geometric nature of the result is underlined in Birkhoff 1960.

After initial doubts, it was established that the Kolmogorov delta variance accounts surprisingly well for turbulence in the ocean, the atmosphere, and all large vessels. (See Grant, Stewart & Moillet 1959.) This verification constitutes a striking triumph of abstract a priori thought over the messiness of raw data. It deserves (despite numerous qualifications, to which Chapter 10 adds fresh ones) to be known outside of the circle of specialists.

The Gaussian function with the Kolmogorov delta variance also rings a bell. In the present context, concerned with a scalar (one-dimensional) temperature, this Gaussian function is a fractional Brown 3-space-to-line function, with $H = 1/3$. Thus the Kolmogorov field involves antipersistence, while Earth's relief favors persistence. A more basic difference is that, while the H required to represent Earth's data is purely phenomenological so far, the Kolmogorov $H = 1/3$ is rooted in the geometry of space.

IN HOMOGENEOUS TURBULENCE, THE ISOSURFACES ARE FRACTALS (MANDELBROT 1975f)

Despite its triumph in predicting that $H = 1/3$, the Kolmogorov approach has a major shortcoming: the distribution of the differences of velocity or of temperature in a fluid remains unknown, except that it cannot be Gaussian.

Such negative results are awkward, but rarely force a convenient assumption to be abandoned. At most, the students of turbulence must be cautious when investigating a Gaussian model: if and when a calculation yields a logical impossibility, they abandon the model. Otherwise, they forge ahead.

In particular—and now we return to temperature— Mandelbrot 1975f combines the Gaussian assumption with the Burgers and the Kolmogorov delta variances. One can hope that the conclusions would remain correct without the Gaussian assumption, because they use little more than continuity and self-similarity.

In the 4-dimensional space of coordinates x,y,z,T, the temperature T defines a function T=T(x,y,z). The graph of a fractional Brown function is a fractal of dimension 4−H, and many of its lower-dimensional sections are the following fractals we know well.

LINEAR SECTIONS. The isotherm for fixed y_0, z_0, and T_0 is made of the points along a spatial axis where a certain value of T is observed. They form a fractional Brown zeroset, and their fractal dimension is 1−H.

PLANAR SECTIONS. For fixed y_0 and z_0, the curve representing the variation of temperature along the x axis is a fractional Brown line-to-line function, and its dimension is 2−H. For fixed z_0 and T_0, the implicit equation $T(z_0,x,y)=T_0$ defines isotherm in a plane. These isotherms are of dimension 2−H. Except for the value of D, they are identical to the coastlines studied in Chapter 28.

SPATIAL SECTIONS. For fixed z_0, the section is the graph of $T(x,y,z_0)$, a fractal of dimension 3−H. For H=½, it is identical in definition to the Brownian relief in the plates of Chapter 28. For H=⅓, it is a fractional Brown relief in the same plates.

EXPLANATION OF PLATES 10-11

For fixed T_0, the isosurface defined by the implicit equation $T(x,y,z)=T_0$ is a three-dimensional generalization of a coastline and introduces us to a new kind of fractal with D=3−H. Thus, D=3−½ in Gauss Burgers nonpersistent turbulence and D=3−⅓ in Gauss Kolmogorov antipersistent turbulence.

Such surfaces are illustrated on Plate 11, whose origin can at long last be explained. For the sake of contrast, Plate 10 adds the isosurface of a persistent function T(x, y, z), with H=.75. Due to the cost of this huge computation, the surfaces had to be smoothed out to excess. The fact that differences due to D affect the overall form less drastically than expected is explained on page 267. ▬

31 ◻ Interval Tremas; Linear Lévy Dusts

The structure of this group of chapters is a bit involved. The two themes of *random trema* and *texture* do not converge until Chapter 35, when it is shown how texture can be controlled. And Chapter 34 introduces texture without much reference to tremas; it describes facts that might have been scattered over several earlier chapters but are better collected together to provide a unified treatment.

As to Chapters 31 to 33, they do not involve texture but use the notion of trema to construct random fractals, many of them new. Like those of the preceding (Brown) chapters, the new fractals are free from time and/or space grids.

This chapter describes random dusts constrained to the line, applies them to the noise problem first tackled in Chapter 8, and grooms them to become the basis of two distinct extensions to the plane and to space described, respectively, in Chapters 32 and 33.

The primary concrete goal of Chapters 32, 33, and 35 is to help model the galaxy clusters, a challenge first described in Chapter 9.

CONDITIONALLY STATIONARY ERRORS (BERGER & MANDELBROT 1963)

We were exhilarated in Chapter 8 to find in Cantor dust a reasonable first model of the principal features of certain excess noises. But we did not even attempt an actual fit of the model to the data. The reason is, obviously, that the fit is expected to be terrible. Cantor dusts are much too regular to be precise models of any irregular natural phenomenon I can think of. In particular, their self-similarity ratios are restricted to values of the form r^k. Furthermore, a Cantor dust's origin plays a privileged role that cannot be justified but has a most unfortunate effect: the set fails to be superposable upon itself by translation; in technical terms, it is not translation invariant.

Irregularity is easy to inject—through randomness. As to invariance by translation, our hoped-for substitute for the Cantor dust will only be required to match up with its translation in a statistical sense. In probabilistic terminology, this means that a set has to be stationary, or at least satisfy a suitably weakened condition of stationarity.

A simple means of accomplishing part of this goal is proposed in Chapter 23. The present chapter takes three further steps forward.

The first step is involved in the earliest realistic random model of intermittency. Berger & Mandelbrot 1963 starts out from a finite approximation of the Cantor dust, with scales satisfying $\epsilon > 0$ and $\Omega < \infty$, and shuffles its gaps at random to make them statistically independent of one another. The intervals of length ϵ between successive gaps are left untouched. Chapter 8 shows that in a Cantor dust the relative number of gaps of length exceeding u is given by a near hyperbolic stairlike function. Randomization reinterprets this function as a tail probability distribution $Pr(U > u)$.

This yields a randomized Cantor dust, with $\epsilon > 0$. Unfortunately, the stairs of $Pr(U > u)$ bear the trace of the original values of N and r. This is why Berger & Mandelbrot 1963 smoothes these stairs out: the successive gaps measured in units of ϵ are taken to be statistically independent integers ≥ 1, the distribution of their lengths being

$$Pr(U > u) = u^{-D}.$$

The model's fit is surprisingly good: the German Federal telephones yield $D \sim 0.3$, and follow-up studies of different channels by various authors find D's from 0.2 to nearly 1.

In the Berger & Mandelbrot model, the durations of successive gaps are independent, hence errors constitute what probabilists call a "renewal" or "recurrent" process (Feller 1950). Each error is a point of recurrence, where the past and the future are statistically independent of each other and follow the same rules as from other errors.

LINEAR LÉVY DUSTS

Unfortunately, the set obtained by shuffling the gaps of the truncated Cantor dust (and smoothing their distribution) remains defective in several ways: (a) the fit of the formula to the data on excess noises remains imperfect in details, (b) the restriction to $\epsilon > 0$ may be acceptable to the physicist but is annoying from the esthetic point of view, (c) the construction is awkward and arbitrary, and (d) it is too far removed in spirit from Cantor's original construction.

Mandelbrot 1965c uses a set due to Paul Lévy to construct a more refined model that avoids defects (a) and (b). Let me call this set the *Lévy dust*. Once D is prescribed, the Lévy dust is the only set that combines two desirable properties. As in the randomized truncated Cantor dust, the past and the future are independent if seen from a point in this set. Like the Cantor dust, it is a self-similar fractal. Better than the Cantor dust, this Lévy dust is statistically identical to itself reduced in an *arbitrary* ratio r between 0 and 1.

The zeroset of Brownian motion, Chapter 25, turns out to be the Lévy dust with D=½.

Unfortunately, the method Lévy uses to introduce this set preserves defects (c) and (d) listed above. And it is technically delicate: Instead of constraining u to be an integer ≥1, one must let it be a positive real number with $Pr(U>u)=u^{-D}$ extended down to u=0. Because $0^{-D}=\infty$, the total "probability" is infinite. The method used to exorcise this seemingly ridiculous implication is important and interesting, but of no other use in this work.

Fortunately, these difficulties vanish if one adopts a more natural "trema" construction proposed in Mandelbrot 1972z.

ACTIVE AND VIRTUAL TREMAS

As a preliminary, I claim it is useful to describe the original Cantor dust by means of a combination of "active" and "virtual" tremas. Again, one starts from [0,1] and cuts out its open mid-third]⅓,⅔[. From then on, the construction's substance remains the same but the formal description changes. One makes believe that the second stage cuts out the mid-thirds of *each* third of [0,1]. While cutting out the mid-third of the already vanished mid-third has no perceivable effect, virtual tremas will momentarily prove convenient. In the same way, one cuts out the mid-third of *each* ninth of [0,1], of *each* 27th, and so on. Note

that the distribution of the number of tremas of length exceeding u is now given by a step function, whose overall behavior is proportional to u^{-1}, instead of u^{-D}. The same dependence upon u holds with different rules of curdling, except that the positions of the steps and the factor of proportionality both depend on the method of construction.

INTERVAL TREMAS & THE RESULTING GAPS (MANDELBROT 1972z)

Next, Mandelbrot 1972z randomizes the Cantor construction by smoothing the steps of the distribution, and selecting the lengths and positions of the tremas at random, independently of one another. Finally, to implement the proportionality to u^{-1}, it is assumed that the number of tremas which are centered in an interval of length Δt and have a length above u has an expectation equal to $(1-D_*)\Delta t/u$ and a Poisson distribution. The reason for the notation $1-D_*$ will soon become clear.

Being independent, the tremas are allowed to intersect, and they do so with gusto: the probability of a trema's being intersected by no other trema is zero. In other words, the notions of *trema* and of *gap* cease to coincide: the term *gap* will be reserved for the intervals created by overlapping tremas. And the question arises of whether all the tremas eventually coalesce into one huge gap, or leave some points uncovered. We state the answer, then the next section justifies it by an intuitive birth process argument, and shows that the uncovered points form *unforced* clusters.

Consider an interval that fails to be wholly covered by tremas of length above ϵ_0, then introduce smaller tremas above a moving threshold ϵ that decreases from ϵ_0 to 0. When $D_* \leq 0$, letting $\epsilon \to 0$ makes it almost certain (the probability tends to 1) that *no* point is left uncovered. When $0 < D_* < 1$, the same outcome *may* also happen, but it ceases to be almost certain. With some positive probability, there is an uncovered "trema fractal" even at the limit. Mandelbrot 1972z proves that it is a Lévy dust of dimension $D = D_*$.

In summary, $D = \max(D_*, 0)$.

A BIRTH PROCESS AND UNFORCED CLUSTERING IN LÉVY DUSTS

By the construction of Chapter 8, Cantor errors come in hierarchical bursts or "clusters," the intensity of clustering being measured by the exponent D. This property is preserved when the gaps are shuffled at random, but the proof is neither perspicuous nor illuminating.

The proof of the same result for the random trema dust is, to the contrary, simple and of intrinsic interest.

The key, again, is to begin with tremas of length above a threshold ϵ, then to multiply ϵ repeatedly by some $r < 1$, say $r = \frac{1}{3}$, so that its value tends to 0. We start with a trema-free intergap interval bounded by two "ϵ-gaps." Adding tremas of length between $\epsilon/3$ and ϵ occasionally has the devastating effect of eras-

ing everything. But there is a good probability of seeing a much milder effect: (a) the bounding ϵ-gaps extend into longer ($\epsilon/3$)-gaps, and (b) small additional ($\epsilon/3$)-gaps appear within our intergap. The newly redefined intergaps are unavoidably perceived as clustered. And in the same fashion, subclusters are generated by replacing $\epsilon/3$ by $\epsilon/9, \ldots 3^{-n}\epsilon, \ldots$

These clusters' evolution as $n \to \infty$ is ruled by a novel birth and death process. As in the classical theory used in Chapter 23, clusters die or multiply independently of other clusters with the same n, and of their family histories. A long intergap has a smaller probability of being erased than a short one, and generates more children on the average. When $1 - D_*$ increases, the intervals between ϵ-gaps become shorter. And in addition some intervals between ($\epsilon/3$)-gaps disappear completely. Therefore, the expected number of offspring decreases in two ways. The value $D_* = 0$ is a critical value in the sense that for $D_* \leq 0$ the family line almost surely dies off, while for $D_* > 0$ there is a positive probability of seeing the family line survive forever.

MEAN NUMBERS OF ERRORS IN THE BERGER & MANDELBROT MODEL

◁ This technical digression proposes to show that the main results relative to the numbers of errors in the Cantor dust model remain valid after randomization. In fact the arguments and conclusions are considerably simplified, particularly if $\Omega = \infty$. The topic exemplifies the uses of conditional expectation in self-similar processes.

◁ Suppose that there is at least one error in the interval [0,R], the value of R being in the range where $R \gg \eta$ and $R \ll \Omega$. This condition reads $M(R) > 0$. The reason why the Berger & Mandelbrot model is called *conditionally stationary* is this: if [t,t+d] is entirely within [0,R], the conditioned number of errors, denoted by $\{M(t+d) - M(t) \mid M(R) > 0\}$, has a distribution independent of t. Hence it suffices to study it for t=0. Also, given that expectations are additive, conditional stationarity alone implies

$$\langle M(d) \mid M(R) > 0 \rangle = (d/R) \langle M(R) \mid M(R) > 0 \rangle.$$

As to self-similarity, it implies that

$$\Pr\{M(d) > 0 \mid M(R) > 0\} = (d/R)^{1-D^*},$$

where D^* is some constant to be determined by the process under study. To prove this assertion, it suffices to introduce an intermediate d' satisfying $d < d' < R$, and to decompose our conditional Pr as

$$\Pr\{M(d) > 0 \mid M(d') > 0\} \Pr\{M(d') > 0 \mid M(R) > 0\}.$$

Combining the last two equalities, we see that

$$\langle M(d) \mid M(d) > 0 \rangle = (d/R)^{D^*} \langle M(R) \mid M(R) > 0 \rangle.$$

Therefore, combining conditional stationarity and self-similarity suffices to show that

$$\langle M(d) \mid M(d) > 0 \rangle d^{-D^*} = \text{constant}.$$

The specific model under study determines the exponent as being $D^* = D$. Furthermore, self-similarity alone implies that the ratios

$$\{\text{instant of first error} \mid M(R) > 0\} / R,$$

$$\text{and } \{M(R) \mid M(R) > 0\} / \langle M(R) \mid M(R) > 0 \rangle$$

are random variables that depend on D but are independent of R and of Ω.

◁ By contrast with the conditional probabilities, the absolute probability of the conditioning event $M(R) > 0$ depends strongly on Ω. However, if the truncation to $\Omega < \infty$ is done properly, one finds that

$$\Pr\{M(R) > 0\} = (R/\Omega)^{1-D}.$$

Since this last expression can be deduced from an expression in the preceding paragraph by replacing R by L and d by R, the event "$M(R) > 0$ knowing that $L < \infty$" can be treated like the event "$M(R) > 0$ knowing that $M(L) > 0$." In the limit $\Omega \to \infty$, the probability that [0, R] falls completely within a very long gap converges to 1, so that the probability of observing an error becomes infinitely small. But the previously derived conditional probability of the number of errors is unaffected.

◁ The preceding argument adds to the discussion of the conditional cosmographic principle in Chapter 22. ▶   ■

Plate 285 ☐ **RANDOM PATTERN OF STREETS**

As noted in Chapter 8, it is regrettable that the Cantor dust should be so hard to illustrate directly. However, it can be visualized indirectly as the intersection of the triadic Koch curve with its base. And in the same way the Lévy dust can be imagined indirectly. On this plate, the black street-like stripes are placed at random, and in particular their directions are isotropic. Their widths follow a hyperbolic distribution and rapidly become so thin that they cannot be drawn. Asymptotically, the white remainder set (the "blocks of houses") is of zero area and of dimension D less than 2.

As long as the remaining blocks of houses have a dimension D>1, their intersection by an arbitrary line is a Lévy dust of dimension D−1. On the other hand, if D<1, the intersection is almost certainly empty. This result is, however, not very apparent here because the construction could not be carried far enough.

Chapter 33 provides a better illustration. When the tremas subtracted from the plane are random discs as exemplified by Plates 306 to 309, the trema fractals' intersections with straight lines are Lévy dusts. ■

285

D=.9000

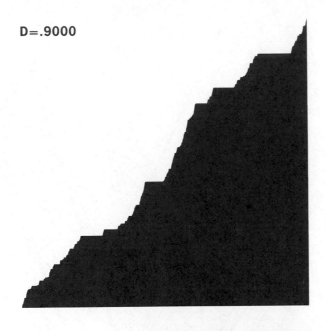

These graphs are randomized analogs of the Cantor function, or Devil's Staircase, in Plate 83. In the largest of these Lévy staircases, the dimension is the same as in the Cantor original, and in the two small ones it is either much smaller or much larger.

To draw a Lévy staircase, one evaluates the abscissa as function of the ordinate. In a first stage, whenever the ordinate increases by an amount Δy (in these instances, $\Delta y = .002$), the abscissa increases by a random amount having the distribution $Pr(\Delta X > u) = u^{-D}$. In a second stage, the abscissa is rescaled so that the staircase terminates at the point of coordinates $(1,1)$. The small staircase for $D = .3$ seems reduced to a small number of steps, due to the overwhelming clustering of the risers' abscissas. ■

D=.3000

D=.6309

32 ¤ Subordination; Spatial Lévy Dusts; Ordered Galaxies

The central concern of this chapter and the next is with galaxy clusters, a topic already touched in Chapters 9, 22, and 23. The underlying techniques generalize last chapter's dusts to the plane and the space. This chapter is primarily concerned with the spatial Lévy dusts. Following Bochner, we introduce these fractals by "processing" Brownian motion by the method of "subordination." Under the Lévy dust, one encounters the Lévy flight, a nonstandard random walk. The chapter begins with an informal preview of random walk clusters. Then subordination is explained and justified, by being extended to a nonrandom context. The claims made in the preview are justified in the final section.

PREVIEW: RANDOM WALK CLUSTERS

The goal of my early model of galaxy clusters was to exhibit a distribution of mass with the following features. (a) The mass $M(R)$ in a sphere centered on the distribution satisfies $M(R) \propto R^D$ with $D < 2$. (b) The distribution satisfies the conditional cosmographic principle in its statistical form.

RAYLEIGH FLIGHT STOPOVERS. A preliminary is provided by a construct that has neither the fractal nor the topological dimension of galaxy clusters. Starting from a point $\Pi(0)$ in space, a Rayleigh flight rocket jumps in an isotropic random direction. The duration of each jump is $\Delta t = 1$, and the distance U to the next stopover $\Pi(1)$ is random Gaussian with $\langle[\Pi(1) - \Pi(0)]^2\rangle = 1$. The rocket then jumps off to $\Pi(2)$, such that

$$U_1 = \Pi(1) - \Pi(0) \text{ and } U_2 = \Pi(2) - \Pi(1)$$

are independent and identically distributed vectors. And so on.

In order to view the rocket as going on forever, we add its previous stopovers $\Pi(-1)$, $\Pi(-2)$.... But a change in the direction of time does not affect a random walk, hence it is sufficient to draw two independent trajectories starting from $\Pi(0)$.

Our rocket's trail (including the "contrails" it leaves as it jumps) is a random set. So is the collection of its stopovers considered without taking into account the order in which they were reached. Both sets follow exactly the same distribution when examined from any of the points $\Pi(t)$. In the terms introduced in Chapter 22, both sets satisfy the conditional cosmographic principle in its proper statistical form.

LOADING. Identically distributed and statistically independent masses are assigned at random to each stopover of a Rayleigh flight, extending conditional stationarity to mass.

THE DIMENSION D=2. As is widely known, the distance the flight covers in K jumps increases like $\sqrt{K}$. A consequence is that in a ball with radius R and center $\Pi(t)$ the number of stopovers is $M(R) \propto R^2$. The exponent in the last formula conforms to the idea that the dimension of the set of stopovers $\Pi(t)$ is D=2. In particular, the global density vanishes.

BROWNIAN MOTION. By interpolating the Rayleigh flight in continuous time, one obtains a Brown trail, which (Chapter 25) is a continuous curve with D=2. Thus the Rayleigh flight model is essentially a fractal curve ($D_T=1$ and D=2) satisfying the conditional, but not the strong, cosmographic principle. The last conclusion is satisfactory, but the values of D_T and D are unacceptable.

GENERALIZED DENSITY. If we load a Brown trail between the points $\Pi(t_0)$ and $\Pi(t)$ by the mass $\delta|t_0-t|$, the mass M(R) becomes the time spent in the ball of radius R, multiplied by the uniform generalized density δ.

EXPANSION OF THE UNIVERSE. In standard discussions, the initial distribution has a uniform density δ. As the Universe expands uniformly, δ decreases, but the distribution remains uniform. On the other hand, it is generally believed that every other distribution changes by expansion. The uniformly loaded Brown trail shows constructively that this conclusion is *incorrect*: again, δ changes with expansion, but it remains defined and uniform.

Therefore Rayleigh stopovers are neutral with respect to the question of whether our Universe does or does not expand. This property is preserved when D is decreased through the use of the Lévy flight, to be surveyed now.

LÉVY FLIGHT STOPOVERS; NONINTEGER DIMENSIONS <2. My random walk model of the distribution of galaxies implements any desired fractal dimension D<2 using a dust, i.e., a set of correct topological dimension $D_T=0$. To achieve this goal, I use a random walk wherein the mathematical expectation $\langle U^2(t)\rangle$ is *infinite*, because U is a hyperbolic random variable, with an inner cutoff at u=1. Thus, for $u \leq 1$, $Pr(u>u)=1$, while for u>1, $Pr(U>u) \propto u^{-D}$, with 0<D<2.

A major consequence is that $\langle M(R)\rangle \propto R^D$ when R>>1. This is the relationship we had set out to implement. It allows any dimension likely to be suggested by fact or theory.

◁ ASIDE ON LÉVY STABILITY. As $t \to \infty$, the mass carried over a time span t (properly scaled) converges to a random variable independent of t, first investigated by Paul Lévy and best called "Lévy stable" (Chapter 39). Hence the term "Lévy flight" proposed for the process underlying my model.

◁ Due to $\langle U^2 \rangle = \infty$, the standard central limit theorem ceases to be valid, and a special central limit theorem must be used instead. This replacement has considerable consequences. The standard theorem is "universal," in the sense that the limit depends only on the quantities $\langle U \rangle$ and $\langle U^2 \rangle$. The nonstandard theorem is *not* universal. Through D, the distribution of M(R) depends explicitly upon the distribution of the jumps. ►

The remainder of this chapter constructs a dust that plays relative to the Lévy flight the same role as Brownian motion plays relative to Rayleigh flight. A direct interpolation is tediously technical, because it must give a meaning to the distribution $Pr(U > u) = u^{-D}$ applied down to $u = 0$, where it diverges. An indirect approach, to the contrary, can be made both simple and precise, through the use of the process of subordination. This process is of independent interest and opens up numerous obvious generalizations.

CAUCHY FLIGHT AND D=1

We introduce subordination through an example. To generate a dimension equal to D=1 starting with the Brown trail of dimension D=2, we must seek to decrease D by 1. In the case of classical shapes from Euclid, such a decrease is easy to achieve. In the plane, it suffices to take the section by a line; in 3-space, it suffices to take the section by a plane; and in 4-space, to take a section by a

3-space. We also saw in Chapter 23 that the same rule holds for random fractal curds, and in Chapter 25 that the Brown line-to-line function has the dimension 3/2, while its zeroset and all sections that are not perpendicular to the t-axis have the dimension ½.

Extended by formal analogy, this method for subtracting 1 from D leads us to suspect that appropriately selected sections of a Brown trail are typically of dimension 2−1=1. This hunch is indeed verified (Feller, 1971, p. 348). Moreover, it should extend to plane sections of a trail in the ordinary 3-space and to 3-dimensional sections of a trail in 4-space, in which the coordinates are x, y, z, and humor.

Starting from a line-to-4-space Brownian trail, consider the points where humor=0. These "humorless" sites can be viewed as generated in the order in which they are visited by the underlying Brownian motion, and the distances between such visits are independent and isotropic. As a result, the humorless sites can be viewed as the stopovers of a random flight whose steps follow rules very different from those of Brownian motion. This walk will be called Cauchy motion or flight. Given two time instants 0 and t, one finds that the probability density of the vector from $\Pi(0)$ to $\Pi(t)$ is a numerical multiple of

$$t^{-E}[1 + |\Pi(t) - \Pi(0)|^2 t^{-2}]^{-E/2}.$$

The formal hunch that D=1 is confirmed in S. J. Taylor 1966, 1967. The Cauchy flight is illustrated in one of the views of Plate 298.

THE IDEA OF SUBORDINATION

Let us ponder the preceding construction. A line-to-E-space Brownian motion hits the humorless points at the times when one of its line-to-line coordinate functions vanishes. But each of the coordinates is a one-dimensional Brownian motion. Not only (Chapter 25) do this function's zerosets form a set of dimension $D=\frac{1}{2}$, but the fact that interzero intervals are mutually independent implies that this zeroset is a linear Lévy dust. In summary, the Cauchy motion is the map on a Brownian motion of a linear Lévy dust. Recalling that decimation was the Romans' charming way of punishing a hostile group by killing every tenth member, we see that Cauchy motion is obtained by a fractal form of decimation. It was pioneered in Bochner 1955, who called it *subordination*. (Feller 1971 includes scattered nonelementary comments on this notion.)

For future reference, let us note that

$$D_{\text{Cauchy trail}} = D_{\text{Brown trail}} \times D_{\text{Brown zeroset}}.$$

SUBORDINATION CAN BE EXTENDED BACK TO NONRANDOM FRACTALS

To elaborate on the nature of fractal subordination, we apply it to some Koch and Peano fractal curves. (Oddly enough, the present discussion seems to be the first mention of subordination in a *non*random context.)

The idea is that one can modify these curves by leaving the initiator unchanged but replacing the generator by a subset of the original. This replaces the limit fractal set, to be called the *subordinand*, by a *subordinate* subset. First we describe examples, then we introduce the important *rule of multiplication of dimensions*.

EXAMPLE WITH D < 2. Take the four-legged generator of the triadic Koch curve, as used in Plate 42. Erasing the second and third legs yields the classic generator of the triadic Cantor dust, Plate 78. Thus, the Cantor dust is subordinate to a third of a snowflake. A different subordinate dust, not restricted to the line, results if one erases the first and third of the N=4 sides of the Koch generator. In either case, subordination changes the dimension from log4/log3 to log2/log3. If only one leg of the generator is erased, the subordinate dust is *not* a subset of the line, although it is of dimension log3/log3=1.

EXAMPLE WITH D = 2. Take the four-legged second stage of the Peano-Cesàro curve of Plate 64, and erase the second and third leg. The new generator is the interval [0,1] itself! Thus, the straight interval is a (most trivial!) subordinate of the Peano-Cesàro curve. Erasing a different set of two legs yields a fractal dust with D=1. Erasing one leg leaves a set of dimension log3/log2.

MULTIPLICATION OF DIMENSIONS

Recall from Chapters 6 and 7 that the Koch and Peano curves can be viewed as the trails of "motions" whose time parameter t lies in [0,1]. This time is defined in such a way that, to take an example, a snowflake generator's four legs are covered during the instants whose expansion in the base 4 begin respectively with 0, 1, 2, and 3. For example, the second fourth of the third fourth is covered during the instants whose expansion in the base 4 begins with 0.21. Viewed as motions, our Koch or Peano curves are themselves "fractal maps" of the interval [0,1]. In this framework, the effect of the first mentioned decimation of generator legs is to eliminate the values of t that include the digits 1 or 2 (or 0 and 3), thus limiting t to belong to a certain Cantor dust of [0,1].

We can therefore describe our subordinate subsets of the Koch or Peano curves as *fractal* maps of a *fractal* subset of time. This subset is clearly a Cantor dust, and it is called *subordinator*. Its dimension is $\log N / \log N' = \log 2 / \log 4 = \frac{1}{2}$. More generally, we find the self-explanatory relation

$$D_{subordinate} = D_{subordinand} \times D_{subordinator}.$$

This generalizes the relation we saw characterizes Cauchy motion. As we know, sums of dimensions occur in the study of sections and intersections. Now we discover a lovely "calculus," giving a meaning to products of dimensions as well as to sums.

Of course, this rule suffers exceptions, analogous to those applicable to the rule that codimensions add under intersection.

LINEAR LÉVY DUST AS SUBORDINATOR

The linear Lévy dust of Chapter 31 is the first subordinate used by Bochner, and it continues to be so widely used as subordinator by pure mathematicians that the related Lévy staircase is often called the stable subordinator function. To obtain self-similar subordinate sets, one uses a self-similar subordinand, e.g., Brownian or fractional Brownian motion.

Observe that, while Brownian motion's intrinsic dimension is 2, Brownian motion restricted to the line is of dimension 1. Therefore, last section's rule is replaced by

$$D_{subordinate} = \min\{E, 2 \times D_{subordinator}\}.$$

More generally, a fractional Brownian motion's intrinsic dimension is $1/H$, but

$$D_{subordinate} = \min\{E, D_{subordinator}/H\}.$$

Thus, the largest space that the subordinate set can fill to the hilt corresponds to E=integer part of $1/H$.

BROWNIAN MOTION AS SUBORDINAND. The most important subordinand is the Brown trail. The Brownian map of time instants restricted to a linear Lévy dust of dimension $D/2$ between 0 and 1 is a spatial dust with arbitrary dimension D between 0 and 2. It deserves to be called spatial Lévy dust.

Granted that the subordinator dust's gaps and the subordinand's increments are both statistically independent, the subordinate process also has statistically independent increments. Granted that the subordinator's gap lengths satisfy $Pr(W>w)=w^{-D/2}$, and that during a gap of duration w the Brownian motion moves by an amount of the order of $u=\sqrt{w}$, the spatial dust's gaps seem to satisfy $Pr(U>u) = Pr(W>u^2) = u^{-D}$. It can be shown that such is indeed the case.

ORDERED GALAXY CLUSTERS

The formula $Pr(U>u)=u^{-D}$ shows that the subordinate dust implements the process previewed at the beginning of this chapter.

DIMENSIONS. The dust itself is of dimension D. If the maps of each linear gap's endpoints are joined by intervals, one obtains a Lévy trail; its dimension is $max(1,D)$—as in the study of trees in Chapter 16.

CORRELATIONS. A Lévy trail induces a linear ordering among the galaxies it generates, implying that each galaxy only interacts with its immediate neighbors. And each couple of neighbors interacts independently of the other couples. In this sense, a Lévy flight is equivalent to the unjustified replacement of an unsolvable N-body problem by a manageable combination of many two-body problems. The result might have been atrociously unrealistic, but is not. Mandelbrot 1975u (described fully in Peebles 1980, pp. 243-249) shows that Lévy flight leads to two- and three-point correlations on the celestial sphere that are identical to those that P. J. E. Peebles and Groth obtained in 1975 by curve fitting; see Peebles 1980. ∎

Plate 293 ⌑ THE COMPUTER "BUG" AS ARTIST, OPUS 2

This plate can be credited in part to faulty computer programming. The "bug" was promptly identified and corrected (but only after its output had been recorded, of course!), and the final outcome was Plate 69.

The change that had been wrought by a single tiny bug in a critical place had gone well beyond anything we had expected.

It is clear that a very strict order is designed into Plate 69. Here, this order is hidden, and no other order is apparent.

The fact that, at least at first blush, this plate could pass for High Art, cannot be an accident. My thoughts on this account are sketched in Mandelbrot 1981l, and are to be presented fully in the near future. ∎

Plates 296 and 297, AFTER PLATE 295 ¤ MANDELBROT EARLY MODEL CLUSTERS OF DIMENSION D=1.2600. LÉVY FLIGHT AND ITS STOPOVERS

A Lévy flight is roughly a sequence of jumps separated by stopovers. Only the latter are of direct interest in this chapter, but jumps are a necessary part of the construction.

Therefore, the top (black on white) figures in these plates include, as part of the motion's trail, the "contrails" created during actual flights. The trail in three-dimensional space is shown through its projections on two perpendicular planes. The original can be visualized by holding the book half open.

To proceed to the bottom (white on black) figures, one wipes away the intervals that represent the jumps. Then one takes a photographic negative. Each stopover is a star, a galaxy, or a more general blob of matter.

More precisely, the straight intervals of the black-on-white top figures have the following characteristics. Their direction in space is random and isotropic (that is, parallel to the vector joining the origin of space to a point chosen at random on a sphere). The different intervals are statistically independent, and their lengths follow the probability distribution $Pr(U>u) = u^{-D}$, except that $P(U>u) = 1$ when $u<1$. The value of $D=1.2600$ is close to the $D\sim1.23$ found for actual galaxies.

The overwhelming majority of the intervals are too small to be perceived. In fact, we lined the plane with a uniform grid and marked the cells containing one or more stopovers. In other words, each point stands for a whole minicluster.

In addition, regardless of D, the miniclusters are themselves clustered. They exhibit such clear-cut hierarchical levels that it is hard to believe that the model involves no explicit hierarchy, only a built-in self-similarity.

Let us elaborate by mentioning that all the plates in the present portfolio represent the beginning of two distinct flights, forward and reverse, and such flights are nothing but two statistically independent replicas of the same process. Clearly, if the origin is displaced to some other stopover, the two halves are again independent. Hence, every stopover has precisely the same claim to be called the Center of the World. This feature is the essence of the *conditional cosmographic principle* I propound in this Essay.

The present method does not claim to account for the way the galaxies had actually been generated, but it brings home my theme that the conditional cosmographic principle is compatible with ostensible multilevel clustering. A great variety of such configurations may be present even when none has been inserted "to measure." ∎

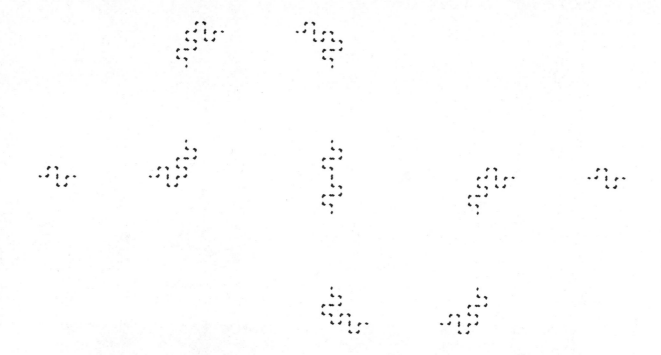

Plate 295 ◻ NONRANDOM SUBORDINATION: CLUSTERED FRACTAL DUST, OF DIMENSION D=1, SUBORDINATE TO A KOCH CURVE OF DIMENSION D=1.5

One can modify the method of recursion basic to the Koch construction so that it breaks a line systematically and leads to a dust that has the same dimension D=1 as the line, but is entirely different in topology and appearance.

Imagine that a rubber band, initially laid along [0,1], is extended to follow the Koch generator that is used in Plate 49 to yield a fractal curve of dimension 3/2. Then the corners are pinned down permanently, and each of the 8 straight intervals of the band is cut in its middle, leaving 16 pieces that snap back to their original lengths 1/16. These pieces' free ends are then pinned down, and the process is repeated. The end result is a self-similar hierarchically clustered dust with r=1/16 and N=16, hence D=1.

This construction amounts to allowing us

to mark a generator's side so it is erased at the next stage of the Koch construction. This process is called subordination in the text. The only points we keep are the positions of a Koch motion when time belongs to a subset of fractal dimension $\log 16/\log 64 = 4/6$. And the fact that $(4/6)\times(3/2) = 1$ is a special case of the rule of multiplication of dimensions discussed in the text.

Note that all the points on this plate are ordered intrinsically by the Koch curve of which the generator is a subset. Furthermore, it is easy to derive the frequency distribution of the snap-back distances between successive pinned-down points. Roughly, the number of distances $\geq u$ is proportional to u^{-D} with D=1, Plates 296 and 297 use the same frequency distribution differently. ∎

D=1.2600. CAPTION IS ON P. 294

D=1.2600. CAPTION IS ON P. 294

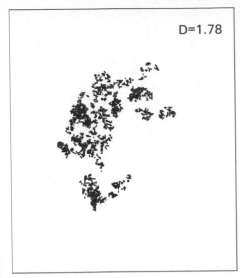

D=1.78

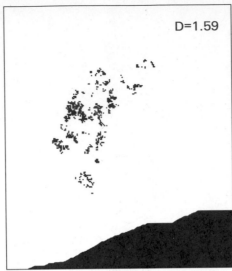

D=1.59

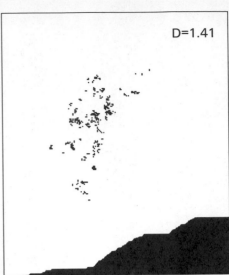

D=1.41

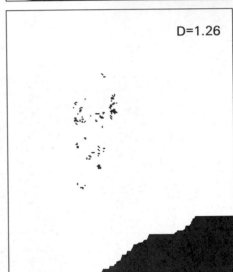

D=1.26

Plate 298 ◻ DECREASE OF D BY SUBORDINATION.
MAKING THE LÉVY CLUSTERS BECOME INCREASINGLY SEPARATE

A planar Lévy dust's degree of clustering depends upon its D. Here, this effect is illustrated by processing a planar Brownian trail, with D=2, using successive linear Lévy subordinations, each riding on its predecessor. Throughout, $D_{subordinator} = 2^{-1/6} = .89$, hence the subordinate dusts have the dimensions: 1.78 (= 2×.89), 1.59, 1.41, 1.26, 1.12, 1, and .89. The Lévy staircases next to most dusts show how time was decimated to generate this dust from the dust with D=1.78. A "ghost" of the subordinand, a continuous Brown trail, is perceived clearly for D close to 2, but becomes increasingly faint as D decreases (see Chapter 35). Increasing clustering is not provoked by the concentration of all points around a few of them but by the disappearance of most points, leading to an increasing number of apparent hierarchic levels. ◼

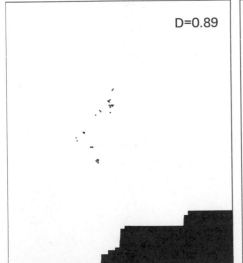

D=0.89

D=1.00

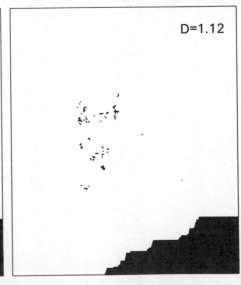

D=1.12

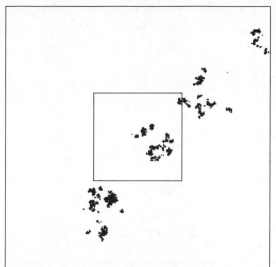
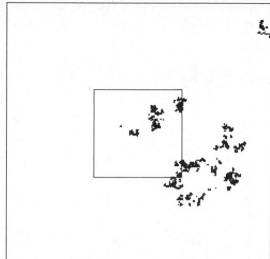

Plate 299 ⌑ ZOOMING TOWARD A LÉVY DUST WITH D=1.2600

The first figure on the top left represents a cluster of 12,500,000 positions of a Lévy motion, as seen through the square window from a far away spaceship. Between each view and the view that follows in clockwise direction, the distance from the spaceship to the center of the cluster and the field of vision are divided by b=3. The structure seen through the window changes in detail, but remains unchanged in broad lines. This is expected, due to the fact that the set is self-similar. ■

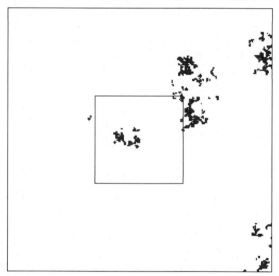

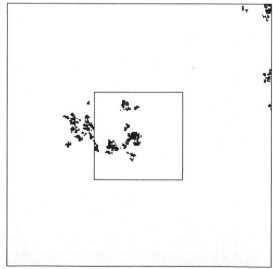

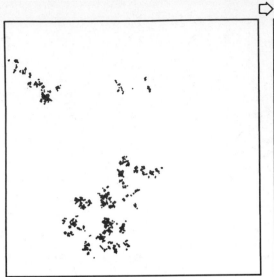

Plates 300 ¤
CIRCUMNAVIGATION OF LÉVY
CLUSTERS OF DIMENSION D=1.3000

The shape of clusters generated as sites of a Lévy flight in the plane is highly sample dependent, meaning that if one simulates clusters again and again while keeping the same dimension, one must expect to obtain a great variety of different shapes.

The same is true of a small isolated spatial Lévy cluster when viewed from many different directions—by following the present "strip" clockwise from the top of this plate. ■

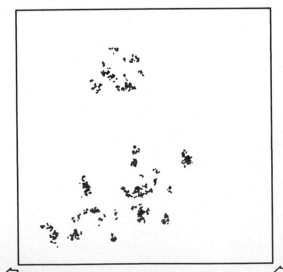

33 ◻ Disc and Sphere Tremas: Moon Craters and Galaxies

Having introduced the linear Lévy dust as a trema fractal, via interval shaped random tremas (Chapter 31), we promptly sidetracked in Chapter 32: we generalized this dust to the plane and the space via the process of subordination. In this chapter and the next, we generalize the random tremas directly.

In this chapter, the planar and spatial tremas are discs and balls, hence the generalization bears directly upon the shapes of Moon craters and of meteorites. But the spatial tremas' most important application is a different and less obvious one. When D is close to 1, a trema fractal is a dust, hence a candidate to replace Lévy flight stopovers in modeling the galaxy clusters. The main novelty, compared with the random walks model, is that here galaxies are *not* ordered along a trail. Hence a gain in a priori verisimilitude, a resulting loss in computational convenience, and an ultimate gain in quality of fit: the predicted covariance properties are even closer to the empirical evidence. The nonspherical tremas in Chapter 35 improve the fit further.

PLANAR AND SPATIAL TREMAS

As background for the random and overlapping tremas, let planar curdling in a grid, Chapters 13 and 14, be restated in terms of virtual tremas. The first cascade stage consists in marking N out of b^2 squares, and keeping them as curds. Alternatively, one may say that the first stage cuts out b^2-N square tremas. The next stage cuts out second-order square tremas numbering $b^2(b^2-N)$, including $N(b^2-N)$ genuinely new tremas and $(b^2-N)^2$ tremas that are "virtual": they eliminate again something that had already been eliminated in the first stage. And so on.

Counting both genuine and virtual tremas, we find that the number of tremas with an *area* in excess of s is proportional to $1/s$. The corresponding result relative to curdling in 3-space is that the number of tremas with a *volume* in excess of v is proportional to $1/v$.

Similarly, the bulk of this chapter and of Chapter 35 concerns the case in which the numbers of independent tremas centered in a box of sides dx and dy, or dx, dy, and dz, is a

Poisson random variable of expectation

$$\langle Nr(area > a)\rangle = (C/2a)dx\,dy,$$
$$\langle Nr(volume > v)\rangle = (C/3v)dx\,dy\,dz.$$

The corresponding expectation in $\mathbb{R}^E$ is

$$(C/Ev)dx_1...dx_E.$$

The fractal properties of the resulting trema set are as simple as in the linear case tackled in Chapter 31. When $C<1$, these properties can be derived from those of the linear case, and the predecessor Essays conjectured they held for all C. This was confirmed by El Hélou 1978.

When $C>E$, the trema set is almost surely empty. When $C<E$, it is a fractal of dimension $D=E-C$.

As to the trema fractals' topology, general principles show that a trema set with $D<1$ is a dust with $D_T=0$. When $D>1$, on the other hand, general principles do not suffice, the topology being determined by the trema's shape. The problem of percolation arises here, in yet another fractal context.

LUNAR CRATERS AND DISC TREMAS

We begin with a side issue that provides an easier two-dimensional preparation and is amusing: the geometric nature of the set that lunar craters leave uncovered. While the Greek $\kappa\rho\alpha\tau\eta\rho$ denotes a bowl or drinking vessel, almost all of Earth's craters are of volcan-

ic origin. But it is generally believed that the craters observed on Earth's Moon, Mars, the Jovian satellite Callisto, and other planets and their satellites, are overwhelmingly due to the impact of meteorites.

The larger the meteorite, the larger and deeper the crater resulting from its impact. Furthermore, a large crater due to the late impact of a heavy meteorite may wipe out several previously formed small craters, while a small crater due to the late impact of a light meteorite may "dent" the rim of a large older crater. As for the sizes, there is solid empirical evidence that at the moment of meteorite impact the crater areas follow a hyperbolic distribution: the number of craters with an area exceeding s km^2 and such that their centers are located within a square of 1 km^2 can be written as C/s, with C a constant. This evidence is discussed in Marcus 1964, Arthur 1954, and Hartmann 1977.

To simplify the argument (with no change in the main result), we approximate the lunar surface by a plane, and the lunar craters by disc-shaped tremas. If the Moon went on perpetually scooping up meteorites from a statistically invariant environment, every point of its surface would obviously be covered again and again ad infinitum. However, it may be that craters are wiped clean every so often, say by volcanic lava, in which case the trema set they fail to cover at a given moment may be nontrivial. Alternatively, it may be that the solar system evolved in such a fashion that our Moon was only bombarded during a finite period of time. The parameter C may measure either the time since the last attrition of craters or the total duration of the bombardment.

To assess its effect upon the trema fractal's shape, let us keep the seed invariant, and vary C. As C increases from 0 to 2, the Moon's surface becomes increasingly saturated, and the result stated in the preceding section shows that D decreases and reaches 0 for $C \geq 2$. The trema fractal's dependence upon D is illustrated by Plates 306 through 309.

APPENZELLER AND EMMENTHALER. When C is very small, other lovers of Swiss cheese may join me in thinking that the shape we deal with resembles a slice of cheese that is almost entirely pierced by very small pin holes. It is a wild extrapolation of the structure of Appenzeller. When C increases, we turn progressively to a wildly extrapolated Emmenthaler, with large overlapping holes.

(Thus, the English nursery rhyme about the Moon being made of green cheese proves correct, except for color.)

TOPOLOGY. CRITICAL D'S. Either of the above extrapolations of cheese must be called "wild," because the trema fractal "cheese slices" are of vanishing area. I conjecture the following. As long as C is small enough, the trema fractal is a σ-cluster, each contact cluster being a web of connected filaments and having the topological dimension $D_T = 1$. When D reaches a certain critical dimension, D_{crit}, the value of D_T drops to 0, and the σ-web collapses into dust.

The next critical dimension is D=0. When C>2, the Moon's surface is oversaturated, every point being almost certainly covered by

at least one crater. In particular, such would be the case if the Moon's surface were never wiped clean and continued scooping up meteorites endlessly.

NONSCALING CRATERS. Some planets other than Earth's Moon are characterized by a density of craters of the form $Ws^{-\gamma}$ with $\gamma \neq 1$. The problem these craters raise is tackled in the appendix to this chapter.

GALAXIES AND INTERGALACTIC VOIDS GENERATED VIA SPHERICAL TREMAS

While the Moon's tremas have an independently recognized existence as craters, ball-shaped tremas with a scaling distribution began as a natural extension of the same geometric device to space. I thought they may yield an alternative to the galaxy model of Chapter 32. Thus, I postulated the existence of intergalactic voids that combine many tremas, and may range up to very large size. The good fit of the resulting model was a very pleasant surprise, and demands further theory (Chapter 35) and experiment.

COVARIANCES. Because the statisticians and the physicists trust correlations and spectra, the first test of the trema fractals as models of galaxy clusters relies upon their correlation properties. The covariance between two points in space is the same as in my random walk model, as it should be, since the latter fitted the data well. The same is true, as it should be, of the covariance between two directions in the sky. The predicted covariances between three and four directions fit better than those predicted by the random walk model, but the improvements are technical and are better discussed elsewhere. Basically, once D is known, the various models give the same correlations.

Now recall that Gaussian phenomena, including Brown or fractional Brown fractals, are fully characterized by the covariance properties. When they are scaling, they are fully characterized by D. Given the influence of the Gaussian phenomena on the statisticians' thinking, one may be tempted to stop at the covariances. But fractal dusts are *not* Gaussian phenomena, and their D fails to specify many important facts about them.

CRITICAL DIMENSIONS. More basic than the correlation is the question of whether the trema fractals have the right topology. To check, it is best, as in the preceding section, to keep a fixed seed and let C increase from 0 to 3. As long as C is small, $D_T=2$, and our fractal is made of ramified veils. When D traverses a certain value D_{2crit}, called upper critical dimension, the veils split into filaments, with $D_T=1$. And when D traverses a smaller value D_{crit} called lower critical dimension, the filaments collapse into dust, with $D_T=0$. Since the modeling of galaxy clusters requires dusts, it is important to verify that D_{crit} exceeds the observed $D\sim1.23$. My computer simulations confirm this inequality.

PERCOLATION. The hope that the world is not more complicated than need be makes me believe that $D>D_{crit}$ is the necessary and sufficient condition for the trema fractal to percolate, in the sense described in Chapter 13.

METEORITES

The mass distribution of Earth impacting meteorites has been studied carefully, for example in Hawkins 1964. Mid-size meteorites are made of stone, and 1 km^3 in space contains roughly $P(v)=10^{-25}/v$ meteorites of volume exceeding v km^3.

This claim is ordinarily expressed differently, using the following very mixed units. During each year, each km^2 of Earth's surface is on the average host to $0.186/m$ meteorites of mass above m grams. Their average density being 3.4 g cm^{-3}, this relation boils down, in more consistent units, to 5.4 $10^{-17}/v$ meteorites of volume exceeding v km^3. Moreover, Earth moves on by roughly 1 km during 10^{-9} years—the inverse of the order of magnitude of Earth's trajectory around the Sun in km. Hence, using consistent units, and keeping to orders of magnitude so that 5.4 becomes 10, we find that while Earth moves on by 1 km in space, each km^2 of Earth's surface is host to $10^{-25}/v$ meteorites of volume exceeding v km^3. Assuming that the meteorites impacting Earth as it sweeps through space are a representative sample of the meteorites' distribution in space, we obtain the result that has been asserted.

This $10^{-25}/v$ law is formally identical to the C/s law for lunar craters, but there is a difference: craters can overlap, while meteorites cannot.

Nevertheless, it is fun to see what would happen if $P(v)=10^{-25}/v$ held down to $v=0$ and—wild thought!—if meteorites could overlap. Adding the innocuous assumption that meteorites are spherical, the trema set can be investigated directly (with no need of the results in El Hélou 1978). The sections of meteorites by straight lines randomly thrown in space are rectilinear tremas, and it can be shown that the number of such intervals centered within 1 km and of length exceeding u km is $C'10^{-25}/u$. (C' is a numerical factor of the order of magnitude of 1, unimportant in this context.) Hence a result in Chapter 32 shows that the dimension of the trema set's linear section is $1-10^{-25}$. Adding 2 when we go back from the linear sections to the full shape, we find $3-D=10^{-25}$.

This result is inane, since it implies in particular that meteorites nearly fill space, even after one allows for overlap. Nevertheless, the codimension $3-D=10^{-25}$ deserves just another glance. Let us assume in a first approximation that the $10^{-25}/v$ relationship holds down to a positive cutoff $\eta>0$ and that there is no meteorite of smaller size. The argument we have sketched asserts that if one could actually pass to the limit $\eta \to 0$, the set outside of all meteorites would converge to a trema set of dimension $D=3-10^{-25}$. Fortunately, this limit set would be attained so extraordinarily slowly that in the observable range allowing meteorite overlap can pose no problem. Unfortunately, the value of D can have no practical importance whatsoever.

APPENDIX: NONSCALING CRATERS

The Moon's crater distribution is best written for the present purpose as $Pr(A>a)=Fa^{-\gamma}$, with $\gamma=1$. The same γ seems to hold for Mars, but for Jovian satellites one finds different values of γ (Soderblom 1980). Similarly, $\gamma<1$ for small volume meteorites. The resulting trema sets are not scaling.

THE CASE WHEN $\gamma>1$. In this first nonscaling case, any given point of planetary surface, regardless of the value of W, almost surely falls into an infinity of craters. The surface texture is overwhelmingly dominated by small craters. The Jovian satellite Callisto has such a texture, and indeed it is characterized by $\gamma>1$. When discussed in predecessors to this Essay, before the Voyager mission, $\gamma>1$ was merely a theoretical possibility.

THE CASE WHEN $\gamma<1$ AND CRATER AREAS ARE BOUNDED. Denoting this bound by 1, the probability for a point to remain outside all craters is positive ◁ because the integral $\int_0^1 Pr(A>a)da$ converges ▶, but it decreases as W increases. The resulting pocked surface is (even more than in the scaling case) reminiscent of a slice of Swiss cheese. The greater the value of γ, the smaller the number of small holes, and the more "chunky" the resulting cheese. However, regardless of the value of γ, the slice is of positive area, hence it is a (non-self-similar) set of dimension 2. On the other hand, I have no doubt that its topological dimension is 1, meaning it is a fractal.

In space (meteorites) this trema fractal's dimensions are $D=3$ and $D_T=2$.

**Plates 306 and 307 ¤ SMALLISH ROUND TREMAS, IN WHITE,
AND RANDOM SLICES OF "SWISS CHEESE" (DIMENSIONS D=1.9900 AND D=1.9000)**

The tremas are white circular discs. Their centers are distributed at random on the plane. For the disc of rank ρ, the area is $K(2-D)/\rho$, the numerical constant then being chosen suitably to fit the trema model described in the text. Plate 306 shows a sort of Appenzeller wherein the black portion is of dimension D=1.9900, and Plate 307 a sort of Emmenthaler with a black portion of dimension D=1.9000. ■

306

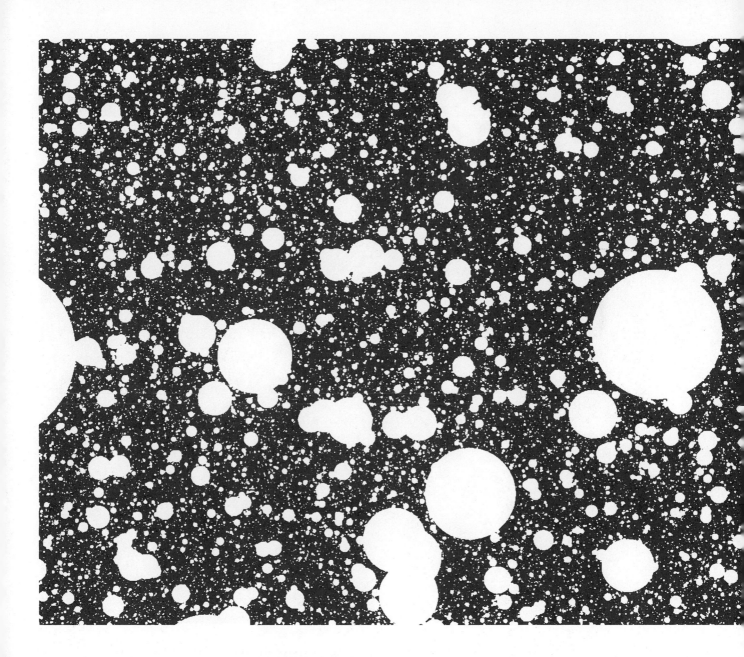

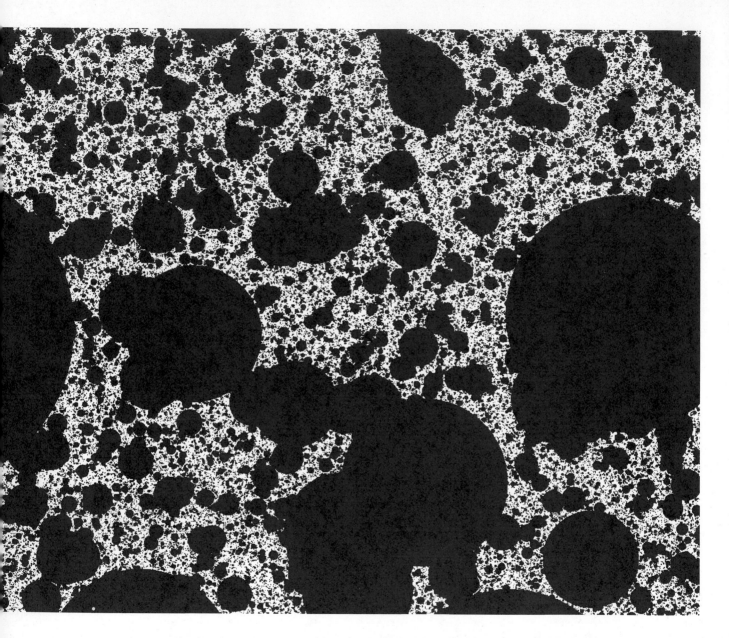

Plates 308 and 309 ¤ **LARGER ROUND TREMAS, IN BLACK, AND RANDOM FORKED WHITE THREADS (DIMENSIONS D=1.7500 AND D=1.5000)**

The construction proceeds as in Plates 307 and 308, but the tremas are bigger, so hardly anything is left out, and they are represented in black. The D's are the dimensions of the remaining white fractal. ▬

308

34 ¤ Texture:
Gaps and Lacunarity; Cirri and Succolarity

Texture is an elusive notion which mathematicians and scientists tend to avoid because they cannot grasp it. Engineers and artists cannot avoid it, but mostly fail to handle it to their satisfaction. There are many indications, however, that several individual facets of texture are about to be mastered quantitatively.

In fact, much of fractal geometry could pass as an *implicit* study of texture. In this and the next chapter, two specific facets are approached *explicitly*, with stress on galaxy clusters. Remarks on texture could have been scattered between earlier chapters, beginning with Chapters 8 and 9, but it seemed preferable (at the cost of interrupting the discussion of tremas!) to collect all my comments on texture in one place.

As stated repeatedly, my search for models of galaxy clusters proceeded by stages. Early ones, described in Chapters 32 and 33, fit the desired D while preserving the conditional cosmographic principle. Later ones, described in Chapter 35, also fit texture.

This chapter's several introductory sections present the basic observations about galaxies that led me to distinguish two aspects of texture, calling them *lacunarity* and *succolarity*. *Lacuna* (related to *lake*) is Latin for *gap*, hence a fractal is to be called *lacunar* if its gaps tend to be large, in the sense that they include large intervals (discs, or balls). And a *succolating* fractal is one that "nearly" includes the filaments that would have allowed percolation; since *percolare* means "to flow through" in Latin (Chapter 13), *succolare* (sub-colare) seems the proper neo-Latin for "to almost flow through."

The remainder of this chapter introduces several measures of lacunarity, but measures of succolarity are beyond the present elementary discussion.

Chapter 35 proceeds to show how lacunarity and succolarity can both be controlled through tremas.

Up to now, a predominant role in measuring fractals was given to the topological and fractal dimensions. Chapter 14 was an exception (without follow up), since the order of ramification injects finer distinctions between fractals that share the same values of D_T and of D. We encountered many different expressions of the form

prefactor x (quantity)^{exponent},

but so far we only considered the exponent. The study of texture forces us to extend our attention to the prefactor. Since it could not be neglected forever, we cannot be surprised that neither Nature (science) nor human thought (mathematics) are simple!

GALAXIES' "CIRRIFORM" FILAMENTS

A mysterious empirical finding was brought to my attention in Paris, in 1974, after my first lecture on the model described in Chapter 32. My sole purpose had been to achieve the desired value of D in a fractal (actually, I had not yet coined the term *fractal*). But an unidentified astronomer pointed out during the discussion that there was a further, unexpected, element of verisimilitude: on the samples generated by my model, the points often seem to fall along nearly straight lines, and more generally seem scattered along narrow "near-streams" or "near-filaments." The unidentified astronomer informed me that the galaxies shared this property in even clearer form and that a "near-stream" of galaxies observed more closely decomposes into thinner "near-streams." The astronomer stressed that *streams* was a very poor term, the structures in question being disconnected.

To avoid being confused by terminology, I recalled that filmy fleecy clouds are called *cirri* by meteorologists, and filed away the information that galaxies have a *cirriform structure*, and that it would be desirable to improve the model to make the cirri even more apparent.

Actual references came forth much later: Tombaugh had observed "cirri" in 1937, in the Perseus Supergalaxy, and de Vaucouleurs had confirmed them in the 1950's, in the Local and Southern Supergalaxies. Further confirmation came from Peterson 1974 (concerning the Zwicky catalog), from Joêveer, Einasto & Tago 1978, and from Soneira and Peebles in 1978 (concerning the Lick Observatory catalog of Shane and Wirtanen; see Peebles 1980).

CIRRIFORM FRACTALS

Clearly, a cirriform structure *can*, but *need not*, be present in a nonrandom fractal dust. It is absent from the Fournier model in Chapter 9, which generates a collection of "lumps." By contrast, cirri are readily created by taking a Sierpiński carpet of Chapter 14 and disconnecting its generator without brutality. Since the resulting fractal's dimension can take essentially any value, we have made the important point that being cirriform is not a matter of dimension. Nevertheless, specifically built-in nonrandom cirri are too artificial to warrant attention.

This is why it was noteworthy that an unintended but unquestionable cirriform structure should be present in a random model for D close enough to 2.

This led me to a careful examination of other families of random fractals. Particularly immediate and interesting configurations are observed in the plates of Chapter 28 and in Plate C15, wherein the archipelagoes, into which many of the islands seem to coalesce, are more often atoll-shaped than clump-shaped.

CIRRI ARE EXPECTED IN FRACTALS THAT "NEARLY" PERCOLATE

Plates 308 to 309 reveal that an accentuated cirriform structure is present in fractals constructed as in Chapter 33, by removing random disc-shaped tremas. It suffices that the dimension be close to, but "slightly below," the critical percolation dimension, D_{crit}. The reason for the cirriform structure is obvious in this case. Let D decrease through D_{crit}, as we go through a sequence of fractals, each imbedded in its predecessors. We know that topological dimension crashes discontinuously from 1 to 0, but this discontinuity is exceptional: most facets of form vary continuously. For example, the out-of-focus picture obtained by replacing each point by a ball of radius ρ varies continuously. This out-of-focus picture is streamlike, not only when $D > D_{crit}$ but also when $D_{crit} - D$ is positive but small.

Observe that D_{crit} can also be said to be defined for the fractals of Chapter 32, but its value is degenerate, equal to max D=2.

GALAXIES' OBSERVED LACUNARITY

A second skeleton rattles in the closets of most models of the distribution of galaxies. To avoid invidious (even when justified) criticism of others, consider either of my own early models, as analyzed in Chapters 32 and 33. When D is matched to experiment (D~1.23), the limited portions of space shown in my plates look reasonable at first glance. But overall sky maps are completely wrong. Their gaps include immense domains (one-tenth of the sky or more) that are totally empty of galaxies within any prescribed distance. In devastating contrast, actual maps (e.g., the processed Lick Observatory map, Peebles 1980) seem fairly homogenous or isotropic, except on rather fine scales. I say that the sky is of low, and the models are of high, *lacunarity*.

APPARENT COSMOLOGICAL IMPLICATION. This last circumstance tempted me, circa 1970, to interpret the sky's appearance wrongly, as due to a D *much* larger than the value D~1.2 suggested by de Vaucouleurs 1970. As to cosmologists, we know they are enamored of a homogeneous Universe, and expect homogeneity, with D=3, to prevail above a small outer cutoff. They might hasten to interpret the above discrepancy as supporting the notion that fractals with D~1.23 (more generally, with D<3) are only applicable to the description of a small region of the universe.

LACUNARITY IS A PARAMETER DISTINCT FROM D. Actually, I am about to show that it is often possible to preserve a fractal's D, while modifying the perceived lacunarity. The main idea is illustrated in Plate 318, by two different Sierpiński carpets of identical D but very different appearance. The one to the left has the bigger gaps, and it is the more lacunar one, both intuitively and according to the measures I shall propose.

COSMOLOGICAL IMPLICATION. The customary inference, that the perceived low lacunarity implies a "small" outer cutoff Ω, may be overly hasty. The Devil's Advocate is prepared to argue that the small scale evidence in favor of D~1.23 and the large scale evidence in favor of near isotropy are *not* incompatible with a properly designed fractal model in which $\Omega=\infty$. To win this argument is not to disprove that $\Omega<\infty$, but merely to demonstrate that the determination of Ω requires additional care and data.

THE LACUNARITY OF TURBULENCE

The issue of whether the outer cutoff Ω is small or large also affects the study of turbulence. As mentioned in Chapter 10, Richardson 1926 proclaims that Ω is extremely large in the atmosphere, while most meteorologists think it is small. Therefore most of the comments in the preceding section have their turbulence counterpart.

There being few vocal living proponents of $\Omega=\infty$, the issue is less acute for turbulence than for galaxies, and is better discussed in the latter context.

A CANTOR DUST'S LACUNARITY

The notion of lacunarity (contrary to the notion of succolarity) makes sense on the line, hence previous sections' claims are most readily justified for linear dusts. We recall from Chapter 8 that a Cantor dust $\mathcal{C}$ on [0,1] may achieve any given D between 0 and 1 (limits excluded) in many different ways, and that the results need not look alike.

This is the case even if $\mathcal{C}$ decomposes into a prescribed number N of equal parts. Indeed, D and N determine $r=N^{-1/D}$, the parts' common length, but *not* the parts' positions within [0,1]. As a result, the same values of D and N (hence of r) are compatible with markedly different distributions for the parts.

At one extreme, one may collect the parts into two clumps terminating near 0 and near 1, respectively. This leaves in the middle a big gap, whose relative length $1-Nr = 1-N^{1-1/D}$ is very close to 1. An example is seen in the horizontal mid section of the Sierpiński gasket to the left in Plate 318. Essentially the same effect is achieved by placing a single big clump anywhere between 0 and 1.

At the other extreme, one may separate the N parts by N−1 gaps of the same length $(1-Nr)/(N-1)$. An example is seen in the horizontal mid section of the Sierpiński gasket to the right in Plate 318. When curdling is random, as in Chapter 23, the gaps are nearly of the same length.

When N >> 1, the outcome of the first extreme construction looks like a few points, hence "mimics" the dimension D=0, while the outcome of the second extreme construction looks like a "full" interval, hence mimics the dimension D=1. And of course, one can mimic any D between 0 and 1, by choosing for the N−1 gaps an appropriate collection of intervals, whose relative lengths add to 1−Nr.

The contrast between the extremes increases with N, 1/r, and b. The fractal dimension is hard to guess from the appearance of a minimally lacunar fractal with large N. However, it is clear-cut for small N. Therefore, the game of guessing D by just looking at a fractal has limitations. It is not an idle game (and we are correct in dwelling upon it in earlier chapters), but for galaxies it is misleading.

◁ This issue is clarified by a topic which necessity "exiled" into Chapter 39. The inspection of a *nonlacunar fractal* reveals its similarity dimension, which we shall see is 1, and not its Hausdorff dimension. In this case, the two dimensions differ, and the latter is the more suitable embodiment of fractal dimension. ▶

GAPS VERSUS CIRRI FOR N≫1 AND D>1

When N≫1 and D>1, a judicious choice of the generator can yield either of four outcomes: lacunarity can be either high or low, and cirri can either be arbitrarily close to percolation or absent. Thus, our two aspects of texture can in principle vary independently of each other.

ALTERNATIVE LACUNARITY MEASURES

In the short time since I started examining lacunarity, several distinct approaches proved worthy of examination. Unfortunately, one *must* not expect the resulting alternative measures to be monotone functions of one another. They are real numbers chosen to summarize the shape of a curve, hence they participate of the notions of "average man" and of "typical value of a chance variable." The fact is sad but unchangeable (notwithstanding many statisticians' willingness to risk everything in defense of their favorite) that typical values are by nature indeterminate.

THE GAP DISTRIBUTION'S PREFACTOR

One is tempted to measure a Cantor dust's degree of lacunarity by the largest gap's relative length. Alternatively, in plane shapes as those in Plate 318, lacunarity tends to vary inversely with the ratio between the trema's perimeter and the square root of its area. But a more promising measurement is deduced from the distribution of gap sizes.

From Chapter 8, a Cantor dust's gap lengths satisfy $Nr(U>u) \propto Fu^{-D}$, in the sense that $\log Nr(U>u)$ viewed as function of $\log u$ has a regular stair-shaped graph. The present discussion changes nothing to this last result, but the prefactor F, which was not significant till now, comes to the fore.

We must face the fact that the definition of F is somewhat arbitrary. For example F may be taken as relative to the line joining either the left, or the right endpoints of the stair's risers, or their midpoints. Fortunately such detail does not matter. As lacunarity goes *up*, one observes that any sensibly defined prefactor goes *down*. The same result holds for the volume or area scale factors relative to the Sierpiński carpets and the fractal foams. In all cases, lacunarity increase is due to the collapse of many gaps into a single bigger one. This makes the graph of the stairs slide toward 4:30 o'clock, a direction that is steeper than the stairs' own overall slope of $-D/E$, provoking the decrease in F that is claimed above.

Thus, we see that, within the broad nevertheless special class of fractals that include the Cantor dusts and Sierpiński carpets, lacunarity can be measured, hence defined, by F.

But this is a definition of limited validity. It already ceases to be compelling when a carpet's large central medallion is interrupted in its midst by a smaller carpet. Hence we need alternative definitions. The best is to substitute for F the more broadly valid prefactor of the relation $M(R) \propto R^D$.

LACUNARITY AS 2ND ORDER EFFECT CONCERNING THE MASS PREFACTOR

When a fractal is *not* constructed recursively (e.g., when it is random) lacunarity stand-ins are needed. Those described in this and the following sections are statistical averages, even in the case of the Cantor dust, which is nonrandom.

First, ponder the Cantor dusts obtained as horizontal mid sections of the two figures in Plate 318. Take the total mass of either dust to be 1, and consider the mass in diverse subintervals of lengths $2R = 2/7$. In the more lacunar example to the left, this mass ranges widely, from 0 to ½, while in the less lacunar example to the right, it ranges only a little around its mean value. Unfortunately, the precise distribution of mass is complicated in the Cantor dust case, and it is best to switch to the simpler case of a fully random Cantor dust, $\mathcal{D}$.

We take it that $\mathcal{D}$ intersects [0,1], and we denote the expected mass in this interval as $\langle W \rangle$ (the reason for this notation will transpire in a moment). When a small interval $[t, t+2R]$ is chosen in [0,1], the expected mass in it is $2R\langle W \rangle$, as it should be. But if one excludes the uninteresting cases where the mass vanishes, the expected mass increases to $(2R)^D \langle W \rangle$. Its value depends on D—but nothing else. (This shows that our dust's probability of intersecting [0,1] is $(2R)^{1-D}$.) In other words, the mass itself comes out as $W(2R)^D$, where W is a random variable: sometimes large and in other cases small, but on the average equal to $\langle W \rangle$, irrespective of lacunarity.

Now let us dig deeper, and seek how far the actual values of $W/\langle W \rangle - 1$ differ from 0. The conventional measure of discrepancy is the expected value of the second order expression $(W/\langle W \rangle - 1)^2$, denoted $\langle (W/\langle W \rangle - 1)^2 \rangle$. This second order lacunarity is small when lacunarity is intuitively viewed as low, and large when lacunarity is intuitively viewed as high. Therefore $\langle (W/\langle W \rangle - 1)^2 \rangle$ is a candidate to *define* lacunarity. Alternatives such as $\langle |W/\langle W \rangle - 1| \rangle$ are tempting, but they are far harder to evaluate than the mean square.

To summarize, we have moved beyond the relation "mass $\propto R^D$" to give individual attention to the prefactor of proportionality of mass to R^D. Observe that the notion of lacunarity has nothing to do with topology, and that it concerns comparisons at given D; its possible use for inter-D comparisons remains unexplored.

LACUNARITY AS 1ST ORDER EFFECT CONCERNING THE MASS PREFACTOR

An alternative approach to lacunarity involves the distribution of the mass in $[t,t+2R]$ when its midpoint $t+R$ is conditioned to belong to $\mathcal{D}$. This condition implies that $[t, t+2R]$ intersects $\mathcal{D}$, but the converse need not be true: if $[t,t+2R]$ intersects $\mathcal{D}$, the midpoint $t+R$ need not be in $\mathcal{D}$. The tighter conditioning we are now imposing on $[t,t+2R]$ has a greater tendency to eliminate the cases where the mass is well below the average, therefore results in an increased expected mass. In other words, W is replaced by W* satisfying $\langle W^* \rangle > \langle W \rangle$. And the ratio $\langle W^* \rangle / \langle W \rangle$ is large for very lacunar $\mathcal{D}$, and small for less lacunar ones. Hence we find an alternative candidate to define and measure of lacunarity: $\langle W^* \rangle / \langle W \rangle$.

CROSSOVER AT CUTOFF, & LACUNARITY

The approaches to lacunarity discussed up to this point are intrinsic, that is, do not involve any external point of comparison. We know, however, that many physical systems involve a finite outer cutoff Ω. These systems allow yet another approach to lacunarity, of slightly decreased generality than the two preceding ones but of very much greater convenience.

Let us indeed replace our fractal set $\mathcal{D}$, for which $\Omega = \infty$, by a fractal set $\mathcal{D}_\Omega$ which is "like $\mathcal{D}$" on scales below Ω and near homogeneous on scales above Ω. An example of Ω is the crossover radius where the galaxies' distribution changes from $D < E = 3$ to $D = 3$. This crossover could be left without precise definition until now, but no longer. The idea is that an observer who sits on a point of $\mathcal{D}$ views Ω as the size of the smallest chunk he must investigate to obtain a fair idea of the whole. To an inhabitant, the less lacunar world should seem to become homogeneous very rapidly, and the more lacunar world should seem to become homogeneous very slowly.

A first impulse is to write

$$\langle M(R) \rangle = \alpha R^D \text{ for } R \ll \Omega$$
$$\text{and } \langle M(R) \rangle = \beta R^E \text{ for } R \gg \Omega,$$

and to argue that the crossover occurs when $\alpha R^D = \beta R^E$, i.e., $\Omega^{E-D} = \alpha/\beta$. Hence

$$\langle M(R) \rangle = \alpha \Omega^{D-E} R^E \text{ for } R \gg \Omega.$$

A minor variant picks the point where the

34 ⌗⌗⌗ TEXTURE: GAPS AND LACUNARITY; CIRRI AND SUCCOLARITY · 317

two formulas have equal derivatives, hence $\Omega*^{E-D} = D\alpha/E\beta$. When lacunarity (i.e., α) increases but β and D remain fixed, Ω and $\Lambda*$ both increase. Both are fresh candidates to define and measure lacunarity.

IMPROVED TRANSLATION INVARIANCE

The fact that a straight line can slide upon itself is expressed by saying it is translation invariant. By contrast, Chapter 22 stressed that Cantor dusts have the eminently *un*desirable property that they are *not* translation invariant. For example, the original triadic dust $\mathcal{C}$ and its translate by $1/3$ do not even intersect. On the other hand, $\mathcal{C}$ and its translate by $2/3$ have one-half of $\mathcal{C}$ in common.

In the case of maximally lacunar Cantor dusts with $N \gg 1$, the only admissible translations yielding a significant overlap are of length close to 1 or close to 0. In the minimally lacunar case, on the other hand, the admissible translation length may be (approximately) any multiple of $1/N$.

In other words, translation invariance must be weakened in order to apply to Cantor dusts, but one gets away with lesser weakening when the lacunarity is low.

The conclusion of Chapter 22 was that one can extend translation invariance and the cosmologic principle to fractals, by making them random and recasting the invariances in "conditional" form. This recasting provides a major reason for introducing random fractals.

FROM STRATIFIED TO NONSTRATIFIED TEXTURE

The process used in this chapter to vary the succolarity in a Sierpiński carpet, and the lacunarity in a Cantor dust and a Sierpiński carpet involves a return to the strata characteristic of the nonrandom and the early random fractals. This method is powerful but artificial. In particular, the restriction of the scaling ratios to the form r^k buys lacunarity by narrowing the scope of self-similarity. With a high value of N (e.g., $N=10^{22}$, see the caption of Plate 114), and a correspondingly low r, the stratification is pronounced and conspicuous.

This way of controlling succolarity and lacunarity is obviously undesirable. Therefore, it is fortunate that I found one can do much better by extending the method of tremas: replacing intervals, discs, and balls by the more general shapes discussed in the chapter that follows.

NONLACUNAR FRACTALS

A fractal may be of vanishing lacunarity, as shown in an entry in Chapter 39. ∎

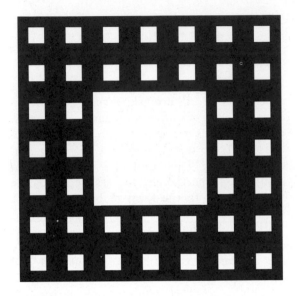
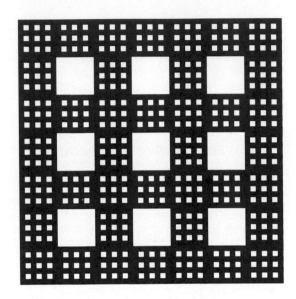

Plate 318 ⌑ **CARPETS' LACUNARITY**

Consider the following Sierpiński carpets constructed using the generators

Both generators satisfy b=1/r=7 and N=40, hence D~1.8957. The fact that N=40 may not be obvious, but it becomes obvious by inspection of the next stages, as shown above on 7 times larger scale.

Clearly, D being the same in both cases is not obvious. This is overwhelmed by the fact that the carpet to the left gives the impression of having definitely larger gaps, that is, of being much more lacunar (*lacuna* = hole, gap). Chapter 34 advances several alternative methods to pin this impression down.

This dimension D~1.8957 is remarkably close to that of Bernoulli percolation (end of Chapter 13), but the resemblance is misleading, because the topologies are very different in these two cases. ■

35 ◘ General Tremas, and the Control of Texture

In agreement with this Essay's method, Chapters 31 and 33 introduced the trema fractals through the simplest examples, based upon intervals, discs, and balls. The results were gratifyingly varied, but the use of more general tremas brings in even greater riches.

It is true that El Hélou (1978) shows that a trema fractal's dimension is solely determined by the distribution of the trema length (area or volume). But the days when D was the sole numerical parameter of a fractal ended when Chapter 34 introduced succolarity and lacunarity. The present chapter shows how these characteristics are affected by the trema shape. Again, the demand from the case studies and the supply from geometry are uncannily matched.

From the viewpoint of succolarity, the tremas' shape affects D_{crit}, hence for a given D, it affects the sign and magnitude of the difference $D-D_{crit}$.

From the viewpoint of lacunarity, the simplest improvements upon earlier chapters are achieved as follows. In the case of linear trema fractals (Chapter 31), the Lévy dusts are the most lacunar, and any lesser degree of lacunarity can be achieved most simply and naturally by taking as trema the union of many intervals. In the case of spatial trema fractals obtained directly (Chapter 33), the simplest is to take each trema to be other than a disc or ball. In the case of spatial trema fractals subordinated to Brownian or fractional Brownian motion (Chapter 32), the simplest is to take as subordinator a fractal dust less lacunar than Lévy dust.

Unfortunately, deadlines are closing in (this being the last chapter of this Essay to be written), and the arguments concerning trema fractals would take much reworking to make them suitable for inclusion in this Essay. Therefore, the chapter must be a mere sketch.

TREMA GENERATORS; ISOTROPY

The term *trema shape* used in the preceding introductory section involves the notion of *trema generator*. Of course, the term *generator* is already used in several early chapters. We remember that the stick generators of the Cantor or Koch shapes, and the trema generator of the Sierpiński shapes, determine *both* a fractal's shape and its D. Here, to the contrary, the trema generator determines *everything except* D.

NONRANDOM TREMA GENERATOR. This is an open set within which an arbitrary point is singled out as the *center*, and whose length (respectively, area or volume) is equal to 2 (respectively, π or $4\pi/3$). The tremas are rescaled versions of this generator. Their positions and sizes are random, with the same distribution as in Chapters 31 and 33.

For example, in the case E=1, the number of tremas having a length above τ and centered in an interval of length Δt continues to be a Poisson random variable of expectation $(E-D_*)\Delta t/\tau$. And the familiar formula for the dimension, $D=\max(D_*, 0)$, is shown in El Hélou 1978 to apply under mild restrictive assumptions about the trema generator's shape. (The question of whether these restrictive assumptions are intrinsic or due to the method of proof deserves investigation.)

BOUNDEDNESS OF THE GENERATOR. Since the philosophical goal of the trema construction is to create global structures from local interaction, it is sensible to include the assumption that the tremas are local, that is, bounded. But unbounded tremas may bring interesting surprises. A further generalized trema model is embodied in Plate 285.

DEFINITION OF GAPS. A gap is no longer the union of tremas, but the union of maximal open components of tremas.

NONRANDOM ISOTROPY. For the generator to be isotropic, one must be able to choose the origin so that the generator is the set of points whose distance from the origin falls within some set of the positive real line (usually, a collection of prescribed intervals). The isotropic case is the simplest and most thoroughly investigated.

However, nonisotropy is not excluded. In particular, we see that a fractal dust can be made *asymmetric* with respect to the past and the future.

RANDOM TREMA GENERATOR. This is a partly or fully random set of length (area or volume) equal to 1. A careful check of the applicability of the theorem in El Hélou 1978 would be welcome.

The least level of randomness consists in picking a single sample from a process that generates random sets, and in making all the tremas identical to this sample (up to displacement and size). The next useful level of randomness adds a random rotation, chosen independently for each trema. Even more generally, the tremas may be obtained by taking independent samples from a process that generates random sets. The sample sets need not all have the same volume, because volume is pinned down during resizing. Then the resized samples are rotated. Nonindependent rotations or samples are conceivable, but I have not used them thus far.

RANDOM ISOTROPY. In the first of the above alternatives, isotropy requires the sample to be rotation invariant. In the second alternative, the rotation sample must be distributed uniformly. In the third alternative, only the process must be rotation invariant.

STRATIFICATION. The preceding definitions would allow the trema length (area, volume) to be stratified, i.e., restricted to values of the form r^k. But this would confuse the distinct effects of stratification and of general trema shapes, there is no stratification.

CONTROL OF SUCCOLARITY THROUGH THE D_{crit} OF GENERAL TREMA FRACTALS

A section of Chapter 34 shows that a cirriform structure is expected if a fractal "nearly" percolates, that is, if it belongs to a family with a well-defined D_{crit}, and if its D is "only a little" below D_{crit}. In other words, D and the intensity of cirriform structure can be fitted jointly if the model involves both D and D_{crit} as parameters.

In a trema fractal, the parameters are the real number D and a function that specifies the trema generator. Let me show that D_{crit} is a function of this last functional parameter: it can be brought arbitrarily close to E, and if E > 2, D_{crit} can be made arbitrarily close to 1.

A CASE WHERE D_{crit} IS ARBITRARILY CLOSE TO E. It suffices to take as generator an arbitrarily thin needle or flat pancake with fixed shape but isotropically oriented axes (Plate 323). To prove this assertion in the plane (E=2) observe that, given an arbitrary D < 2, the trema centers, sizes, and direction can be selected inspective of the generator's flatness ratio. Next, consider a square of side L, and subdivide the tremas into 3 ranges: a mid range with areas below $\pi L^2 / 10$ and above

$\pi\eta^2$, a high range, and a low range. When D is much above the D_{crit} relative to disc shaped tremas, and the tremas are barely flattened discs, the situation is as in Chapter 33: the mid range tremas mostly form separate holes surrounded by a highly connected set. But if the tremas nearly flatten into lines, they almost surely cut up our square into small disconnected polygons. The added effect of flattened low range tremas can only be to cut these polygons further. Adding high range tremas can erase our square, or dissect it into pieces, or leave it alone. When it is left alone, it can no longer percolate. In other words, I showed that flattening the tremas can force D_{crit} to become larger than any prescribed D < 2.

The generalization to E > 2 is obvious.

The same effect is achieved for E ≥ 2, and also extends to E = 1, by taking as trema generator the domain contained between a ball (or sphere) of radius well above 1, and a suitably smaller ball (or sphere).

A CASE WHERE D_{crit} IS ARBITRARILY CLOSE TO 1. A heuristic argument suggests that when E ≥ 3 and the tremas are nearly needle shaped, D_{crit} is arbitrarily close to 1.

CONTROL OF LACUNARITY THROUGH THE L OF GENERAL TREMA FRACTALS

A section in Chapter 34 shows how one can control lacunarity where the trema lengths are stratified. Now let us put into the record (without detail) the fact that the same goal can be achieved via the trema generator. We focus on the measure of lacunarity that is mentioned last in Chapter 34, and involves an outer cutoff Ω.

As a matter of fact, we first go a step further and perform a double cutoff by constraining the linear scale of the trema to lie between $\epsilon > 0$ and $\Lambda < \infty$.

It is easy to see that an arbitrarily picked point continues to have the probability $(\epsilon/\Lambda)^{E-D}$ of belonging to the resulting truncated trema fractal. Next spread mass on this set with the density ϵ^{D-E}. We find that the prefactor $\beta = \alpha \Omega^{D-E}$ of Chapter 34 becomes Λ^{D-E}. Performing the passage to $\epsilon \to 0$ properly, this expression continues to hold for $\epsilon = 0$. Hence, $\Omega = \Lambda \alpha^{1/(E-D)}$.

(If Ω is defined through the variant definition, $\Omega = \Lambda \alpha^{1/(E-D)} (D/E)^{1/(E-D)}$.)

It remains to evaluate α. One finds that it depends on the trema generator's whole shape. It is largest when the generator is an interval (disc, ball) and can take arbitrarily low values. The threshold Ω is correspondingly low.

When the trema is contained between concentric spheres of radii $\alpha \gg 1$ and $\beta \gg 1$, the result is very simple: $\Omega \propto 1/\alpha$.

Thus, it is possible to arrange for $\langle M(R) \rangle$, hence, for the covariance of the distribution of mass, to go over arbitrarily fast to its behavior in the asymptotic region, meaning that the densities at two points separated by more than Ω become effectively independent.

It is odd that a decrease in lacunarity, through a decrease in α, should be accomplished by spreading out the generator. We would rather expect an increasingly spread generator to lead to an *increase* in the size of the pre-asymptotic region. This fact underlines again that the behavior of $\langle M(R) \rangle$, hence of the relative covariance of a distribution of mass, gives but a partial view of a set's structure. Higher moments of $M(R)$ carry much additional information, but we cannot dwell on this issue.

CONTROL OF LACUNARITY IN DUSTS SUBORDINATED TO BROWN TRAILS

Once we control a linear dust's lacunarity, we can map the result into space, via the process of subordination examined in Chapter 32. Working in the plane, and using as subordinand a Brown net as in Plate 243, one can achieve a dust that is arbitrarily close to seeming itself to be net-like, and to having an infinite order of ramification. Starting with $E=2$, let the subordinand be a fractional Brown net with $H > \frac{1}{2}$ whose gaps are smaller than for $H = \frac{1}{2}$. When, in addition, the subordinator's dimension satisfies $D/H < E = 2$, and the subordinator is of low lacunarity, the subordinate can be made to seem arbitrarily close to plane-filling. When $E = 3$ and $H = \frac{1}{3}$, the subordinand is a space-filling curve. When $D/H < E$, and the subordinator is of low lacunarity, the subordinate dust can be made to fill *space* to as low a degree of lacunarity as one wishes, irrespective of D.   ∎

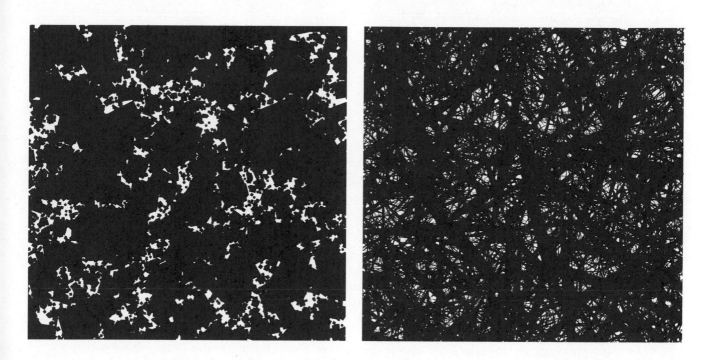

Plate 323 ⌑ **EFFECT OF THE TREMA GENERATOR
UPON THE LACUNARITY OF A TREMA FRACTAL**

These two illustrations ought to give an idea of the effect of the trema generator's shape upon lacunarity. While both trema generators are diamond-shaped, one is nearly a square, and the other is a sharp needle. Isolated small black diamonds are seen against the white areas.

Both constructions involve the same parameter D and the same areas for the largest and the smallest diamonds. One can show that it follows that the white remainders have identical areas in both cases, except for statistical variability. Nevertheless, it is obvious by inspection that one of the white remainders spreads out very much more than the other. The measures of lacunarity that I introduced attribute to the more spread out remainder a much lower value of the coefficients lacunarity. ■■

Plate 325 □ FRACTAL DUST OBTAINED WHEN THE TREMAS ARE NONSPHERICAL: ONE OCTANT'S PROJECTION ON A SPHERICAL SKY

For a most embarrassing reason, not only is this not the illustration intended for this spot, but the precise specification of this plate escapes me at the moment. Indeed, illustrations of fractals with $D \sim 1.23$ and varying and controlled degrees of lacunarity and succolarity had been produced by us in large numbers around January 1, 1979. But the file containing the bulk of the output is misplaced (or lost), and the few preliminary runs that survive in other files carry inadequate labels. Lacking time to reactivate the program, I can only show what is available.

As I recall it, the computation begins with a periodic pattern, whose period is a 600^3 cubic lattice. In other words, the computation is carried on a 600^3 lattice, whose opposite faces are identified to create a torus. The distribution of trema volumes is truncated. The tremas having been removed, the origin is moved to a nonremoved point, chosen either arbitrarily or within a region of high density.

Points close to the origin are not plotted, and other points are sorted into shells defined by $R_1^2 < x^2 + y^2 + z^2 < R_2^2$ corresponding to decreasing brightness ranges. Each shell is projected on the spherical sky.

The goal is to process the available data, so as to extract the maximum of *independent* information. For small R_2, one can map the whole sky, but for larger R_2, one must not process more than some suitable fraction of one period of the initial periodic pattern. The value of R_2 in the outer-most shell is greatest when the map is limited to a single octant of the sky, for example the domain where $x > 0$, $y > 0$, $z > 0$. In spherical coordinates, one can define this octant as corresponding to positive latitudes (northern hemisphere) and longitudes between $-45°$ and $45°$. Under the Hammer projection used here, this octant maps on the "gothic ogive window" in the following diagram.

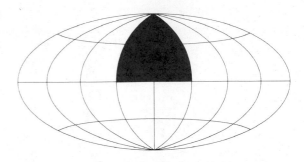

When R_2 reaches 600, the data in the neighborhoods of the three vertices become statically dependent, and the neighborhoods of the bottom vertices are best disregarded. In this fashion, the data beyond $R_2 = 600$, and the data near $x = z = 0$, $y = 600$, and near $y = z = 0$, $x = 600$, are sacrificed to the task of avoiding the statistical dependence induced by periodicity. On the other hand, to plot the antipodal region $x < 0$, $y < 0$, $z < 0$, i.e., southern latitudes and longitudes θ satisfying $|\theta - 180°| < 45°$, does not require a fresh computation, and the outcome may look sufficiently different to be viewed as providing additional information.

In a final stage of processing, meant to erase the trace of the original cubic lattice, every point is moved along a vector whose coordinates are uniformly distributed on $[0,1]$. Unfortunately, this procedure generates solid grey areas of various degrees of blackness, which misrepresent the underlying fractal: what we see are smoothed out versions of areas of great nonuniformity.

In the present Plate, $R_2 = 600$ and $R_1 = R_2/1.5$, hence the magnitudes lie in a narrow range of width $2.5 \log_{10}(1.5)^2 \sim .88$.

Figure 7 of Mandelbrot 1980b shows another fractal dust (also incompletely labeled) obtained via a different choice of f tremas.

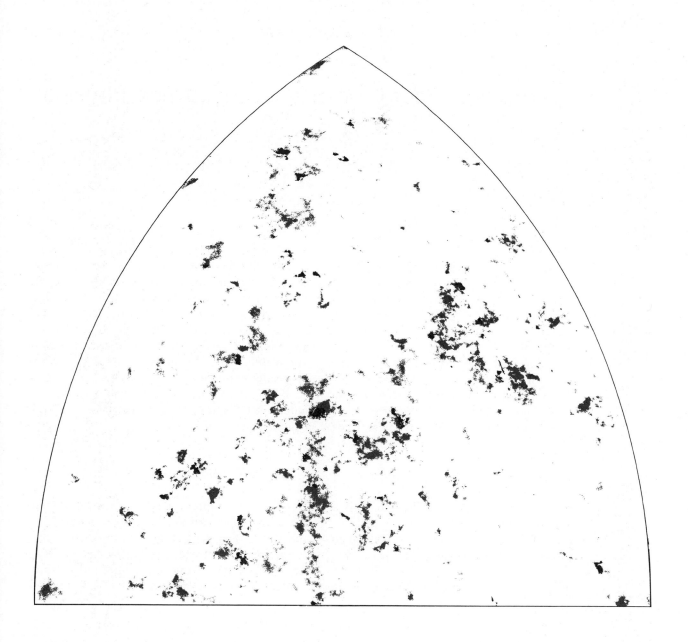

36 ◻ Logic of Fractals in Statistical Lattice Physics

From the viewpoint of fractals, most problems of physics are not specifically different from those raised by other fields, which is why case studies from physics are scattered throughout this Essay, with only a few kept aside to be discussed in this chapter.

Some readers, however, may have started on this Essay with the present chapter, because its title is the only one containing the word *physics*. Let me encourage these readers to scan the index, but first let me draw their attention to the following extensive case studies in physics that do not show in chapter titles.

Chapters 13 and 14 include a case study of percolation.

The Apollonian "soap" in Chapter 18 is a smectic liquid crystal.

Texture (Chapters 34 and 35) is bound to find many new applications in physics in the very near future.

Finally, a few references will be of inter-est. The term *diffractals* was coined in Berry 1979 to denote waves that are either reflected by a fractal surface or refracted by a slab of transparent material with fractally turbulent refractive index. Diffractals are a new wave regime in which ever finer levels of structures are explored and geometrical optics is never applicable. Berry calculates some of their properties explicitly.

Berry 1978 calculates the distribution of modes of fractal drums: resonators whose boundary is a fractal.

ON TWO KINDS OF CONVERGENCE

Now to this chapter's goal. By scattering the preceding topics around, a very important is-sue was either neglected or swept under the rug when encountered. In many areas of phys-ics, a basic step in the construction of mathe-matical fractals is impossible as a matter of principle.

As a prelude, let us recall again that the bulk of this Essay is devoted to fractals that involve recursive interpolation, either as a matter of definition, or at least through an after-the-fact explicit construction. Each construction stage begins with a geometrically standard shape, for example a broken line "teragon," and interpolates it a bit further. The fractal is the limit of these teragons, in the sense that the distance between the teragon and the limit (defined by suitable generalization of the usual notion of distance between points) tends to zero. Such a limit is called "strong" by mathematicians.

By contrast, other limits that occur in the statistical context, are called "weak" (or "vague"). As ordinarily presented, the distinction between the two sorts of limit seems yet another hairsplitting nicety. But the theme of weak convergence permeates all the cases, both old and new ones, where random fractals enter into "lattice physics," which is the customary practice of present statistical physics.

The discussion hangs on some fresh examples of fractals in physics, and on an important problem in lattice hydrology that falls into the same mold.

THE RANDOM WALK'S FRACTAL LIMIT

As a prelude, let us note the role of weak convergence in the context of Brownian motion. As mentioned fleetingly in Chapter 25, a random walk on a lattice (for example, on the points whose coordinates are integers) can be "downsized" as it walks on, until the lattice step becomes invisible, and its effects on the observables become negligible.

Everybody knows that this procedure "generates" Brownian motion, but the term "generates" has a new meaning here. The sequence of teragons which Chapter 6 uses to generate a Koch curve behaves like a picture to which detail is continually added by sharper focusing. By contrast, a sequence of downsized random walks moves around, first seeming to be at a small distance from some Brownian motion, then even closer to a different one, then still closer to yet another one, and so on...without ever settling down. There is good reason for mathematicians to describe this process as a *weak* or *vague* convergence. And there is good reason for considering a finitely downsized random walk as a fractal curve with an inner cutoff equal to the lattice spacing. But this is a novel kind of cutoff. In earlier chapters, an inner cutoff was superposed *after the fact* upon defined geometric constructions which in theory involve no cutoff, and can be interpolated to infinitesimal scales and generate fractals. On the contrary, there is no way of interpolating random walk.

FRACTALS IN "LATTICE PHYSICS"

The preceding description's scope goes well beyond Brownian motion. Indeed, statistical physics has imperative reasons for replacing many of the actual problems it faces by analogs constrained to a lattice. One may therefore describe the bulk of statistical physics as forming a part of "lattice physics."

As my early Essays pointed out, and many writers have confirmed, lattice physics is rife with fractals and almost fractals. The former are shapes in a parameter space, such as the Devil's Staircase shapes mentioned in the caption of Plate 83. The latter are shapes in real space that are *not* fractals, because they cannot conceivably be interpolated to the infinitely small, but are fractal-like insofar as their medium and large scale properties are those of fractals. A notable example is encountered in Chapters 13 and 14, when we tackle Bernoulli percolation.

Needless to say, I am utterly convinced that these shapes' downsized versions converge weakly to fractal limits. And the arguments in Chapters 13 and 14 are based on this conviction. Physicists find it totally persuasive, despite the fact that, insofar as I know, the only case where a full mathematical proof is available is Brownian motion. Thus, I tend to think of these *non*fractal shapes with presumed fractal limits as *latticed* fractals. Major additional examples are discussed later in the chapter.

A related but different inference is that the actual problems, of which lattice physics is a manageable simplification, involve the same (or nearly the same) fractals. In the case of polymers (to be studied momentarily), Stapleton, Allen, Flynn, Stinson & Kurtz 1980 supports this inference.

LOCAL INTERACTION/GLOBAL ORDER

A fascinating discovery of lattice physics, and one that deserves to be known widely, is that under certain conditions it happens that purely local interactions snowball into global effects. To take a basic example, interactions between neighboring elementary spins can generate magnets one can hold in one's hand.

One should be allowed to dream that the phenomena that I represented by fractional Brownian fractals will one day be explained in this manner.

A FICTITIOUS EXAMPLE

Let me describe an example that differs in fundamental ways from the physical mechanism of ordering, but has the virtue of being simple and of bringing back our old fractal friend the Sierpiński gasket (Chapter 14) as example of demonstrable weak limit. Spins are placed at the points with integer valued coordinate, so that at even (resp., odd) times, they sit on even (resp. odd) points. The rule of change is that the spin $S(t,n)$ at time t and position n is -1 if $S(t-1, n-1)$ and $S(t-1, n+1)$ are identical, and is $+1$ otherwise.

A line uniformly covered with -1 spins is left invariant by this process. Now let us follow the effects of the introduction of an $+1$ "impurity" at $n=0$ and $t=0$. The spins $S(1, n)$ are all -1 except for $n=-1$ and $n=+1$, and later configurations are as follows:

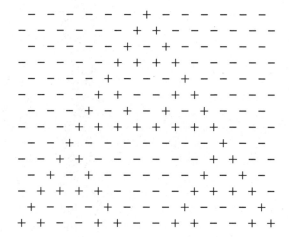

Many readers recognize here a Pascal triangle that has been summarized by marking by + the positions of odd valued binomial coefficients. The tth line of the complete Pascal triangle gives the coefficients in the development of the binomial $(a+b)^t$.

And everyone who read Chapter 14 sees that if we join each + to the neighboring +'s, we obtain a graph with obvious kinship to the Sierpiński gasket (Rose 1981). In fact, by downsizing this graph, we make it converge to the Sierpiński gasket.

SELF-AVOIDING RANDOM WALK AND LINEAR POLYMERS' GEOMETRY

We now return to an important specific problem. The self-avoiding random walk (SARW) goes forward with no regard to its past positions, except that it is prohibited from passing through a point more than once, and from entering a region from which it will find it impossible to exit. All the permissible directions are given equal probabilities.

On the straight line, such a motion poses no problem: it necessarily continues in either direction and never reverses itself.

In the plane and space, to the contrary, the problem is interesting and very difficult, so difficult that to date no analytical study has been successful. Yet its practical importance in the study of macromolecules (polymers) is such that it has become the object of careful heuristics and detailed computer simulations. The result that interests us most, due to C. Domb and described in Barber & Ninham 1970, is as follows:

After $n \gg 1$ steps, the root mean square displacement R_n is of the order of magnitude of n raised to a power we denote by $1/D$.

This result strongly suggests that, within a circle or sphere with radius R surrounding a site, the number of other sites is approximately R^D. This is a good reason for checking whether or not D is a fractal dimension.

Its value on a straight line is (trivially) $D=1$. A theoretical argument due to Flory, and computer simulations for $E=2$ and 3, agree on $D=(E+2)/3$ (de Gennes 1979, Section 1.3, which denotes D by $1/\nu$, is a good survey). The fractal dimension $D_B=2$ of Brownian motion exceeds this value for $E=2$ and 3, but coincides with it for $E=4$.

Only in the limit $E \to \infty$ does a limit argument due to Kesten establish that $D \to 2$. However, the value $D=2$ for $E \geq 4$ is suggested by delicate physics arguments, and also by the simple fractal argument that runs as follows: When $E \geq 4$, the codimension of Brownian motion is 2, hence the codimension of its double points is 0, meaning that Brownian motion has no double points. Therefore it is self-avoiding with no further ado.

The values of D happen to be sensitive to the details of the underlying assumptions. If a polymer in 3-space is made of two different types of atoms (so that the walk is not constrained to a lattice), Windwer finds that $D = 2/1.29$, which he claims is significantly below Domb's value $D = 1.67 \sim 2/1.2$. In a polymer dissolved in a reacting solvent, the imbedding space is even less inert, in particular, D becomes interaction-dependent. The Θ-point is defined as the point when D takes the Brownian value $D_B = 2$. In good solvents $D < 2$, and D decreases with the solvent's quality; in particular, a perfect solvent yields $D = 2/1.57$ if $E = 2$ and $D = 2/1.37$ if $E = 3$. Even the worst solvent in 2-space can never lead beyond $D = 2$, but a bad solvent in 3-space yields $D > 2$. Coagulation and phase separation set in, and a nonbranching chain is no longer a satisfactory model.

The preceding paragraphs do nothing but transcribe known results into fractal terminology, but I feel that this transcription helps to clarify their statement. Nevertheless, it must be restated that, by calling D a dimension, we assume that repeatedly downsized SARW converge weakly toward some family of fractals with the empirically observed D as their dimension. The physicist has no doubt on this account, but a demanding mathematician insists that, as of now, this assertion remains a conjecture. The following section sketches a direction the proof may take.

Observe that the downsized fractal limit is *not* expected to be self-avoiding, because points where a SARW is "reflected" on its distant past become double points. Their dimension is indeed $(4-E)/3 > 0$. However, triple points can be expected *not* to occur, and indeed their dimension is $\max(0, 2-E) = 0$.

Sequences that converge strongly to fractals are incomparably easier to study, both analytically and computationally, than downsized SARW. Hence it is useful—so to say—to "shadow" SARW by a sequence blessed with ordinary (strongly) convergent approximations. This goal is achieved by my "squig curves," Chapter 25. One striking result is that the least contrived and most isotropic squigs have a dimension extraordinarily close to the value $D = 4/3$ characteristic of plane SARW. A second "shadow" is the self-avoiding Brownian motion, defined in Plate 243 as the boundary of the hull of a bounded Brown trail. It may be recalled that it also yields $D = 4/3$. This bunching of values can hardly be a coincidence; it must tell us something profound about the structure of the plane.

It is interesting to sidetrack here to examine whether a self-avoiding random walk satisfies the cosmological principle (Chapter 22). Its first few steps do not. A conditionally cosmographic steady state seems certain to prevail (but I do not know of a proof).

RENORMALIZATION ARGUMENTS

The *analytic* study of scaling in lattice physical systems (pursued in a tradition distinct from mine) relies greatly on the powerful tool called (inaccurately) the "renormalization group (RG) technique." Wilson 1979 is a readily available survey by the originator. When an early version of this Essay and an early RG paper were still in preprint form, H. G. Callen drew my attention to an obvious conceptual kinship between them.

To investigate this kinship, let us ponder the following quotes from Wilson 1975, p. 774: (a) "The crucial feature of the statistical continuum limit is the absence of characteristic length or energy or time scales." (b) "[The RG] is the tool that one uses to study the statistical continuum limit in the same way that the derivative is the basic procedure for studying the ordinary continuum limit.... [Universality, an additional hypothesis] has an analogue in the case of an ordinary derivative. Normally, there are many different finite difference approximations to a single derivative." (c) "One is still a long way from the simple and yet explicit nature of the derivative." (d) "A divergent integral is a typical...symptom of a problem lacking a characteristic scale." (e) "[An earlier] renormalization theory...eliminates the divergences of quantum electrodynamics.... [Its] worst feature...is that it is a purely mathematical technique for subtracting out the divergent parts of integrals." (f) "The basic physical idea underlying the RG approach is...that there is

a cascade effect.... [The first] principal feature of the cascade picture is scaling." (g) "[The second principal feature] is amplification or deamplification."

Now a few comments. Quote (a) states that RG and fractals address the same class of concrete problems, and quote (d) that they encounter the same first difficulty. Quote (b) becomes by far more accurate when applied to the theory of fractals. In the fractal context, the complaint stated in quote (c) is unwarranted: there is now a simple and explicit replacement for the derivative, the first element of which is fractal dimension. Quote (d) brings back familiar memories to the reader of this Essay: we start in Chapter 5 by arguing that the integral that is supposed to give the length of a coastline is divergent. Elsewhere we manage to live with infinite variance, infinite expectation, or infinite probability (as when we deal with the distribution $\Pr(U>u)=u^{-D}$ for $0<u<\infty$, despite the fact that $0^{-D}=\infty$). Quote (e) gives us a cosy feeling: we always manage to avoid divergences without recourse to purely mathematical techniques. Quote (f) is also totally familiar.

In sum, there is indeed no question that RG and fractals draw on the same inspiration, and lead to the analytic and the geometric face of the same coin. But there is no fractal counterpart of (g), hence the parallelism is not complete.

◁ One of the outputs of RG is a fixed-point Hamiltonian $\mathcal{H}_0$. To be a physicist is to believe that a physical system's Hamiltonian $\mathcal{H}$ implies in principle everything there is to know concerning the system's structure. If so, one should also be able to use Hamiltonians to derive the various random shapes' joint probability distributions. The finitely renormalized $\mathcal{H}$ should yield a downsized shapes' distributions, and the fixed point $\mathcal{H}_0$ should yield the limits' distribution—and in particular their D. The program of research implied in this sketch may be hard to implement, but I am fully confident it will work out. ▶

SELF-AVOIDING POLYGONS

Let a polygon be chosen at random among all the n-sided self-avoiding polygons whose sides are links of a plane (E=2) square lattice. Sometimes it is squarish, with an area about $(n/4)^2$. Sometimes it is spindly and skinny, with an area about n/2. If one averages by giving to each polygon the same weight, numerical simulations indicate that the average area is about $n^{2/D}$ with D~4/3 (Hiley & Sykes 1961). Hence from the fractal viewpoint a polygon behaves like a self-avoiding random walk biting its tail.

BACK TO COASTLINE MODELS

Their dimension being D~4/3 seems to qualify self-avoiding polygons as models of coastlines of above average irregularity. We may rejoice at this finding, but the question concerning the shape of coastlines, raised in Chapter 5, is not settled thereby.

First of all, there is the problem of islands. The concept of dimension should at the same time account for the coastlines' irregularity, their fragmentation, and the relationship between irregularity and fragmentation. But self-avoiding polygons have no offshore islands.

Second, I think that no single value of D could suffice for all of Earth's coastlines.

Last but not least, when a very large self-avoiding random walk or polygon is downsized, so that the lattice step decreases from 1 to a small value η, two points that used to be distant by 1 converge to the same limit point. The limit downsized walk or polygon is therefore no longer self-avoiding: it does not self-intersect but it self-contacts. I do not wish to see such points in a model of a coastline. For example, they imply the existence of a strict interpretation of the etymology of *peninsulas* (almost-islands that touch the mainland at a single point) and the existence of almost-lakes.

RIVERS' FAILURE TO RUN STRAIGHT

Chapter 12 mentions Hack's empirical finding that it is typical for a river's length to increase like the power of D/2 of its drainage area. Were it true that rivers flow straight through round drainage areas, stream lengths would be proportional to the square root of the drainage area, so the value of D would be D=1. In fact, D is the range from 1.2 to 1.3. In response, Chapter 12 describes a model based on a plane-filling rivers network, in which the rivers are fractal curves.

In a very different stochastic attempt to explain the Hack effect, Leopold & Langbein 1962 reports computer simulations of the development of drainage patterns in regions of uniform lithology. The model involves an original two-dimensional random walk in a square lattice, which ought to interest the physicist. It is assumed that both source locations and directions of propagation are chosen by chance. The source of the first stream is a square chosen at random, and a channel is generated by a SARW into adjacent squares, until it goes off the boundary area. Then a second source is chosen at random and another stream, generated as before, terminates either by going off the boundary or by joining the first stream. Often the second stream, the "Missouri," is longer than the portion of the Mississippi above their junction. It is possible that the junction should occur at the first stream's source. The same procedure continues until all squares are filled in. In addition to these general rules, various rather arbitrary decisions avoid loops, snags, and inconsistencies.

Computer simulation indicates that, in this random walk model, river length increases as the 0.64 power of the drainage area. Hence $D \sim 1.28$. The discrepancy between this value and Domb's $D \sim 4/3$ may be a statistical variation due to insufficiently extensive simulations. But I am tempted to view the discrepancy as genuine: the cumulative interference from other streams seems more accentuated

than the interference from a SARW's past values, hence we would expect a small D for the Leopold & Langbein model.

Compared to actual maps, Leopold & Langbein rivers wander excessively. To avoid this defect, numerous alternatives have been proposed. The model due to Howard 1971 postulates headward growth, according to various perfectly artificial schemes, from mouths placed on the boundary of a square toward sources placed inside. This procedure generates rivers that are markedly straighter than in the Leopold & Langbein scheme, and hence presumably involve a smaller D.

Thus far, the study of random networks such as those of Leopold & Langbein and of Howard is limited to a few computer simulations. It is a shame, and I wish to bring these very interesting problems to the attention of mathematicians. The fact that the SARW has proven extremely resistant to analysis should serve to warn off seekers of easy problems with a large payoff, but the Leopold & Langbein variant might be easier.

To repeat: the mathematical difficulties encountered in the study of SARW are rooted in the fact that local changes may have global effects. Similarly, a local change in a Leopold & Langbein network may result in a big river breaking through a dividing line into the neighboring basin. One would be happy to be able to measure the intensity of the resulting long-term interaction macroscopically. Naturally, I expect this parameter to be a fractal dimension. ◼

37 ◻ Price Change and Scaling in Economics

Only half in jest, the variation of prices on stock and commodity exchanges may be said to raise a geometric problem, since the newspapers' financial pages are full of advertisements by self-styled "chartists," who plot the past suitably and proclaim they can predict the future from those charts' geometry.

The basic counterclaim, first asserted in 1900 by Louis Bachelier, is that charting is useless. The most assertive statement is that successive price changes are statistically independent. A milder statement is that every price follows a "martingale" stochastic process, meaning that the market is "perfect": everything in its past has been discounted fully. An even milder statement is that imperfections remain only as long as they are smaller than transaction costs; such markets are called "efficient." Bachelier's notion of efficiency has proved extraordinarily accurate.

A more specific assertion by Bachelier is that any competitive price follows, in the first approximation, a "one-dimensional Brownian motion" B(t). The fact that a process so fundamental to physics has been invented by a maverick mathematician is worth remember-

ing, see Chapter 40. Sad to report, when actual data became available, B(t) turned out to represent them very poorly. The present chapter describes an alternative description, which I constructed on the basis of a scaling assumption (one of the earliest to be made in *any* field). It proves astonishingly accurate.

THE DISCONTINUITY OF PRICES

My simplest anti-Brown argument is based on an experimental observation that is so plain and direct that one may be surprised that it should prove fundamental. But the arguments which earlier chapters use to show that $D > 2$ for galaxies and $D > 2$ for turbulence are also surprisingly plain and direct. The unsophisticated observation is that a continuous process *cannot* account for a phenomenon characterized by *very sharp* discontinuities. We know that Brownian motion's sample functions *are* ◁ almost surely, almost everywhere ▶ continuous. But prices on competitive markets *need not* be continuous, and they are conspicuously *discontinuous*. The only reason for assuming continuity is that many sciences tend, knowingly or not, to copy the procedures that prove successful in Newtonian physics. Continuity should prove a reasonable assumption for diverse "exogenous" quantities and rates that enter into economics but are defined in purely physical terms. But prices are different: mechanics involves nothing comparable, and gives no guidance on this account.

The typical mechanism of price formation involves both knowledge of the present and anticipation of the future. Even when the exogenous physical determinants of a price vary continuously, anticipations change drastically, "in a flash." When a physical signal of negligible energy and duration, "the stroke of a pen," provokes a brutal change of anticipations, and when no institution injects inertia to complicate matters, a price determined on the basis of anticipation can crash to zero, soar out of sight, do anything.

FALLACIES OF FILTER TRADING (MANDELBROT 1963b)

The idea that price can be discontinuous hardly seems to have any predictive value by itself. But it proved basic to the fall and the burial of the method of trading using "filters," due to Alexander 1961. In principle, a p% filter is a device that monitors price continuously, records all the local maxima and minima, gives a buy signal when price first reaches a local minimum *plus exactly* p%, and gives a sell signal when a price first reaches a local maximum *minus exactly* p%. Since continuous monitoring is impractical, Alexander monitored the sequence of *daily* highs and lows. He took it for granted that a price record can be handled like a continuous function. The algorithm seeks the days when the high *first exceeds* an earlier day's low plus p%. The assumption is that at some time during the day d, the price was *exactly* equal to said low plus p%, at which point the filter triggered a buy signal. Similarly for sell signals. Alexander's empirical conclusion is that a filter's buy or sell signals bring higher returns than "buy and hold."

In fact, Mandelbrot 1963b, p. 417, points out that the 24-hour days on which the filter gives a buy signal are very likely to be days of strong overall upward price motion. On many such days, price actually jumps, either overnight or while trading is stopped on the Exchange's initiative. Thus, as the moments when Alexander's filter ought to emit a buy signal, it is likely to be turned off! It will emit a buy signal as soon as it turns on again, but the resulting buy price is often significantly higher than Alexander assumed.

Additional possibility: on many days, price variation is made reasonably continuous by the deliberate actions of a market specialist, performing his assigned functions of matching buyers and sellers, *and* of "insuring the continuity of the market" by buying or selling from his own holdings. Whenever the specialist fails to insure continuity, he must file a written explanation, and he often prefers to smooth out the discontinuity artificially. Obviously, the resulting bargains are reserved to friends, while the bulk of customers has to buy at a higher price.

Third possibility: certain daily price changes are subjected to limits, and can move the daily limit for several days with no trading, preventing "stop limit" moves from being executed.

Theoretical and experimental studies (to be described momentarily) convinced me that the above biases are significant, and that the computed advantage of filter trading over buy and hold was spurious. Upon rechecking, Alexander 1964 found my prediction to be correct, and the method of filters to be no better than "buy and hold." Fama and Blume 1966 carried out a thorough "post-mortem" check, replacing Alexander's price indices by individual price series; the method of filters is now buried for good. This episode underlines the risk of error inherent in what I call the "fallacy of price continuity."

Winning "martingales" resemble perpetual motion machines. It is to the credit of Bachelier's efficient market hypothesis that it had predicted in advance that filters *should* not work, but to the discredit of Bachelier's Brownian motion model that it could not explain why filters *seemed* to work. Therefore, it is to the credit of my specific models that they permit an analysis, and pinpoint the flaws of this and other explicitly described paths to sure wealth.

STATISTICAL "FIXES"

The failure of the Brownian motion as a model of price variation elicited two very different responses. On the one hand, there is a plethora of ad hoc "fixes." Faced with a statistical test that rejects the Brownian hypothesis that price changes are Gaussian, the economist can try one modification after another until the test is fooled.

A popular fix is censorship, hypocritically labeled "rejection of statistical outliers." One distinguishes the ordinary "small" price changes from the large price changes that defeat Alexander's filters. The former are viewed as random and Gaussian, and treasures of ingenuity are devoted to them..., as if anyone cared. The latter are handled separately as "nonstochastic." A second popular fix is a mixture of several random populations: when X is not Gaussian, maybe it is a mixture of two, three, or more Gaussian variables. Yet another fix is nonlinear transformation: when X is positive and grossly non-Gaussian, maybe $\log X$ is Gaussian; when X is symmetric and non-Gaussian, maybe $\tan^{-1}X$ will fool the test. Yet another procedure (which I view as suicidal coming from a statistician) proclaims that price follows Brownian motion, but the motion's parameters vary uncontrollably. This last fix can never be falsified, hence the philosopher Karl Popper proclaims it cannot be a scientific model.

A SCALING PRINCIPLE IN ECONOMICS
(MANDELBROT 1963b)

At the opposite of the fixes stands my own work. It applies to diverse data of economics, but the principle is best expressed in the context of price.

SCALING PRINCIPLE OF PRICE CHANGE. When $X(t)$ is a price, $\log X(t)$ has the property that its increment over an arbitrary time lag d, $\log X(t+d) - \log X(t)$, has a distribution independent of d, except for a scale factor.

Before exploring this principle's consequences, let us run through a checklist of properties.

A scientific principle *must* yield predictions that can be checked against the evidence. *This one does so,* as will be seen momentarily, and the fit is very good.

It is *nice* for scientific principles to be deducible from other theoretical considerations in their respective fields. The scaling principle of price change can be based on the general (not necessarily standard) form of the probabilistic "central limit" argument, but it has not yet been deduced from standard economics. The only "explanatory" arguments (Mandelbrot 1966b, 1971e) support it as the consequence of scaling in exogenous physical variables. These arguments are less well established than the result they claim to justify.

Finally, even when no actual explanation is available, it is *pleasant* if a scientific principle does not actually *clash* with earlier presuppositions. The present scaling principle seems innocent enough. The question it answers has not previously been raised, so that contrary opinions could not be expressed. All that scaling seems to assert is that in competitive markets no time lag is really more special than any other. It asserts that the obvious special features of the day and the week (and the year in case of agricultural commodities) are compensated or arbitrated away. Since all the usual "fixes" of Brownian motion involve privileged time scales, my principle seems simply to affirm that there is no "sufficient reason" to assume that any time scale is more privileged than any other.

THE INFINITE VARIANCE SYNDROME

However, we want the actual implementation of the scaling principle to yield a result distinct from Brownian motion. To achieve this goal, I took the radical step of assuming that $\log X(t+d) - \log X(t)$ has an infinite variance. Before my papers, no one hesitated to write "denote the variance by V." The underlying assumption that V is finite was not even mentioned...rightly so, because science writing collapses when one lists every assumption irrespective of *established* significance. My reasons for taking the opposite view are discussed later in this section. Needless to say, the success of assuming $V = \infty$ made it easier for me to allow curves to have infinite length and surfaces to have infinite area.

THE OBSERVED MISBEHAVIOR OF THE SAMPLE VARIANCE OF PRICE CHANGES. "Typical values" used to summarize data are the least sophisticated level of descriptive statistics, but in the case of price changes, the usual summaries turn out to be tricky and wholly unreliable. Indeed, the motivation for using a sample average to measure location, and a sample root mean square to measure dispersion, resides in the belief that these are "stable" characteristics which eventually converge to population values. But the figure in Mandelbrot 1967b shows that their behavior in the case of prices proves extraordinarily elusive:

(A) Values of the mean square corresponding to different long subsamples often have different orders of magnitude.

(B) As sample size increases, the mean square fails to stabilize. It goes up and down, with an overall tendency to increase.

(C) The mean square tends to be influenced predominantly by a few contributing squares. When these so-called outliers are eliminated, the estimate of dispersion often changes in its order of magnitude.

THE HYPOTHESIS OF NONSTATIONARITY. These properties taken together, or even any one of them taken singly, used to suggest to everyone that the process is nonstationary. My preliminary counterproposal is that the process is in fact stationary but the unknown theoretical second moment is extremely large. Under the assumption of large but finite moment, sample moments converge according to the law of large numbers, but the convergence is extremely slow and the value of the limit matters very little in practice.

THE INFINITE VARIANCE PRINCIPLE. My further counterproposal is that the population mean square is infinite. The choice between "very large" and "infinite" is of course familiar to anyone who has plowed this far in this Essay, but those who are starting here may be differently disposed, and so were all my readers of 1962. To anyone with the usual training in statistics, an infinite variance seems at best scary and at worst bizarre. In fact, "infinite" does not differ from "very large" by any effect one could detect through the sample moments. Also, of course, the fact that a variable X has an infinite variance in no way denies that X is finite with a probability equal to 1. For example, the Cauchy variable of density $1/\pi(1+x^2)$ is almost surely finite, but has an infinite variance and an infinite expectation. Thus, the choice between variables with very large and infinite variances should not be decided a priori, and should hinge solely on the question of which is the more convenient to handle. I accept infinite variance because it makes it possible to preserve scaling.

STABLE LÉVY MODEL (MANDELBROT 1963b)

Mandelbrot 1963b combines the scaling principle with the acceptable idea that successive price changes are independent with vanishing expectation, and furthermore allows the variance of price changes to be infinite. A brief mathematical argument leads to the conjecture that price change is ruled by a Lévy stable distribution, which also enters in Chapters 32, 33, and 39.

This conjecture proves to be of very broad validity. The first tests (Mandelbrot 1963b, 1967b) applied to many commodity prices, some interest rates and some 19th century security prices. Later, Fama 1963 studied recent security prices, and Roll 1970 studied other interest rates. Here we must be content with a single illustration, Plate 340.

THE MODEL'S PREDICTIVE POWER

The predictive value of the scaling principle of price change resides in the following finding. One starts with the distribution of *daily* price changes over a period of five years of middling price variability. And one finds that if this distribution is extrapolated to *monthly* price changes, its graph goes right through the data from various recessions, depression, etc. It accounts for all the most extreme events of nearly a century in the history of an essential and most volatile commodity.

In particular, Plate 340, the process that rules the changes in the price of cotton has remained approximately stationary over the very long period under study. This amazing finding is best presented in two steps.

FIRST TEST OF STATIONARITY. Plate 340 indicates that the analytic form of the process of price changes, and the value of D, both remain constant. There is no disputing the major changes in the value of currency, etc., but overall trends are negligible in comparison with the fluctuations with which we deal here.

SECOND TEST OF STATIONARITY: CORREC-TION OF AN ERROR IN PLATE 340. A fluke gave rise to a second test of stationarity. On the plate, the curves (a$^+$) and (b$^+$) (similarly, (a) and (b$^-$)) differ by a horizontal translation. Since translation on doubly logarithmic coordinates corresponds to a change of scale in natural coordinates, this discrepancy led Mandelbrot 1963b to concur with the economists' opinion that the price change distribution had changed between 1950 and 1900. I thought the distribution preserved the same shape, but its scale had become smaller.

However, this concession to opinion turns out to have gone beyond necessity. The data behind the curves (a$^+$) and (a$^-$) had been read incorrectly (Mandelbrot 1972b). Once this error is corrected, one is led to curves near identical with the curves (b$^+$) and (b$^-$).

One cannot deny that the data give at casual glance the impression of being grossly nonstationary, but this is so only because casual impressions are formed against the background of a belief that the underlying process is Gaussian. My alternative to the nonstationary but Gaussian process is a stationary but non-Gaussian stable process.

CONCLUSION

I know of no other comparably successful prediction in economics.   ▬

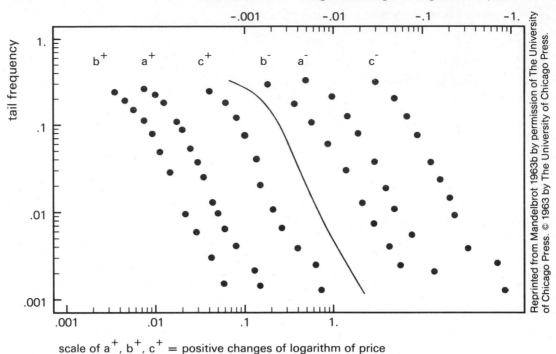

scale of a⁻, b⁻, c⁻ = negative changes of logarithm of price

scale of a⁺, b⁺, c⁺ = positive changes of logarithm of price

Plate 340 ¤ ORIGINAL EVIDENCE FOR SCALING IN ECONOMICS

This old plate, reproduced from Mandelbrot 1963b, is one to which I am attached (as I later became attached to Plate 271). It combines doubly logarithmic graphs of positive and negative tails for the recorded changes in the logarithm of cotton price, together with the cumulated density function of the symmetric stable distribution of exponent $D=1.7$, which is actually a slightly overestimated value of D. The ordinate gives the relative frequency of cases where the change of one of the quantities X defined below exceeds the change in abscissa.

Copy this plate on a transparency, and move it horizontally. You will see that the theoretical curve superimposes on either of the empirical graphs with slight discrepancies of general shape. This is precisely what my scaling criterion postulates!

The discrepancies are largely due to a slight asymmetry in the distribution. This is an important observation that requires skew variants of the stable distribution.

The following series of data are plotted, the positive and negative values of X being treated separately in both cases.

(a) $X = \log_e Z(t + 1 \text{ day}) - \log_e Z(t)$, where Z is the daily closing price at the New York Cotton Exchange, 1900–1905 (Data communicated by the U.S. Department of Agriculture).

(b) $X = \log_e Z(t + 1 \text{ day}) - \log_e Z(t)$ where Z is an index of daily closing prices of cotton on various Exchanges in the U.S., 1944–1958 (communicated by Hendrik S. Houthakker).

(c) $X = \log_e Z(t + 1 \text{ month}) - \log_e Z(t)$, where Z is the closing price on the 15th of each month at the New York Cotton Exchange, 1880–1940 (communicated by the U. S. Department of Agriculture). ■

38 ¤ Scaling and Power Laws Without Geometry

If monographs or textbooks on fractals come to be written, the discussion of random geometric shapes, which is mathematically delicate, will come after the less difficult topic of random functions, and these books will begin with random variables. On the other hand, this Essay plunges straight into the most complicated topic, because it is the most interesting one, and gives play to geometric intuition.

Most closely related to fractals are the hyperbolic probability distributions. Many examples of their use are encountered in earlier chapters, beginning with hyperbolic functions $Nr(U>u)$. But much remains to be said. This chapter begins with general comments and continues with certain phenomena of linguistics and economics in which empirical evidence, both abundant and sound, is very well represented by hyperbolic laws. The argument is the same in both cases and spotlights scaling and similarity dimension in wholly "disincarnated" forms.

The example from linguistics was the object of my first paper (Chapter 42). It familiarized me with certain manipulations that are straightforward but of wide applicability. The example from linguistics also has a thermodynamic facet, involving my independent discovery of a counterpart to negative temperature.

MORE ON HYPERBOLIC DISTRIBUTIONS

In a definition we know well, a random variable (r.v.) U is called hyperbolic when $P(u)=Pr(U>u)=Fu^{-D}$. This definition is bizarre, insofar as every finite prefactor σ leads to the conclusion that $P(0)=\infty$, which seems absurd, and certainly indicates that special care must be taken—as we know well. For example, we saw in Chapter 12 that when a Koch generator includes an island the resulting curve includes an infinity of islands, with those of area above a numbering $Nr(A>a)=Fa^{-B}$. Let us rank them by decreasing area, islands of identical areas being ordered arbitrarily. To select such an island at random with uniform probability is to select the island's rank at random. Achieving this goal would authorize the replacement of $Nr(A>a)$ by $Pr(A>a)$. But in fact an island's rank is a positive integer, and it is *not* possible to choose a positive integer at random.

Another familiar story: the hyperbolic distribution leads to straightforward *conditional* distributions. For example, the *conditional* r.v. $\{U, \text{ knowing that } U > u_0\}$, written as $\{U | U > u_0\}$, satisfies

$$\Pr\{U > u | U > u_0\} = \begin{cases} 1 & \text{when } u < u_0 \\ (u/u_0)^{-D} & \text{when } u > u_0 \end{cases}$$

EXPECTATION PARADOXES

When $D > 1$, the corresponding expectation is

$$\langle U | U > u_0 \rangle = D(D-1)^{-1} u_0$$

This result suggests endless paradoxical stories. Sober readers are urged to forge ahead.

THE LINDY EFFECT. The future career expectation of a television comedian is proportional to his past exposure. *Source*: *The New Republic* of June 13, 1964.

For a key, see the next story.

PARABLE OF THE YOUNG POETS' CEMETERY. In the cemetery's most melancholy section, among the graves of poets and scholars who had fallen unexpectedly in the flower of their youth, each monument is surmounted by a symbol of loss: one half of a book, of a column, or of a tool. The old groundskeeper, himself a scholar and a poet in his youth, urges visitors to take these funereal symbols most literally: "Anyone who lies here," he proclaims, "had accomplished enough to be viewed as full of promise, and some monuments' sizes reflect the accomplishments of those whose remains they shelter. But how can we assess their broken promise? A few of my charges may have lived to challenge Leonhard Euler or Victor Hugo in fecundity, if perhaps not in genius. But most of them, alas, were about to be abandoned by their Muses. Since promise and accomplishments are precisely equal in young life, we must view them as equal at the moment of sudden death."

The key. ANYONE WHO STOPS YOUNG STOPS IN THE MIDDLE OF A PROMISING CAREER. "Proof." According to A. Lotka, the distribution of the number of scientific papers due to any single author is hyperbolic with the exponent $D = 2$. This rule incorporates the qualitative fact that most people write nothing or little, but a few write an awful lot. If so, however long a person's past collected works, it will on the average continue for an equal additional amount. When it eventually stops, it breaks off at precisely half of its promise.

Comments. The only way of avoiding apparent disappointment is to be so old that age corrections must be considered when computing the expected future. The coefficient of proportionality in the Lindy Effect is doubtless equal to 1.

PARABLE OF THE RECEDING SHORE. Far, far away, there is a country called the Land of Ten Thousand Lakes, affectionately known as Big, Second Biggest,..., Nth Biggest, etc., down to 10,000th Biggest. Big is an uncharted sea, nay, a wide ocean at least 1600 miles across, the width of N-th Biggest $1600N^{-0.8}$, and so the smallest has a width of 1 mile. But each lake is always covered with a haze that makes it impossible to see beyond a mile to identify its width. The land is unmarked, and has no inhabitants to help the traveler. As a traveler who believes in mathematical expectation stands on an unknown shore, he knows he has before him a stretch of water of expected width equal to 5 miles. If he sails on for a few miles m, finds he has not yet reached his goal and calculates the new expected distance to the next shore, he obtains the value 5m. Do spirits inhabit these lakes, and *actually move the shore away?*

The key. The above distribution of lake widths merely restates the Korčak distribution encountered in Chapters 12 and 30.

SCALING PROBABILITY DISTRIBUTIONS

Now we return to serious matters. To be able to speak of scaling random variables, the term *scaling* must be defined without geometry. The reason is that the only geometric shape associated with a random variable is a point, which cannot be subdivided. As a substitute, say that a random variable X is *scaling under the transformation* $\mathcal{T}(X)$ if the distributions of X and $\mathcal{T}(X)$ are identical except for scale.

Transformation is understood here in a broad sense: e.g., the sum of two independent realizations of X is viewed as a transform of X. The corresponding variables should be called *scaling under addition*, but are called

Lévy stable (they enter in Chapters 31, 32 and 39). Chapter 39 (pp. 373a and 379a) goes on to *scaling under weighted addition.*

ASYMPTOTIC SCALING. ASYMPTOTICALLY HYPERBOLIC R.V. Most fortunately, the above definition is less indeterminate than may seem the case. For many transformations, invariance turns out to demand an *asymptotically hyperbolic distribution.* This means that there must exist an exponent D > 0 such that

$$\lim_{u \to \infty} \Pr(U < u)u^D \text{ and } \lim_{u \to \infty} \Pr(U > u)u^D$$

are defined and finite, and one of the limits is positive.

PARETO DISTRIBUTION. "Asymptotically hyperbolic" can be viewed as synonymous with a term familiar to economic statisticians, *Paretian.* Vilfredo Pareto was an Italian economist who hoped to translate the laws of mechanical equilibrium into terms of economic equilibrium, but is likely to be remembered more durably for having discovered a basic statistical regularity: he found that in certain societies the number of individuals with a personal income U exceeding a large value u is approximately hyperbolically distributed, i.e., proportional to u^{-D}. (We turn to the distribution of income later in this chapter.)

"NEW METHODS OF STATISTICAL ECONOMICS" (MANDELBROT 1963e)

Hyperbolic laws similar to Pareto's were later discovered in many areas of economics, and many efforts have been directed at explaining their prevalence. But let us first describe a heretical approach to this problem.

A field such as economics can never forget that its "data" are an awfully mixed bag. Therefore, the distribution of the data is the joint effect of a fixed underlying "true distribution," and of a highly variable "filter." Mandelbrot 1963e observes that asymptotically hyperbolic distributions with $D < 2$ are very "robust" in that respect, meaning that a wide variety of filters leave their asymptotic behavior unchanged. On the other hand, practically all other distributions are highly *non*robust. Therefore, an hyperbolic true distribution *can* be observed with consistency: different sets of distorted data suggest the same distribution with the same D. But the same treatment applied to most other distributions leads to "chaotic" incompatible results. In other words, the practical alternative to the asymptotically hyperbolic distribution is not any other distribution but chaos. Since chaotic results tend not to be published or noticed, the fact that asymptotically hyperbolic distributions are very widespread was to be expected, and tells us little about their actual prevalence in nature.

ZIPF LAW OF WORD FREQUENCIES

A word is simply a sequence of proper letters terminating with an improper letter called space. We rank the words in a text by decreasing frequency in a sample of one individual's discourse, the words of identical frequency being ordered arbitrarily. In this classification, ρ designates the rank assumed by a word of probability P, and the term *distribution of word frequencies* denotes the relationship between ρ and P.

One might expect this relationship to vary wildly according to the language and the speaker, but in fact it does not. An empirical law made known by Zipf 1949 (on G. K. Zipf, see Chapter 40) asserts that the relation between ρ and P is "universal," i.e., parameter-free, and takes the form

$$P \propto 1/\rho.$$

And in a second approximation, which I obtained theoretically during an unsuccessful attempt to derive the parameter free law $P \propto 1/\rho$, all the differences between languages and subjects boil down to

$$P = F(\rho + V)^{-1/D}.$$

Since $\Sigma P = 1$, the three parameters D, F and V are related by $F^{-1} = \Sigma(\rho + V)^{-1/D}$.

Together, these parameters measure how rich is a subject's vocabulary use.

The main parameter is D. It is sensible to measure how rich is a subject's use of vocabulary through the relative frequency of his use of rare words; for example, through the frequency of the word of rank $\rho = 1000$ compared to that of the word of rank $\rho = 10$. This relative frequency increases with D.

Why is the above law of such universality?

Since it is near perfectly hyperbolic, and granted all we have learned so far in the Essay, it is eminently sensible to try and relate Zipf's law to some underlying scaling property. (The procedure seemed highly nonobvious in 1950, when I first tackled this topic.) As suggested by the notation, the exponent plays the usual role of dimension. The prefactor F (recall Chapter 34) comes second.

LEXICOGRAPHIC TREES

An "object" that could be scaling does indeed exist in the present case: it is a lexicographical tree. We first define it and describe what scaling means in its context. Then we prove that when the lexicographical tree is scaling, word frequencies follow the two-parameter law written above. We discuss the validity of the explanation. Then we point out the interpretation of D as a dimension.

TREES. A lexicographical tree has N+1 trunks, numbered from 0 to N. The first trunk corresponds to the "word" constituted by the improper letter "space" taken by itself, and each of the other trunks corresponds to one of N proper letters. The "space" trunk is barren, but each of the other trunks carries N+1 leaders corresponding to the space and to N proper letters. The next generation space leader is barren and the others branch out into N+1 as before. Hence the barren tip of each space leader corresponds to a word made of proper letters followed by a space. And the construction continues ad infinitum. Each

barren tip is inscribed with the corresponding word's probability. And the tip of a nonbarren branch is inscribed with the total probability of the words that begin with the sequence of letters that determines said branch.

SCALING TREES. A tree can be termed *scaling* if each branch taken by itself is in some way a reduced-scale version of the whole tree. To truncate such a tree is near-literally to cut a branch from it. Hence our first conclusion is that a scaling tree must branch out without bound. In particular, contrary to untrained intuition, the total number of different words is not a sensible way of measuring richness of vocabulary. (Nearly everyone "knows" so many more words than he uses that his vocabulary is practically infinite.) A further argument (which we skip) determines the form that must be observed for the probability P of a barren branch at the kth level, i.e., growing on top of k live ones.

DERIVATION OF THE GENERALIZED ZIPF LAW IN THE SIMPLEST CASE. (Mandelbrot 1951, 1965z, 1968p.) The simplest scaling tree corresponds to discourse that is a sequence of statistically independent letters, the probability of each proper letter being $r < 1/N$, and that of the improper letter "space" being the remainder $(1-Nr)$. In this case, the kth level has the following properties

$$P=(1-Nr)r^k=P_0r^k,$$

and ρ varies between the bound

$$1+N+N^2+...+N^{k-1}=(N^k-1)/(N-1)$$

(excluded) and the bound

$$(N^{k+1}-1)/(N-1)$$

(included). Writing

$$D = \log N / \log(1/r) < 1 \text{ and } V = 1/(N-1),$$

and inserting

$$k = \log(P/P_0)/\log r$$

in each bound, we have

$$P^{-D}P_0{}^D - 1 < \rho/V \leq N(P^{-D}P_0{}^D) - 1.$$

The desired result is obtained by approximating ρ through the average of its bounds.

GENERALIZATION. Less simple scaling trees correspond to letter sequences generated by other stationary random processes, for example by Markov chains, and later cut into words by the recurrences of the space. The argument becomes more complex (Mandelbrot 1955b), but the final result is the same.

CONVERSE. Does it conversely follow from Zipf's data that the lexicographical tree using ordinary letters is scaling? Of course not: many short sequences of letters never occur and many long sequences are fairly common, hence actual lexicographical trees are far from being strictly scaling, but it is generically felt that the above argument suffices to explain why the generalized Zipf law holds. Also, one might mention that it had originally been hoped that Zipf's law would contribute

to the field of linguistics, but my explanation shows this law is linguistically very shallow.

◁ The generalized Zipf law also holds within certain restricted vocabularies. For example, the esoteric discipline styling itself hagioanthroponymy, which investigates the uses of names of saints as surnames of humans (Maître 1964), establishes that the Zipf law applies to such surnames. Also, Tesnière 1975 finds it applies to family names. Does this suggest that the corresponding trees are scaling? ▶

D IS A FRACTAL DIMENSION. The new observation that D is formally a similarity dimension is not as shallow as one may fear. Indeed, if one precedes it with a decimal point, a *word* as we defined it is nothing but a number between 0 and 1 written in the counting basis (N+1) and containing no zero except at the end. Mark such numbers on the interval [0,1] and add the limit points of this set. The construction amounts in effect to cutting out of [0,1] all the numbers that include the digit 0 otherwise than at the end. One finds that the remainder is a Cantor dust, the fractal dimension of which is precisely D.

As to the scaling lexicographical trees other than the simplest ones, to which we have alluded as providing a generalized proof of the Zipf law, they correspond in the same way to generalized Cantor dusts of dimension D. The equation for D in Mandelbrot 1955b is a matrix generalization of the definition of the similarity dimension through $Nr^D = 1$.

FURTHER GENERALIZATION: THE CASE D > 1. Curiously, the condition D < 1 is not universally fulfilled. The instances where the generalized Zipf law holds but the estimated D satisfies D>1, are rare but unquestionable. To describe the role of the special value D=1 let us assume that the law $P = F(\rho + V)^{-1/D}$ holds only up to $\rho = \rho^* \leq \infty$. If D<1, there is no difficulty with the infinite dictionaries suggested by the theoretical argument. But the infinite series $\Sigma(\rho + V)^{-1/D}$ diverges when D≥1. Hence $\Sigma P = 1$ and F>0 demand that $\rho^* < \infty$: the dictionary must contain a *finite* number of words.

It turns out, indeed, that D>1 is only encountered in cases where the vocabulary is unnaturally limited by artificial extraneous means (e.g., Latin inserts in a non-Latin text). These special cases are discussed in my papers on this subject. Since a construction limited to a finite number of points never leads to a fractal, D>1 is not interpretable as a fractal dimension.

TEMPERATURE OF DISCOURSE

The above deviations allow for a second, very different interpretation, patterned after statistical thermodynamics. The counterparts of physical energy and physical entropy are a cost of coding and Shannon's information. And D is the "temperature of discourse." The "hotter" the discourse, the higher the probability of use of rare words.

The case D<1 corresponds to the standard case where there is no upper bound for the formal equivalent of energy.

On the other hand, the case when words are so "hot" as to lead to D>1 involves the highly unusual imposition of a finite upper bound on the energy.

Shortly after I described this sharp dichotomy in terms of language statistics, a counterpart was independently recognized in physics. The inverse physical temperature $1/\theta$ is smallest—it vanishes—when a body is hottest, and Norman Ramsey recognized that if the body is to become hotter still, $1/\theta$ must become negative. See Mandelbrot 1970p for a discussion of this parallelism.

Thermodynamics deduces the bulk properties of objects from microcanonial equiprobability. Since molecules are not known individually, assumptions about their possible states elicit little emotion, but we do have an individual knowledge of words, so in the study of language the assumption of equiprobability is hard to swallow.

◁ The preceding analogy becomes particularly natural within certain more general approaches to thermodynamics. At the risk of overquoting items that are peripheral to this Essay, one such formalism is given in Mandelbrot 1962t, 1964t. ▶

THE PARETO LAW FOR SALARIES

Another example of a scaling abstract tree is found in the organization charts of hierarchic human groups. We deal with the simplest scaling hierarchy if (a) its members distribute among levels in such a way that (except on the lowest level) each member has the same number N of subordinates and (b) all his subordinates have the same "weight" U, which is equal to r<1 times the weight of the immediate superior. It is most convenient to consider this weight to be a salary.

When diverse hierarchies are to be compared from the point of view of the inequality of incomes, one can classify their members in the order of decreasing income (the order within a level is arbitrary), designate each individual by his rank ρ, and evaluate the rate of decrease of income as a function of rank, or vice versa. The more rapid the decrease in income when rank increases, the greater the inequality.

The formalism used for Zipf's law applies without change: the rank ρ of an individual of income U is approximately

$$\rho = -V + U^{-D} F^D.$$

The present derivation is due to Lydall 1959.

The degree of inequality is mostly determined by

$$D = \log N / \log(1/r),$$

which does not seem to have any fractal interpretation worth writing down. The greater the formal D, the greater the value of r and the lower the degree of inequality.

It is possible (as in the case of word frequencies) to generalize the model by assuming that within a given level k the value of U varies between individuals, so that U is equal to the product of r^k by a random factor, the same for everyone. This generalization modifies the parameters V and P_0 and hence D, but it leaves the basic relationship unchanged.

Note that the empirical D is ordinarily near 2. In cases where it is exactly 2, let inverse income be plotted on an axis pointing downward. One obtains an exact pyramid (base equal to the square of the height). In this case, the income of a superior is the geometric mean of the income of all his subordinates taken together and of that of each subordinate taken separately.

CRITIQUE. When D=2, the smallest $1/r$ occurs for N=2 and equals $1/r=\sqrt{2}$. This value seems unrealistically high, suggesting that Lydall's model can only hold in hierarchies in which D>2. If so, the fact that the overall D in a population is about 2 may mean that income differences *within* hierarchies pale in comparison with the differences *between* hierarchies, and the differences within groups that involve no hierarchical trees.

OTHER INCOMES' DISTRIBUTION

A broader study of the distribution of income in Mandelbrot 1960i, 1961e, 1962q, inspired the work described in Chapter 37. ■■

39 ☒ Mathematical Backup and Addenda

Complicated formulas and mathematical definitions and references, avoided elsewhere, are brought together in this chapter, together with several mathematical and other addenda.

AFFINITY AND SELF-SIMILARITY (SELF-)

In the text, the terms *self-similar* and *self-affine* (a neologism) are applied to either bounded or unbounded sets (without, I hope, introducing ambiguity). Many discussions of turbulence, and earlier papers of mine, also use *self-similar* in a "generic" sense that incorporates self-affine, but in this Essay the generic meaning is implemented by *scaling*.

1. SELF-SIMILARITY

In the Euclidean space $\mathbb{R}^E$, a real ratio $r > 0$ determines a transformation called *similarity*. It transforms the point $x = (x_1, \dots x_\delta, \dots x_E)$ into the point $r(x) = (rx_1, \dots rx_\delta, \dots rx_E)$, and hence transforms a set S into the set $r(S)$. See Hutchinson 1981.

BOUNDED SETS. A bounded set S is *self-similar*, with respect to the ratio r and an integer N, when S is the union of N nonoverlapping subsets, each of which is congruent to $r(S)$. *Congruent* means identical except for displacement and/or rotation.

A bounded set S is *self-similar*, with respect to the array of ratios $r^{(1)}...r^{(N)}$, when S is the union of N nonoverlapping subsets, respectively congruent to $r^{(n)}(S)$.

A bounded random set S is *statistically self-similar*, with respect to the ratio r and an integer N, when S is the union of N nonoverlapping subsets, each of which is of the form $r(S_n)$ where the N sets S_n are congruent in distribution to S.

UNBOUNDED SETS. An unbounded set S is *self-similar* with respect to the ratio r, when the set $r(S)$ is congruent to S.

2. SELF-AFFINITY

In the Euclidean space of dimension E, a collection of positive real ratios $r = (r_1...r_\delta...r_E)$ determines a *affinity*. It transforms each point $x = (x_1...x_\delta...x_E)$ into the point

$$r(x)=r(x_1...x_\delta...x_E)=(x_1 r_1...x_\delta r_\delta...x_E r_E),$$

hence transforms a set S into the set $r(S)$.

BOUNDED SETS. A bounded set S is *self-affine*, with respect to the ratio vector r and an integer N, when S is the union of N nonoverlapping subsets, each of which is congruent to $r(S)$.

UNBOUNDED SETS. An unbounded set S is *self-affine*, with respect to the ratio vector r, when the set $r(S)$ is congruent to S.

The preceding definition is often applied under the following conditions: (a) S is the graph of a function X(t) from scalar time t to an E−1 dimensional Euclidean vector; (b) $r_1 = ...r_\delta... = r_{E-1} = r$; (c) $r_E \neq r$. In this case, a direct definition runs as follows: A time-to-vector function X(t) is *self-affine*, with respect to the exponent α and the focal time t_0, if there exists an exponent $\log r_E / \log r = \alpha > 0$ such that for every h > 0 the function $h^{-\alpha}X[h(t-t_0)]$ is independent of h.

LAMPERTI SEMI-STABILITY. Random unbounded self-affine sets are called *semi-stable* in Lamperti 1962, 1972.

ALLOMETRY. Chapter 17 observes that when a botanical tree's height changes by r, its trunk diameter changes by $r^{3/2}$. In fact, the representative points whose coordinates are diverse linear measures of trees relate to each other by an affinity. Biologists call such figures *allometric*.

BROWN FRACTAL SETS

Due to the proliferation of different Brown sets, the terminology is necessarily pedantic, and sometimes even leaden.

1. BROWN LINE-TO-LINE FUNCTION

This term denotes the *classical ordinary Brownian motion*, also called *Wiener function*, *Bachelier function*, or *Bachelier-Wiener-Lévy function*. The following cumbersome definition allows an easy classification of various generalizations.

ASSUMPTIONS. (a) The time variable t is a

real number. (b) The space variable x is a real number. (c) The parameter H is H=½. (d) The probability Pr(X<x) is given by the error function erf(x), which is the distribution of the reduced Gaussian random variable with $\langle X \rangle = 0$ and $\langle X^2 \rangle = 1$.

DEFINITION. The line-to-line Brownian function B(t) is a random function such that, for all t and Δt,

$$Pr([B(t+\Delta t)-B(t)]/|\Delta t|^H < x) = erf(x).$$

WHITE GAUSSIAN NOISE REPRESENTATION. The function B(t) is continuous but nondifferentiable, meaning that B'(t) does not exist as an ordinary but as a generalized function (Schwartz distribution). This B'(t) is called *white Gaussian noise*. One can write B(t) as the integral of B'(t).

SELF-AFFINITY. The notion of probability distribution extends from random variables to random functions. Setting B(0)=0, the rescaled function $t^{-\frac{1}{2}}B(ht)$ has a probability distribution independent of t. This property of scaling is an example of *self-affinity*.

SPECTRUM. In terms of spectral or harmonic analysis, the spectral density of B(t) is proportional to f^{-1-2H}, that is, to f^{-2}. However, the meaning of the spectral density f^{-2} requires a special argument, because the function B(t) is *not* stationary, while the customary Wiener-Khinchin theory of covariance and spectrum is relative to stationary functions. This discussion is therefore postponed to the entry WEIERSTRASS.

NONDIFFERENTIABILITY. The function B(t) is continuous but is not differentiable. Again, the topic is best analyzed in the entry WEIERSTRASS.

REFERENCES. Lévy 1937-1954 and 1948-1965 have a well-deserved reputation for cryptic elegance and very personal style (see Chapter 40). However, they are unmatched for intuitive depth and simplicity.

Businesslike recent references, tailored to the needs of very diverse groups of mathematicians, scientists, and engineers, are too numerous to list, but the recent Knight 1981 looks promising. (Unfortunately, it chooses *not* to include "results on the Hausdorff dimension or measure of the sample paths, however elegant they may be, because they do not seem to have any known applications [!] and [do] not...seem really necessary for a general understanding of the directly applicable material. On the other side...such topics as the everywhere nondifferentiability of the sample paths...do seem to say something definite about the extreme irregularity of the paths".)

2. GENERALIZED BROWN FUNCTIONS

Every assumption in the preceding section has a natural generalization, and every process obtained by generalizing one assumption or more is significantly different from the original B(t), and has significant applications.

(a) The real (scalar) time t may be replaced by a point in Euclidean space $\mathbb{R}^E$, with E>1, on a circle, or on a sphere.

(b) The real (scalar) X may be replaced by

a point P in Euclidean space $\mathbb{R}^E$, with $E > 1$, a point on a circle, or on a sphere.

(c) The parameter H may be given a value other than ½. With the Gaussian distribution erf, H can be allowed to lie anywhere in the range $0 < H < 1$.

(d) The Gaussian distribution erf may be replaced by a non-Gaussian distribution discussed in the entry LÉVY STABLE.

Furthermore, B(t) can be generalized via its white noise representation. This procedure yields substantially different results.

3. DETRENDING

The variation of the Brown line-to-line function B(t) between $t = 0$ and $t = 2\pi$ decomposes into (a) the trend defined by $B^*(t) = B(0) + (t/2\pi)[B(2\pi) - B(0)]$, and (b) an oscillatory remainder $B_B(t)$. In the case of the Brownian B(t), these terms happen to be statistically independent.

THE TREND. The graph of the trend $B^*(t)$ is a straight line with a random Gaussian slope.

BROWN BRIDGE. The "detrended" oscillatory term $B_B(t)$ is identical in distribution to a *Brown bridge*, defined as a Brown line-to-line function that is constrained to satisfy $B(2\pi) = B(0)$.

ABUSE OF DETRENDING. Faced with a samples of unknown origin, many applied statisticians, working in economics, meteorology and the like, hasten to decompose it into a trend and an oscillation (and added periodic terms).

They assume implicitly that the addends are attributable to distinct generating mechanisms, and are statistically independent.

This last implicit assumption is quite unwarranted, except when the sample is generated by B(t).

4. BROWN CIRCLE-TO-LINE FUNCTIONS

LOOPED BROWN BRIDGE. Take the periodic function of t which coincides over the time span $0 < t \leq 2\pi$ with a Brown bridge $B_B(t)$, and select Δt at random (uniformly) over $[0, 2\pi[$. The function $B_B(t + \Delta t)$ is statistically stationary (see the entry STATIONARITY...), and can be represented as a random *Fourier-Brown-Wiener series*. The coefficients are independent Gaussian random variables, having wholly random phases and moduli proportional to n^{-1}. In other words, the discrete spectrum is proportional to n^{-2}, that is, to f^{-2}, and the cumulative spectral energy above the frequency f is $\sim f^{-1}$.

PRACTICAL CONSEQUENCE. The simulation of B(t) is necessarily carried out over a finite time span. If this span is viewed as $[0, 2\pi[$, the simulation can rely on discrete finite Fourier methods. One computes a Brown bridge using a fast Fourier transform, and the required random trend is added.

REFERENCES. Paley & Wiener 1934 deserves its reputation for relentless algebra. However, the profound expository paragraphs in its Chapters IX and X still deserve to be read. Kahane 1968 is recommended, but only to mathematicians; the results are never stated in their simple original context.

ODD LOOPED BROWN BRIDGE. The functions $B_0(t)=\frac{1}{2}[B_B(t)-B_B(t+\pi)]$ and $B_E(t) = \frac{1}{2}[B_B(t)+B_B(t+\pi)]$, are (respectively) the sums of the odd and the even numbered harmonic components of the bridge function $B_B(t)$. The odd component has the virtue of being obtained directly in terms of a white Gaussian noise $B'(t)$ laid along the circle:

$$B_0(t)=\int_{-\pi}^{0} B'(t-s)ds-\int_{0}^{\pi}B'(t-s)ds.$$

BROWN LINE-TO-CIRCLE FUNCTION. Starting with $B(t)$, drop its integer part and multiply the fractional remainder by 2π. The result determines a point's position on the unit circle. This *Brownian line-to-circle* function is mostly mentioned to warn against confusing it with either of the preceding, very different, functions.

5. FRACTIONAL BROWN LINE-TO-LINE FUNCTIONS

To define this function, denoted by $B_H(t)$, start with the ordinary Brown line-to-line function and change the exponent from $H=\frac{1}{2}$ to any real number satisfying $0<H<1$. Cases where $H\neq\frac{1}{2}$ are *properly fractional*.

All the $B_H(t)$ are continuous and nondifferentiable. The earliest mention of them I could locate is in Kolmogorov 1940. Other scattered references and various properties are listed in Mandelbrot & Van Ness 1968. See also Lawrance & Kottegoda 1977.

CORRELATION AND SPECTRUM. Clearly, $\langle[B_H(t+\Delta t) - B_H(t)]^2\rangle = |\Delta t|^{2H}$. The spectral density of $B_H(t)$ is proportional to f^{-2H-1}. The exponent is not an integer, which is one of several reasons why I proposed to denote $B_H(t)$ as being *fractional*.

DISCRETE FRACTIONAL GAUSSIAN NOISE. It is defined as the sequence of increments of $B_H(t)$ over successive unit time spans. Its correlation is

$$2^{-1}[|d+1|^{2H}-2|d|^{2H}+|d-1|^{2H}].$$

LONG-RUN CORRELATIONS. PERSISTENCE AND ANTIPERSISTENCE. Set $B_H(0)=0$, and define the past increment as $-B_H(-t)$ and the future increment as $B_H(t)$. One has:

$$\langle-B_H(-t)B_H(t)\rangle$$

$$=2^{-1}\{\langle[B_H(t)-B_H(-t)]^2\rangle-2\langle[B_H(t)]^2\rangle\}$$

$$=2^{-1}(2t)^{2H}-t^{2H}.$$

Dividing by $\langle B_H(t)^2\rangle = t^{2H}$, one obtains the correlation, which one finds to be independent of t: it equals $2^{2H-1}-1$. In the classical case $H=\frac{1}{2}$, the correlation vanishes, as expected. For $H>\frac{1}{2}$, the correlation is positive, expressing *persistence*, and it becomes 1 when $H=1$. For $H<\frac{1}{2}$, the correlation is negative, expressing *antipersistence*, and it becomes $-\frac{1}{2}$ when $H=0$.

The fact that this correlation should be independent of t even in cases when it does not vanish is an obvious corollary of the self-affinity of $B_H(t)$.

However, most students of randomness begin by being surprised and/or disturbed that the correlation of past and future may be independent of t without reducing to 0.

PRACTICAL CONSEQUENCE CONCERNING SIMULATION. To generate a random function for all integer times between t=0 and t=T, it is customary to select an algorithm in advance with no regard to T, and then to let it run for a time T. The algorithms needed to generate the fractional Brown functions are very different: they necessarily depend on T.

A fast generator for the discrete increments of $B_H(t)$ is described in Mandelbrot 1971f. (That paper is marred by a potentially very disturbing misprint: in the first fraction on p. 545, 1 must be subtracted from the numerator and added to the whole fraction.)

FRACTAL DIMENSIONS. For the graph, one has D=2−H. For the zeroset and other level sets, one has D=1−H. See Adler 1981.

6. FRACTIONAL BROWN CIRCLE-
OR TORUS--TO-LINE FUNCTION

Fractional Brown circle-to-line functions are far less intrinsic than the functions of subentry 4. The simplest is the sum of the *fractional Fourier-Brown-Wiener* series, defined as having independent Gaussian coefficients, wholly random phases, and coefficient moduli proportional to $n^{-H-\frac{1}{2}}$. A fractional Brown torus-to-line function is the sum of a double

Fourier series with the same properties.

WARNING. A superficial analogy would suggest that the fractional Brown circle-to-line function might be obtained by the process that is applicable in the nonfractional case: by forming the trend $B_H^*(t)$ of a fractional Brown line-to-line function, then detrending $B_H(t)$ and forming a periodic function by repetition.

Unfortunately, the periodic function obtained in this fashion and the sum of the Fourier series with coefficients $n^{-H-\frac{1}{2}}$ are *different* random functions. In particular, the Fourier series *is stationary*, while the repeated detrended $B_H(t)$ *is not*. For example, over a small interval on both sides of t=0, the repeated detrended bridge joins together two nonconsecutive subpieces of $B_H(t)$. The pinning down involved in the definition of the bridge is sufficient to make the combined piece continuous but is not sufficient to make it stationary. For example, it is *not* identical in distribution to a small piece made of consecutive subpieces on both sides of $t=\pi$.

REMARKS ON SIMULATION. To compute a fractional Brown line-to-line function by finite discrete Fourier methods is theoretically impossible and in practice is workable but very tricky. The most straightforward procedure is to (a) compute the appropriate circle-to-line function, (b) discard it except for a limited portion corresponding to a small subinterval of the period 2π, say from $0 < t < t^*$, and (c) add a separately computed very low frequency component. As $H \to 1$, this t^* must tend to 0.

FRACTAL DIMENSIONS. For the whole graph, D=2−H (Orey 1970). When the level set is nonempty, $D=1-H$. This result is in Marcus 1976 (strengthening Theorem 5, p. 146, in Kahane 1968).

CRITICAL TRANSITION THROUGH $H=1$. The fractional Fourier-Brown-Wiener series with independent Gaussian coefficients proportional to $n^{-\frac{1}{2}-H}$ converges to a continuous sum for all $H>0$. When H crosses the value $H=1$, the sum becomes differentiable. By contrast, the fractional Brownian process is only defined up to $H=1$. This difference in the range of admissible values of H confirms that these two processes are quite different. It also suggests that critical transition phenomena of physics might be modeled by the line-to-line, but not the circle-to-line, Brown function.

7. FRACTIONAL BROWN LINE (OR CIRCLE)-TO-SPACE TRAILS

In the circle-to-space case with $H<1$, the trail's dimension is $\min(E, 1/H)$. This is part of Theorem 1, p. 143, in Kahane 1968.

8. DIFFERENT FORMS OF FRACTIONAL INTEGRO-DIFFERENTIATION

To transform the Brown line-to-line function $B(t)$ into $B_H(t)$, the simplest is to write

$$B_H(t)=[\Gamma(H+\tfrac{1}{2})]^{-1} \int_{-\infty}^{t} (t-s)^{H-\frac{1}{2}} dB(s).$$

This integral diverges, but increments like $B_H(t)-B_H(0)$ are convergent. This is a moving average of kernel $(t-s)^{H-\frac{1}{2}}$. A classical, albeit rather obscure transformation, it is known to pure mathematicians as the *Riemann Liouville fractional integral or differential* of order $H+\frac{1}{2}$.

HEURISTICS. The idea that the order of integration and/or differentiation need not be an integer is best understood in spectral terms. Indeed, ordinary integration of a periodic function is equivalent to the multiplication of the function's Fourier coefficients by $1/n$, and ordinary integration of a nonperiodic function is equivalent to the multiplication of its Fourier transform (when defined) by $1/f$. Hence, the operation that multiplies the Fourier transform by the *fractional power* $(1/f)^{H+\frac{1}{2}}$ can reasonably be called *fractional integro-differentiation*. Since the spectrum of white noise is f^{-0}, the spectrum of $B_H(t)$ is $(1/f)^{2(H+\frac{1}{2})} = f^{-2H-1}$ (as announced).

REFERENCES. The Riemann-Liouville transform has many other scattered applications (Zygmund 1959, II, p. 133, Oldham & Spanier 1974, Ross 1975, Lavoie, Osler & Tremblay 1976). The less well-known application to probability (with references back to Kolmogorov 1940) is discussed in Mandelbrot & Van Ness 1968.

EFFECT ON SMOOTHNESS. When its order $H-\frac{1}{2}$ is positive, the Riemann Liouville transform is a fractional form of integration, because it increases a function's smoothness. Smoothness equals local persistence, but smoothness obtained by integration extends to function's global properties. When $H-\frac{1}{2}<0$, the Riemann Liouville transform is a fractional form of differentiation, because it enhances irregularity that depends on local behavior.

APPLICATION TO BROWN FUNCTIONS. For a fractional Brown circle-to-line function, H has no upper bound. Fractional integration of order $H-\frac{1}{2}>\frac{1}{2}$ applied to a Brown circle-to-line function creates a differentiable function. On the contrary, Brown line-to-line functions, $H-\frac{1}{2}$ can at most equal $\frac{1}{2}$, and $B_H(t)$ is not differentiable.

For both the Brown circle-to-line and line-to-line functions, local irregularity prohibits differentiation beyond H=0, hence beyond the order $-\frac{1}{2}$.

BILATERAL EXTENSION OF FRACTIONAL INTEGRO-DIFFERENTIATION. The fact that the classical Riemann Liouville definition is strongly asymmetric in t is quite acceptable when t is time. But cases when the coordinate t may "run" in either direction require a symmetric definition. I propose

$$B_H(t)=[\Gamma(H+\tfrac{1}{2})]^{-1}.$$
$$\left[\int_{-\infty}^{t}(t-s)^{H-\frac{1}{2}}dB(s)\ -\int_{t}^{\infty}|t-s|^{H-\frac{1}{2}}dB(s)\right]$$

9. BROWN SPACE-TO-LINE FUNCTIONS

Lévy 1948, 1957, 1959, 1963, 1965 introduced Brown functions from a space Ω to the real line, where Ω is *either* $\mathbb{R}^E$ with the ordinary distance $|PP_0|$, *or* a sphere in $\mathbb{R}^{E+1}$ with distance defined along a geodesic, *or* a Hilbert space. For each of these Brown functions, $B(P)-B(P_0)$ is a Gaussian random variable of zero mean and variance $G(|PP_0|)$, with $G(x)=x$. The literature includes McKean 1963 and Cartier 1971.

WHITE GAUSSIAN NOISE REPRESENTATION WHEN Ω IS A SPHERE. This B(P) is constructed as described in Chapter 28: throw a white noise blanket on the sphere, and take for B(P) this integral of this noise over the half sphere whose North Pole is P. Actually, I prefer the variant that takes $\frac{1}{2}$ of the integral over the half sphere, minus $\frac{1}{2}$ of the integral on the other half sphere. This generalizes the second process in subentry 4 above.

WHITE GAUSSIAN NOISE REPRESENTATION WHEN Ω IS $\mathbb{R}^E$ (CHENTSOV 1957). This case involves a more complicated algorithm, due to Chentsov, which is easiest to visualize when Ω is $\mathbb{R}^2$ and B(0,0)=0. Take an auxiliary cylinder of radius 1 and coordinates u and θ, and place a blanket of white noise on it. As modified in Mandelbrot 1975b, the algorithm begins by integrating this noise over the rectangle from θ to $\theta+d\theta$ and from 0 to u. One obtains a line-to-line Brown function that vanishes for u=0 and that will be denoted by $B(u,\theta,d\theta)$. For each (x,y) in the plane, the line-to-line Brown components $B(x\cos\theta + y\sin\theta,\ \theta,\ d\theta)$ are statistically independent, and their integral over θ is B(x,y).

10. FRACTIONAL BROWN SPACE-TO-LINE FUNCTIONS

Gangolli 1967, anticipated on certain points by Yaglom 1957, generalizes B(P) to the case where $G(x) = x^{2H}$ in the preceding subentry. But he fails to include an explicit algorithm to construct the resulting function. To do so, Mandelbrot 1975b generalizes the Chentsov construction by replacing each $B(u, \theta, d\theta)$ by a bilaterally defined fractional Brown line-to-line function.

For D, see Yoder 1974, 1975.

For simulation via FFT, see Voss, 1982.

11. NONLINEAR TRANSFORMS OF FRACTIONAL GAUSSIAN NOISES

Given a $G(x)$ different from $G(x) = x$, form $\Sigma_{t=1}^{T} G\{B_H(t) - B_H(t-1)\}$, and interpolate linearly for noninteger T's. The result, to be denoted by $B_G(T) - B_G(0)$, is asymptotically scaling if there exists a function $A(T)$ such that $\lim_{T\to\infty} A(T)\{B_G(hT) - B_G(0)\}$ is non-degenerate for every $h\epsilon(0,1)$. Murray Rosenblatt had studied the case $G(x) = x^2 - 1$. Taqqu 1975 shows the problem hinges on the Hermite rank of G, defined as the order of the lowest term in the development of G in Hermite series. More recent results along these lines are in Taqqu 1979 and Dobrushin 1979.

DIMENSION AND COVERING OF A SET (OR OF ITS COMPLEMENT) BY BALLS

The fractal dimension I advocate and all its acceptable variants are not topological but *metric* notions. They involve a metric space Ω, that is, a space in which the distance between any two points is defined suitably. A closed (respectively, open) ball of center ω and radius ρ is the set of all points whose distance to ω is $\leq\rho$ (respectively $< \rho$). (Balls are solids, and spheres are their surfaces.)

Given a bounded set S in Ω, there are many methods of covering it with balls of radius ρ. Often, as in the examples examined in this entry, these methods involve naturally a notion of dimension. In the basic case studies, these notions yield identical values. In other instances, however, their values differ.

1. CANTOR AND MINKOWSKI

The crudest method of covering, pioneered by Cantor, centers a ball on every point in S, and uses these balls' union as a smoothed-out version of S, to be called $S(\rho)$.

Add the assumption that Ω is an E-dimensional Euclidean space. In this case, the concept of volume (vol) is defined, and

vol$\{$d-dimensional ball of radius $\rho\} = \gamma(d)\rho^D$,

with

$$\gamma(d) = [\Gamma(\tfrac{1}{2})]^d / \Gamma(1 + d/2).$$

When S is a cube of volume much greater than ρ^3,

$$\text{vol}\,[S(\rho)] \sim \text{vol}\,[S].$$

When S is a square having an area much greater than ρ^2,

$$\mathrm{vol}[\mathcal{S}(\rho)] \sim 2\rho \ \mathrm{area} \ [\mathcal{S}].$$

When $\mathcal{S}$ is an interval having a length much greater than ρ,

$$\mathrm{vol}[\mathcal{S}(\rho)] \sim \pi\rho^2 \ \mathrm{length} \ [\mathcal{S}].$$

More precisely, let "content" stand for either volume, area or length, whichever is appropriate, and let d be the standard dimension. Letting V denote the expression

$$V = \mathrm{vol}[\mathcal{S}(\rho)] / \gamma(E-d)\rho^{E-d},$$

we see that cubes, squares, and lines satisfy

$$\mathrm{content} \ [\mathcal{S}] = \lim_{\rho \to 0} V.$$

This formula is not, as might seem, a hair-splitting relation between equally innocuous notions. An example due to H. A. Schwarz (reported in 1882) shows that, when a circular cylinder is triangulated and the triangulation is made increasingly fine, the sum of the triangles' areas *does not necessarily* converge to the cylinder's area. To avoid this paradoxical behavior, Minkowski 1901 sought to reduce length and area to the sound and simple concept of volume, through the above method of covering $\mathcal{S}$ by balls.

However, a slight complication enters from the outset: the expression V may fail to have a limit as ρ tends to 0.

When such is the case, the notion of lim is replaced by the twin notions of lim sup and lim inf. To every real number A in the open interval] lim inf, lim sup [, corresponds at least one sequence $\rho_m \to 0$ such that

$$\lim_{m \to \infty} \mathrm{vol}\{[\mathcal{S}(\rho_m)] / \gamma(E-d)\rho_m^{E-d}\} = A.$$

But no such sequence exists if either A < lim inf or A > lim sup. These definitions being granted, Minkowski 1901 calls

$$\lim \sup_{\rho \to 0} \mathrm{vol}[\mathcal{S}(\rho)] / \gamma(E-d)\rho^{E-d}$$

$$\mathrm{and} \quad \lim \inf_{\rho \to 0} \mathrm{vol}[\mathcal{S}(\rho)] / \gamma(E-d)\rho^{E-d}$$

the upper and the lower d-content of $\mathcal{S}$. When they are equal, their value is the d-content of $\mathcal{S}$. Minkowski observes that for standard Euclidean shapes there exists a D such that if d > D the upper content of $\mathcal{S}$ vanishes, and if d < D the lower content of $\mathcal{S}$ is infinite.

2. BOULIGAND

The extension of Minkowski's definition to noninteger d's is due to Bouligand 1928, 1929. In particular, the above lim inf, which may be a fraction, deserves to be called the *Minkowski-Bouligand dimension* D_{MB}.

Bouligand recognized that D_{MB} is sometimes counterintuitive, and more generally is less desirable than the Hausdorff Besicovitch D. But it is often identical to D and easier to evaluate, hence is useful. The case E = 1 is discussed in Kahane & Salem 1963, p. 29, which confirms that D_{MB} is often equal to D, cannot be smaller, and can be greater.

3. PONTRJAGIN & SCHNIRELMAN; KOLMOGOROV & TIHOMIROV

Among all collections of balls of radius ρ that cover a set S in the metric space Ω, the most economical is by definition one that requires the smallest number of balls. When S is bounded, this smallest number is finite and can be denoted by $N(\rho)$. Pontrjagin & Schnirelman 1932 advances the expression

$$\lim \inf_{\rho \to 0} \log N(\rho) / \log (1/\rho)$$

as an alternative definition of dimension.

    This approach is developed in Kolmogorov & Tihomirov 1959, whose authors were inspired by Shannon's information theory to label $\log N(\rho)$ the ρ-entropy of S. Hawkes 1974 calls the corresponding dimension the *lower entropy dimension*, and the variant obtained by replacing lim inf by lim sup the *upper entropy dimension*. Hawkes shows that the Hausdorff Besicovitch dimension is at most equal to the lower entropy dimension; they often coincide but may fail to do so.

    Kolmogorov & Tihomirov 1959 also studies $M(\rho)$, defined as the largest number of points in S, such that their mutual distances exceed 2ρ. For sets on the line, $N(\rho)=M(\rho)$. But for other sets,

$$\lim \inf_{\rho \to 0} \log M(\rho) / \log (1/\rho)$$

is still another dimension.

    ◁ Kolmogorov & Tihomirov 1959 calls $\log M(\rho)$ a *capacity,* which is most unfortunate, because of an entirely different, older, and better-justified meaning for this term in potential theory. In particular, one must avoid the temptation of designating the dimension in the preceding paragraph as a capacity dimension. See POTENTIALS, 3. ▶

4. BESICOVITCH & TAYLOR; BOYD

When Ω is [0,1] or is the real line, we saw in Chapter 8 that a dust S is fully determined by its complement, which is the union of the maximal open intervals, the gaps (in some constructions, each gap is a trema).

TRIADIC CANTOR DUST C IN [0,1]. The lengths of the gaps add up to 1, and follow the hyperbolic distribution $\Pr(U>u)=Fu^{-D}$. Hence, the length λ_n of the nth gap by decreasing size has an order of magnitude of $n^{-1/D}$.

GENERAL LINEAR SETS OF ZERO LEBESGUE MEASURE. The behavior of the λ_n for $n \to \infty$ is studied in Besicovitch & Taylor 1954. There exists a real exponent D_{BT} such that the series $\Sigma \lambda_n^d$ converges when $d > D_{BT}$ (and in particular converges to 1 when $d=1$). Thus D_{BT} is the infimum of the real numbers d such that $\Sigma \lambda_n^d < \infty$. It can be shown that $D_{BT} \geq D$. Hawkes 1974 (p. 707) proves that D_{BT} coincides with the upper entropy dimension, but may be easier to evaluate.

WARNING. When S is not of zero measure,

D_{BT} is not a dimension. It relates to an exponent in Chap. 15, and to the Δ in Chap. 17.

APOLLONIAN PACKING EXPONENT. D_{BT} has a counterpart in the case of Apollonian packing (Chapter 18). It was introduced in 1966 by Z. A. Melzak, and Boyd 1973b showed that (as expected) it is the residual set's Hausdorff-Besicovitch dimension.

DIMENSION (FOURIER) AND HEURISTICS

Let $\mu(x)$ be a nondecreasing function of $x \epsilon [0,1]$. If the maximal open intervals where μ is constant add up to the complement of the closed set S, $d\mu(x)$ is called *supported by* S. The Fourier-Stieltjes transform of μ is

$$\hat{\mu}(f) = \int \exp(ifx)d\mu(x).$$

The smoothest μ's yield the fastest possible rate of decrease of $\hat{\mu}$. Let D_F be the largest real, such that at least one function $\mu(x)$ supported by S satisfies

$$\hat{\mu}(f) = o(|f|^{-D_F/2+\epsilon}), \text{ as } f \to \infty, \text{ for all } \epsilon > 0,$$

but no $\mu(x)$ satisfies

$$\hat{\mu}(f) = o(|f|^{-D_F/2-\epsilon}), \text{ as } f \to \infty, \text{ for some } \epsilon > 0.$$

Here, "a=o(b), as $f \to \infty$," means that $\lim_{f \to \infty}(a/b)=0$. When S is the whole interval $[0,1]$, D_F is infinite. On the contrary, when S is a single point, $D_F=0$. More interestingly, whenever S is of zero Lebesgue

measure, D_F is finite and at most equal to the Hausdorff Besicovitch dimension D of S. The inequality $D_F \le D$ shows that the fractal and harmonic properties of a fractal set are related, but are not necessarily identical.

To prove that these dimensions can differ, suppose that S is a set on a line for which $D=D_F$. When the same S is viewed as a set in the plane, D is unchanged but D_F becomes 0.

DEFINITION. A convenient way of summarizing some of the harmonic properties of S is to call D_F the *Fourier dimension* of S.

SALEM SETS. The equality $D_F=D$ characterizes a category of sets called sets of unicity or Salem sets (Kahane & Salem 1963, Kahane 1968).

RULE OF THUMB AND HEURISTICS. The fractals of interest in case studies tend to be Salem sets. Since D_F is often easy to estimate from data, it can serve as estimate of D.

NONRANDOM SALEM SETS. A nonrandom Cantor dust is a Salem set only if r satisfies certain number theoretic properties.

RANDOM SALEM SETS. A random Cantor dust is a Salem set when randomness is sufficient to break up every arithmetic regularity.

The original example, due to R. Salem, is very complex. The next example is the Lévy dust: Denoting the Lévy staircase (Plate 286) by L(x), Kahane & Mandelbrot 1965 shows that the spectrum of dL(x) is near identical on the average to the spectrum of the fractional Brownian line-to-line function, and is a smoothed form of the spectrum of the Gauss-Wierstrass function.

◁ Kahane 1968 (Theorems 1, p. 165 and

5, p. 173) shows that the image of compact set S of dimension δ by a fractional Brown line-to-line function of exponent H is a Salem set with $D=\min(1,\delta/H)$. ▶

THE CANTOR DUST IS NOT A SALEM SET. The triadic Cantor dust had originally emerged out of Georg Cantor's search for a set of unicity (See Zygmund 1959, I, p. 196), but this search failed. (Cantor then abandoned harmonic analysis, and—as second best!—founded the theory of sets). For example, denote the Cantor staircase (Plate 83) by $C(x)$. The spectrum of $dC(x)$ has the same overall shape as that of $dL(x)$, but it includes occasional sharp peaks of nondecreasing size, which implies that $D_\Gamma=0$. See Hille & Tamarkin 1929.

These peaks make all the difference in the theory of sets of unicity, but in practice they are unlikely to be significant. Most estimators of spectral density will tend to miss the peaks and to pick the background ruled by D.

FRACTALS (ON THE DEFINITION OF)

Although the term *fractal* is defined in Chapter 3, I continue to believe that one would do better without a definition (my 1975 Essay included none).

The immediate reason is that the present definition will be seen to exclude certain sets one would prefer to see included.

More fundamentally, my definition involves D and D_T, but it seems that the notion of fractal structure is more basic than either D or D_T. Deep down, the importance of the notions of dimension is increased by their unexpected new use!

In other words, one should be able to define fractal structures as being invariant under some suitable collection of smooth transformations. But this task is unlikely to be an easy one. To exemplify the difficulty in a standard context, let us recall that certain definitions of *complex number* fail to exclude the real numbers! At the present stage, the main need is to differentiate the basic fractals from the standard sets in Euclid. This is a need my definition does satisfy.

My obvious lack of enthusiasm must be shared by numerous prominent mathematicians who failed to notice the definition in my 1977 Essay. Nevertheless, let us elaborate on it.

1. DEFINITION

A *fractal set* was defined for the first time in the Introduction of my 1977 Essay as a set in a metric space, for which

Hausdorff Besicovitch dimension D
        > topological dimension D_T.

With one exception, the fractals in this book are sets in a Euclidean space of dimension $E<\infty$. They may be called Euclidean fractals. The exception is in Chapter 28: the Brown coastline on the sphere may be viewed as a Riemannian fractal.

2. CRITIQUE. PARTLY ARITHMETIC VERSUS PURELY FRACTAL DIMENSIONS

The above mathematical definition is *rigorous but tentative*, and it would be nice to improve it, but several seemingly natural changes would be ill-inspired.

Long ago, when fumbling for a measure for the properties later to be called fractal, I had settled on the Hausdorff Besicovitch dimension D, because it has been studied most carefully. The fact that treatises such as Federer 1969 find it necessary to introduce innumerable variants, separated from D by details, is unsettling. Nevertheless, there is good reason at present to postpone the examination of these details.

Furthermore, given several possible dimensions to choose between, one *must* avoid those that involve clearly extraneous features. Most important, D involves no arithmetic facet, by contrast with either the Fourier dimension D_F (p. 360) or the Besicovitch & Taylor exponent (p. 359 and Kahane 1971, p. 89).

3. HAUSDORFF BORDERLINE CASES

Borderline cases are always a problem. A priori, a nonrectifiable curve for which $D=1$ may be called either fractal or nonfractal, and the same holds for any set such that $D=D_T$, but the Hausdorff measure using the test function $h(\rho) = \gamma(D)\rho^D$ is infinite (it cannot vanish). More irritatingly, the Devil's staircase of Cantor (Plate 83) is intuitively a frac-

tal, since it exhibits many length scales in obvious fashion. Hence, one hates to have to call it nonfractal, even though $D=1=D_T$. See p. 373. Lacking other criteria, I set the border in such a way as to achieve a short definition. If and when a good reason arises, this definition ought to be changed. See HAUSDORFF, 8.

4. RESTATED DEFINITION

The "capacitary dimension" (see the subentry POTENTIALS, 4) satisfies the criteria set in subentry 2 above, merely because its value is identical to D. Hence, a fractal can be defined alternatively as a set for which

Frostman capacitary dimension
> topological dimension.

5. FRACTAL TIMES, INTRINSIC AND LOCAL

Some raw material on this topic is found in Chapter XII of the 1977 *Fractals*.

HAUSDORFF MEASURE AND HAUSDORFF BESICOVITCH DIMENSION

Convenient general references on this topic are Hurewicz & Wallman 1941, Billingsley 1967, Rogers 1970, and Adler 1981.

1. CARATHÉODORY MEASURE

The thought that "the general notion of volume or magnitude is indispensable in investigations on the dimensions of continuous sets" occurred to Cantor, in passing. Given the problem's difficulty, Lebesgue doubts that Cantor could reach any significant result. The idea is furthered in Carathéodory 1914 and implemented in Hausdorff 1919.

A classical method for evaluating the area of a planar shape begins by approximating S by a collection of very small squares and by adding these squares' sides raised to the power $D=2$. Carathéodory 1914 extends this traditional approach. It avoids reliance on coordinate axes by replacing squares by discs, and strives *not* to use in advance the knowledge that S is a standard Euclidean shape of known dimension imbedded in a known $\mathbb{R}^E$.

Observe therefore that when a planar shape imbedded in three-space is covered by discs, it is a fortiori covered by balls of which these discs are equators. Hence, to avoid prejudging the fact that S is planar, it suffices to cover it by balls instead of discs. When S is indeed a surface, one obtains its approximate contents by adding expressions of the form $\pi\rho^2$ corresponding to all the covering balls. More generally, a d-dimensional standard shape requires us to add expressions of the form $h(\rho) = \gamma(d)\rho^d$, where the function $\gamma(d) = [\Gamma(\frac{1}{2})]^d / \Gamma(1+d/2)$ is defined early in this chapter as the contents of a ball of unit radius. On this basis, Carathéodory 1914 extends the ideas of "length" or "area" to some nonstandard shapes.

2. HAUSDORFF MEASURE

Hausdorff 1919 goes beyond Carathéodory by allowing d to be fractional (the function $\gamma(d)$ was written in such a way that it continues to be meaningful). Thus, instead of limiting oneself to powers of ρ, one can use any positive *test function* $h(\rho)$ that tends to 0 with ρ.

Furthermore, a ball being merely the set of points whose distance from a center ω does not exceed a prescribed radius ρ continues to be defined when the space Ω is not Euclidean, as long as a distance is defined. As has been noted, such spaces are called metric, hence the Hausdorff measure is a metric concept.

Given a test (or "gauge") function $h(\rho)$, a finite covering of the set S by balls of radii ρ_m can be said to have the measure $\Sigma h(\rho_m)$. To achieve economy in covering, one considers all the coverings by balls of radius less than ρ, and one forms the infimum

$$\inf_{\rho_m < \rho} \Sigma h(\rho_m).$$

As $\rho \to 0$, the constraint $\rho_m < \rho$ becomes increasingly stringent. Hence the expression $\inf \Sigma h(\rho_m)$ can only increase; it has a limit

$$\lim_{\rho \to 0} \inf_{\rho_m < \rho} \Sigma h(\rho_m).$$

This limit may be either finite and positive, or infinite, or zero. It defines the h-measure of the set S.

When $h(\rho)=\gamma(d)\rho^d$, the h-measure is called d-dimensional. More precisely, due to the prefactor $\gamma(d)$, it is the normalized d-dimensional measure.

When $h(\rho)=1/\log|\rho|$, the h-measure is called logarithmic.

3. A SET'S INTRINSIC TEST FUNCTION

The function $h(\rho)$ may be called *intrinsic for* S and denoted by $h_S(\rho)$ if the h_S-measure of S is positive and finite. This measure may be called the *fractal measure* of S.

For the standard shapes in Euclid, the intrinsic test function is always of the form $h_S(\rho)=\gamma(D)\rho^D$, with some integer value of D. Hausdorff showed that $h_S(\rho)=\gamma(D)\rho^D$, with noninteger D's, are intrinsic for the Cantor dusts and the Koch curves.

On the other hand, in the case of typical random fractals, even where they are statistically self-similar, the intrinsic $h_S(\rho)$ exists but is more complicated, for example of the form $h_S(\rho)=\rho^D|\log\rho|$. If so, the h-measure of S with respect to $h(\rho)=\gamma(D)\rho^D$ vanishes, hence the shape has less "substance" than if it were D-dimensional, but more substance than if it were $(D-\epsilon)$-dimensional. An example is provided by Brownian motion in the plane, for which Lévy finds $h_S(\rho) = \rho^2\log\log(1/\rho)$. See Taylor 1964.

The 2-dimensional measure of any bounded set in the plane being finite, test functions such as $\rho^2/\log(1/\rho)$ are not intrinsic for any planar set.

Much work on determining $h_S(\rho)$ for random sets is coauthored or authored by S. J. Taylor; a reference is Pruitt and Taylor 1969.

4. HAUSDORFF BESICOVITCH DIMENSION: DEFINITION

If one knows that S is two-dimensional, it suffices to evaluate the Hausdorff h-measure for $h(\rho)=\pi\rho^2$. However, the definition of Hausdorff measure is formulated to insure that advance knowledge of D is not needed. If one deals with a standard shape of unknown dimension, one will evaluate the measure for all test functions $h(\rho)=\gamma(d)\rho^d$ with d an integer. If length is infinite and volume is zero, the shape can only be two-dimensional.

Besicovitch extended the core of this last conclusion to cases where d is not an integer and S is not standard shape. He showed that for every set S there exists a real value D such that the d-measure is infinite for d<D and vanishes for d>D.

This D is called the Hausdorff Besicovitch dimension of S.

To a physicist, this definition means that D is a *critical* dimension.

The D-dimensional Hausdorff measure of a D-dimensional set S may be either zero, or infinite, or positive and finite. Hausdorff had considered only this third and simplest category and showed it includes the Cantor sets and the Koch curves. If, in addition, the set S is self-similar, it is easy to see that its similarity dimension must equal D. On the other hand, we saw that typical random sets have zero measure in their intrinsic dimension.

For a long time, Besicovitch was the author or the co-author of nearly every paper on this subject. While Hausdorff is the father of nonstandard dimension, Besicovitch made himself its mother.

CODIMENSION. When Ω is the space $\mathbb{R}^E$, $D \leq E$, and $E-D$ is called *codimension*.

5. DIRECT PRODUCTS OF SETS (ADDITIVE DIMENSIONS)

Let S_1 and S_2 belong respectively to an E_1-space and an E_2-space, and denote by S the set in E-space, with $E=E_1+E_2$, which is obtained as the product of S_1 and S_2. (If $E_1=E_2=1$, S is the set of points (x,y) in the plane, such that $x \epsilon S_1$ and $y \epsilon S_2$.)

The rule of thumb is that if S_1 and S_2 are "independent" the dimension of S is the sum of the dimensions of S_1 and S_2.

The notion of "independence" embodied in this rule proves unexpectedly difficult to state and prove generally. See Marstrand 1954a, 1954b; Hawkes 1974; Mattila 1975. Luckily, intuition is usually a good guide in the case studies for example in those tackled in this Essay.

6. INTERSECTIONS OF SETS (ADDITIVE CODIMENSIONS)

The rule of thumb is the following. When S_1 and S_2 are independent sets in E-space and

$$\text{codimension}(S_1)+\text{codimension}(S_2)<E,$$

the term on the left is almost surely equal to

$$\text{codimension}(S_1 \cap S_2).$$

When the sum of codimensions is $>E$, one finds typically that the intersection is almost surely of dimension 0.

In particular, two sets of the same dimension miss one another if $D \leq E/2$. The dimension $E=2D$ can be called *critical*.

Most notably, given that Brown trails have the dimension $D=2$, two Brown trails hit one another when $E<4$, and miss when $E \geq 4$.

The rule extends in obvious fashion to the intersection of more than two sets.

SELF-INTERSECTIONS. The set of k-multiple points of S can be viewed as the intersection of k replicas of S. One is tempted to test the assumption that, from the viewpoint of the intersection's dimension, said k replicas can be viewed as independent. In one example at least, this guess turns out to be correct. S. J. Taylor 1966 (generalizing upon results by Dvoretzky, Erdös & Kakutani) studies the trails of Brownian and Lévy motion in $\mathbb{R}^1$ and $\mathbb{R}^2$. The trail's dimension is D, and the sets of its k-multiple points are of dimension $\max[0,E-k(E-D)]$. Taylor's guess is that the result holds in $\mathbb{R}^E$ for all k's up to $k=\infty$.

7. PROJECTIONS OF SETS

The rule of thumb is that, when a fractal S of dimension D is projected along a direction independent of S upon a Euclidean subspace of dimension E_0, the projection S^* satisfies

$$\text{dimension } S = \min (E_0, D).$$

APPLICATION. Let $x_1 \in S_1$ and $x_2 \in S_2$, where S_1 and S_2 are two fractals in $\mathbb{R}^E$, of dimensions D_1 and D_2. Let a_1 and a_2 be nonnegative real numbers and define the set S as made up of the points of the form $x = a_1 x_1 + a_2 x_2$. This set S has a D satisfying

$$\max(D_1, D_2) \leq D \leq \min(E, D_1 + D_2).$$

The proof consists in taking a direct product of $\mathbb{R}^E$ by $\mathbb{R}^E$, then projecting.

In case of independence, the upper bound tends to apply. When $D = E = 1$, S may be either a fractal or a set that includes intervals.

8. SUBORDINATION OF SETS (MULTIPLICATIVE DIMENSIONS)

See Chapter 32.

9. SUBDIMENSIONAL SEQUENCE

When the intrinsic test function of S is $h_S(\rho) = \gamma(D)\rho^D$, the fractal properties are fully described by its D. When

$$h_S(\rho) = \rho^D [\log (1/\rho)]^{\Delta_1} [\log\log (1/\rho)]^{\Delta_2},$$

the description of the fractal properties of S is more cumbersome. It requires the sequence D, Δ_1, Δ_2. The Δ_m may be called *subordinate dimensions*, or *subdimensions*.

The subdimensions may bear on the question of whether the borderline sets discussed in the subentry FRACTAL, 3 are or are not to be called fractal. It may become useful to include among the fractals all the S such that $D = D_T$ but at least one Δ is nonzero.

INDICATOR/COINDICATOR FUNCTIONS

Given a set S, the *indicator* function J(x) is classically defined as being such that J(x)=1 when $x \in S$, and J(x)=0 when $x \notin S$. When S is a Cantor set, a Sierpiński lattice (gasket or carpet), a fractal net, or any one of several other classes of fractals, J(x) is inconvenient. I often find it more convenient to replace J(x) by a different function C(x), which I introduced and now propose to call *coindicator*.

C(x) is a randomly weighted average of the indicator functions of the gaps of S. In other words, C(x) is constant in each gap, and its values in different gaps are independent random variables of identical distribution.

Under the older (and misleading) term, *core function*, C(x) is introduced and investigated in Mandelbrot 1965c, 1967b, and 1967i.

LÉVY STABLE
RANDOM VARIABLES AND FUNCTIONS

The hyperbolic distribution is of unbeatable formal simplicity, and is invariant under truncation (see the entry SCALING UNDER TRUNCATION). But the other transformations that leave it invariant are not important. Far more important are the distributions invariant under addition. They are only asymptotically hyperbolic, and Paul Lévy burdened them with a frightfully overworked term: "stable distributions." He also introduced stable processes, in which both the hyperbolic and the stable distributions play a role.

Until my work, the stable variables were deemed "pathological" or even "monstrous," with the sole exception of the Holtsmark random vector discussed in subentry 9. My main applications are discussed in Chapters 31, 32, and 37, and an application to genetics is mentioned below, in subentry 4.

REFERENCES. They are numerous, but none is satisfactory. In Feller 1966 (Volume II), the material on stability is complete but scattered, hence hard to find when needed. Lamperti 1966 is a good introduction. Gnedenko & Kolmogorov 1954 is still recommended. Lukacs 1970 collects many useful details. The original great treatises, Lévy 1925, 1937-1954, are not to everyone's taste, as they exhibit the distinctive characteristics of their author's style (see Chapter 40).

1. THE GAUSSIAN R.V. IS SCALING UNDER ADDITION

The Gaussian distribution is known to have the following property. Let G_1 and G_2 be two independent Gaussian random variables, with

$$\langle G_1 \rangle = \langle G_2 \rangle = 0; \ \langle G_1{}^2 \rangle = \sigma_1{}^2, \langle G_2{}^2 \rangle = \sigma_2{}^2.$$

Their sum $G_1 + G_2$ satisfies

$$\langle G_1 + G_2 \rangle = 0; \ \langle (G_1 + G_2)^2 \rangle = \sigma_1{}^2 + \sigma_2{}^2.$$

More important, $G_1 + G_2$ is itself Gaussian. Thus the Gaussian property is invariant under the addition of independent random variables. In other words, the functional equation

(L) $$s_1 X_1 + s_2 X_2 = sX,$$

combined with the auxiliary relation

(A:2) $$s_1{}^2 + s_2{}^2 = s^2,$$

has the Gaussian as a possible solution. In fact, except for scale, the Gaussian is the only distribution satisfying both (L) and (A:2).

Furthermore, if (L) is combined with the alternative auxiliary relation $\langle X^2 \rangle < \infty$, the Gaussian is again the unique solution.

(L) was the object of profound study in Lévy 1925, which calls it *stability*. Whenever ambiguity threatens, I use the cumbersome *Lévy stability*.

2. THE CAUCHY RANDOM VARIABLE

Since practically-minded scientists tend to take $\langle X^2 \rangle < \infty$ for granted, the Gaussian is widely believed to be the only stable distribution. Such is definitely not the case, as first recognized in Cauchy 1853, p. 206. Cauchy's example is a certain random variable first considered by Poisson and now called "reduced Cauchy variable." It satisfies

$$Pr(X > -x) = Pr(X < x) = \frac{1}{2} + \pi^{-1} \tan^{-1} x,$$

hence

$$\text{Cauchy density} = 1/[\pi(1 + x^2)].$$

Cauchy showed this variable to be a solution of the combination of (L) with the alternative auxiliary relation

(A:1) $s_1 + s_2 = s.$

For the Cauchy variable, $\langle X^2 \rangle = \infty$, in fact $\langle X \rangle = \infty$. Hence, in order to express the obvious notion that the scale of the product of X by a nonrandom s equals s times the scale of X, one must measure scale by some quantity other than the root mean square. One candidate is the distance between the quartiles Q and Q', where $Pr(X < Q') = Pr(X > Q) = \frac{1}{4}$.

The Cauchy variable most often serves as a counterexample, as in Bienaymé 1853, pp. 321-323. See also Heyde & Seneta 1977.

GEOMETRIC GENERATING MODEL. The above formula $Pr(X < x) = \frac{1}{2} + \pi^{-1} \tan^{-1} x$ is implemented geometrically by positioning the point W with a uniform probability distribution over the circle $u^2 + v^2 = 1$ and defining X as the abscissa of the point where the line from O to W intersects the line $v = 1$. By the same token, the variable Y, defined as the ordinate of the point where the line from O to W intersects the line $u = 1$, has the same distribution as X. Since $Y = 1/X$, we find that the inverse of Cauchy is Cauchy.

Furthermore, whenever OW = (X,Y) is an isotropically distributed random vector in the plane, Y/X is a Cauchy variable. In particular, the ratio of two independent Gauss variables is a Cauchy variable.

3. BROWNIAN MOTION'S RECURRENCES

Now combine the equation (L) with

(A:0.5) $s_1^{0.5} + s_2^{0.5} = s^{0.5}.$

The solution is the random variable whose density is 0 for $x < 0$, and otherwise equals

$$p(x) = (2\pi)^{-\frac{1}{2}} \exp(-1/2x)x^{-3/2}.$$

The quantity $p(x)dx$ is the probability of finding that a Brown function satisfying $B(0) = 0$ also satisfies $B(t) = 0$ for some t in $[x, x+dx]$.

4. GENERAL LEVY STABLE VARIABLES

Cauchy also considered the generalized auxiliary relation

(A:D) $\qquad s_1{}^D + s_2{}^D = s^D.$

SYMMETRIC SOLUTIONS. Cauchy asserted on the basis of formal calculations that for every D the combination of (L) with (A:D) has one solution, the random variable of density

$$\pi^{-2} \int_0^\infty \exp(-u^D)\cos(ux)du.$$

Pólya and Lévy showed that in the case $0 < D \leq 2$, Cauchy's assertion is indeed justified, the Gauss and Cauchy distributions being two special cases. But in the case $D > 2$, Cauchy's assertion is invalid because the above-written formal density takes on negative values, which is an absurdity.

EXTREME NONSYMMETRIC SOLUTIONS. Lévy showed moreover that the combination of (L) and (A:D) allows for *non*symmetric solutions. For the most extremely asymmetric ones, the generating function (Laplace transform) is defined and equal to $\exp(g^D)$.

OTHER NONSYMMETRIC SOLUTIONS. The general solution of the combination of (L) and (A:D) is a weighted difference of two independent and identically distributed solutions with extreme asymmetry. The custom is to denote the weights by $\frac{1}{2}(1+\beta)$ and $\frac{1}{2}(1-\beta)$.

FINAL GENERALIZATION OF (L). Leaving (A:D) unchanged, replace the condition (L) by

(L*) $\quad s_1 X_1 + s_2 X_2 = sX + \text{constant.}$

When $D \neq 1$, this change makes no difference,

but when $D = 1$ it allows for additional solutions, called asymmetric Cauchy variables.

BACTERIAL MUTANTS. Mandelbrot 1974d shows that the total number of mutants in an old culture of bacteria (the Luria-Delbrück problem) is a Lévy stable variable with extreme asymmetry.

5. LEVY STABLE DENSITIES' SHAPE

Aside from three exceptions: $D = 2$ with $\beta = 0$, $D = 1$ with $\beta = 0$, and $D = \frac{1}{2}$ with $\beta = 1$, Lévy stable distributions are *not* known in closed analytic form, but the properties of the three simple exceptions generalize to other cases.

In all extreme asymmetric cases with $0 < D < 1$, the density vanishes for $x < 0$.

The fact that the Gaussian density is $\exp(-\frac{1}{2}x^2)$ generalizes to the short tail of all the extreme asymmetric cases with $1 < D < 2$. The density is $\propto \exp(-c|x|^{D/(D-1)})$.

For $x \to \infty$, the Cauchy density is $\propto (\pi)^{-1} x^{-D-1}$, and the Brown recurrence density is $\propto (2\pi)^{-\frac{1}{2}} x^{-D-1}$. More generally, for all $D \neq 2$, the density in the long tail(s) is $\propto x^{-D-1}$.

Otherwise, the behavior of p(u) must be obtained numerically. Graphs for $1 < D < 2$ are given in Mandelbrot 1960e for the extreme asymmetric case, with added comment in Mandelbrot 1962p concerning the values of D very close to 2, and in Mandelbrot 1963b for the symmetric case. Fast Fourier transform techniques make this task much lighter, see Dumouchel 1973, 1975.

6. INEQUALITY BETWEEN ADDENDS, AND THE RESULTING CLUSTERING

Let X_1 and X_2 be independent random variables with the same probability density $p(u)$. The probability density of $X = X_1 + X_2$ is

$$p_2(u) = \int_{-\infty}^{\infty} p(y)p(u-y)dy,$$

If the sum u is known, the conditional density of either addend y is $p(y)p(u-y)/p_2(u)$. Let us examine this density's shape in detail.

EXAMPLES. When $p(u)$ is Gaussian of unit variance, hence a unimodal function (= it has a single maximum), the conditional distribution is Gaussian centered on $\frac{1}{2}u$ and has the variance $\frac{1}{2}$, which is independent of u (see Brown fractal sets, 3). As $u \to \infty$, the addends become increasingly close to being equal in relative values.

When $p(u)$ is a reduced Cauchy density, which is again unimodal, two very different cases must be distinguished. When $|u| \leq 2$, which happens half of the time, the conditional distribution is again unimodal, and the most likely value is also $\frac{1}{2}u$. On the contrary, when $|u| > 2$, the value $\frac{1}{2}u$ becomes the *least* likely (locally). For $(u) = 2$, the conditional distribution bifurcates into *two* separate "ogives" centered respectively near $y = 0$ and $y = u$. As $u \to \pm\infty$, these ogives become increasingly hard to distinguish from Cauchy ogives centered on 0 and u.

When $p(u)$ is a Brown recurrence density, the situation is like in the Cauchy case but even more extreme, the conditional density being bimodal with a probability $> \frac{1}{2}$.

Corollary: consider three successive zero crossings of a random walk T_{k-1}, T_k and T_{k+1}. If $T_{k+1} - T_{k-1}$ is large, the middle crossing is *most* likely to cluster *extremely* close to either T_{k-1} or T_{k+1}, and *least likely* to fall halfway in between. ◁ This result is related to a celebrated counterintuitive result of probability, Lévy's arc sine law. ▶

Next, consider the conditional distribution of U, given that the sum of M variables U_g takes a very large value u. In the Gaussian case, the most likely outcome is that each addend U_g is nearly u/M. In the Cauchy case, and in the Brown recurrence case, on the contrary, the most likely outcome is that all addends, except one, are smallish.

THE HIDDEN PITFALL IN THE IDEA OF "IDENTICAL" CONTRIBUTIONS TO A SUM. The addends being *a priori identical,* in the sense of having the same distribution, allows their a posteriori values to be either near equal (as in the Gaussian case), or unequal to varying degrees (as in the stable Lévy case when the sum is very large).

7. NONSTANDARD CENTRAL LIMITS. ROLE OF HYPERBOLIC VARIABLES

Given an infinite sequence X_n of independent and identically distributed random variables, the central limit problem inquires whether or not it is possible to select the weights a_n and b_n so that the sum $a_N \Sigma_1^N X_n - b_N$ has a nontrivial limit for $N \to \infty$.

In the standard case $\langle X_n^2 \rangle < \infty$, the answer is standard and affirmative: $a_N = 1/\sqrt{N}$ and $b_N \sim \langle X_n \rangle \sqrt{N}$, and the limit is Gaussian.

The nonstandard case $\langle X_n^2 \rangle = \infty$ is by far more complex: (a) The selection of a_N and b_N is not always possible. (b) When it is possible, the limit is stable non-Gaussian. (c) In order that the exponent of the limit be D, a *sufficient* condition on the X_n is that the distribution be asymptotically hyperbolic of exponent D (Chapter 38). (d) The necessary and sufficient condition is found in the references at the beginning of this entry.

8. LEVY STABLE LINE-TO-LINE FUNCTIONS

These are random functions having stationary independent increments and such that the incremental random variable $X(t)-X(0)$ is Lévy stable. The scaling factor $a(t)$ that makes $[X(t)-X(0)]a(t)$ independent of t must take the form $a(t)=t^{-1/D}$. This process generalizes the ordinary Brownian motion to $D \neq 2$.

The most striking property of $X(t)$ is that it is discontinuous and includes jumps.

THE CASE $D < 1$. Here, $X(t)$ includes nothing but jumps; the number of those occurring between times t and $t+\Delta t$ and having an absolute value exceeding u is a Poisson random variable of expectation equal to $|\Delta t|u^{-D}$.

The relative numbers of positive and negative jumps are $\frac{1}{2}(1+\beta)$ and $\frac{1}{2}(1-\beta)$. The extreme asymmetric case $\beta=1$ involves positive jumps only; it is called *stable subordinator* and serves to define the Lévy staircase in

Plates 286 and 287.

PARADOX. Since $u^{-D} \to \infty$ as $u \to 0$, the total expected number of jumps is infinite, however small the length of Δt. The fact that the associated probability is infinite seems paradoxical. But this feeling ceases when one notes that the jumps for which $u < 1$ add to a finite cumulative total. This conclusion becomes natural after it is noted that a small jump's expected length is finite. It is

$$\propto \int_0^1 Du^{-D-1}u \ du = D\int_0^1 u^{-D}du < \infty.$$

THE CASE $1 < D < 2$. Now, the last-written integral diverges, hence the total contribution of the small jumps is infinite. As a result, $X(t)$ includes a continuous term and a jump term; both are infinite but they have a finite sum.

9. STABLE LEVY VECTORS AND FUNCTIONS

Let the functional equation (L) in the definition of stability be changed by making X into a random vector X. Given a unit vector V, it is clear that the combined equations (L) and (A:D) have an elementary solution that is the product of V by a scalar stable variable.

Lévy 1937-1954 shows that the general solution is merely the sum of elementary solutions that correspond to all directions in space and are weighted by a distribution over the unit sphere. These contributions may be either discrete (finite or denumerably infinite), or infinitesimal. In order that the vector X be isotropic, the elementary contributions must be distributed uniformly over all directions.

STABLE LÉVY VECTOR FUNCTIONS OF TIME. These functions admit the same sort of decomposition as a stable scalar function, into a sum of jumps following the hyperbolic distribution. The jumps' sizes and directions are ruled by a distribution over the unit sphere.

HOLTSMARK DISTRIBUTION. Holtsmark's work in spectroscopy survived by being restated in terms of Newtonian attraction (Chandrasekhar 1943); until my work, it involved the only concrete occurrence of a Lévy stable distribution. Suppose there is a star at O and other stars of unit mass are distributed throughout space, independently of each other and with the expected density δ. What is the total attraction these stars exert upon O? Soon after Newton's discovery of the r^{-2} law of attraction, the Reverend Bentley wrote to him to point out (in effect) that the attraction of the stars within a thin pencil $d\Omega'$ with its apex at O has an infinite expectation, and so does the attraction of the stars within the pencil $d\Omega''$ symmetric of $d\Omega'$ with respect to O. Bentley concluded that the difference between these infinities is undetermined.

The Holtsmark problem, as it is usually restated, avoids this difficulty by concerning itself with the excess of the actual attractions over their expectations. We begin with the stars within a domain bounded by the above pencil of angular angle $d\Omega$ and the spheres of radii r and r+dr. Each exerts the attraction $u=r^{-2}$ and their number is a Poisson variable of expectation $\delta|d\Omega|d(r^3) = \delta|d\Omega|\ |d(u^{-3/2})|$. Hence the attraction in excess of the expecta-

tion has the characteristic function

$$\exp\{\,\delta|d\Omega|\ \int_0^\infty [\exp(i\zeta u)-1-i\zeta u]|d(u^{-3/2})|\,\}.$$

This turns out to correspond to a Lévy stable variable of exponent $D=3/2$ and $\beta=1$. By the above subentry 6, a large positive u is very likely to be due to the presence of a *single* star near O, irrespective of the density of stars elsewhere; and the distribution of U behaves for very large u as the distribution of the attraction of the nearest star.

The overall excess attraction is therefore an isotropic Lévy stable vector with $D=3/2$.

The meaning of stability is that if there are two uniformly distributed clouds of red and blue stars, the forces exerted on O by red stars alone, or by blue stars alone, or by both together, differ only by a scale factor and not in the analytic form of their distributions.

10. SPACE-TO-LINE STABLE RANDOM FUNCTIONS

The construction of the space-to-line Brown function given by Chentsov 1957 was generalized to the stable case in Mandelbrot 1975b.

11. DIMENSIONS

In the non-Gaussian case, the earliest calculations of the dimension of a stable process are found in McKean 1955 and Blumenthal & Getoor 1960c, 1962. A reference with full bibliography is Pruitt & Taylor 1969.

12. SCALING UNDER WEIGHTED ADDITION (MANDELBROT 1974c,f)

As discussed in this chapter in the subentry NONLACUNAR FRACTALS 4, Mandelbrot 1974c,f advances a family of generalizations of the Lévy stable variables. They involve a generalization of Lévy's stability condition (L), wherein the weights $s_i\mu$ become random.

LIPSCHITZ-HÖLDER HEURISTICS

Fractal dimension is originally a local property, notwithstanding the fact that in this Essay the local properties are reflected in the global ones. Therefore, in the case of the graph of an otherwise arbitrary continuous function $X(t)$, D must be related to other local properties. One of the most useful is the Lipschitz Hölder (LH) exponent α. The LH condition at $t+$ is a way of expressing that

$$X(t)-X(t_0) \sim |t-t_0|^\alpha, \text{ for } 0 < t-t_0 < \epsilon,$$

and similarly for $t-$. The global LH exponent in $[t',t'']$ is $\lambda[t',t''] = \inf_{t' \leq t \leq t''}\alpha$. Unless $X(t)$ is a constant, $\lambda \leq 1$.

LH HEURISTICS AND D. Given α, the number of squares of side r necessary to cover the graph of X between times t and $t+r$ is roughly equal to $r^{\alpha-1}$. In this fashion, one can cover the graph of $X(t)$ for $t \in [0,1]$ by N squares,

and a rough dimensional argument yields $D = \log N / \log(1/r)$. This way of guessing D will be denoted here as *Lipschitz-Hölder heuristics*. It is robust and effective.

EXAMPLES. When X is differentiable for every t between 0 and 1, and we neglect the points where $X'(t)=0$, one has $\alpha=1$ throughout, and the number of squares needed to cover the graph is $N \sim r^{\alpha-1}(1/r) = r^{-1}$. It follows that $D=1$, as is of course the case.

When $X(t)$ is a Brown function, ordinary or fractional, one can show that $\alpha \equiv \lambda = H$. The heuristic N is $N \sim r^{H-1-1}$, hence $D=2-H$, which again agrees with the known D.

◁ For the functions in the entry WEIERSTRASS..., Hardy 1916 shows that $\alpha \equiv H$. Hence the conjecture that the Hausdorff Besicovitch dimension is $2-H$. ▶

The case of the Cantor staircase (Plate 101) is quite different. Here X varies only for t's that belong to a fractal dust with fractal dimension $\delta < 1$, and α depends on t. Divide [0,1] into $1/r$ time spans of length r. In $r^{-\delta}$ of these spans, $\alpha=\delta$, and in the other spans, α is undefined, but if the coordinate axes are rotated a bit, one finds $\alpha=1$. Hence the heuristic value of the number of covering squares is $r^{-1}+r^{\delta-1}r^{-\delta} = 2r^{-1}$, and the heuristic dimension is $D=1$. Such is indeed the case, as noted in the caption of Plate 101.

Furthermore, the sum of a Brown function and a Cantor staircase with $\delta < H$ yields $D = 2-H$ and $\lambda=\delta$, hence $1 < D < 2-\lambda$.

SUMMARY. The heuristic inequality $1 \leq D \leq 2-\lambda$. This guess is confirmed in Love & Young 1937 and Besicovitch & Ursell 1937.

See also Kahane & Salem 1963, p. 27.

ON THE DEFINITION OF "FRACTAL". The entry FRACTALS... mentions that it would be desirable to extend the scope of the term *fractal* to include the Cantor staircase. Should we say a curve is fractal when $\lambda < 1$ and α is near λ for "sufficiently many" t's? I prefer not to follow this path, because such extensions are cumbersome and distinguish between $D_T = 0$ and $D_T > 0$.

LINE-TO-PLANE FUNCTIONS. Let $X(t)$ and $Y(t)$ be continuous functions with the LH exponents λ_1 and λ_2. The heuristic suggests that covering the graph for $t \in [0,1]$ of the vector function of coordinates $X(t)$ and $Y(t)$ requires at most $r^{\lambda_1 + \lambda_2 - 3}$ cubes of side r, hence $1 \leq D \leq 3 - (\lambda_1 + \lambda_2)$. For the ordinary Brown line-to-plane trail, this yields the correct $D = 2$.

PROJECTIONS. Now, form a continuous trail, by projecting $\{X(t), Y(t)\}$ on the (x,y) plane. When $\lambda_1 = \lambda_2 = \lambda$, the heuristic suggests that one needs up to $1/r$ squares of size r^{λ}, hence $1 \leq D \leq \min(2, 1/\lambda)$. Similarly, consider the continuous trail of a function $\{X(t), Y(t), Z(t)\}$ whose coordinates have identical LH exponents λ. The heuristic suggests $1 \leq D \leq \min(3, 1/\lambda)$. When $\lambda_1 \neq \lambda_2$, the continuous trail of $\{X(t), Y(t)\}$ must be covered by squares of side $r^{\max \lambda}$, hence

$$1 \leq D \leq 2 - \max\{0, (\lambda_1 + \lambda_2 - 1)/\max(\lambda_1, \lambda_2)\}.$$

All this is confirmed by Love & Young 1937.

MEDIAN AND SKIP POLYGONS

Material on this topic (related to Peano curves) is found in Chapter XII of the 1977 *Fractals*.

MUSIC: TWO PROPERTIES OF SCALING

Music has at least two scaling properties worth mentioning.

TEMPERED MUSICAL SCALES AND THEIR RELATIONSHIP WITH THE FREQUENCY SPECTRUM OF THE MODIFIED WEIERSTRASS FUNCTION. The most widespread use of the Latin root *scala* = ladder is of course *not* found in the term *scaling* encountered throughout this Essay, but in the notion of *musical* scale, which implies a discrete spectrum that is preserved by multiplication of the frequencies. In a tempered scale, the frequencies are spread *logarithmically*. For example, the twelve-tone scale corresponds to the base $b = 2^{1/12}$. As a result, the fundamental notes of each musical instrument make up a high proportion of the low frequencies within its overall frequency band, but a low proportion of the high frequencies.

Extrapolated to inaudibly high and low frequencies, such a frequency spectrum becomes identical to that of the Weierstrass (modified) function (p. 389b) with the same value of b. Consequently, in order to add low frequencies to a piece of music, it suffices to add new instruments capable of the desired low tones.

Since the Euler-Fourier theorem represents the most general periodic function as a series of *linearly* spaced harmonics, the functions that represent the sequence of the fundamental notes in the most general piece of music are very restricted functions.

MUSIC AS A SCALING (1/f) NOISE (R. F. VOSS). A second scaling facet of music concerns the variation in time of diverse measures of the audio signal: for example its power (measured by square of its intensity), or its instantaneous frequency (measured by the rate of zero crossings of the audio signal). Voss & Clarke 1975 and Voss 1978 (see also Gardner 1978) observe that in the works of such diverse composers as Bach, Beethoven, and the Beatles, both of the above measures of the audio signal are scaling noises, 1/f noises, as described on p. 254.

Conversely, if random music is triggered by an outside physical noise source, with $1/f^B$ spectrum and varying scaling exponents, Voss & Clarke 1975, and Voss 1978 find that the resulting sound is closest to being "music-like" when the trigger is an 1/f noise.

This was a totally unexpected finding, but, like many findings in the body of this Essay, it becomes "natural" *after the fact*. The argument I favor is that musical compositions are, as indicated by their name, composed: First, they subdivide into movements characterized by different overall tempos and/or levels of loudness. The movements subdivide further in the same fashion. And teachers insist that every piece of music be "composed" down to the shortest meaningful subdivisions. The re-

sult is bound to be scaling!

However, this scaling range does not extend below time spans of the order of one note. Higher frequencies are ruled by entirely different mechanisms (including the resonance of lungs, of fiddle bodies, and of woodwind pipes), therefore the high energy spectrum is more like f^{-2} than f^{-1}.

NONLACUNAR FRACTALS

Given the definitions of lacunarity in Chapter 34, a nonlacunar set in the space $\mathbb{R}^E$ should intersect every cube or sphere in said space. In mathematical terms, it should be everywhere dense, hence nonclosed. (The only everywhere dense closed set is $\mathbb{R}^E$ itself!) This entry shows that such fractals do exist, but "feel" very different from the closed fractals in the rest of this Essay. A key symptom is that the Hausdorff Besicovitch dimension remains workable, but the similarity and Minkowski Bouligand dimensions are equal to E, rather than to the Hausdorff Besicovitch D.

1. RELATIVE INTERMITTENCY

The phenomena to which nonlacunar fractals are addressed are scattered throughout this Essay, in the sense that many of my case studies of natural fractals negate some unquestionable knowledge about Nature.

We forget in Chapter 8 that the noise that causes fractal errors weakens between errors but does not desist.

We neglect in Chapter 9 our knowledge of the existence of interstellar matter. Its distribution is doubtless *at least* as irregular as that of the stars. In fact, the notion that it is impossible to define a density is stronger and more widely accepted for interstellar than stellar matter. To quote deVaucouleurs 1970, "it seems difficult to believe that, whereas visible matter is conspicuously clumpy and clustered on all scales, the invisible intergalactic gas is uniform and homogeneous...[its] distribution must be closely related to...the distribution of galaxies." Other astronomers write of intergalactic *wisps* and *cobwebs*.

And in Chapter 10 the pastrylike sheets of turbulent dissipation are an obviously oversimplified view of reality.

The end of Chapter 9 mentions very briefly the fractal view of the distribution of minerals. Here, the use of closed fractals implies that, between the regions where copper can be mined, the concentration of copper vanishes. In fact, it is very small in most places, but cannot be assumed to vanish anywhere.

In each case, some areas of less immediate interest were artificially emptied to make it possible to use *closed* fractal sets, but eventually these areas must be filled. This can be done using a fresh hybrid, *nonlacunar fractals*. To take an example, a nonlacunar mass distribution in the cosmos will be such that no portion of space is empty, but, for every set of small thresholds θ and λ, a proportion of mass at least $1-\lambda$ concentrates on a portion of space of relative volume at most θ.

2. QUOTE FROM DE WIJS, AND COMMENT

The basic intuitive circumstances that call for nonlacunar fractals are described in de Wijs 1951, which makes a *"Working Hypothesis"* that is worth summarizing.

"Consider a [body of ore] with a tonnage W and an average grade M. With an imaginary cut we slash this body into two halves of equal tonnage ½W, differing in average grade. Accepting for the grade of the richer half $(1+d)M$, the grade of the poorer half has to be $(1-d)M$ to satisfy the condition that the two halves together average again M.... A second imaginary cut divides the body into four parts of equal tonnage ¼W, averaging $(1+d)^2M$, $(1+d)(1-d)M$, $(1+d)(1-d)M$, and $(1-d)^2M$. A third cut produces $2^3=8$ blocks, namely 1 block with an average grade of $(1+d)^3M$, 3 blocks of $(1+d)^2(1-d)M$, 3 blocks of $(1+d)(1-d)^2M$, and one block of $(1-d)^3M$. One can visualize the continued division into progressively smaller blocks....

"The coefficient d as a measure of variability adequately replaces the collective intangibles [dear to those who feel that ore estimation is an art rather than a science], and statistical deductions based upon this measure can abolish the maze of empirical rules and intuitive techniques."

COMMENT. De Wijs did not even begin to explore the geometric aspects of this model, and neither he nor his otherwise notable followers (including G. Matheron) had an inkling of fractals. However, if one assumes that the ore density is independent of grade, making tonnage equivalent to volume, precisely the same scheme had been investigated for totally different purposes by the pure mathematician A.S. Besicovitch and his disciples.

Anticipating the next subentry, if the (reinterpreted) scheme of de Wijs is continued ad infinitum, the ore curdles into a nonlacunar fractal. To write its dimension in the customary form, $D = \log N^*/\log 2$, it is necessary to define $\log N^*$ by

$$\log N^* = -\Sigma \pi_i \log \pi_i,$$

where $\pi_1 = (1+d)^3$, $\pi_8 = (1-d)^3$, $\pi_2 = \pi_3 = \pi_4 = (1+d)^2(1-d)$, and $\pi_5 = \pi_6 = \pi_7 = (1+d)(1-d)^2$.

CONCLUSION. De Wijs's hunch was well-inspired, but the coefficient d is an unsuitable measure, because it only applies to one model. The proper measure of ore variability is D.

3. BESICOVITCH WEIGHTED CURDLING

To appreciate the results of Besicovitch, it is best to restate them on [0,1] with b=3.

ASSUMPTIONS. We start with mass distributed over [0,1] with density equal to 1, and share it among the thirds through nonrandom multiplication by three weights W_0, W_1, W_2, satisfying the following conditions.

(A) $\frac{1}{3}W_0 + \frac{1}{3}W_1 + \frac{1}{3}W_2 = 1$. This expresses the conservation of mass, and implies that each W_i is bounded by b. The quantity $\frac{1}{3}W_i$, which is the mass in the ith third, will be denoted by π_i.

(B) The uniform distribution $W_i \equiv \frac{1}{3}$ is excluded.

(C) $W_0 W_1 W_2 > 0$. In particular, the Cantor construction—corresponding to $W_0 = \frac{1}{2}$, $W_1 = 0$ and $W_2 = \frac{1}{2}$—is excluded.

Further stages of the cascade proceed similarly; for example, the densities over the subeddies are W_0^2, $W_0 W_1$, $W_0 W_2$, $W_1 W_0$, W_1^2, $W_1 W_2$, $W_2 W_0$, $W_2 W_1$, W_2^2.

CONCLUSIONS. Iterating ad infinitum, we reach the following results, mostly due to Besicovitch and Eggleston. (Billingsley 1965 is a valuable exposition.)

(A) *Singularity. The Besicovitch fractal.* The density at almost every point is asymptotically zero. The set of points where the asymptotic density is *not* zero (it is infinite there) is to be called *Besicovitch fractal*, $\mathcal{B}$. It is the set of points of [0,1] whose ternary development is such that the ratio

k^{-1} (number of i's in the first k "digits")

converges to π_i. Such points form an open set: the limit of a sequence of such points need not be in the set.

(B) *Nonlacunarity.* The limit distribution of mass is everywhere dense: even asymptotically, no open interval (however small) is entirely empty. The mass between 0 and t strictly increases with t. While the points where ΠW fails to converge to 0 are very few in relative numbers, their absolute number insures that the mass within any interval [t',t"] has a nonzero limit for $k \to \infty$.

(C) *The Hausdorff Besicovitch dimension*

of $\mathcal{B}$. It is

$$D = -(\pi_1 \log \pi_1 + \pi_2 \log \pi_2 + \pi_3 \log \pi_3).$$

Formally, D is an "entropy" as defined in thermodynamics, or else an "information" as defined by Shannon (see Billingsley 1965).

(D) *The similarity dimension of $\mathcal{B}$.* It is 1. Indeed, $\mathcal{B}$ is self-similar with $N=3$ and $r=\frac{1}{3}$, hence $D_S = \log 3 / \log 3 = 1$; the reason for adding the index S will transpire momentarily. Similarly, three-dimensional variants have the dimension 3. In this instance D_S cannot have much physical significance: firstly, it does not depend on the W_i's, as long as they fulfill the conditions we impose on them; secondly it jumps from 1 to $\log 2 / \log 3$ if $\mathcal{B}$ is replaced by its Cantor dust limit.

Furthermore, a fractally homogeneous distribution can no longer be founded upon self-similarity. Indeed, if we attribute equal weights to all pieces of length 3^{-k}, the resulting distribution is uniform on [0,1]. It is unrelated to the values of the W_i's, and it differs from the measure by which the set itself has been generated. Also, passing to the Cantor dust limit, this uniform distribution changes discontinuously into a very nonuniform one.

(E) *The similarity dimension of the "set of concentration" of $\mathcal{B}$.* It is D. The point is that the Besicovitch measure is closely approximated by a fractally homogeneous measure whose similarity dimension is equal to the Hausdorff Besicovitch D. To be precise, after a large number k of cascade stages, the overwhelming bulk of an initially uniform mass becomes concentrated upon 3^{kD} triadic intervals of length 3^{-k}. These intervals' distribution is not uniform over [0,1], but its largest gap tends to 0 as $k \to \infty$.

COMMENT. One must distinguish between the "full set" necessary to include the whole mass and the "partial set" in which the bulk of the mass concentrates. Both are self-similar, but their self-similarity dimensions D_S and D are different. See subentry 5 below.

4. RANDOM WEIGHTED CURDLING (MANDELBROT 1974f c)

A natural and rich generalization of the Besicovitch scheme is introduced in Mandelbrot 1974f c and developed in Kahane & Peyrière 1976.

The effect of each cascade stage is to multiply the densities in the b^3 eddies of each eddy by identically distributed statistically independent random weights W_i.

After k stages of a weighted curdling cascade, the overwhelming bulk of the mass becomes concentrated in a number of eddies of the order of magnitude of b^{kD*} out of a total of b^{3k}, where

$$D^* = -\langle W \log_b(r^3 W)\rangle = 3 - \langle W \log_b W \rangle.$$

In particular, if W is discrete and its possible values w_i have the respective probabilities p_i, one has

$$D^* = 3 - \Sigma p_i w_i \log_b w_i.$$

IN THE CASE $D^* > 0$, $D = D^*$. The measure generated by weighted curdling is approximated by a fractally homogeneous measure of dimension $D = D^*$, obtained as in Chapter 23.

IN THE CASE $D^* < 0$, $D = 0$. The number of nonempty cells tends asymptotically to 0, therefore the limit is almost surely empty.

In summary, the carrier of mass is approximated by a closed set with $D = \max(0, D^*)$.

SECTIONS. Similarly, the mass within the planar or linear sections concentrates in relatively small numbers of eddies, respectively $b^{(D^* - 1)}$ out of a total of b^2, and $b^{(D^* - 2)}$ out of a total of b. Therefore, the sections are nondegenerate if, respectively, $D^* > 1$ or $D^* > 2$, and they are approximated by fractals having the respective dimensions $D^* - 1$ or $D^* - 2$. Thus, the dimensions of the sections follow the same rule as for lacunar fractals.

NEW RANDOM VARIABLES, INVARIANT UNDER WEIGHTED ADDITION. Denote by X the random variable that rules the asymptotic mass within an eddy of any order k, or its section by a line or plane of dimension Δ. I showed that the X satisfy the functional equations

$$(1/C)\Sigma_{g=0}^{C-1} X_g W_g = X,$$

where $C = b^\Delta$, the r.v. W_g and X_g are independent, and the equality expresses identity of distribution. This equation generalizes the equation (L) discussed in the subentry LEVY STABLE.... The solutions generalize the Lévy stable variables; they are discussed in the op. cit. of Mandelbrot and Kahane & Peyrière.

5. LIMIT LOGNORMAL RANDOM CURDLING AND FUNCTION (MANDELBROT 1972j)

Mandelbrot 1972j gives up the eddy grid which both absolute and weighted curdling borrow from Cantor. The eddies are not prescribed in advance, but are generated by the same statistical mechanism as the distribution of mass within them. And in addition the discrete eddy strata merge into a continuum.

LIMIT LOGNORMAL FUNCTION, MOTIVATED. We proceed by successive modifications of weighted curdling, performed (for simplicity) on a function L(t) of one variable.

After the nth stage, the density of weighted curdling is a function $Y_n(t)$ such that $\Delta \log Y_n(t) = \log Y_{n+1}(t) - \log Y_n(t)$ is a step function; it varies when t is an integral multiple of $b^{-n} = r^n$, and its values between such instants are independent random variables of the form $\log W$. Now let $\Delta \log W$ be lognormal with the mean $-\frac{1}{2}(\log b)$ and the variance $\mu \log b$. One finds that the covariance between $\Delta \log Y_n(t)$ and $\Delta \log Y_n(t+\tau)$ takes the value $\mu(\log b)(1 - |\tau|/r^n)$ in the interval $|\tau| < r^n$, and vanishes outside of this interval. This $\Delta \log Y_n(t)$ is not Gaussian, because the joint distribution of its values for two (or more) t's fails to be a multidimensional Gaussian random variable.

First modification. Replace each $\Delta \log Y_n(t)$ by $\Delta \log Y_n^*(t)$, defined as the Gaussian random function with the barely different covariance $\mu(\log b)\exp(-|\tau|/r^n)$. The result retains the same "range of dependence" as the original, but it breaks up the discrete boundaries between eddies of duration r^n.

Second modification. Replace the discrete parameter $n \log b$ by a continuous parameter λ. The sum of finite differences $\Delta \log Y_n^*(t)$ changes to an integral of infinitesimal differentials $d \log L_\lambda(t)$, of mean $-\frac{1}{2} \mu d\lambda$ and variance $\mu d\lambda$, and the eddies become continuous.

DEFINITION OF L(t). Consider the limit

$$L(t) = L_\infty(t) = \lim_{\lambda \to \infty} L_\lambda(t).$$

The random variable $\log L_\lambda(t)$ is Gaussian with the mean $\langle \log L_\lambda(t) \rangle = -\frac{1}{2} \lambda \mu$ and the variance $\sigma^2 \log L_\lambda(t) = \lambda \mu$. This insures that $\langle L_\lambda(t) \rangle = 1$ for all λ. But the limit of $L_\lambda(t)$ may be either nondegenerate or almost surely vanishing. This question has not been settled mathematically, but the following heuristic arguments can doubtless be made rigorous. They are stated for the more interesting functions $L(x)$ of a three dimensional variable.

THE CONCENTRATION SET OF A LIMIT LOGNORMAL MEASURE. In order to obtain an idea of the set where $L_\lambda(x)$ it is not small but extremely large, it is convenient to use reference squares of side r^n. They are not imposed subeddies, merely a measuring device. When $n \gg 1$ and x is fixed, the lognormal $L_{n \log b}(x)$ has an extremely high probability of being extremely close to 0, hence, is extremely small over most of its domain.

Since $L_{n \log b}(x)$ is continuous, it varies little over a cell of side r^n, hence the derivation of the set of concentration for weighted curdling with a lognormal W also applies to the

present model. Neglecting logarithmic terms, the number of cells that contribute the bulk of the integral of $L_{n \log b}(x)$ has the expectation $Q = (r^{-n})^{D^*}$, with $D^* = 3 - \mu/2$.

When $\mu > 6$, so that $D^* < 0$, $Q \to 0$ as $\lambda \to \infty$, $L(x)$ is almost surely degenerate.

When $4 < \mu < 6$, so that $0 < D^* < 1$, $L(x)$ is nondegenerate with $D = D^*$, but its traces on planes and straight lines are almost surely degenerate.

When $2 < \mu < 4$, so that $1 < D^* < 2$, $L(x)$ and its traces on planes are nondegenerate with dimensions D^* and $D^* - 1$, but its traces on straight lines are almost surely degenerate.

When $0 < \mu < 2$, so that $2 < D^* < 3$, $L(x)$ and its traces on both planes and straight lines are nondegenerate with the dimensions D^*, $D^* - 1$ and $D^* - 2$.

6. DIMENSION OF A MEASURE'S CONCENTRATE

The study of relative intermittency suggests yet other definitions of dimension. Instead of a set in a metric space, consider a measure $\mu(S)$ that is defined over a bounded subspace Ω (in a suitable σ-field including the balls), and has the following properties. (A) When S is a ball, $\mu(S) > 0$ and also $\mu(\Omega) = 1$, hence "the set in which $\mu > 0$" is identical to Ω. (B) However, intuition suggests that μ "concentrates" over a very small portion of Ω. We seek fresh ways of quantifying B).

Given $\rho > 0$ and $0 < \lambda < 1$, consider the sets Σ_λ for which $\mu(\Omega - \Sigma_\lambda) < \lambda$. Let $N(\rho, \Sigma_\lambda)$ denote the inf of the number of balls of radius ρ needed to cover Σ_λ. Define

$$N(\rho,\lambda)=\inf N(\rho,\Sigma_\lambda).$$

The dimensionlike expressions

$$\lim\inf_{\alpha\downarrow 0}\log N(\alpha,\alpha)/\log(1/\alpha)$$
$$\lim\inf_{\rho\downarrow 0}\log N(\rho,\lambda)/\log(1/\rho)$$
$$\lim\inf_{\lambda\downarrow 0}\lim\inf_{\rho\downarrow 0}\log N(\rho,\lambda)/\log(1/\rho)$$

lurk behind certain heuristic estimates I found useful, and a rigorous exploration would be welcome. Of course, heuristic estimates replace $\inf N(\delta,\lambda)$ by the actual $N(\delta,\Sigma_\lambda)$ relative to some sensible covering Σ_λ.

PEANO CURVES

Additional material on this topic, and on noninteger counting bases, is found in Chapter XII of the 1977 *Fractals*.

POTENTIALS & CAPACITIES. FROSTMAN DIMENSION

The Hausdorff Besicovitch dimension D plays the central role in the modern theory of classical potentials and of generalized (Marcel Riesz) potentials using kernels of the form $|u|^{-F}$, where $F\neq E-2$. Among recent nonelementary mathematical treatments of potential theory, I favor duPlessis 1970, Chapter 3, and the more detailed Landkof 1966–1972.

1. CONJECTURE

We shall see that the special value $D=1$ is intimately linked with the Newtonian potential in $\mathbb{R}^3$. This link underlies the comments in Chapter 10 concerning the various cosmological theories that predict $D=1$, such as the Fournier and Jeans-Hoyle theories.

It should be possible to rephrase these theories as corollaries of Newtonian gravitation.

Thus, the departure of the observed value $D\sim 1.23$ from 1 should be traceable to non-Newtonian (relativistic) effects.

2. DIMENSION & POTENTIALS: HEURISTICS

As mentioned in Chapter 10, Bentley and Newton knew that Kepler's Blazing Sky Effect ("Olbers paradox") has a counterpart in terms of gravitational potential. Suppose that $E=3$, that the mass $M(R)$ within a sphere of radius R around the origin ω is $\propto R^D$ with $D=3$, and that the potential's kernel is the Newtonian R^{-F} with $F=1$. The mass in a shell of thickness dR and radius R is $\propto R^{D-1}$, hence the total potential at ω, which is given by $\propto\int R^{-F}R^{D-1}\,dR=\int R\,dR$, diverges at infinity. There is no divergence at infinity when $D=3$ but $F>3$, implying a non-Newtonian potential. The same result is achieved in the Fournier-Charlier model with $F=1$ and $D<1$.

For the general integral $\int R^{D-1-F}\,dR$, the condition of convergence at infinity is clearly $D<F$. And the condition of convergence at the origin is $D>F$. This argument establishes a one-to-one link between D and F, and in particular it relates D=1 to F=1.

3. POTENTIAL AND CAPACITY

This link was tightened by G. Pólya and G. Szegö and put in final form in Frostman 1935. The major advance is that the argument goes beyond a single origin ω to all points in a (compact) set S. Consider a unit mass distributed on S so that the domain du contains the mass $d\mu(u)$. At the point t, the kernel $|u|^{-F}$ yields the potential function

$$\Pi(t) = \int |u-t|^{-F} d\mu(u).$$

The physical concept of electrostatic capacity was used by de la Vallée Poussin to measure the "contents" of sets. The idea is that if S has a high capacity $C(S)$, the total mass μ can be shuffled to insure that the maximum potential is as small as possible.

Definition: Take the supremum of the potential over all points t, then the infimum of the result with respect to all the distributions of a unit mass over S, and finally set

$$C(S) = \{\inf[\sup_t \Pi(t)]\}^{-1}.$$

If the 1/r kernel is used, this minimal potential is actually achieved by electric charges on a conducting set.

Equivalent definition: $[C(S)]^{-1}$ is the infimum, among all the distributions of mass supported by S, of the energy defined by the double integral

$$\iint |t-u|^{-F} d\mu(s) d\mu(t).$$

4. D AS THE FROSTMAN DIMENSION

There is a simple relationship between $C(S)$ and F. When the exponent F used in defining $C(S)$ is greater than the Hausdorff Besicovitch D, the capacity of $C(S)$ vanishes, meaning that even the "most efficient" distribution of mass over S leads to a potential that is infinite somewhere. When F is less than D, on the other hand, the capacity of S is positive. Thus the Hausdorff Besicovitch dimension is also a capacity dimension in the sense due to Pólya and Szegö. This identity is proved in full generality in Frostman 1935.

The detailed relation between the capacity measure and the Hausdorff measure in the dimension D is involved; see Taylor 1961.

5. "ANOMALOUS" DIMENSION

The kernels $|u|^{-F}$ with $F \neq E-2$ are associated in the physicist's mind with an imbedding space having the "anomalous Euclidean" dimension 2−F. (I do not believe this usage is meant to imply any actual generalization of E to positive reals other than integers.) Given (a) the link between D and F (Frostman), and (b) the role of D in describing galaxy clusters (established in Chapter 10 of this Essay), the terminology of anomalous dimension leads to the following statements. A fractal dimension $D=1$ for galaxies is not anomalous, but the observed fractal dimension $D \sim 1.23$ seems to involve an imbedding space of anomalous dimension.

SCALING UNDER TRUNCATION

Underlying its link with scaling, the hyperbolic distribution is the only distribution such that the rescaled truncated variable "U/u_0, knowing that $U/u_0 > 1$" has a distribution independent of u_0.

PROOF. Assume that there is an underlying distribution $P(u)$, with the rescaled truncated r.v. $W = U/u_0$ following the usual conditional distribution $P(wu_0)/P(u_0)$. We want this conditional distribution to be the same for $u_0 = h'$ and $u_0 = h''$. Write $v' = \log h'$ and $v'' = \log h''$ and consider $R = \log P(u)$ as a function of $v = \log u$. The desired identity $P(uh')/P(h') = P(uh'')/P(h'')$ demands that $R(v'+v) - R(v') = R(v''+v) - R(v'')$ for all choices of v, v', and v''. This requires that R be a linear function of v.

SIMILARITY DIMENSION: ITS PITFALLS

Certain open sets (not containing their limit points) involve a serious discrepancy between dimensions.

The set of trema endpoints of the Cantor dust set is self-similar with the same N and r as the whole Cantor dust, hence it has the same similarity dimension. But it is denumerable, hence its Hausdorff Besicovitch dimension is 0. By adding the limit points of this dust, one falls back on the Cantor dust, and the discrepancy vanishes "to the benefit" of similarity dimension, which is the more important characteristic for this set.

A second simplest example, which I call Besicovitch set, is investigated in the entry NONLACUNAR FRACTALS, 3.

STATIONARITY (DEGREES OF)

Ordinary words used in scientific discourse combine (a) diverse intuitive meanings, dependent on the user, and (b) formal definitions, each of which singles out one special meaning and enshrines it mathematically. The terms *stationary* and *ergodic* are fortunate in that mathematicians agree about them. But my experience indicates that many engineers, physicists, and practical statisticians pay only lip service to the mathematical definition, and hold narrower views. And I prefer an even broader view. These misunderstandings or preferences are revealing.

THE MATHEMATICAL DEFINITION. A process $X(t)$ is stationary if the distribution of $X(t)$ is independent of t, the joint distribution of $X(t_1+\tau)$ and $X(t_2+\tau)$ is independent of τ, and similarly—for all k—for the joint distributions of $X(t_1+\tau) \dots X(t_k+\tau)$.

FIRST MISUNDERSTANDING (PHILOSOPHY). It is a platitude that there can be no science, except of phenomena that follow unchanging rules. Stationarity is often misunderstood in this light: many think that it merely demands that the rules governing the process be invariant in time. But this summary is invalid. For example, the Brownian motion's increment $B(t_1+\tau)-B(t_2+\tau)$ is Gaussian with mean and variance independent of τ. This rule, and also the rule of the Brownian motion's set of zeros, are independent of τ. However, stationarity refers specifically to the rules governing the values of the process itself. For Brownian motion, those rules *are not* time invariant.

SECOND MISUNDERSTANDING (PRACTICAL STATISTICIANS). Numerous techniques (and canned computer programs) that are billed as "*the* analysis of stationary time series" are far narrower in their scope than indicated by this label. This is unavoidable, due to the fact that mathematical stationarity is too general a notion for any single technique to apply in all cases. But as a result the statisticians foster among their customers the opinion that the notion of "stationary time series" is identical to the much narrower notions grasped by the current techniques. Even when they take the trouble of checking that their techniques are "robust," they envision minimal departures from the simplest hypothesis, not the drastic departures that stationarity does allow.

THIRD MISUNDERSTANDING (ENGINEERS AND PHYSICISTS). Many investigators (partly due to the previous misunderstanding) believe that stationarity asserts that the sample processes "may move up and down, but sort of stay statistically the same." This summary applied at an early informal stage, but now it is also invalid. The mathematical definition refers specifically to the generating rules, not to the objects they generate. When mathematicians first encountered stationary processes having extremely erratic samples, they marvelled that the notion of stationarity could encompass such wealth of unexpected behavior. Unfortunately, this is a kind of behavior that many practitioners insist is *not* stationary.

A GRAY ZONE. There is no question that the boundary between stationary and nonstationary processes lies somewhere between white Gaussian noise and Brownian motion, but its precise location is disputed.

SCALING NOISES AS A BENCHMARK. The Gaussian scaling noises of Chapter 27 are a convenient benchmark to refine this boundary, their spectral density being of the form f^{-B} with $B \geq 0$. For white noise, $B=0$; for Brownian motion, $B=2$; and for different purposes, the boundary between stationary and nonstationary processes falls at different values of B.

Mathematicians seeking to avoid the "infrared catastrophe" place the boundary at $B=1$, because $\int_0^1 f^{-B}df < \infty$ is equivalent to $B<1$.

But the behavior of a sample of scaling noise changes continuously at $B=1$. As a matter of fact, there is more of a visible change between $B=0$ and $B>0$, so much so that practitioners faced with *any* sample for which $B>0$ tend to call it *non*stationary. And they tend to be consistent, and claim that data that look like a sample with $B>0$ require a *non*stationary model to represent them.

On the other hand, I found that excluding $B > 1$ makes the definition of stationarity *insufficiently* general in many case studies.

CONDITIONALLY STATIONARY SPORADIC PROCESSES. For example, the theory of fractal noises (Chapter 8) suggests that the process of Brownian zeros is stationary in a weakened form. Indeed, assume that there is at least one zero anywhere between $t=0$ and $t=T$. The result is a random process depending on T as an additional extrinsic parameter. I observed that the joint distribution of the values $X(\tau + t_m)$ is independent of t as long as all the instants $\tau + t_m$ lie between 0 and T. Thus, the nonstationary Brownian zeros process incorporates latently a whole class of random processes, each satisfying a *conditional* form of stationarity, which often suffices.

The processes in this class are so intimately interrelated that Mandelbrot 1967b argues that they must be viewed as one *generalized* stochastic process, to be called *sporadic process*. Compared to a standard random process, the novelty is that the measure of the whole sample space Ω is $\mu(\Omega) = \infty$. Thus, it cannot be normalized to $\mu(\Omega) = 1$. The acceptance of $\mu(\Omega) = \infty$ for random *variables* goes back at least to Rényi 1955. To prevent $\mu(\Omega) = \infty$ from leading to catastrophe, the theory of generalized variables assumes they never observed directly, only as conditioned by some event C such that $0 < \mu(C) < \infty$.

While Rényi random variables are of limited importance, sporadic functions are important: In particular, they allow Mandelbrot 1967b to exorcize some instances of infrared

catastrophe, thus accounting for certain scaling noises with $B \in [1,2]$.

ERGODICITY, MIXING. A second notion that is subject to differing interpretations is ergodicity. In the mathematical literature, ergodicity splits into multiple forms of *mixing*. Some processes *mix strongly*, while others *mix weakly*. As presented in books of mathematics, the distinction hardly seems to affect the study of nature. But in fact it does, with a vengeance! In particular, scaling noises with $0 < B < 1$ are weakly but *not* strongly mixing.

FOURTH MISUNDERSTANDING (CONCERNING THE VALIDITY OF LIMIT CONVERGENCE TO B(t)). It is widely believed that to say that $X(t)$ is stationary is the same as to say that its running sum $X^*(t) = \Sigma_{s0}^{t} X(s)$ can be normalized so as to converge to Brownian motion. Mathematicians have long known that this belief is unwarranted (Grenander & Rosenblatt, 1957). And many of the case studies in this Essay involve functions $X(t)$ that contradict this belief, because of either the Noah Effect ($\langle X^2(t) \rangle = \infty$) or the Joseph Effect (infinite dependence, as in f^{-B} noises with $B > 0$). However, nearly all of my case studies have at some stage been dismissed a priori by an "expert" maintaining that the underlying phenomena are patently *non*stationary, hence my stationary models are foredoomed. This argument is wrong, but is psychologically significant.

CONCLUSION. The frontier between mathematically stationary and nonstationary process encourages disputes over semantics. In practice, this frontier is straddled by processes that differ from the intuitively stationary ones, nevertheless can be the object of science. They also happen to be needed throughout the present Essay and the rest of my research work.

QUESTIONS OF VOCABULARY: "LAPLACIAN," "BENIGN," OR "SETTLED" VS. "VAGRANT". Again, new terms become indispensable. Let me hereby recommend *settled* as (a) a synonym of what mathematicians call "stationary and such that $X^*(t)$ converges to $B(t)$," and (b) a term for the intuitive idea certain practitioners tend to call "stationarity." The alternative antonyms would be *unsettled* and *vagrant*.

An early paper, Mandelbrot 1973f, uses (instead of *settled*) the terms *Laplacian* and *benign*. The latter means "harmless, readily controllable;" it applies because this kind of chance can be trusted *not* to generate any of the wild and varied configurations which make vagrant chance so much more difficult, and so much more interesting.

STATISTICAL ANALYSIS USING R/S

Two assumptions concerning time series were a matter of course in practical statistics: that $\langle X^2 \rangle < \infty$, and that X is short-run dependent. However, I showed (Chapter 37) that long-tailed empirical records are often best interpreted by accepting $\langle X^2 \rangle = \infty$. And the question of whether a record is weakly (short-run) or strongly (long-run) dependent was first faced where I injected long-run dependence to

interpret the Hurst phenomenon (Chapter 27).

The mixture of long-tailedness and very long-run dependence might have been statistically unmanageable, because the standard second-order techniques geared towards dependence (correlation, spectra) invariables assume $\langle X^2 \rangle < \infty$. But there is an alternative.

One can disregard the distribution of X(t), and tackle its long-run dependence with the help of rescaled range analysis, also called R/S analysis. This statistical technique, introduced in Mandelbrot & Wallis 1969c and given mathematical foundation in Mandelbrot 1975w, concerns the distinction between the short and the very long run. The constant it introduces is denoted by J and called *Hurst Coefficient or* R/S *Exponent*, and can lie anywhere between 0 and 1.

Even before defining J, one can describe its significance. The special value $J = \frac{1}{2}$ is characteristic of independent, Markov and other short-run dependent random functions. Therefore, the absence of very long-run nonperiodic statistical dependence in empirical records or in sample functions can be investigated by *testing* whether the hypothesis that $J = \frac{1}{2}$ is statistically acceptable. If not, the intensity of very long-run dependence is measured by $J - \frac{1}{2}$, whose value can be *estimated* from the data.

The principal virtue of this approach is that the exponent J is *robust* with respect to the marginal distribution. That is, not only is it effective when the underlying data or random functions are near Gaussian, but it continues to be effective when X(t) is so far from Gaussian that $\langle X^2(t) \rangle$ diverges, in which case all second order techniques are invalid.

DEFINITION OF THE STATISTIC R/S. In continuous time t, define $X^*(t) = \int_0^t X(u)du$, $X^{2*}(t) = \int_0^t X^2(u)du$, and $X^{*2} = (X^*)^2$. In discrete time i, define $X^*(0) = 0$, $X^*(t) = \Sigma_{i=1}^{[t]} X(i)$, with [t] the integer part of t. For every d>0, called the lag, define the *adjusted range* of $X^*(t)$ in the time interval 0 to d, as

$$R(d) = \max_{0 \leq u \leq d} \{X^*(u) - (u/d)X^*(d)\}$$
$$- \min_{0 \leq u \leq d} \{X^*(u) - (u/d)X^*(d)\}.$$

Then evaluate the *sample standard deviation* of X(t),

$$S^2(d) = X^{2*}(d)/d - X^{*2}(d)/d^2.$$

The expression Q(d) = R(d)/S(d) is the R/S *statistic*, or *self-rescaled self-adjusted range* of $X^*(t)$.

DEFINITION OF THE R/S EXPONENT J. Suppose there exists a real number J such that, as $d \to \infty$, $(1/d^J)[R(d)/S(d)]$ converges in distribution to a nondegenerate limit random variable. Mandelbrot 1975w proves this implies that $0 \leq J \leq 1$. The function X is then said to have the R/S *exponent* J with a constant R/S *prefactor*.

Suppose, more generally, that the ratio $[1/d^J L(d)][R(d)/S(d)]$ converges in distribution to a nondegenerate random variable, where L(d) denotes a slowly varying function at infinity, that is, a function which satisfies

$L(td)/L(d) \to 1$ as $d \to \infty$ for all t>0. The simplest example is L(d) = log d. The function X is then said to have the R/S exponent J, and the R/S prefactor L(d).

PRINCIPAL RESULTS (MANDELBROT 1975w). When X(t) is a white Gaussian noise, one finds $J = \frac{1}{2}$ with a constant prefactor. More precisely, $e^{-\delta J} R(e^\delta)/S(e^\delta)$ is a stationary random function of $\delta = \log d$.

More generally, $J = \frac{1}{2}$ whenever S(d) $\to \langle X^2 \rangle$ and the rescaled $a^{-\frac{1}{2}} X^*(at)$ converges weakly to B(t) as $a \to \infty$.

When X(t) is the discrete fractional Gaussian noise, that is the sequence of increments of $B_H(t)$ (see p. 353), one finds J=H, with $H \in]0,1[$.

More generally, in order to obtain $J = H \neq \frac{1}{2}$ with a constant prefactor, it suffices that S(d) $\to \langle X^2 \rangle$ and that $X^*(t)$ be attracted by $B_H(t)$, with $\langle X^*(t) \rangle \sim t^{2H}$.

Still more generally, $J = H \neq \frac{1}{2}$ with the prefactor L(d) prevails if S(d) $\to \langle X^2 \rangle$, and $X^*(t)$ is attracted by $B_H(t)$ and satisfies $\langle X^*(t)^2 \rangle \sim t^{2H} L(t)$.

Finally, $J \neq \frac{1}{2}$ when S(d) $\to \langle X^2 \rangle$, and $X^*(t)$ is attracted by a non-Gaussian scaling random function of exponent H=J. Examples are given in Taqqu 1975, 1979a,b.

On the other hand, when X is a white Lévy stable noise, hence $\langle X^2 \rangle = \infty$, one finds $J = \frac{1}{2}$.

When X becomes stationary when differenced (or differentiated), one finds J=1.

WEIERSTRASS FUNCTIONS AND KIN. ULTRAVIOLET AND INFRARED CATASTROPHES

The complex Wierstrass function is the sum of the series

$$W_0(t) = (1-w^2)^{-1/2} \sum_0^\infty w^n \exp(2\pi i b^n t),$$

where b is a real number >1, and w is written either as $w = b^H$, with $0 < H < 1$, or as $w = b^{D-2}$, with $1 < D < 2$. The real and imaginary parts of $W_0(t)$ are called Weierstrass cosine and sine functions.

The function $W_0(t)$ is continuous but nowhere differentiable. But its formal extension to $D < 1$ is continuous *and* differentiable.

In addition to $W_0(t)$, this entry discusses several variants that I found it indispensable to introduce, due to the new role the theory of fractals gives to $W_0(t)$.

FREQUENCY SPECTRUM OF $W_0(t)$. The term "spectrum" is overloaded with meanings. *Frequency spectrum* designates the set of admissible values of the frequency f, irrespective of the corresponding terms' amplitudes.

A periodic function's frequency spectrum is the sequence of positive integers. A Brown function's frequency spectrum is $\mathbb{R}^+$. And Weierstrass function's frequency spectrum is the discrete sequence b^n from $n = 1$ to $n = \infty$.

ENERGY SPECTRUM OF $W_0(t)$. *Energy spectrum* designates the set of admissible values f together with the corresponding energies (amplitudes squared). For each frequency of the form $f = b^n$, $W_0(t)$ has a spectral line of energy $(1-w^2)^{-1} w^{2n}$. Hence, the cumulative energy in frequencies $f \geq b^n$ is convergent and $\propto w^{2n} = b^{-2nH} = f^{-2H}$.

COMPARISON WITH FRACTIONAL BROWNIAN MOTION. The cumulative energy is also f^{-2H} in several previously encountered cases. (A) The fractional Fourier-Brown-Wiener periodic random functions, for which the acceptable frequencies are of the form $f = n$, and the corresponding Fourier coefficients are $n^{H-1/2}$. (B) The random processes with the continuous population spectral density $\propto 2H f^{-2H-1}$. These are the fractional Brown functions $B_H(t)$ in Chapter 27. For example, if $H = 1/2$, the Weierstrass cumulative spectrum $\propto f^{-1}$ is encountered for ordinary Brownian motion $B(t)$, whose spectral density is f^{-2}. An essential difference is that the Brown spectrum is absolutely continuous, while the Fourier-Brown-Wiener and Weierstrass spectra are discrete.

NONDIFFERENTIABILITY. To prove that $W_0(t)$ does not have a finite derivative for any value of t, Weierstrass had to add two conditions: (a) b is an odd integer, hence $W_0(t)$ is a Fourier series, and (b) $\log_b (1+3\pi/2) < D < 2$. The necessary and sufficient conditions, $b > 1$ and $1 < D < 2$, are from Hardy 1916.

DIVERGENCE OF ENERGY. To a physicist accustomed to spectra, Hardy's conditions are intuitively obvious. Applying the rule of thumb that a function's derivative is obtained by multiplying its kth Fourier coefficient by k, the physicist finds for the formal derivative of $W_0(t)$ that the Fourier coefficient with $k = b^n$ has an amplitude squared equal to $(1-w^2)^{-1} w^{2n} b^{2n}$. The cumulative energy in frequencies $\geq b^n$ being *infinite*, the physicist agrees that $W_0'(t)$ cannot be defined.

It is interesting to note that Riemann's search for a counterexample to differentiability led him to $R(t) = \Sigma_1^\infty n^{-2}\sin(2\pi n^2 t)$, whose energy in frequencies $\geq f=n^2$ is $\propto n^{-3} = f^{-2H}$, with $H=\frac{3}{4}$. Thus, the same heuristic argument suggests that $R'(t)$ is not definable, hence $R(t)$ is not differentiable. This conclusion is "almost" correct, but $R'(t)$ does exist for certain t's (Gerver 1970, Smith 1972).

ULTRAVIOLET DIVERGENCE/CATASTROPHE. The term "catastrophe" first entered physics around 1900, after Rayleigh and Jeans devised a theory of blackbody radiation predicting that the frequency band of width df near the f contains an energy proportional to f^{-4}. The implication, that the total high frequency energy is infinite, is catastrophic *for the theory*. Since the trouble comes from frequencies beyond the ultraviolet, it was described as an *ultraviolet (UV) catastrophe.*

Everyone knows that Planck built quantum theory upon the ruins created by the UV catastrophe of radiation.

HISTORICAL ASIDE. Note (the point *must* have been made by others, but I have no reference) that the same divergence killed the old physics ($\neq 1900$), and the old mathematics ($\neq 1875$) that believed that continuous functions must be differentiable. The physicists' reaction was to change the rules of the game, but the mathematicians' reaction was to learn to live with nondifferentiable functions and their formal differentials. (The latter are the only examples of Schwartz distribution of fre-

quent use in physics.)

SEARCH FOR A SCALING DISCRETE SPECTRUM. INFRARED DIVERGENCE. While the frequency spectrum of the Brown function is continuous, is scaling, and extends to $f=0$, the frequency spectrum of the Weierstrass function fitted to the same H is discrete and is bounded below by $f=1$. The presence of this lower bound is solely due to the fact that Weierstrass's original b was an integer and the function was periodic. Now we would like to eliminate this feature, and the obvious procedure is to allow n to run from $-\infty$ to $+\infty$. To let the scaling property extend to the energy spectrum, it suffices to attribute to the component of frequency b^n the amplitude w^n.

Unfortunately, the resulting series is divergent, due to low frequency components. This defect is called *infrared (IR) divergence (or "catastrophe")*. However, the divergence must be faced, because the lower bound $f=1$ clashes with the self-similarity otherwise embodied in the energy spectrum f^{-2H}.

WEIERSTRASS FUNCTION, MODIFIED TO BE SELF-AFFINE WITH RESPECT TO THE FOCAL TIME $t=0$. To extend the Weierstrass frequency spectrum f^{-2H} down to $f=0$ without dreadful consequences, it is simplest to *first* form the expression $W_0(0)-W_0(t)$, and *then* to let n run from $-\infty$ to ∞. The added terms corresponding to $n<0$ converge if $0<H<1$, and their sum is continuous and differentiable. The function thus modified,

$$W_1(t)-W_1(0) = (1-w^2)^{-\frac{1}{2}}\Sigma_{-\infty}^\infty w^n[\exp(2\pi i b^n t)-1],$$

is still continuous but nowhere differentiable. In addition it is scaling in the sense that

$$W_1(tb^m) - W_1(0) = (1-w^2)^{-1/2}$$
$$\Sigma_{-\infty}^{\infty} \; w^{-m}w^{n+m}[\exp(2\pi ib^{n+m}t - 1]$$
$$= w^{-m}[W_1(t) - W_1(0)].$$

Thus, the function $w^m[W_1(b^m t) - W_1(0)]$ is independent of m. Alternatively, as long as $r = b^m$, $r^{-H}[W_1(rt) - W_1(0)]$ is independent of h. That is, $W_1(r) - W_1(0)$, and its real and complex parts, are self-affine with respect to r's of the form b^{-m} and the focal time t=0.

An extensive study of the Weierstrass (modified) functions $W_1(t)$, with enlightening graphics, is found in Berry & Lewis 1980.

GAUSS RANDOM FUNCTIONS WITH AN EXTENDED WEIERSTRASS SPECTRUM. The next step toward realism and applicability is taken when the extended Weierstrass function is randomized. The simplest and most intrinsic method consists in multiplying its Fourier coefficients by independent complex Gaussian factors of zero mean and unit variance. The real and imaginary parts of the result deserve to be called *Weierstrass (modified)-Gauss functions*. In several ways, they are approximate fractional Brown functions. When the values of H match, their spectra are as close to being the same as allowed by one being discrete and the other continuous. Moreover, the result of Orey 1970 and Marcus 1976 remains applicable and shows that their level sets have the same fractal dimension.

FRACTAL PROPERTIES. By a theorem in Love & Young 1937 and Besicovitch & Ursell 1937 (see LIPSCHITZ...,), the graph of a function satisfying for all x the Lipschitz condition of exponent H has a fractal dimension between 1 and 2−H. For the fractional Brown function having the same cumulative spectrum f^{-2H}, the dimension is known to take the largest possible value 2−H=D. I conjecture that the same holds for the Weierstrass curve. And that its zeroset is of dimension 1−H.

ZEROSETS OF RELATED FUNCTIONS. The Rademacher functions are squared-off variants of the sine functions of the form $\sin(2\pi b^n t)$ in which b=2. Where the sine is positive (respectively, negative or vanishing), the Rademacher function is equal to 1 (respectively, to −1, or 0) (Zygmund 1959 I, p. 202.) The natural generalization of the Weierstrass function is a series in which the nth term is the product of w^n by the nth Rademacher function. This function is discontinuous, but its spectral exponent continues to be 2H. Intuitively, the precedent of fractional Brownian motion suggests that the zerosets of the Weierstrass-Rademacher function are of dimension 1−H. This is confirmed in Beyer 1962, but only under the restriction that 1/H is an integer.

Singh 1935 refers to numerous other variants of the Weierstrass function. In some cases, the zeroset's D is easy to evaluate. The topic deserves a fresh look. ▬

40 ¤ Biographical Sketches

As a prelude to this chapter devoted to bits of biography, note that a life story that is interesting to tell is rarely the reward (or is it a punishment?) of those who keep to the mainstream of science. As an example, take John William Strutt, third Baron Rayleigh. A steady flow of triumphs made his name recognizable in almost every province of science. Yet, with one exception, his life appears uneventfully subordinated to his evolution as a scientist. The unexpected occurs when, having been admitted to Trinity College as a birthright, being the elder son of a landowning lord, he decides to become a scholar.

Science does have its great Romantic, Evariste Galois, whose story fits the canons of French Court tragedy since it combines within the confines of one day his eclosion as scientist and his death in a duel. But most scientists' stories are like Rayleigh's: hardly touched even by extreme uprooting (witness A. S. Besicovitch), and ultimately almost pre-

dictable, except for the occasionally colorful circumstances of the revelation of their talent, and of their entry into the mainstream. The three-year-old Carl Friedrich Gauss corrects the arithmetic of his father. The adolescent Srinivasa Ramanujan reinvents mathematics. Harlow Shapley, upon finding that he must wait out a term before he can register in a school of journalism, selects a department from an alphabetical list. He skips archaeology because he does not know the word's meaning, proceeds to astronomy, and meets his fate. More atypical is the story of Felix Hausdorff. Until the age of 35, he devotes most of his time to philosophy, poetry, writing and directing plays, and similar endeavors. Then he settles down to mathematics and soon produces his masterpiece, Hausdorff 1914.

Tales according to the typical pattern are legion, but the stories selected for this chapter are entirely different. Entry into the mainstream is postponed, and in many cases it is even posthumous. Strong feelings persist of really belonging to other times. The hero is a loner. Like certain painters, he might be called a *naive* or a *visionary,* but there is a better term in American English: *maverick.* When the curtain falls on the prologue of his life story, he is still, by choice or by chance, unbranded.

Mavericks' work frequently exhibits a peculiar freshness. Even those who fail to achieve greatness tend to share with the giants a sharply personal style. The key seems to be time to spare. In the words of the daughter of D'Arcy Thompson, speaking about his book *On Growth and Form* (Thompson 1917), "It is a matter of speculation whether [such a work] would ever have been written if [its author] had not spent thirty years of his early life in the wilderness." Indeed, he was 57 when he published it, and many other mavericks do their best very late: the cliché that science is very largely a young man's game is definitely not true in their case.

I find such stories appealing and wish to share the emotions a few of them evoke.

As mavericks should, our heroes differ greatly from one another. Paul Lévy lived long enough to set his mark deeply in his province of science, but his admirers (and I am one) think that he deserves even better; call it true fame. (So did D'Arcy Wentworth Thompson, who would not be out of place in this company, but whose life is fully documented in the abridged edition of his book, Thompson 1962.) Lewis F. Richardson also made it—barely. But Louis Bachelier's story

was sadder; no one read through his books and papers, and he stood as a perennially unsuccessful applicant until all his work had been duplicated by others. Hurst had better luck, and his story is intriguing. And Fournier d'Albe and Zipf deserve lasting footnotes. Thus each of the stories in this chapter brings some insight into the psychology of a peculiar kind of strong mind.

In cases where standard biographies exist, they are not repeated unless necessary. The great *Dictionary of Scientific Biography* (Gillispie 1970-1976) includes bibliographies. Its omissions are also significant.

LOUIS BACHELIER (1870-1946)

The story of the beginnings of the theory of Brownian motion is worth knowing and is touched upon in the next chapter. However, physics might have been preceded in this context by mathematics—and also (a most unusual sequence of events) by economics.

The fact is that a truly incredible proportion of the results of the mathematical theory of Brownian motion had been described in detail five years before Einstein. The precursor was Louis Bachelier (*Dictionary of Scientific Biography,* I, 366-367).

Our story centers on a doctoral dissertation in the mathematical sciences, defended in Paris on March 19, 1900. Sixty years later it received the rare compliment of an English translation, with extensive comments. However, it started badly: the committee that examined it was not overly impressed and gave it the unusual and near-insulting *mention honorable* at a time when no one stood for the French doctorate unless he foresaw an academic opening and felt sure of receiving the required *mention très honorable*.

It is not surprising therefore that this dissertation had no direct influence on anyone else's work. Bachelier, in turn, was not influenced by anything written in this century, even though he remained active and published (in the best journals) several papers filled with endless algebraic manipulations. In addition, his popular book, Bachelier 1914, enjoyed several printings and even now bears being read. It is not to be recommended to just anyone, because its subject matter has changed profoundly, and because it is not clear whether short sentences summarize established knowledge or outline problems yet to be explored. The cumulative effect of such ambiguity is rather disconcerting. Only very late, after repeated failures, was Bachelier finally appointed to a University professorship, in the tiny University of Besançon.

In view of his slow and mediocre career and of the thinness of the personal trace he left (my search, though diligent, has discovered only some odd scraps of recollections by students and colleagues, and not a single photo), the posthumous fame of his dissertation makes him an almost romantic personality. Why the sharpness of this contrast?

To begin with, his life might have been brighter were it not for a certain mathematical error. The story is told in Lévy 1970 (pp. 97–98) and in greater detail in a letter Paul Lévy wrote me on January 25, 1964.

"I first heard of him a few years after the publication of my *Calcul des Probabilités,* that is, in 1928, give or take a year. He was a candidate for a professorship at the University of Dijon. Gevrey, who was teaching there, came to ask my opinion of a work Bachelier published in 1913 (*Annales de l'Ecole Normale*). In it, he had defined Wiener's function (prior to Wiener) as follows: In each of the intervals $[n\tau,(n+1)\tau]$, he considered a function $X(t|\tau)$ that has a constant derivative equal to either $+v$ or $-v$, the two values being equiprobable. He then proceeded to the limit (v constant, and $\tau \rightarrow 0$), and claimed he was obtaining a proper function $X(t)$! Gevrey was scandalized by this error. I agreed with him and confirmed it in a letter which he read to his colleagues in Dijon. Bachelier was blackballed. He found out the part I had played and asked for an explanation, which I gave him and which did not convince him of his error. I shall say no more of the immediate consequences of this incident.

"I had forgotten it when in 1931, reading Kolmogorov's fundamental paper, I came to 'der Bacheliers Fall.' I looked up Bachelier's works, and saw that this error, which is repeated everywhere, does not prevent him from obtaining results that would have been correct if only, instead of $v=$constant, he had written $v=c\tau^{-\frac{1}{2}}$, and that, prior to Einstein and prior to Wiener, he happens to have seen some important properties of the so-called Wiener or Wiener-Lévy function, namely, the diffusion equation and the distribution of $\max_{0 \leq \tau \leq t} X(t)$.

"We became reconciled. I had written him that I regretted that an impression, produced by a single initial error, should have kept me from going on with my reading of a work in which there were so many interesting ideas. He replied with a long letter in which he expressed great enthusiasm for research."

That Lévy should have played this role is tragic, for his own career, as we will see very soon, also nearly foundered because his papers were not sufficiently rigorous.

We now reach the second and deeper reason for Bachelier's career problems. It is revealed by the title of his dissertation, which (on purpose) I have not yet mentioned: "Mathematical theory of speculation." The title did not by any means refer to (philosophical) speculation on the nature of chance, rather to (money-grubbing) speculation on the ups and downs of the market for consolidated state bonds (*"la rente"*). The function X(t) mentioned by Lévy stood for the price of these bonds at time t.

The professional difficulties that Bachelier was to experience as a result were foreshadowed in the delicately understated comment by Henri Poincaré, who wrote the official report on this dissertation, that "the topic is somewhat remote from those our candidates are in the habit of treating." One may argue that Bachelier should have avoided seeking the judgment of unwilling mathematicians (the idea of assigning thesis subjects was totally foreign to French professors of that peri-

od), but he had no choice: his lower degree was in mathematics and, while Poincaré did little research in probability, he was in charge of teaching it.

Bachelier's tragedy was to be a man of the past and of the future but not of his present. He was a man of the past because he worked on the historical roots of probability theory: the study of gambling. He chose to introduce continuous time stochastic processes through the continuous form of gambling, *La Bourse*. He was a man of the future, both in mathematics (witness the above letter by Lévy) and in economics, where he is acknowledged as the creator of the probabilistic concept of "martingale" (this is the proper formulation of the notion of a *fair game* or of an *efficient market*, see Chapter 37), and he was well ahead of his time in understanding many specific aspects of uncertainty as related to economics. He owes his greatest fame to the concept that prices follow the Brownian motion process. Unfortunately, no organized scientific community of his time was in a position to understand and welcome him. To gain acceptance for his ideas would have required supreme political skills that he evidently did not possess.

To survive and go on producing new works under these circumstances, Bachelier had to feel strongly about the importance of his work. In particular, he knew very well that he was the originator of the theory of the diffusion of probability. In an unpublished *Notice* that he wrote in 1921 (while applying for some unspecified academic position), he stated that his principal scholarly contribution had been to provide "images taken from natural phenomena, like the theory of radiation of probability, in which [he] likens an abstraction to energy—a strange and unexpected linkage and a starting point for great progress. It was with this concept in mind that Henri Poincaré had written, 'Mr. Bachelier has evidenced an original and precise mind.'"

The preceding sentence is taken from the already-mentioned report on the dissertation, which deserves further excerpting: "The manner in which the candidate obtains the law of Gauss is most original, and all the more interesting as the same reasoning might, with a few changes, be extended to the theory of errors. He develops this in a chapter which might at first seem strange, for he titles it 'Radiation of Probability.' In effect, the author resorts to a comparison with the analytical theory of the propagation of heat. A little reflection shows that the analogy is real and the comparison legitimate. Fourier's reasoning is applicable almost without change to this problem, which is so different from that for which it had been created. It is regrettable that [the author] did not develop this part of his thesis further."

Poincaré, therefore, had seen that Bachelier had advanced to the threshold of a general theory of diffusion. However, Poincaré was notorious for lapses of memory. A few years later, he took an active part in discussions concerning Brownian diffusion, but had forgotten Bachelier's 1900 dissertation.

Other comments in Bachelier's *Notice* are also worth summarizing: "1906: *Théorie des probabilités continues*. This theory has no relation whatsoever with the theory of geometric probability, whose scope is very limited. This is a science of another level of difficulty and generality than the calculus of probability. Conception, analysis, method, everything in it is new. 1913: *Probabilités cinématiques et dynamiques*. These applications of probability to mechanics are the author's own, absolutely. He took the original idea from no one; no work of the same kind has ever been performed. Conception, method, results, everything is new."

The hapless authors of academic *Notices* are not called upon to be modest, and Louis Bachelier did exaggerate to some extent. Moreover, he gave no evidence of having read anything written in the twentieth century. Unfortunately, his contemporaries discounted everything he said and refused him the position he was seeking!

Does anyone know more about him?

Poincaré's statements are paraphrased, with permission, from a report filed in the Archives of the Pierre and Marie Curie University (Paris VI), heir to the archives of the former Faculty of Sciences of Paris. This fascinating document, in the lucid style characteristic of Poincaré's popular writings, suggests that more extensive selections from Poincaré's letters and confidential reports to universities and academies ought to be made available. As of today, a broad and intriguing aspect of his personality is near absent from his books and his *Collected Works*.

EDMUND EDWARD
FOURNIER D'ALBE (1868-1933)

Fournier d'Albe (*Who's Who in Science,* p. 593) chose to live as a free-lance science journalist and inventor: he constructed a prosthesis to enable the blind to "hear" letters and was the first to transmit a television signal from London.

His name was witness to Huguenot ancestry. Despite his partly German education and his eventual residence in London, where he obtained his A.B. by attending evening college, a stint in Dublin transformed him into an Irish patriot and a militant in a Pan-Celtic movement. He was a believer in spiritualism and a religious mystic.

He is remembered for his book *Two New Worlds.* It received very good reviews in *Nature,* which called its arguments "simple and reasonable," and *The Times,* which called its speculations "curious and attractive." However, the obituaries for Fournier d'Albe that appeared in *Nature* and *The Times* somehow failed to mention this book. It has become almost impossible to find and is rarely mentioned without sarcastic comments.

True, it is the kind of work in which a physicist is surprised to find anything of permanent technical value. In fact, I had been advised against attracting attention to it, lest the disputable bulk of the material be taken seriously. But should one use against Fournier an argument one would not consider using against Kepler? This is not to say Fournier was a Kepler; he hardly rose to the level of accomplishment of others in this chapter. Yet a critic's claim that "scientifically the work of the self-styled 'Newton of the soul' is worthless" is too sweeping by far.

Indeed, Fournier was the first to restate an old intuition about galactic clustering (dating back to Kant and to Kant's contemporary Lambert) in terms sufficiently precise to allow us today to conclude that the galaxies should satisfy $D=1$. Thus, we are indebted to him for something of lasting value.

HAROLD EDWIN HURST (1880-1978)

Hurst, hailed as perhaps the foremost Nilologist of all time and spoken of as "Abu Nil," the Father of the Nile, spent the bulk of his career in Cairo as a civil servant of the British Crown, then of Egypt. (*Who's Who, 1973,* p. 1625, and *Who's Who of British Scientists 1969/70,* pp. 417-418.)

His early training, as he and Mrs. Marguerite Brunel Hurst described to me, is worth retelling. The son of a village builder of limited means, whose family had lived near Leicester for almost three centuries, he left school at age 15. He had been trained mostly in chemistry, and also in carpentry by his father. He then started as a pupil teacher at a school in Leicester, attending evening classes to continue his own education.

At age 20, he won a scholarship that enabled him to go to Oxford as a noncollegiate student. After a year, he became an undergraduate at the recently reestablished Hertford College, and soon switched to a major in physics and worked at Clarendon Laboratory.

His lack of preparation in mathematics was a handicap, but thanks to the interest that Professor Glazebrook took in an unusual candidate who was very strong in practical work, he won a first-class honors degree, to everyone's surprise, and was asked to stay for three years as a lecturer and demonstrator.

In 1906, Hurst went to Egypt for a short stay that was to last 62 years, of which the most fruitful were after he had turned 65. His first duties included transmitting standard time from the Observatory to the Citadel of Cairo, where a gun was to be fired at midday. However, he became increasingly fascinated with the Nile, and his study and exploration of the Nile basin made him well known internationally. He traveled extensively by river and on land—on foot with porters, using a bicycle, later by car, and later still by plane. The low Aswan Dam had been build in 1903, but he realized how important it was to Egypt that provision should be made not only for the dry years but for a series of dry years. Irrigation storage schemes should be adequate for every situation, very much, as in the Old Testament, Joseph stored grain for the lean years. He was one of the first to realize the need for the "Sudd el Aali," the High Dam and Reservoir at Aswan.

Hurst's name is likely to survive because of a statistical method he initiated and used to discover a major empirical law concerning long run dependence in geophysics. At first, it seems surprising that anything of the kind could come from an author so poorly prepared in mathematics and working so far from any major center of learning, but at second thought these circumstances may have been vital to both the birth of his idea and its survival. He investigated the Nile using a peculiar method of analysis of his own design, one that might be termed narrow and ad hoc, but in fact has turned out to be eminently intrinsic. Not being pressed by time and having exceptionally abundant data at his disposal, he was in a position to compare them with the standard model of stochastic variability (white noise) through their respective effects upon the design of the High Dam. This led him to the expression Chapters 28 and 39 (p. 387) denote by $R(d)/S(d)$.

One can imagine the amount of hard work implied in such research before the advent of computers—but of course the Nile is sufficiently important to Egypt to justify comparatively large expenditures (and to preclude forcing Hurst to retire).

Hurst adamantly maintained that his finding was significant, despite the fact that no test existed by which such significance could be assessed objectively. Finally, at the ages of 71 and 75, he read two long papers on his discovery, and its potential importance became recognized.

In E. H. Lloyd's words (but my notation), Hurst put us "in one of those situations, so salutary for theoreticians, in which empirical discoveries stubbornly refuse to accord with theory. All the researches described above lead to the conclusion that in the long run

R(d) should increase like $d^{0.5}$, whereas Hurst's extraordinarily well-documented empirical law shows an increase like d^H, where H is about 0.7. We are forced to the conclusion that either the theorists' interpretation of their work is inadequate or their theories are falsely based; possibly both conclusions apply." Similarly, in the words of Feller 1951: "We are here confronted with a problem which is interesting from both a statistical and a mathematical point of view."

My fractional Brownian motion model (Chapter 28) arose as a direct response to the Hurst phenomenon, but this is not the end of the Hurst story. It is hard to quibble with the glowing comments in the last paragraph...but both were unwittingly based upon an incorrect reading of Hurst's claims. Lloyd neglected the division of R by S, and Feller knew Hurst's work from a third party's verbal report (as he acknowledged), and failed to realize that a division by S had been performed. The value of Feller's work was not affected. For the importance of the division by S, see Mandelbrot & Wallis 1969c and Mandelbrot 1975w.

We see again in this instance that when a result is truly unexpected it is hard to comprehend, even by those best disposed to listen.

PAUL LÉVY (1886–1971)

Paul Lévy, who acknowledged no pupil but came closest to being my mentor, achieved goals that Bachelier only saw from afar. Lévy lived long enough to gain recognition as possibly the greatest probabilist of all time, and (when nearly 80 years old) he finally came to occupy, at the Paris Académie des Sciences, the seat that had been Poincaré's, then Hadamard's. See *World Who's Who in Science*, p. 1035.

And yet, almost to the end of his active life, Lévy had been kept at arm's length by the Establishment. Not only did Poincaré's former University chair elude him repeatedly, but his repeated offers to give noncredit lectures were accepted with reluctance, for fear they might disrupt the curriculum.

His life, thoughts, and opinions are documented at length in Lévy 1970, a book well worth reading because of a lack of self-conscious attempt to appear better or worse than life. The end is best skipped, but the best passages are splendid. In particular, he describes in touching terms both his fear of being "a mere survivor of the last century," and his feeling of being a mathematician "unlike all the others." This feeling was widely shared. I recall John von Neumann saying in 1954, "I think I understand how every other mathematician operates, but Lévy is like a visitor from a strange planet. He seems to have his own private methods of arriving at the truth, which leave me ill at ease."

He had few obligations to distract him, aside from a score of lectures each year as a professor of mathematical analysis at the Ecole Polytechnique. Working alone, he transformed probability theory from a small collection of odd results into a discipline in which rich and varied results could be obtained through methods so direct as to be classical. He became interested in the topic when asked for a lecture on errors in the firing of guns. He was near 40 at the time, a brilliant man short of fulfilling his promise and a professor at Polytechnique at a time when the school's appointments favored him as an alumnus. His major books were written at ages 50 and 60, and much of his work on Hilbert space-to-line Brownian functions came much later.

Of the countless interesting tales in his autobiography, one relates to the short paper he devoted to the Bentley paradox, relative to the Newtonian gravitation potential (Chapter 9). In 1904, when a 19-year-old student, Lévy independently discovered the Fournier model of the universe. However, he believed that "the argument was so simple that I would not have thought of publishing it if, 25 years later, chance had not made me overhear a conversation between Jean Perrin and Paul Langevin. These illustrious physicists agreed that one could only escape the paradox by assuming the universe to be finite. I spoke up to point out their error. They did not seem to see my point, but Perrin was shaken by my self-assurance and asked me to write down my ideas, which I did."

Apropos of results being "too simple to publish," the phrase appears often in Lévy's recollections. Many creative minds overrate their most baroque works, and underrate the simple ones. When history reverses such judgments, prolific writers come to be best remembered as authors of "lemmas," of propositions they had felt to be "too simple" in

themselves and had published solely as preludes to forgotten theorems.

The remarks that follow paraphrase part of what I said at a ceremony in Lévy's memory: "The trace left in my memory by his spoken lectures at Polytechnique has become very blurred, because chance had assigned me to the rear of a large lecture hall, and Lévy's voice was weak and not amplified. The most vivid recollection is that of the resemblance some of us noticed between his figure—long, gray, and well groomed—and the somewhat peculiar way he had of tracing on the blackboard the symbol of integration.

"But his written course notes were quite another matter. They were not the traditional well-ordered procession, beginning with a regiment of definitions and of lemmas followed by theorems, every assumption being clearly stated, this majestic flow being perhaps interrupted by the statement of a few unproven results, clearly emphasized as such. Rather, the recollection I have is of a tumultuous flood of remarks and observations.

"In his autobiography, Lévy suggests that in order to interest children in geometry, one should proceed as quickly as possible to theorems they are not tempted to consider evident. His method at Polytechnique was not all that different. To give an account of it, we are irresistibly attracted to images borrowed from geography and mountaineering. We are thus reminded of an old review of an earlier great *Cours d'Analyse de l'Ecole Polytechnique*. The course had been taught by Camille Jordan, and the reviewer was Henri Lebesgue.

Because Lebesgue's disdain for Lévy's work was strong and public, it is ironic that his comments in praise of Jordan apply so well to Lévy. He was unlike 'a person who would attempt to reach the peak of an unknown region, but who would not allow himself to look around before reaching his goal. If led there by someone else, perhaps he may be able to look down upon many things, but he could not know what they are. In fact, one cannot generally see anything from a very high peak; mountaineers climb them only for the sake of the effort.'

"Needless to say, Lévy's course notes were not popular. To many excellent Polytechnique students, they were a source of worry when cramming for the general examination. In the ultimate rewrite, which I had to study in 1957 as his Maître de Conférences, all those features had become even more strongly accentuated. For example, the treatment of the theory of integration was frankly no more than an approximation. No one, he had written, can do a good job by trying to force his talent. It would seem that in his last course notes, his talent had been forced.

"But my recollection of the course he had taught to the class admitted in 1944 remains extraordinarily positive. Intuition, though it cannot be taught, can only too easily be thwarted. I believe that this is what Lévy was trying above all to avoid, and I think he had mostly succeeded.

"At Polytechnique, I had heard many references to his creative work. One would praise it as being very important, then promptly add the comment that it did not contain a single faultless mathematical proof, and included infuriatingly many arguments of uncertain footing. In conclusion, the most urgent thing was to make everything rigorous. This task has been performed, and today the intellectual grandchildren of Lévy rejoice in being accepted as full-fledged mathematicians. As one of them put it a moment ago, they see themselves as 'probabilists turned bourgeois.'

"I fear that far too much may have been paid for this acceptance. In every branch of knowledge, there seem to be many successive levels of precision and generality. Some are unsuited to attack any but the most trivial problems. More and more, however, and in almost every branch of knowledge, one is able to push precision and generality to excess. For example, a hundred pages of preliminaries may be needed to prove one theorem in a form that is hardly more general than its predecessors, and discloses no new horizon. But some fortunate branches of knowledge allow an intermediate level of precision and generality that may be termed classical. Paul Lévy's almost unique greatness lies in the fact that he was, at the same time, a forerunner and *the* classic in his field.

"Lévy rarely concerned himself with anything but pure mathematics. Also, those who have to solve a problem that has already been well-posed rarely find in his work a formula ready to serve them with no further effort. On the other hand, if I can believe my personal experience, Lévy's approach to more basic issues of formulation of chance makes him stand out more and more as a giant.

"Whether in the diverse topics to which the present Essay is devoted or in those I examine in other works, a proper mathematical formalization seems to demand very quickly either a conceptual tool that Lévy had provided or a tool wrought in the same spirit and possessing the same degree of generality. More and more, the inner world which Lévy explored as if he were its geographer reveals itself as sharing with the world which surrounds us a kind of premonitory accord that is, without doubt, a token of his genius."

LEWIS FRY RICHARDSON
(1881–1953)

Even by the standards of the present chapter, the life of L. F. Richardson is unusual, its strands failing to become integrated in any predominant direction. He was, incidentally, the uncle of Sir Ralph Richardson, the actor. See *World Who's Who in Science,* p. 1420, *Obituary Notices of Fellows of the Royal Society,* 9, 1954, 217–235—summarized in Richardson 1960a and 1960s, and a story by M. Greiser in *Datamation*, June 1980. Personal tidbits were kindly contributed by a relation of Richardson, David Edmundson.

In the words of his influential contemporary G. I. Taylor, "Richardson was a very interesting and original character who seldom thought on the same lines as did his contemporaries, and often was not understood by them." To paraphrase E. Gold, his scientific work was original, sometimes difficult to follow, sometimes illuminated by lucid unexpected illustrations. In his studies of turbulence and in the publication that led to Richardson 1960a and 1960s, he was occasionally, and not unnaturally, groping, perhaps with a little confusion. He was breaking new ground and had to find his way with the assistance of a knowledge of advanced mathematics gained as he went—not drawn from a stock obtained in his university career. In view of his inclination to explore new subjects—or even "bits of subjects"—his achievement might seem surprising if one did not realize his amazing and orderly industry.

Richardson attended Cambridge on a scholarship and earned his B.A. in physics, mathematics, chemistry, biology, and zoology, for he was uncertain as to the career he should follow. Helmholtz, who had been a physician before becoming a physicist, seemed to Richardson to have partaken of the feast of life in reverse order.

For some reason he had quarreled with Cambridge, and when he wanted a Doctor's degree many years later, he refused to proceed to his M.A., which cost 10 pounds. Instead, he matriculated at London University, where he was then lecturing, sat with his own pupils, and obtained his doctorate at age 47, in mathematical psychology.

He had begun his career at the Meteorological Office, but being an austere Quaker and a conscientious objector during the War of 1914–1918, he resigned when the Meteorological Office joined the new Air Ministry after the War.

Weather prediction by numerical process is the topic of Richardson 1922–1965, clearly the work of a practical visionary. It was reprinted after 33 years as a classic, but for 20 years it was viewed as disreputable. It turns out that, while approximating the differential equations of the evolution of the atmosphere with equations of finite differences, Richardson had selected unsuitable values of the elementary steps of space and time. Since the need for care in the selection of these steps had not yet been perceived, his mistake was hardly avoidable.

Nevertheless, this work soon won him election to the Royal Society. And five lines from Richardson 1922, p. 66, are widely quoted:

> *Big whorls have little whorls,*
> *Which feed on their velocity;*
> *And little whorls have lesser whorls,*
> *And so on to viscosity*
> *(in the molecular sense),*

In fact, these lines reach the highest level of fame by being often quoted anonymously. Seeing them, a scholar of English literature pointed out to me their kinship to some classics. It is clear that Richardson parodied the following verse from Jonathan Swift 1733, lines 337–340:

> *So, Nat'ralists observe, a Flea*
> *Hath smaller Fleas that on him prey,*
> *And these have smaller Fleas to bit 'em,*
> *And so proceed ad infinitum.*

But Richardson avoided the alternative statement in deMorgan 1872, p. 377:

> *Great fleas have little fleas*
>   *upon their backs to bite 'em*
> *And little fleas have lesser fleas,*
>   *and so ad infinitum,*
> *And the great fleas themselves,*
>   *in turn, have greater fleas to go on,*
> *While these again have greater still,*
>   *and greater still, and so on.*

The difference between these variants is not as slight as it may seem. In fact, it gives one a nice feeling to believe that Richardson was careful in matching his literary models to his notions about physics. Indeed he thought that turbulence only involves a "direct" cascade of energy from large to small eddies—hence Swift. Had he also believed in an "inverse" cascade of energy from small to large eddies—as some believe today—one hopes he might have parodied De Morgan!

In a somewhat analogous light vein, the second section of Richardson 1926 is titled "Does the Wind Possess a Velocity?" and begins as follows: "The question, at first sight foolish, improves on acquaintance." He then goes on to show how wind diffusion may be studied with no need to mention its velocity. In order to give an idea of the degree of irregularity of the motion of air, a fleeting mention is made of the Weierstrass function (which is continuous but has nowhere a derivative; it is mentioned in Chapter 2 and studied in Chapters 39 and 41). Unfortunately, the matter is dropped immediately. What a pity he failed to notice that the Weierstrass function is scaling. Also, as pointed out by G. I. Taylor, Richardson defined the law of turbulent mutual dispersion of particles, but missed the Kolmogorov spectrum by a hair's breadth. However, each fresh glance at his papers seems to show some angle that had passed unnoticed.

Richardson was also a careful and thrifty experimenter. His earliest experiments consisted in measuring wind velocity within clouds by shooting into them diverse steel marbles ranging from the size of a pea to that of a cherry. A late experiment in turbulent diffusion (Richardson & Stommel 1948) required a large number of buoys, which had to be highly visible, hence preferably whitish in color, while remaining almost totally immersed so as not to catch the wind. His solution was to buy a large sack of parsnips, which were thrown from one bridge on the Cape Cod Canal while he made his observations from another bridge downstream.

He spent many years as teacher or administrator off the beaten path. Then an inheritance enabled him to retire early to devote himself fully to the study of the psychology of armed conflicts between states, which he had been pursuing on the side since 1919. Two books appeared after his death, Richardson 1960a,s (Newman 1956, pp. 1238–1263 reprints the author's summaries). Posthumous articles include Richardson 1961, the investigation of the length of coastlines that is described in Chapter 5 and that had such an influence on the genesis of the present Essay.

GEORGE KINGSLEY ZIPF
(1902–1950)

Zipf, an American scholar, started as a philologist but came to describe himself as a statistical human ecologist. He was for twenty years a Lecturer at Harvard, and died just after having published, apparently at his own expense, *Human Behavior and the Principle of Least Effort,* (Zipf 1949-1965).

This is one of those books (Fournier 1907 is another) in which flashes of genius, projected in many directions, are nearly overwhelmed by a gangue of wild notions and extravagance. On the one hand, it deals with the shape of sexual organs and justifies the Anschluss of Austria into Germany because it improved the fit of a mathematical formula. On the other hand, it is filled with figures and tables that hammer away ceaselessly at the empirical law that, in social science statistics, the best combination of mathematical convenience and empirical fit is often given by a scaling probability distribution. Some examples are studied in Chapter 38.

Natural scientists recognize in "Zipf's laws" the counterparts of the scaling laws which physics and astronomy accept with no extraordinary emotion—when evidence points out their validity. Therefore physicists would find it hard to imagine the fierceness of the opposition when Zipf—and Pareto before him—followed the same procedure, with the same outcome, in the social sciences. The most diverse attempts continue to be made, to discredit in advance all evidence based on the use of doubly logarithmic graphs. But I think this method would have remained uncontroversial, were it not for the nature of the conclusion to which it leads. Unfortunately, a straight doubly logarithmic graph indicates a distribution that flies in the face of the Gaussian dogma, which long ruled uncontested. The failure of applied statisticians and social scientists to heed Zipf helps account for the striking backwardness of their fields.

Zipf brought encyclopedic fervor to collecting examples of hyperbolic laws in social sciences, and unyielding stamina to defending his findings and analogous findings by others. However, the present Essay makes it obvious that his basic belief was without merit. It is *not* true that frequency distributions are always hyperbolic in the social sciences, and always Gaussian in the natural sciences. An even more serious failing was that Zipf tied his findings together with empty verbal argument, and came nowhere close to integrating them into a body of thought.

At a critical point in my life (Chapter 42), I read a wise review of *Human Behavior* by the mathematician J. L. Walsh. By only mentioning what was good, this review influenced greatly my early scientific work, and its indirect influence continues. Therefore, I owe a great deal to Zipf through Walsh.

Otherwise Zipf's influence is likely to remain marginal. One sees in him, in the clearest fashion—even in caricature—the extraordinary difficulties that surround any interdisciplinary approach. ■

My fondest hope,
        Dear reader,
    is that you will ask
many further questions
        to my answers.

This drawing,
    dated January 30, 1964,
        is reproduced
    by kind permission
of Monsieur Jean Effel.

—Veuillez, messieurs les journalistes, fournir vos questions à mes réponses.

41 ◻ Historical Sketches

Gauss's dictum, "when a building is completed no one should be able to see any trace of the scaffolding," is often used by mathematicians as an excuse for neglecting the motivation behind their own work and the history of their field. Fortunately, the opposite sentiment is gaining strength, and numerous asides in this Essay show to which side go my own sympathies. However, I am left with several longer stories with which to educate and entertain the reader. They include odds and ends gathered in library forays prompted by my current passion for Leibniz and Poincaré.

ARISTOTLE AND LEIBNIZ, GREAT CHAIN OF BEING, CHIMERAS AND FRACTALS

A reference to Aristotle and Leibniz has long ceased to be required in serious books. But the present entry is not a joke, however unexpected it may be even to its author. Several basic ideas of fractals might be viewed as mathematical and scientific implementations of loose but potent notions that date back to Aristotle and Leibniz, permeate our culture, and

affect even those who think they are not subject to philosophical influences.

My first clue came from a remark in Bourbaki 1960: the idea of fractional integro-differentiation, described in Chapter 27, had occurred to Leibniz, as soon as he had developed his version of calculus and invented the notations $d^k F/dx^k$ and $(d/dx)^k F$. In free translation of Leibniz's letter to de l'Hospital dated September 30, 1695 (Leibniz 1849–, II, XXIV, 197ff.): "John Bernoulli seems to have told you of my having mentioned to him a marvelous analogy which makes it possible to say in a way that successive differentials are in geometric progression. One can ask what would be a differential having as its exponent a fraction. You see that the result can be expressed by an infinite series. Although this seems removed from Geometry, which does not yet know of such fractional exponents, it appears that one day these paradoxes will yield useful consequences, since there is hardly a paradox without utility. Thoughts that mattered little in themselves may give occasion to more beautiful ones." Further elaborations were communicated to John Bernoulli on

December 28, 1695 (Leibniz 1849–, III.1, 226ff.).

While Leibniz devoted much thought to such matters, they never enter in Newton's thoughts about calculus, and there was good reason for this difference of approach. Indeed (see *The Great Chain of Being*, Lovejoy 1936), Leibniz believed deeply in what he called the "principle of continuity" or of "plenitude." Aristotle had already believed that the gap between any two living species can be bridged continuously by other species. He was therefore fascinated by "in-between" animals, which he denoted by a special term (of which I heard from G. E. R. Lloyd), επαμφοτεριζειν. See also this chapter's entry on NATURA NONFACIT SALTUS.

This principle of continuity reflected (or justified?) the belief in "missing links" of all sorts, including chimeras in the sense this term had in Greek mythology: beasts having a lion's head and a goat's body—and also having the tail of a dragon and spitting fire from their mouths!. (Should chimeras be mentioned in this book? If I come to read that it is a fractally written account of chimeric notions, I shall know whom to blame.)

Of course, modern atomic theory's search for distant origins has tended to draw greater attention to the opposite tradition of Greek philosophy, that of Democritus. And the tension between these two contrary forces continues to play a central creative role in our thought. Note that the Cantor dust may be seen as defusing an ancient paradox: it is divisible without end but is not continuous. Inci-

dentally, in the ancient Hebrew cultural tradition, chimeras were either ignored or rejected, as demonstrated from a surprising angle in Soler 1973.

The belief in biological chimeras became discredited, but this does not matter. In mathematics, Aristotle's idea finds an application in the interpolation of the sequence of integers by ratios of integers, then by limits of ratios of integers. In such a tradition, every phenomenon defined by a sequence of integers is a candidate for interpolation. Thus, Leibniz's haste to talk about fractional differentials was spurred on by an idea that sat at the very core of his thought (and underlay his packing of the circle, Chapter 18).

Now what about Cantor, Peano, Koch, and Hausdorff? In creating their monster sets, were not the first three genuinely engaged in actual implementation of mathematical chimeras? And should we not view the Hausdorff dimension as a scale allowing chimeras to be ordered? Today's mathematicians do not read Leibniz or Kant, but the scholars of 1900 did. Thus, having read the verse by Jonathan Swift in the preceding chapter's entry on RICHARDSON, we can fantasize Helge von Koch constructing his snowflake curve to the following tune. He defines a "big flea" as being the original triangle drawn in Plate 36. Then he centers a smaller triangular "flea" on the middle of each back of the big flea; then smaller triangular fleas wherever possible on the backs of old or new fleas. And thus he too proceeds ad infinitum. This fantasy is not based on evidence, but it should make my point. Koch could not fail to be nourished by the cultural currents that descend from Leibniz. And Swift's parody reflects some popular expositions of Leibniz's thought.

Next, we turn from mathematicians concerned with Art for Art's sake (and convinced, in Cantor's words, that "the essence of mathematics is freedom") to men who celebrate Nature by trying to imitate it.

They would not dream of chimeras, would they? In fact, many among them do. Chapter 10 refers to practical students of turbulence, thwarted in efforts to decide whether the process they study concentrates on "peas, spaghetti, or lettuce," irritated that different ways of asking the question should seem to yield different answers, and ending with a call for "in-between" shapes whose nature partakes of both lines and surfaces. Chapter 35 mentions a different band of seekers of the "in-between," found among students of galactic clustering who have to describe the texture of certain shapes that "look streamlike" even though they are clearly composed of isolated points. Would it be artificial to proclaim to these sober seekers, unaware of being concerned with ancient scribblings and old Greek nightmares, that they follow the well-worn path toward chimeras?

Yet another clue pointing to common roots between the Cantorians and the Richardsonians is found in the study of stellar and galactic clustering. Here is a sensitive topic for those who search for conceptual roots, because professional astronomers are loath to acknowledge any influence from the stargazing riffraff, "however attractive their conceptions may be in their grandeur" (to quote Simon Newcomb). This disinclination may explain why it is customary to credit the first fully described hierarchical model to Charlier, an astronomer, instead of Fournier d'Albe (discussed in Chapter 40) or Immanuel Kant.

Kant's comments on the lack of homogeneity in the distribution of matter are eloquent and clear-cut. Witness these highlights (which should encourage one to savor Kant 1755-1969 or Munitz 1957): "That part of my theory which gives it its greatest charm...consists of the following ideas...It is...natural...to regard [the nebulous] stars as being...systems of many stars...[They] are just universes and, so to speak, Milky Ways... It might further be conjectured that these higher universes are not without relation to one another, and that by this mutual relationship they constitute again a still more immense system...which perhaps, like the former, is yet again but one member in a new combination of numbers! We see the first members of a progressive relationship of worlds and systems; and the first part of this infinite progression enables us already to recognize what must be conjectured of the whole. There is no end but an abyss...without bound."

Kant brings us back to Aristotle and Leibniz, and the above case stories may explain why Cantor and Richardson so often sound alike, at least to me. To heighten the drama, allow me to paraphrase, from Verdi's opera *Il Trovatore*, some of the last words of Azucena to Luna *Egl'era tuo fratello*.

These great traditions' leaders grew scorning and fighting each other, but in their intellectual roots *they were brothers.*

Of course, history cannot explain the mystery of the unreasonable effectiveness of mathematics, Chapter 1. The mystery merely moves on and changes character. How can it be that the mixture of information, observation, and search for introspectively satisfying structures that characterize our ancient scribblers should repeatedly yield themes so potent that, long after many details have been found to contradict better observation and the themes themselves have seemingly faded away, they continue to inspire effective developments in both physics and mathematics?

BROWNIAN MOTION AND EINSTEIN

Natural Brownian motion is "the chief of those fundamental phenomena which the biologists have contributed or helped to contribute to the science of physics" (Thompson 1917). A biologist discovered this phenomenon (well before 1800), and another biologist, Robert Brown, found in 1828 that this phenomenon is not biological but physical in nature. This second step was vital, hence the adjective *Brownian* is not as undeserved as some critics make it appear.

Brown had other claims to fame, and Brownian motion is not mentioned in his biography in the *Encyclopaedia Britannicas* ninth edition, 1878. In the eleventh to thirteenth editions, 1910 to 1926, it receives a few words in passing. It is of course treated fully in the editions published since Perrin's 1926 Nobel Prize. The slow acceptance of the physical nature of Brownian motion is recounted in Brush 1968 and Nye 1972. Outlines are given in recent *Britannica's*, Perrin 1909 and 1913, Thompson 1917, and Nelson 1967.

The developments started by Brown culminated in 1905–1909 with theories mostly due to Einstein, and with experiments mostly due to Perrin. One could think that Einstein set out to explain old nineteenth century observations, but in fact he did not.

Einstein 1905 (reprinted in Einstein 1926) begins with the words: "In this paper, it will be shown that according to the molecular-kinetic theory of heat, bodies of microscopically visible size suspended in a liquid will perform movements of such magnitude that they can be easily observed in a microscope, on account of the molecular motions of heat. It is possible that the movements to be discussed here are identical with the so-called ' Brownian molecular motion'; however, the information available to me regarding the latter is so lacking in precision, that I can form no judgment in the matter."

Then we read in Einstein 1906 (reprinted in Einstein 1926): "Soon after the appearance of [Einstein 1905, I was] informed [that] physicists—in the first instance, Gouÿ (of Lyons)—had been convinced by direct observation that the so-called Brownian motion is caused by the irregular thermal movements of the molecules of the liquid. Not only the qualitative properties of the Brownian motion but also the order of magnitude of the paths described by the particles correspond completely with the results of the theory. I will not attempt here a comparison [with] the slender experimental material at my disposal."

Much later, in a January 6, 1948, letter to Michele Besso, Einstein reminisces that he had "deduced [Brownian motion] from mechanics, without knowing that anyone had already observed anything of the kind."

"CANTOR" DUSTS AND HENRY SMITH

A wit observed that crediting Brownian motion to Roger Brown violated a basic law of eponymy, because fame is incompatible with a plain name like Brown. This may be why I had been writing on Cantor dusts for twenty years before chancing on the fact that they should be credited to a Henry Smith.

H. J. S. Smith (1826-1883) was long the Savilian professor of geometry at Oxford, and his *Scientific Papers* were published and reprinted, Smith 1894. In a bizarre episode managed by Hermite, he starred posthumously by sharing a prize with Hermann Minkowski. He also became an early critic of Riemann's theory of integration. A (different) wit observed that, while the integration theories of Archimedes, Cauchy and Lebesgue are God-given, Riemann's theory is unmistakably an awkward human invention. Indeed, Smith 1875 (Chapter XXV of Smith 1894) showed it fails to apply to functions whose discontinuities fall on certain sets. Which counterexam-

ples did he invoke? He invoked the Cantor dust used in Chapter 8, and the dust of positive measure used in Chapter 15.

Vito Volterra (1860-1940) reconstituted Smith's second counterexample in 1881.

Of course, Smith and Volterra did not do much with their examples, but neither did Cantor! All this being described in Hawkins 1970, why is Smith never (to my knowledge) mentioned as claimant for the honor of inventing the "Cantor" dusts?

DIMENSION

EUCLID. (circa 300 B.C.) Dimension underlies the definitions that begin Euclid's Book I on plane geometry:
1. A point is that which has no part.
2. A line is breadthless length.
3. The extremities of a line are points...
5. A surface is that which has length and breadth only.
6. The extremities of a surface are lines.

The theme is developed in the definitions that begin his short Book XI on spatial geometry:
1. A solid is that which has length, breadth, and depth.
2. An extremity of a solid is a surface.

(Heath 1908 comments on this topic.)

These ideas' roots are murky indeed. Guthrie (1971-I) sees traces of the notion of dimension in Pythagoras (582–507 B.C.), but van der Waerden thinks that these traces must be discounted. On the other hand, Plato (427–347 B.C.) comments to Socrates, in Book

VII of *The Republic*, that "after plane surfaces...the right way is next in order after the
second dimension to take the third..., the dimension of cubes and of everything that has
depth." It would be good to know more about
other studies of dimension before Euclid.

RIEMANN. The lack of any study of the
concept of dimension was noted by Riemann
in his 1854 dissertation, "On the Hypotheses
which Form the Foundations of Geometry."

CHARLES HERMITE. Hermite's reputation of
being a mathematical arch-conservative (as
documented by his letter to Stieltjes quoted in
Chapter 6) is confirmed by his letters to
Mittag-Leffler (Dugac 1976c).

April 13, 1883: "To read Cantor's writings
seems a veritable torture...and no one among
us is tempted to follow.... The mapping between a line and a surface leaves us absolutely
indifferent and we think that this observation,
as long as one will not have deduced something from it, results from considerations of
such arbitrariness that the author would have
been better inspired to wait...[But Cantor
may] find readers who will study him with
interest and a pleasure, which we do not."

May 5, 1883: "The translation [of a paper
by Cantor] was edited with utmost care by
Poincaré...[His] view is that almost all French
readers will be alien to investigations which
are at the same time philosophical and mathematical, and in which there is too much arbitrariness, and I think this view is correct."

POINCARÉ. An eloquent and ultimately very
fruitful elaboration of Euclid's views was giv-
en by Poincaré in 1903 (Poincaré 1905, Chap-

ter III, Section 3) and 1912 (Poincaré 1913,
Part 9). Here is a free translation:

"When we say that space has the dimension three, what do we mean? If to divide a
continuum C it suffices to consider as cuts a
certain number of distinguishable elements,
we say that this continuum is of *dimension
one*.... If, on the contrary,...to divide a continuum it suffices to use cuts which form one or
several continua of dimension one, we say that
C is a continuum of *dimension two*. If cuts
which form one or several continua of at most
dimension two suffice, we say that C is a continuum of *dimension three*; and so on.

"To justify this definition it is necessary to
check how geometers introduce the notion of
dimension at the beginning of their works.
Now, what do we see? Usually they begin by
defining surfaces as the boundaries of solids
or pieces of space, curves as the boundaries of
surfaces, points as the boundaries of curves,
and they state that the same procedure cannot
be carried further.

"This is just the idea given above: to divide
space, cuts that are called surfaces are necessary; to divide surfaces, cuts that are called
curves are necessary; and a point cannot be
divided, not being a continuum. Since curves
can be divided by cuts which are not continua,
they are continua of dimension one; since surfaces can be divided by continuous cuts of
dimension one, they are continua of dimension
two; and finally space can be divided by continuous cuts of two dimensions, it is a continuum of dimension three."

◁ The preceding words are inapplicable

to fractal dimension. For the interiors of the various islands in this Essay, D and D_T coincide and both equal two, but the coastlines are an entirely different matter: they are topologically of dimension 1, but fractally of dimension above 1. ▶

BROUWER TO MENGER. Now to a free quote from Hurewicz & Wallman 1941: "In 1913 Brouwer constructed on Poincaré's intuitive foundation a precise and topologically invariant definition of dimension, which for a very wide class of spaces is equivalent to the one we use today. Brouwer's paper remained unnoticed for several years. Then in 1922, independently of Brouwer and of each other, Menger and Urysohn recreated Brouwer's concept, with important improvements.

"Before then mathematicians used the term dimension in a vague sense. A configuration was called E-dimensional if the least number of real parameters needed to describe its points, in some unspecified way, was E. The dangers and inconsistencies in this approach were brought into clear view by two celebrated discoveries in the last part of the 19th century: Cantor's one-to-one correspondence between the points of a line and the points of a plane, and Peano's continuous mapping of an interval on the whole of a square. The first exploded the feeling that a plane is richer in points than a line, and showed that dimension can be changed by a one-to-one transformation. The second contradicted the belief that dimension can be defined as the least number of continuous real parameters required to describe a space, and

showed that dimension can be raised by a one-valued continuous transformation.

"An extremely important question was left open: Is it possible to establish a correspondence between Euclidean space of dimensions E and E_0 combining the features of both Cantor's and Peano's constructions, that is, a correspondence which is both one-to-one *and* continuous? The question is crucial since the existence of a transformation of the stated type between Euclidean E-space and Euclidean E_0-space would signify that dimension (in the natural sense that Euclidean E-space has dimension E) has no topological meaning whatsoever! The class of topological transformations would in consequence be much too wide to be of any real geometric use.

"The first proof that Euclidean E-space and Euclidean E_0-space are not homeomorphic unless E equals E_0 was given by Brouwer in 1911 [Brouwer 1975– **2**, pp. 430–434; the special case $E \leq 3$ and $E_0 > E$ had previously been settled in 1906 by J. Lüroth.] However, this proof did not explicitly reveal any simple topological property of Euclidean E-space distinguishing it from Euclidean E_0-space and responsible for the nonexistence of a homeomorphism between the two. More penetrating, therefore, was Brouwer's procedure in 1913 when he introduced an integer-valued function of a space which was topologically invariant by its very definition. In Euclidean space, it is precisely E (and therefore deserves its name).

"Meanwhile Lebesgue had approached in another way the proof that the dimension of a Euclidean space is topologically invariant. He had observed in 1911 [Lebesgue 1972–, 4, 169–210] that a square can be covered by arbitrarily small 'bricks' in such a way that no point of the square is contained in more than three of these bricks; but that if the bricks are sufficiently small at least three have a point in common. In a similar way a cube in Euclidean E-space can be decomposed into arbitrarily small bricks so that not more than $E+1$ of these bricks meet. Lebesgue conjectured that this number $E+1$ could not be reduced further; that is, for any decomposition in sufficiently small bricks there must be a point common to at least $E+1$ of the bricks. [The proof was given by Brouwer in 1913.] Lebesgue's theorem also displays a topological property of Euclidean E-space distinguishing it from Euclidean E_0-space and therefore it also implies the topological invariance of the dimension of Euclidean spaces."

Concerning the relative contributions of Poincaré, Brouwer, Lebesgue, Urysohn, and Menger, see the notes by H. Freudenthal in Brouwer 1975–, **2**, Chapter 6, and a response in Menger 1979, Chapter 21.

FRACTIONAL DIMENSION AND DELBOEUF. The story of fractal dimension is much simpler: it emerges near fully armed from the work of Hausdorff. But a bit of mystery is present anyhow. Indeed, Russell 1897, p. 162 ignores the raging controversies aroused by Cantor and Peano, but includes the following footnote: "Delboeuf, it is true, speaks of Geometries with m/n dimensions, but gives no reference (Rev. Phil. T. xxxxvi, p. 450)." Delboeuf turns out to deserve attention (see the

entry on SCALING IN LEIBNIZ AND LAPLACE); but my search (done with the assistance of F. Verbruggen) through his works uncovers no further lead about fractional dimension.

BOULIGAND. The Cantor-Minkowski-Bouligand definition of dimension (Chapters 5 and 39) is much less satisfactory than the Hausdorff Besicovitch definition, but I would like to include here a word in praise of Georges Bouligand (1889–1979). His many books are not read much today, even in Paris, but they were prominent when I was a student and was examined by him. Skimming through his works, I am reminded that they initiated me to "modern" mathematics. I wonder whether other presentations, less soft and humane though perhaps pedagogically more durable, would have provided equal intuitive understanding, to be filed away for use when the need arose. I think not. Had Bouligand lived to witness the present conquests of the geometry he loved so gently, I hope that he would view them as personally fulfilling.

NATURA NON FACIT SALTUS AND "THE TRUE STORY OF THEUTOBOCUS"

Natura non facit saltus is the best known statement of the "principle of continuity," which is discussed in this chapter's first entry, and was viewed by Leibniz as being "one of [his] best and best verified." And it is the tenuous distant precursor of the "in-between" geometric shapes: fractals. However, Bartlett 1968 credits this statement to Linné. Surprised by a credit that seemed unfair, I investigated and unearthed a few facts and a story.

True, the celebrated eighteenth century botanist and taxonomist Linné did write this phrase, but only in passing, not as a weighty new pronouncement but as conventional wisdom. He was translating *La nature ne fait jamais de sauts,* due to Leibniz. The latter also penned innumerable variants, including: *Nulla mutatio fiat per saltum, Nullam transitionem fieri per saltum, Tout va par degrés dans la nature et rien par saut.* But Linné's exact Latin words may not be in Leibniz.

Secondly, funny and intriguing, Linné's exact Latin had been anticipated well before Leibniz, in 1613, in the phrase, *Natura in suis operationibus non facit saltum.* (The singular *saltum* instead of the plural *saltus* is preferred by the surly minority for whom *zero* is singular.) Who wrote this phrase? Stevenson 1956, p. 1382, No. 18, credits Jacques Tissot. Who was Tissot? The fact that no one seemed to know gave me an excuse to crash the Bibliothèque Nationale in Paris.

The phrase is found in a fifteen page pamphlet with a very long title that begins thus: *True Story of the Life, Death, and Bones of the Giant Theutobocus, King..., who was Defeated in 105 (*B.C.*) by Marius the Roman Consul and Buried...near Romans.* The account follows, in French intermixed with Latin, of the discovery near Grenoble of bones of gigantic size, and of reasons for attributing them to said King Theutobocus, a human.

There is a reprint of the *True Story* in *Variétés historiques et littéraires, recueil de pièces volantes rares et curieuses,* annotées par M. Edouard Fournier, Tome IX, 1859, pp. 241–257. My curiosity was rewarded. In an extremely long footnote, Fournier describes the following durable imposture. On January 11, 1613, workers digging under 17 or 18 feet of sand unearth a number of very large bones, and rumors circulate that the pit was the tomb of a giant, and was marked by a medal of Marius and a stone bearing the name of Theutobocus. The bones are "authenticated" by two local worthies, featured in newspapers, and shown to King Louis XIII. Controversy ensues concerning their origin, then peters out, to resume only at a time when other old bones were being credited to vanished species. Paleontologists enter the discussion, and identify "King Theutobocus" as a mastodon.

The footnote also says that *no* Jacques Tissot was in fact involved, the "True Story" having been published under a pseudonym by the two worthies mentioned above...as the prospectus for a proposed circus attraction.

But the *Natura non...* remains mysterious. Its being first uttered by small-town charlatans pretending to quote Aristotle would be anticlimactic. More likely, they were merely repeating a standard phrase of their time, and the question of origins is not yet closed.

POINCARÉ AND FRACTAL ATTRACTORS

Contrary to the other entries in this chapter, the present one is devoted to findings that were not merely amusing but had an immediate and durable effect upon my work. Certain texts by Henri Poincaré (1854-1912) came to my attention when the 1977 *Fractals* was in proof, and led to new lines of research sketched in Chapters 18 to 20, and scheduled to be fully presented elsewhere. Let me answer some questions inevitably raised by these and related works of Poincaré.

Yes and No: He definitely was the first student of fractal ("strange") attractors. But nothing I know of his work makes him even a distant precursor of the fractal geometry of the visible facets of Nature.

Yes: The fact had been forgotten, but within a year of Cantor 1883 sets close to the triadic dust and the Weierstrass function arose in orthodox mathematics, well before the creation of the revolutionary theories of sets and of functions of a real variable.

No: Those applications did not go unnoticed in their time. The first was in the theory of automorphic functions (Chapter 18), which made Poincaré and Felix Klein famous. Those applications were pursued by Paul Painlevé (1863–1933), a scholar influential well beyond the realm of pure mathematics. He was fascinated with engineering (he was Wilbur Wright's first passenger after Orville Wright's accident) and eventually entered politics, rising to the post of Prime Minister of France. Incidentally, finding that Perrin had been a close friend of Painlevé, the "daydream" described in Chapter 2 seems less isolated.

Yes: Cantor and Poincaré ended on opposite sides of various intellectual battles, with Cantor, as Peano, the victim of Poincaré's

sarcasm, such as the famous comment that "Cantorism [promises] the joy of a doctor called to follow a fine pathological case." See also the subentry HERMITE. It is useful therefore to know that, when need arose, Poincaré recognized that the classic monsters could enter, not into descriptions of visible Nature, but into abstract mathematical physics. I translate freely from *New Methods in Celestial Mechanics*, Poincaré 1892-III, pp. 389–390.

"Let us try to visualize the pattern formed by the two curves [C' and C''] which correspond to a doubly asymptotic solution [to the three-body problem]. Their intersection points form a sort of infinitely tight...grid. Each curve never intersects itself, but must fold upon itself in very complex fashion so as to intersect infinitely often each apex of the grid.

"One must be struck by the complexity of this shape, which I do not even attempt to illustrate. Nothing can give us a better idea of the complication of the three-body problem, and in general of all problems of dynamics for which there is no uniform integral...

"Diverse hypotheses come to mind:

"1) [The set S' (or S'') defined as C' (or C'') plus the limit points of this curve] fills a half-plane. If so, the solar system is unstable.

"2) [S' or S''] is of [positive and] finite area, and occupies a bounded region of the plane, with possible 'gaps'...

"3) Finally, [S' or S''] is of vanishing area. It is the analog of a [Cantor dust]."

To bolster the impression left by these undeservedly neglected comments, here are free translations of excerpts from Hadamard 1912, Painlevé 1895, and Denjoy 1964, 1975.

First Hadamard: "Poincaré was a precursor of set theory, in the sense that he applied it even before it was born, in one of his most striking and most justly celebrated investigations. Indeed he showed that the singularities of the automorphic functions form either a whole circle or a Cantor dust. This last category was of a kind which his predecessors' imagination could not even conceive. The set in question is one of the most important achievements of set theory, but Bendixson and Cantor himself did not discover it until later.

"Examples of curves without tangent are indeed classical since Riemann and Weierstrass. Anyone can grasp, however, that deep differences exist between, on the one hand, a fact established under circumstances arranged for the enjoyment of the mind, with no other aim and no interest other than to show its possibility, an exhibit in a gallery of monsters, and on the other hand, the same fact as encountered in a theory that is rooted in the most usual and the most essential problems of analysis."

Now to Painlevé: "I must insist on the relations that exist between function theory and Cantor dusts. The latter kind of research was so new in spirit that a mathematical periodical had to be bold to publish it. Many readers viewed it as philosophical rather than scientific. However, the progress of mathematics soon invalidated this judgment. In the year 1883 (which will remain doubly memorable in the history of mathematics in this cen-

tury), *Acta Mathematica* alternated between Poincaré's papers on Fuchsian and Kleinian functions and Cantor's papers."

Cantor's papers, found on pp. 305–414 of Vol. 2 of the *Acta* (the Cantor set on p. 407), were French translations that Mittag-Leffler, the editor of *Acta*, sponsored to help Cantor fight for recognition. Some (see the subentry HERMITE on p. 410) were edited by Poincaré. However, Poincaré's results had already been sketched in *Comptes Rendus* before Cantor's work appeared in German. Poincaré adopted one of Cantor's innovations so promptly that in his first *Acta* paper he denoted *sets* by the German *Mengen*, without taking time to seek a French equivalent.

Next to Denjoy 1964: "Some scientists view certain truths as being in good taste, well-educated, and properly brought up, while to others the gentleman's door must forever remain closed. I think mostly of set theory, which is a whole new universe, incomparably vaster and less artificial, simpler and more logical, apter to model the physical universe; in a word, truer than the old universe. The Cantor dust shares many properties of continuous matter, and seems to correspond to a very deep reality."

In Denjoy 1975, p. 23, we read the following: "I think it obvious that discontinuous models account in a much more satisfactory manner and more successfully than the present ones for a host of natural phenomena. Therefore, the laws of the discontinuous being much less well elucidated than those of the continuous, they should be investigated broadly and in depth. Insuring that the degrees of knowledge of the two orders are comparable will enable the physicist to use one or the other approach according to need."

Unfortunately, Denjoy could not buttress this "daydream" by any specific development beyond the broad hints by Poincaré and Painlevé. An exception involves Denjoy's 1932 paper on differential equations on the torus. Answering a question raised by Poincaré, he shows that the intersection between a solution and a meridian could be the whole meridian or any prescribed Cantor dust. The former behavior, but not the latter, agrees with the physicist's notion of ergodic behavior. An analogous example had been given by Bohl in 1916.

Jacques Hadamard (1865–1963) was a famous mathematician and mathematical physicist, and Arnaud Denjoy (1884–1974) a prominent very pure mathematician, but one to whom no physicist would think of listening. In any event, their remarks found no echo in their time. Both occur in eulogies for Poincaré and Painlevé, and revive ideas the originators had never refreshed by repetition.

POINCARÉ
AND THE GIBBS DISTRIBUTION

The current Poincaré revival may serve as an excuse for referring here to a technical tidbit unrelated to the rest of this Essay.

It concerns what is known to physicists as the Gibbs canonical distribution and to statis-ticians as distribution of exponential type. Poincaré 1890 seeks the probability distributions such that the maximum likelihood estimate of a parameter p, based on the M sample values $x_1, \ldots, x_m, \ldots, x_M$, is of the form $G[\Sigma_{m=1}^{M} F(x_m)/M]$. In other words, they are such that the scale of x and p can be changed by the functions $F(x)$ and $G^{-1}(p)$, so that the maximum likelihood estimate of p is the sample average of the x. This is of course the case if p is the expectation of a Gaussian variable, but Poincaré gives a more general solution, now called Gibbs distribution.

This fact was rediscovered independently by Szilard in 1925. Then, around 1935, Koopman, Pitman, and Darmois asked the same question concerning the most general estimation procedure, without being restricted to maximum likelihood. This property of the Gibbs distribution, called *sufficiency* by statisticians, plays a central role in the Szilard-Mandelbrot axiomatic presentation of statistical thermodynamics, Mandelbrot 1962t, 1964t. In this approach, the arbitrariness that is intrinsic to statistical inference is present in the definition of a closed system's temperature, but is absent from the derivation of the canonical distribution. (A later axiomatic presentation based on the "maximum information precept" grounds the canonical distribution itself in statistical inference, which I think misrepresents its significance.)

SCALING: OLD EMPIRICAL EVIDENCE

SCALING IN ELASTIC SILK THREADS. The earliest empirical observation that can now be reinterpreted as evidence of scaling in a physical system was made, extraordinarily enough, *a hundred and fifty years ago*. On the urging of Carl Friedrich Gauss, Wilhelm Weber set out to investigate the torsion of the silk threads used to support moving coils in electric and magnetic instruments. He found that applying a longitudinal load provokes an immediate extension which is followed by a further lengthening with time. On removal of the load, an immediate contraction equal to the initial immediate extension takes place. This is followed by a gradual further decrease of length until the original length is reached. The aftereffects of a perturbation follow a law of the form $t^{-\gamma}$: they decay hyperbolically in time, not exponentially as everyone expected then, and expects to this day.

The next work on this topic is Kohlrausch 1847, and the elastic torsion of glass fibers is further studied by William Thomson, later Lord Kelvin, in 1865, by James Clerk Maxwell in 1867, and by Ludwig Boltzmann, in a 1874 paper that Maxwell viewed as important enough to discuss in the ninth (1878) edition of *Encyclopaedia Britannica*.

These names and dates should be pondered carefully. They prove that, in order to make a problem worth studying, a show of interest by the likes of Gauss, Kelvin, Boltzmann, and Maxwell is not enough. A problem that fascinated but defeated them could fall into extreme obscurity.

SCALING IN ELECTROSTATIC LEYDEN JARS. The background, in the words of E.T. Whittaker, is as follows: "In 1745 Pieter van Musschenbrock (1692–1761), Professor at Leyden, attempted to find a method of preserving electric charges from the decay which was observed when the charged bodies were surrounded by air. With this purpose he tried the effect of surrounding a charged mass of water by an envelope of some nonconductor, for example, glass. In one of his experiments, a phial of water was suspended from a gun barrel by a wire let down a few inches into the water through the cork; and the gun barrel, suspended on silk lines, was applied so near an excited glass globe that some metallic fringes inserted into the gun barrel touched the globe in motion. Under these circumstances a friend named Cunaeus, who happened to grasp the phial with one hand, and touch the gun barrel with the other, received a violent shock; and it became evident that a method of accumulating or intensifying the electric power had been discovered. This discovery was named *Leyden phial* by Nollet."

Kohlrausch 1854 found for the speed of discharge by the Leyden jar the same result as in his work on silk threads: the charge decays hyperbolically in time. Dielectrics other than glass are investigated in detail in the Ph.D. thesis of Jacques Curie (Pierre Curie's brother and his first collaborator), who finds that in some dielectrics the decay is exponential, but in others it is hyperbolic, with varying values of the exponent γ.

SCALING: DURABLE ANCIENT PANACEAS

Innumerable explanations of the scaling decays or noises are scattered over a hundred years of the most diverse journals. All make for sad reading. Their lack of success is consistent and monotonous, since dead-ends recognized in the 1800's keep being explored again and again, in different contexts and words.

HOPKINSON'S MIXTURE PANACEA. Faced with the hyperbolic decay of the charge of a Leyden jar, Hopkinson (a student of Maxwell) advances in 1878 the "rough explanation [that] glass may be regarded as a mixture of a variety of different silicates that behave differently." It would follow that a decay function that seems to be a hyperbola, is in fact a mixture of two or more different exponentials of the form $\exp(-s/\tau_m)$, each of them characterized by a different relaxation time τ_m. However, even the early data suffices to show that two to four exponentials do not suffice, and the argument is abandoned.

But it keeps popping out wherever data are not sufficiently abundant to disprove it.

DISTRIBUTED RELAXATION TIMES PANACEA. When data cover many decades and cannot be fitted unless the mixture involves exponentials in ridiculous number, say 17 or 23, one is tempted to go all the way to a mixture of an *infinite* number of exponentials. The definition of Euler's gamma function yields

$$t^{-\gamma} = [\Gamma(\gamma)]^{-1} \int_0^\infty \tau^{-(\gamma+1)} \exp(-t/\tau) d\tau.$$

This identity shows that, if the exponential relaxation time τ has the "intensity" $\tau^{-(\gamma+1)}$, the mixture is hyperbolic. However, this argument is logically circular. A scientific explanation's output is supposed to be less obvious a priori than its input, but $t^{-\gamma}$ and $\tau^{-(\gamma+1)}$ are functionally identical.

TRANSIENT BEHAVIOR PANACEA. Upon hearing of the diverse symptoms of scaling listed in the preceding entry, a second near-universal first reaction is this: Surely, these hyperbolic functions $t^{-\gamma}$ are only transient complications that will be cut off exponentially when decays are observed long enough. The first systematic search for the cutoff is in von Schweidler 1907, who measured Leyden jars' decay, first at intervals of 100 seconds, then less frequently, for a total time of 16 million seconds (200 days, through summer and winter!). The hyperbolic decay continues on the dot. More recent experiments on electric $1/f$ noises had started by lasting a few hours, then a night, then a weekend, then a short vacation. In surprisingly many cases, the $1/f$ behavior continues on the dot.

Earlier chapters, for instance the study of galaxy clusters in Chapter 9, note that scientists can become so engrossed in the search for a cutoff as to neglect the need for describing and explaining the phenomena characteristic of the scaling range. Oddly, an overinvolvement with the cutoff can be even stronger among engineers. To take an example, discussed in Chapter 27, many hydrologists hesitate to use my model because it involves an infinite cutoff to scaling. In an engineering project, the finiteness of the cutoff is immaterial, nevertheless a finite cutoff is fervently desired by presumably practical people.

SCALING IN LEIBNIZ AND LAPLACE

To sample Leibniz's scientific works is a sobering experience. Next to the calculus, and to other thoughts that have been carried out to completion, the number and variety of premonitory thrusts is overwhelming. We saw examples in "packing," Chapter 17, and in this chapter's first entry. In addition, Leibniz started formal logic, and was the first (in a 1679 letter to Huygens) to suggest that geometry should include the branch that came to be called topology. (On a less exalted level, he pioneered Hebrew letters in mathematical notation..., in addition to Zodiac symbols!)

My Leibniz mania is further reinforced by finding that for one moment its hero attached importance to geometric scaling. In "Euclidis $\pi\rho\omega\tau\alpha$" (Leibniz 1849—II.1, pp. 183–211), which is an attempt to tighten Euclid's axioms, he states, on p. 185, "IV(2): I have diverse definitions for the straight line. The *straight line* is a curve, any part of which is similar to the whole, and it alone has this property, not only among curves but among sets." This claim can be proved today. Later Leibniz describes the more restricted self-similarity properties of the plane.

The same thought occurred independently in 1860 to Joseph Delboeuf (1831–1896), a Belgian writer whose views Russell 1897 criticizes kindly. He turns out to have been a truly unusual scientific personality, moving his amateurish enthusiasm from classics to the philosophy of geometry. However, his "similitude principle" adds little mathematically to the above Leibniz quote (which he did not know when he did his work, and to which he refers—and steered me—with a nice mixture of generosity and pride). Delboeuf also stars (dimly) on p. 412.

A different encounter with scaling may be read (by those ready to be generous toward the very rich) into Maxims 64 and 69 of Leibniz's *Monadology*, where it is stated that minute portions of the world are precisely as complex and organized as large portions.

A thought related to scaling also occurred to Laplace. In the fifth edition of his *System of the World*, published in 1842 and translated into English (but not in the fourth edition of 1813), one finds in Chapter V of Book V the following remark (Laplace 1879, Vol. VI). "One of [the] remarkable properties [of Newtonian attraction] is, that if the dimensions of all the bodies in the universe, their mutual distances and their velocities were to increase or diminish proportionately, they would describe curves entirely similar to those which they at present describe; so that the universe reduced to the smallest imaginable space would always present the same appearance to observers. The laws of nature therefore only permit us to observe relative dimensions...[The text continues in footnote] Geometers' attempts to prove Euclid's axiom about parallel lines have been hitherto unsuccessful.... The notion of...a circle does not involve anything which depends on its absolute magnitude. But if we diminish its radius, we are forced to diminish also in the same proportion its circumference, and the sides of all inscribed figures. This proportionality seems to be much more natural an axiom than that of Euclid. It is curious to observe this property in the results of universal gravitation."

WEIERSTRASS FUNCTIONS

The continuous but nowhere differentiable functions of Weierstrass had such an impact on the development of mathematics, that one is curious to know whether their story followed the pattern which Farkas Bolyai described to his son, János: "There is some truth in this, that many things have an epoch, in which they are found at the same time in several places, just as violets appear on every side in spring." One also expects to see the co-inventors rush to print.

But in the present case, events unfolded very differently. The nearly unbelievable fact is that Weierstrass never published his discovery, though he read it at the Berlin Academy on July 18, 1872. The talk's manuscript did make it to the *Collected Works,* Weierstrass 1895, but the world was informed, and the claim staked in Weierstrass's name, in DuBois Reymond 1875. Thus, 1875 is but a convenient symbolic date for the beginning of the great crisis of mathematics.

DuBois Reymond wrote that "the meta-physics of these functions seems to hide many puzzles, as far as I am concerned, and I cannot get rid of the thought that [they] will lead to the limit of our intellect." However, one gets the distinct feeling that no one was in a hurry to explore those limits. Some contemporaries who dabbled with this task for a moment (for example, Gaston Darboux) promptly turned back to extreme conservatism, but the others were hardly bolder. One is also forcibly reminded of the more famous story of Gauss hiding his discovery of non-Euclidean geometry, as he wrote to Bessel on January 27, 1829, "for fear of the uproar of the Boeotians." (But later he revealed it to János Bolyai—with disastrous consequences on the latter's mind—after this son of a friend had published his independent discovery). Finally, one thinks of the advice Mittag-Leffler was later to give to Cantor, that he should not fight editors, but withhold his more daring findings until the world is ready for them. Rarely has the avant-garde been so extraordinarily reluctant as in these various cases.

In addition to Weierstrass, three names must be mentioned here. It has long been rumored, and is documented in Neuenschwander 1978, that Riemann told his students around 1861 that $R(t) = \Sigma\, n^{-2} \cos (n^2 t)$ is a continuous and nondifferentiable function. But no precise statement and proof is known. In fact, if "nondifferentiable" meant "nowhere differentiable," any purported proof had to be flawed, since Gerver 1970 and Smith 1972 show that $R(t)$ *does* have positive and finite derivatives at certain points. Kronecker also was concerned about the Riemann function, an interest that underlines the importance the question held at the time. (Manheim 1969, T. Hawkins 1970, and Dugac 1973, 1976 add to our knowledge of this background.)

Bolzano, whose name is hyphenated with that of Weierstrass in a different and better-known context, also enters in this story. Bernhard Bolzano (1781–1848) was one of the few underground heroes of mathematics, most of whose work lay dormant until the 1920's. He discovered in 1834 a close analog of the Weierstrass function, but he failed to notice the property that makes the function of interest to us (Singh 1935, p. 8).

The third man, unknown in his lifetime as in ours, matters more in the present story than anyone but Weierstrass. Charles Cellérier (1818–1890) had taught in Geneva and published little of note, but the files opened after his death included a "revelation." An undated folder marked "Very important and I think new. Correct. Can be published as is" contained a text in his hand describing the limit case D=1 of the Weierstrass function and using it for the familiar purpose. The yellowed pages were shown to a scholar named Cailler, who added a footnote (from which the preceding comments are excerpted) and promptly published the paper as Cellérier 1890. Scattered evidence of interest ensued, especially on the part of Grace C. Young. Raoul Pictet remembered in 1916 that Cellérier had mentioned this work in class when Pictet was among his students, around 1860. But no written evidence came forth. And eventually Cellérier's claim proved to be flawed.

Thus Weierstrass remains alone and unchallenged in the claim made in his name, but we are left with truly odd events to ponder. A certain expression was actually published by Bolzano who thought it innocuous, but the two later scholars who knew better, the modest provincial with no reputation that could be tarnished and the great master who might have felt untarnishable, both chose to sit, wait, and see. "Publish or Perish" could not be farther from their minds.

Since the Weierstrass function is often used to argue for a divorce by mutual consent between mathematics and physics, it may be of interest to mention its discoverer's attitude toward the relationship between these two endeavors. His name found its way into geometric optics (through the Young-Weierstrass points of a spherical lens). Also, in his inaugural lecture for 1857 (quoted in Hilbert 1932, **3**, pp. 337-338), Weierstrass stressed that the physicist should not see in mathematics a simple auxiliary discipline, and the mathematician should not consider the physicist's questions as a simple collection of examples for his methods. "To the question of whether it is really possible to extract something useful from the abstract theories that modern [=1857] mathematics seems to favor, one could answer that it was only on the basis of pure speculation that Greek mathematicians derived the properties of conic sections, long before one could guess that they represent the planets' orbits." AMEN.

42 ¤ Epilog: The Path to Fractals

The Essays on fractals I wrote in 1975 and 1977 start without Preface and end without conclusion. The same is true of the present work, but a few things remain on my mind. Now that fractal geometry is taking ominous steps towards becoming organized, it is a good time to sketch its improbable genesis on the record. And to add a few words on its relative contributions to scientific understanding, description and explanation. As the new geometry marches on all fronts from description to explanation (either generic, as in Chapters 11 and 20, or geared to specific case studies), it is good to recall why it had long benefited from an uncommon (and unpopular) disregard for explanation through "models."

By now, the reader knows well that the probability distribution characteristic of fractals is hyperbolic, and that the study of fractals is rife with other power law relationships. By accepting the validity of scaling and exploring its geometric-physical implications with care, we find so much to occupy us, that it seems strange indeed that, as of yesterday, I felt I had this rich new land all for myself. Many populated clearings surrounded it, and many authors had peeked in once, but no one else had stayed in.

This lifetime involvement was triggered in 1951, by a casual side interest in Zipf's law (Chapters 38 and 40). This empirical regularity concerning word frequencies had come to my attention through a book review. The event seems too symbolic to be true, but the review in question had been retrieved from a "pure" mathematician's wastebasket, for light reading on the Paris subway. Zipf's law proved easy to account for, and the birth of mathematical linguistics was helped along by my work. But the study of word frequencies was a self-terminating enterprise.

However, its aftereffects linger on. Having recognized that (using present vocabulary) my work had been a case study of the usefulness of scaling assumptions, I became sensitive to analogous empirical regularities in diverse fields, beginning with economics. Though astonishingly numerous, these regularities were viewed as of little consequence to established fields. The more successful I was in accounting for them, the more they loomed as visible symptoms of a widespread phenomenon which the sciences had failed to face, and to which I could devote my energies for a while.

My way of investigating these regularities started with the usual search for generating models but gradually changed, because I kept observing instances where minor changes in seemingly insignificant assumptions of a model provoked drastic changes in its predictions. For example, many occurrences of the Gaussian distribution are customarily "explained" through the standard central limit theorem of probability, as resulting from the addition of many independent contributions. This argument's explanatory value hinged on the fact that the numerous other central limit theorems were not even known to the research scientists, and were viewed by Paul Lévy and the other pioneers as "pathological." But the study of scaling laws led me to recognize that nonstandard central limit behavior is in fact part of nature. Unfortunately, as soon as the central limit theorem argument is recognized to have more than one possible outcome, it ceases to be persuasive. An explanation hardly brings understanding, if it is more complicated than its outcome, and if equally plausible variants yield totally different predictions.

Anyhow, exploring the *consequences* of self-similarity was proving full of extraordinary surprises, helping me to understand the fabric of nature. By contrast, the muddled discussion of the *causes* of scaling had few charms. It seemed, on certain days, hardly better than Zipf's ravings on the principle of least effort (p. 403).

This mood was strengthened by a spike of renewed interest in the model of near scaling in taxonomy presented in Yule 1922. The revival's claim of providing an all-purpose explanation of every occurrence of scaling in the social sciences was based upon a technical error (as I showed), but many of my readers of that time somehow became convinced that the scaling relationships in social science have a universal and straightforward explanation, hence (!) do not deserve attention.

My existing bent towards stressing consequences before causes was reinforced as a result. It soon proved a godsend, and in particular helped the full strength of scaling methods become apparent, when (in 1961) I turned to the variation in time of commodity prices on competitive markets (Chapter 37). Economists complain about the paucity and low quality of their data, but data about prices and incomes come in a flood. However, economic theory and econometrics, which claim they can elucidate the relationships between hundreds of ill-defined variables, dare make no prediction about the structure of price records. And the common statistical techniques prove incapable of extracting any order from the data. This illustrates W. Leontief's observation: that "in no field of empirical enquiry has so massive and sophisticated a statistical machinery been used with such indifferent results." But descriptions deduced by scaling methods worked amazingly well. The scaling property incorporates the two most striking characteristics of competitive market prices: their being highly discontinuous, and their being "cyclic" but nonperiodic. This investigation may well be the only example of the use in economics of an invariance-symmetry in the style of physics.

In 1961, I extended the notion of scaling to tackle several noise phenomena. All these diverse efforts were carried out in near total isolation from physicists and mathematicians. But, during my visiting professorship at Harvard in 1962-1964, Garrett Birkhoff pointed out analogies between my approach and the theory of turbulence pioneered by Richardson and highlighted by Kolmogorov 1941. While I had heard of this theory as a student, its influence was not necessarily stronger than that of the philosophical tradition described in Chapter 40, in the entry on ARISTOTLE. In any event, all this was happening well before physicists became enamored of scaling!

Furthermore, G. W. Stewart's lectures on the intermittency of turbulence introduced me to Kolmogorov 1962. The preprint of that work and of Berger & Mandelbrot 1963 had come out within weeks of each other! While Kolmogorov tackled a more interesting problem, my tools were more powerful, and in no time I adapted them to turbulence, obtaining the substance of Chapters 10 and 11.

Finally, I became aware of 1/f noises, of Hurst 1951, 1955, of Richardson 1961, and of the issue of galactic clustering. Again, I felt in each instance that understanding was helped by a good description and the exploration of its consequences. By contrast, the early models I conceived seemed but idle decorations added to the description. They distracted from the basic geometric ideas I was in the process of formulating, and actually hindered understanding, in my opinion. I kept withholding them, even when my papers failed to be accepted for publication. Again, the explanations in Chapters 11 and 20, and *passim*, are an entirely different story, and I rejoice in them.

Thus, the pursuit of scaling kept being revitalized, and enriched by fresh tools and ideas, thanks to changes in the field of study, and led to the gradual emergence of an overall theory. In no way did this theory follow the "top-to-bottom" pattern of being first revealed and formulated and then "applied." It kept surprising everyone, including myself, by growing from a modest bottom to an increasingly (dizzyingly!) ambitious top. Early overviews were given at the International Congress of Logic and Philosophy of Science (1964), in Trumbull Lectures at Yale (1971), and at the Collège de France (1973 & 1974).

The geometric face of this theory of scaling grew in importance, and gave size to fractal geometry. Given the strong geometric flavor of the early studies of turbulence and of critical phenomena, one may have expected a theory of fractals to develop in either of these contexts. But none developed.

Instances where new concepts and techniques come into science through branches of low competitiveness are rare today, hence anomalous. Fractal geometry is a new example of such an historical anomaly.   ■

LIST OF REFERENCES

Each entry in this list includes an author's or an editor's name, and a date. A date followed by—refers to the first volume in a multivolume collection. In cases of ambiguity, the date is followed by a letter, which is most often related to the publication's title or to the name of the serial in which it appeared. This new convention is intended as a mnemonic device.

Because the serials referenced in this list belong to different disciplines, their titles are less abbreviated than is customary.

Few general references are included, and the list does not remotely attempt a balanced or complete coverage of the diverse fields touched in this work.

ABELL, G. O. 1965. Clustering of galaxies. *Annual Reviews of Astronomy and Astrophysics* **3**, 1-22.

ABBOT, L. F. & WISE, M. B. 1981. Dimension of a quantum-mechanical path. *American J. of Physics* **49**, 37-39.

ADLER, R. J. 1981. *The geometry of random fields*, New York: Wiley.

ALEXANDER, S. S. 1961. Price movements in speculative markets: or random walks. *Industrial Management Review of M.I.T.* **2**, Part 2 7-26. Reprint in *The random character of stock market prices*. Ed. P. H. Cootner, 199-218. Cambridge MA: MIT Press, 1964.

ALEXANDER, S. S. 1964. Price movements in speculative markets: No. 2. *Industrial Management Review of M.I.T.* **4**, Part 2, 25-46. Reprint in Cootner (preceding ref.) 338-372.

ALLEN, J. P., COLVIN, J. T., STINSON, D. G., FLYNN, C. P. & STAPLETON, H. J. 1981. Protein conformation from electron spin relaxation data (preprint). Champaign, Illinois.

APOSTEL, L., MANDELBROT, B. & MORF, A. 1957. *Logique, langage et théorie de l'information*. Paris: Presses Universitaires de France.

ARTHUR, D. W. G. 1954. The distribution of lunar craters. *J. of the British Astronomical Association* **64**, 127-132.

AUBRY, S. 1981. *Many defect structures, stochasticity and incommensurability*. *Les Houches 1980*. Ed. R. Balian and M. Kléman. New York: North-Holland, 1981.

AVRON, J. E. & SIMON, B. 1981. Almost periodic Hill's equation and the rings of saturn. *Physical Review Letters* **46**, 1166-1168.

AZBEL, M. YA. 1964. Energy spectrum of a conduction electron in a magnetic field. *Soviet Physics JETP* **19**, 634-645.

BACHELIER, L. 1900. *Théorie de la spéculation*. Thesis for the Doctorate in Mathematical Sciences (defended March 29, 1900). *Annales Scientifiques de l'Ecole Normale Supérieure* **III-17**, 21-86. Translation in *The random character of stock market prices*. Ed. P. H. Cootner, 17-78. Cambridge, MA: MIT Press, 1964.

BACHELIER, L. 1914. *Le jeu, la chance et le hasard*. Paris: Flammarion.

BALMINO, G., LAMBECK, K. & KAULA, W. M. 1973. A spherical harmonic analysis of the Earth's topography. *J. of Geophysical Research* **78**, 478-481.

BARBER, M. N. & NINHAM, B. W. 1970. *Random and restricted walks: theory and applications*. New York: Gordon & Breach.

BARRENBLATT, G. I. 1979. *Similarity, self-similarity, and intermediate asymptotics*. New York: Plenum.

BARTLETT, J. 1968. *Familiar quotations* (14th ed.) Boston: Little Brown.

BATCHELOR, G. K. 1953. *The theory of homogeneous turbulence.* Cambridge University Press.

BATCHELOR, G. K. & TOWNSEND, A. A. 1949. The nature of turbulent motion at high wave numbers. *Pr. of the Royal Society of London* **A 199**, 238-255.

BATCHELOR, G. K. & TOWNSEND, A. A. 1956.

Turbulent diffusion. *Surveys in Mechanics* Ed. G. K. Batchelor & R. N. Davies. Cambridge University Press.

BERGER, J. M. & MANDELBROT, B. B. 1963. A new model for the clustering of errors on telephone circuits. *IBM J. of Research and Development* **7**, 224-236.

BERMAN, S. M. 1970. Gaussian processes with stationary increments: local times and sample function properties. *Annals of Mathematical Statistics* **41**, 1260-1272.

BERRY, M. V. 1978. Catastrophe and fractal regimes in random waves & Distribution of nodes in fractal resonators. *Structural stability in physics*. Ed. W. Güttinger & H. Eikemeier, New York: Springer.

BERRY, M. V. 1979. Diffractals. *J. of Physics* **A12**, 781-797.

BERRY, M. V. & HANNAY, J. H. 1978. Topography of random surfaces. *Nature* **273**,573.

BERRY, M. V. & LEWIS, Z. V. 1980. On the Weierstrass-Mandelbrot fractal function. *Pr. of the Royal Society London* **A370**, 459-484.

BESICOVITCH, A. S. 1934. On rational approximation to real numbers. *J. of the London Mathematical Society* **9**, 126-131.

BESICOVITCH, A. S. 1935. On the sum of digits of real numbers represented in the dyadic system (On sets of fractional dimensions II). *Mathematische Annalen* **110**, 321-330.

BESICOVITCH, A. S. & TAYLOR, S. J. 1954. On the complementary interval of a linear closed set of zero Lebesgue measure. *J. of the London Mathematical Society* **29**, 449-459.

BESICOVITCH, A. S. & URSELL, H. D. 1937. Sets of fractional dimensions (V): On dimensional numbers of some continuous curves. *J. of the London Mathematical Society* **12**, 18-25.

BEYER, W. A. 1962. Hausdorff dimension of level sets of some Rademacher series. *Pacific J. of Mathematics* **12**, 35-46.

BIDAUX, R., BOCCARA, N., SARMA, G., SÈZE, L., DE GENNES, P. G. & PARODI, O. 1973. Statistical properties of focal conic textures in smectic liquid crystals. *Le J. de Physique* **34**, 661-672.

BIENAYMÉ, J. 1853. Considérations à l'appui de la découverte de Laplace sur la loi de probabilité dans la méthode des moindres carrés. *Comptes Rendus* (Paris) **37**, 309-329.

BILLINGSLEY, P. 1967. *Ergodic theory and information*. New York: Wiley.

BILLINGSLEY, P. 1968. *Convergence of probability measures*. New York: J. Wiley.

BIRKHOFF, G. 1950-1960. *Hydrodynamics* (1st and 2nd eds.). Princeton University Press.

BLUMENTHAL, L. M. & MENGER, K. 1970. *Studies in geometry*. San Francisco: W.H. Freeman.

BLUMENTHAL, R. M. & GETOOR, R. K. 1960c. A dimension theorem for sample functions of stable processes. *Illinois J. of Mathematics* **4**, 308-316.

BLUMENTHAL, R. M. & GETOOR, R. K. 1960m. Some theorems on stable processes. *Tr. of the American Mathematical Society* **95**, 263-273.

BLUMENTHAL, R. M. & GETOOR, R. K. 1962. The dimension of the set of zeros and the graph of a symmetric stable process. *Illinois J. of Mathematics* **6**, 370-375.

BOCHNER, S. 1955. *Harmonic analysis and the theory of probability*. Berkeley: University of California Press.

BONDI, H. 1952; 1960. *Cosmology*. Cambridge University Press.

BOREL, E. 1912-1915. Les théories moléculaires et les mathématiques. *Revue Générale des Sciences* **23**, 842-853. Translated as Molecular theories and mathematics. *Rice Institute Pamphlet* **1**, 163-193. Reprint in Borel 1972–, **III**, 1773-1784.

BOREL, E. 1922. Définition arithmétique d'une distribution de masses s'étendant à l'infini et quasi périodique, avec une densité moyenne nulle. *Comptes Rendus* (Paris) **174**, 977-979.

BOREL, E. 1972–. *Oeuvres de Emile Borel*. Paris: Editions du CNRS.

BOULIGAND, G. 1928. Ensembles impropres et nombre dimensionnel. *Bulletin des Sciences Mathématiques* **II-52**, 320-334 & 361-376.

BOULIGAND, G. 1929. Sur la notion d'ordre de mesure d'un ensemble plan. *Bulletin des Sciences Mathématiques* **II-53**, 185-192.

BOURBAKI, N. 1960. *Eléments d'historie des mathématiques*. Paris: Hermann.

BOYD, D. W. 1973a. The residual set dimension of the Apollonian packing. *Mathematika* **20**, 170-174.

BOYD, D. W. 1973b. Improved bounds for the disk packing constant. *Aequationes Mathematicae* **9**, 99-106.

BRAGG, W. H. 1934. Liquid crystals. *Nature* **133**, 445-456.

BRAY, D. 1974. Branching patterns of individual sympathetic neurons in culture. *J. of Cell Biology* **56**, 702-712.

BRODMANN, K. 1913. Neue Forschungsergebnisse der Grossgehirnanatomie... *Verhandlungen der 85 Versammlung deutscher Naturforscher und Aertze in Wien*, 200-240.

BROLIN, H. 1965. Invariant sets under iteration of rational functions. *Arkiv för Matematik* **6**, 103-

144.

BROUWER, L. E. J. 1975–. *Collected works.* Ed. A. Heyting and H. Freudenthal. New York: Elsevier North Holland.

BROWAND, F. K. 1966. An experimental investigation of the instability of an incompressible separated shear layer. *J. Fluid Mechanics* **26**, 281-307.

BROWN, G. L. & ROSHKO, A. 1974. On density effects and large structures in turbulent mixing layers. *J. of Fluid Mechanics* **64**, 775-816.

BRUSH, S. G. 1968. A history of random processes. I. Brownian movement from Brown to Perrin. *Archive for History of Exact Sciences* **5**, 1-36. Also in Brush 1976, 655-701.

CANTOR, G. 1872. Uber die Ausdehnung eines Satzes aus der Theorie der Trigonometrischen Reihen. *Mathematische Annalen* **5**, 123-132.

CANTOR, G. 1883. Grundlagen einer allgemeinen Mannichfältigkeitslehre. *Mathematische Annalen* **21**, 545-591. Also in Cantor 1932. Trans. H. Poincaré, as Fondements d'une théorie générale des ensembles. *Acta Mathematica* **2**, 381-408.

CANTOR, G. 1932. *Gesammelte Abhandlungen mathematischen und philosophischen Inhalts.* Ed. E. Zermelo. Berlin: Teubner. Olms reprint.

CANTOR, G. & DEDEKIND, R. 1937. *Briefwechsel.* (=*Selected Letters*) Ed. E. Noether & J. Cavaillès. Paris: Hermann.

CANTOR, G. & DEDEKIND, R. 1962. *Correspondence.* (=French translation of the 1937 *Briefwechsel*, by Ch. Ehresmann). Insert in Cavaillès 1962.

CANTOR, G. & DEDEKIND, R. 1976. *Unveröffentlicher Briefwechsel.* (=unpublished letters) Appendice XL of Dugac 1976a.

CARATHÉODORY, C. 1914. Über das lineare Maß von Punktmengen — eine Verallgemeinerung des Längenbegriffs. *Nachrichten der K. Gesellschaft der Wissenschaften zu Göttingen. Mathematisch-physikalische Klasse* 404-426. Also in Carathéodory 1954: *Gesammelte mathematische Schriften.* Munich: Beck, **4**, 249-275.

CARLESON, L. 1967. *Selected problems on exceptional sets.* Princeton, NJ: Van Nostrand.

CARTAN, H. 1958. Sur la notion de dimension. *Enseignement Mathématique,* Monographie No. **7**, 163-174.

CARTIER, P. 1971. Introduction à l'étude des mouvements browniens à plusieurs paramètres. *Séminaire de Probabilités V* (Strasbourg). *Lecture Notes in Mathematics* **191**, 58-75. New York: Springer.

CAUCHY, A. 1853. Sur les résultats les plus probables. *Comptes Rendus* (Paris) **37**, 198-206.

CAVAILLÈS, J. 1962. *Philosophie mathématique.* Paris: Hermann.

CELLÉRIER, CH. 1890. Note sur les principes fondamentaux de l'analyse. *Bulletin des Sciences Mathématiques* **14**, 142-160.

CESÀRO, E. 1905. Remarques sur la courbe de von Koch. *Atti della Reale Accademia delle Scienze Fisiche e Matematiche di Napoli* **XII**, 1-12. Also in Cesàro 1964, **II**, 464-479.

CESÀRO, E. 1964–. *Opere scelte.* Rome: Edizioni Cremonese.

CHANDRASEKHAR, S. 1943. Stochastic problems in physics and astronomy. *Reviews of Modern Physics* **15**, 1-89. Reprinted in *Noise and Stochastic Processes.* Ed. N. Wax. New York: Dover.

CHARLIER, C. V. L. 1908. Wie eine unendliche Welt aufgebaut sein kann. *Arkiv för Matematik, Astronomi och Fysik* **4**, 1-15.

CHARLIER, C. V. L. 1922. How an infinite world may be built up. *Arkiv för Matematik, Astronomi och Fysik* **16**, 1-34.

CHENTSOV, N. N. 1957. Lévy's Brownian motion for several parameters and generalized white noise. *Theory of Probability and its Applications* **2**, 265-266.

CHORIN, A. J. 1981. Estimates of intermittency, spectra, and blow up in developed turbulence. *Communications in Pure and Applied Mathematics* **34**, 853-866.

CHORIN, A. J. 1982. The evolution of a turbulent vortex (to appear).

CLAYTON, D. D. 1975. *Dark night sky, a personal adventure in cosmology.* New York: Quadrangle.

COLLET, P. & ECKMANN, J. P. 1980. *Iterated maps on the interval as dynamical systems.* Boston: Birkhauser.

COMROE, J. H., Jr., 1966. The lung. *Scientific American* (February) 56-68.

COOTNER, P. H. (Ed.) 1964. *The random character of stock market prices.* Cambridge, MA: MIT Press.

CORRSIN, S. 1959d. On the spectrum of isotropic temperature fluctuations in isotropic turbulence. *J. of Applied Physics* **22**, 469-473.

CORRSIN, S. 1959b. Outline of some topics in homogeneous turbulence flow. *J. of Geophysical Research* **64**, 2134-2150.

CORRSIN, S. 1962. Turbulent dissipation fluctuations. *Physics of Fluids* **5**, 1301-1302.

COXETER, H. S. M., 1979. The non-Euclidean symmetry of Escher's picture "Circle Limit III". *Leonardo* **12**, 19-25.

DAMERAU, F. J. & MANDELBROT, B. B. 1973. Tests of the degree of word clustering in samples of written English. *Linguistics* **102**, 58-75.

DAUBEN, J. W. 1971. The trigonometric background to Georg Cantor's theory of sets. *Archive for History of Exact Sciences* **7**, 181-216.

DAUBEN, J. W. 1974. Denumerability and dimension: the origins of Georg Cantor's theory of sets. *Rete* **2**, 105-133.

DAUBEN, J. W. 1975. The invariance of dimension: problems in the early development of set theory and topology. *Historia Mathematicae* **2**, 273-288.

DAUBEN, J. W. 1978. Georg Cantor: The personal matrix of his mathematics. *Isis* **69**, 534-550.

DAVIS, C. & KNUTH, D. E. 1970. Number representations and dragon curves. *J. of Recreational Mathematics* **3**, 66-81 & 133-149.

DE CHÉSEAUX, J. P. L. 1744. Sur la force de la lumière et sa propagation dans l'éther, et sur la distance des étoiles fixes. *Traité de la comète qui a paru en décembre 1743 et en janvier, février et mars 1744.* Lausanne et Genève: Chez Marc-Michel Bousquet et Compagnie.

DE GENNES, P. G. 1974. *The physics of liquid crystals.* Oxford: Clarendon Press.

DE GENNES, P. G. 1976. La percolation: un concept unificateur. *La Recherche* **7**, 919-927.

DE GENNES, P. G. 1979. *Scaling concepts in polymer physics.* Ithaca, NY: Cornell University Press.

DENJOY, A. 1964. *Hommes, formes et le nombre.* Paris: Albert Blanchard.

DENJOY, A. 1975. Evocation de l'homme et de l'œuvre. *Astérisque* **28-28**. Ed. G. Choquet. Paris: Société Mathématique de France.

DE VAUCOULEURS, G. 1956. The distribution of bright galaxies and the local supergalaxy. *Vistas in Astronomy* **II**, 1584-1606. London: Pergamon.

DE VAUCOULEURS, G. 1970. The case for a hierarchical cosmology. *Science* **167**, 1203-1213.

DE VAUCOULEURS, G. 1971. The large scale distribution of galaxies and clusters of galaxies. *Publications of the Astronomical Society of the Pacific* **73**, 113-143.

DE WIJS, H. J. 1951 & 1953. Statistics of ore distribution. *Geologie en Mijnbouw* (Amsterdam) **13**, 365-375 & **15**, 12-24.

DHAR, D. 1977. Lattices of effectively nonintegral dimensionality. *J. of Mathematical Physics* **18**, 577.

DICKSON, F. P. 1968. *The bowl of night; the physical universe and scientific thought.* Cambridge, MA: MIT Press.

DIEUDONNÉ, J. 1975. L'abstraction et l'intuition mathématique, *Dialectica* **29**, 39-54.

DOBRUSHIN, R. L. 1979. Gaussian processes and their subordinated self-similar random generalized fields. *Annals of Probability* **7**, 1-28.

DOMB, C. 1964. Some statistical problems connected with crystal lattices. *J. of the Royal Statistical Society* **26B**, 367-397.

DOMB, C. & GREEN, M.S. (Eds.) 1972–. *Phase transitions and critical phenomena.* New York: Academic.

DOMB, C., GILLIS, J. & WILMERS, G. 1965. On the shape and configuration of polymer molecules. *Pr. of the Physical Society* **85**, 625-645.

DOUADY, A. & OESTERLE, J. 1980. Dimension de Hausdorff des attracteurs, *Comptes Rendus* (Paris), **290A**, 1136-1138.

DUBOIS REYMOND, P. 1875. Versuch einer Classification der willkürlichen Functionen reeller Argument nach ihren Änderungen in den kleinsten Intervallen. *J. für die reine und angewandte Mathematik* (Crelle) **79**, 21-37.

DUGAC, P. 1973. Elements d'analyse de Karl Weierstrass. *Archive for History of Exact Sciences* **10**, 41-176.

DUGAC, P. 1976a. *Richard Dedekind et les fondements des mathématiques.* Paris: Vrin.

DUGAC, P. 1976b. Notes et documents sur la vie et l'œuvre de René Baire. *Archive for History of Exact Sciences* **15**, 297-384.

DUGAC, P. 1976c. Des correspondances mathématiques du XIXe et XXe siècles. *Revue de Synthèse* **97**, 149-170.

DUMOUCHEL, W. H. 1973. Stable distributions in statistical inference: 1. Symmetric stable distributions compared to other symmetric long-tailed distributions. *J. of the American Statistical Association* **68**, 469-482.

DUMOUCHEL, W. H. 1975. Stable distributions in statistical inference: 2. Information of stably distributed samples. *J. of the American Statistical Association* **70**, 386-393.

DUPLESSIS, N. 1970. *An introduction to potential theory.* New York: Hafner.

DUTTA, P. & HORN, P. M. 1981. Low-frequency fluctuation in solids: 1/f noise. *Reviews of Modern Physics* **53**, 497-516.

DVORETZKY, A., ERDÖS, P. & KAKUTANI, S. 1950. Double points of Brownian motion in n-space. *Acta Scientiarum Mathematicarum* (Szeged) **12**, 75-81.

DYSON, F. J. 1966. The search for extraterrestial technology, *Perspectives in Modern Physics: Essays in Honor of Hans A. Bethe.* Ed. R. E. Marshak, 641-655, New York: Interscience.

EGGLESTON, H. G. 1949. The fractional dimension of a set defined by decimal properties. *Quarterly J. of Mathematics, Oxford Series* **20**, 31-36.

EGGLESTON, H. G. 1953. On closest packing by equilateral triangles. *Pr. of the Cambridge Philosophical Society* **49**, 26-30.

EINSTEIN, A. 1926. *Investigations on the theory of the Brownian movement.* Ed. R. Fürth. Tr. A. D. Cowper. London: Methuen (Dover reprint).

EL HÉLOU, Y. 1978. Recouvrement du tore par des ouverts aléatoires et dimension de Hausdorff de l'ensemble non recouvert. *Comptes Rendus* (Paris) **287A**, 815-818.

ELIAS, H. & SCHWARTZ, D. 1969. Surface areas of the cerebral cortex of mammals. *Science* **166**, 111-113.

ESSAM, J. W. 1980. Percolation theory. *Reports on the Progress of Physics* **43**, 833-912.

FAMA, E. F. 1963. Mandelbrot and the stable Paretian hypothesis. *J. of Business* (Chicago) **36**, 420-429. Reproduced in *The Random Character of Stock Market Prices*, Ed. P. H. Cootner. Cambridge, MA: MIT Press.

FAMA, E. F. 1965 The behavior of stock-market prices. *J. of Business* **38**, 34-105. Based on a Ph.D. thesis, University of Chicago: *The distribution of daily differences of stock prices: a test of Mandelbrot's stable paretian hypothesis.*

FAMA, E. F. & BLUME, M. 1966. Filter rules and stock-market trading. *J. of Business* (Chicago) **39**, 226-241.

FATOU, P. 1906. Sur les solutions uniformes de certaines équations fonctionnelles. *Comptes rendus* (Paris) **143**, 546-548.

FATOU, P. 1919-1920. Sur les équations fonctionnelles. *Bull. Société Mathématique de France* **47**, 161-271; **48**, 33-94, & **48**, 208-314.

FEDERER, H. 1969. *Geometric measure theory.* New York: Springer.

FEIGENBAUM, M. J. 1978. Quantitative universality for a class of nonlinear transformations. *J. of Statistical Physics* **19**, 25-52.

FEIGENBAUM, M. J. 1979. The universal metric properties of nonlinear transformations. *J. of Statistical Physics* **21**, 669-706.

FEIGENBAUM, M. 1981. Universal behavior in nonlinear systems. *Los Alamos Science* **1**, 4-27.

FELLER, W. 1949. Fluctuation theory of recurrent events. *Tr. of the American Mathematical Society* **67**, 98-119.

FELLER, W. 1951. The asymptotic distribution of the range of sums of independent random variables. *Annals of Mathematical Statistics* **22**, 427.

FELLER, W. 1950-1957-1968. *An Introduction to Probability Theory and Its Applications,* Vol. 1. New York: Wiley.

FELLER, W. 1966-1971. *An Introduction to Probability Theory and Its Applications,* Vol. 2. New York: Wiley.

FEYNMAN, R. P. 1979 in *Pr. of the Third Workshop on Current Problems in High Energy Particle Theory,* Florence, Ed. Casalbuoni, R., Domokos, G., & Kovesi-Domokos, S. Baltimore: Johns Hopkins University Press.

FEYNMAN, R. P. & HIBBS, A. R. 1965. *Quantum mechanics and path integrals.* New York: McGraw-Hill.

FISHER, M. E. 1967. The theory of condensation and the critical point. *Physics* **3**, 255-283.

FOURNIER D'ALBE, E. E. 1907. *Two new worlds: I The infra world; II The supra world.* London: Longmans Green.

FRÉCHET, M. 1941. Sur la loi de répartition de certaines grandeurs géographiques. *J. de la Société de Statistique de Paris* **82**, 114-122.

FRICKE, R. & KLEIN, F. 1897. *Vorlesungen über die Theorie der automorphen Functionen.* Leipzig: Teubner (Johnson reprint).

FRIEDLANDER, S. K. & TOPPER, L. 1961. *Turbulence: classic papers on statistical theory.* New York: Interscience.

FRIEDMAN, J. B. 1974. The architect's compass in creation miniatures of the later middle ages. *Traditio, Studies in Ancient and Medieval History, Thought, and Religion,* 419-429.

FROSTMAN, O. 1935. Potentiel d'équilibre et capacité des ensembles avec quelques applications à la théorie des fonctions. *Meddelanden fran Lunds Universitets Mathematiska Seminarium* **3**, 1-118.

FUJISAKA, H. & MORI, H. 1979. A maximum principle for determining the intermittency exponent μ of fully developed steady turbulence. *Progress of Theoretical Physics* **62**, 54-60.

GAMOW, G. 1954. Modern cosmology. *Scientific American* **190** (March) 54-63. Reprint in Munitz (Ed.) 1957, 390-404.

GANGOLLI, R. 1967. Lévy's Brownian motion of several parameters. *Annales de l'Institut Henri Poincaré* **3B**, 121-226.

GARDNER, M. 1967. An array of problems that can be solved with elementary mathematical techniques. *Scientific American* **216** (March, April and June issues). Also in Gardner 1977, pp. 207-209 & 215-220.

GARDNER, M. 1976. In which "monster" curves

force redefinition of the word "curve." *Scientific American* **235** (December issue), 124-133.

GARDNER, M. 1977. *Mathematical magic show.* New York: Knopf.

GEFEN, Y., MANDELBROT, B. B. & AHARONY, A. 1980. Critical phenomena on fractals. *Physical Review Letters* **45**, 855-858.

GEFEN, Y., AHARONY, A., MANDELBROT, B. B. & KIRKPATRICK, S. 1981. Solvable fractal family, and its possible relation to the backbone at percolation. *Physical Review Letters.* **47**, 1771-1774.

GELBAUM, B. R. & OLMSTED, J. M. H. 1964. *Counterexamples in analysis.* San Francisco: Holden-Day.

GERNSTEIN, G. L. & MANDELBROT, B. B. 1964. Random walk models for the spike activity of a single neuron. *The Biophysical J.* **4**, 41-68.

GERVER, J. 1970. The differentiability of the Riemann function at certain rational multiples of π. *American J. of Mathematics* **92**, 33-55.

GILLISPIE, C. C. (Ed.) 1970-1976. *Dictionary of scientific biography.* Fourteen volumes. New York: Scribner's.

GISPERT, H. 1980. Correspondance de Fréchet....et....théorie de la dimension. *Cahiers du Séminaire d'Histoire des Mathématiques* (Paris) **1**, 69-120.

GNEDENKO, B. V. & KOLMOGOROV, A. N. 1954. *Limit distributions for sums of independent random variables.* Trans. K.L. Chung. Reading, MA: Addison Wesley.

GOLITZYN, G. S. 1962. Fluctuations of dissipation in a locally isotropic turbulent flow (in Russian). *Doklady Akademii Nauk SSSR* **144**, 520-523.

GRANT, H. L., STEWART, R. W. & MOILLIET, A. 1959. Turbulence spectra from a tidal channel. *J. of Fluid Mechanics* **12**, 241-268.

GRASSBERGER, P. 1981. On the Hausdorff dimension of fractal attractors (preprint).

GREENWOOD, P. E. 1969. The variation of a stable path is stable. *Z. für Wahrscheinlichkeitstheorie* **14**, 140-148.

GRENANDER, U. & ROSENBLATT, M. 1957 & 1966. *Statistical analysis of stationary time series.* New York: Wiley.

GROAT, R. A. 1948. Relationship of volumetric rate of blood flow to arterial diameter. *Federation Pr.* **7**, 45.

GROSSMAN, S. & THOMAE, S. 1977. Invariant distributions and stationary correlation functions of one-dimensional discrete processes. *Z. für Naturforschung* **32A**, 1353-1363.

GUREL, O. & RÖSSLER, O. E. (Eds.) 1979. Bifurcation theory and applications in scientific disciplines. *Annals of the New York Academy of Sciences* **316**, 1-708.

GURVICH, A. S. 1960. Experimental research on frequency spectra of atmospheric turbulence. *Izvestia Akademii Nauk SSSR; Geofizicheskaya Seriia* 1042.

GURVICH, A. S. & YAGLOM, A. M. 1967. Breakdown of eddies and probability distribution for small scale turbulence. *Boundary Layers and Turbulence.* (Kyoto International Symposium, 1966), *Physics of Fluids* **10**, S59-S65.

GURVICH, A. S. & ZUBKOVSKII, S. L. 1963. On the experimental evaluation of the fluctuation of dissipation of turbulent energy. *Izvestia Akademii Nauk SSSR; Geofizicheskaya Seriia* **12**, 1856-.

GUTHRIE, W. K. C. 1950. *The Greek philosophers from Thales to Aristotle.* London: Methuen (Harper paperback).

GUTHRIE, W. K. C. 1971-. *A history of Greek philosophy.* Cambridge University Press.

HACK, J. T. 1957. Studies of longitudinal streams in Virginia and Maryland. *U.S. Geological Survey Professional Papers* **294B**.

HADAMARD, J. 1912. L'œuvre mathématique de Poincaré. *Acta Mathematica* **38**, 203-287. Also in Poincaré 1916-, **XI**, 152-242. Or in Hadamard 1968, **4**, 1921-2005.

HADAMARD, J. 1968. *Oeuvres de Jacques Hadamard.* Paris: Editions du CNRS.

HAGGETT, P. 1972. *Geography: a modern synthesis.* New York: Harper & Row.

HAHN, H. 1956. The crisis in intuition, Translation in *The world of mathematics*, Ed. J. R. Newman. New York: Simon & Schuster, Vol. III, 1956-1976. Original German text in *Krise und Neuaufbau in den Exakten Wissenschaften* by H. Mark, H. Thirring, H. Hahn, K. Menger and G. Nöbeling, Leipzig and Vienna: F. Deuticke, 1933.

HALLÉ, F., OLDEMAN, R. A. A., & TOMLINSON, P. B., 1978. *Tropical trees and forests.* New York: Springer.

HALLEY, J. W. & MAI, T. 1979. Numerical estimates of the Hausdorff dimension of the largest cluster and its backbone in the percolation problem in two dimensions. *Physical Review Letters* **43**, 740-743.

HANDELMAN, S. W. 1980 A high-resolution computer graphics system. *IBM Systems J.*, **19**, 356-366.

HARDY, G. H. 1916. Weierstrass's non-differentiable function. *Tr. of the American Mathematical Society* **17**, 322-323. Also in Hardy 1966-, **IV**, 477-501.

HARDY, G. H. 1966–. *Collected papers.* Oxford: Clarendon Press.

HARRIS, T. E. 1963. *Branching processes.* New York: Springer.

HARRISON, E. R. 1981. *Cosmology.* Cambridge University Press.

HARISON, R. J., BISHOP, G. J. & QUINN, G. P. 1978. Spanning lengths of percolation clusters. *J. of Statistical Physics* **19**, 53-64.

HARTER, W. G. 1979-1981. Theory of hyperfine and superfine links in symmetric polyatomic molecules. I Trigonal and tetrahedral molecules. II Elementary cases in octahedral hexafluoride molecules *Physical Review*, **A19**, pp. 2277-2303 & **A24**, pp. 192-263.

HARTMANN, W. K. 1977. Cratering in the solar system. *Scientific American* (January) 84-99.

HARVEY, W. 1628. *De motu cordis.* Trans. Robert Willis, London, 1847, as *On the motion of the heart and blood in animals.* Excerpt in *Steps in the scientific tradition: readings in the history of science.* Ed. R.S. Westfall et al. New York: Wiley.

HAUSDORFF, F. 1919. Dimension und äusseres Mass. *Mathematische Annalen* **79**, 157-179.

HAWKES, J. 1974. Hausdorff measure, entropy and the independence of small sets. *Pr. of the London Mathematical Society* (3) **28**, 700-724.

HAWKES, J. 1978. Multiple points for symmetric Lévy processes. *Mathematical Pr. of the Cambridge Philosophical Society* **83**, 83-90.

HAWKINS, G. S. 1964. Interplanetary debris near the Earth. *Annual Review of Astronomy and Astrophysics* **2**, 149-164.

HAWKINS, T. 1970. *Lebesgue's theory of integration: Its origins and development.* Madison: University of Wisconsin Press.

HEATH, T. L. 1908. *The thirteen books of Euclid's elements translated with introduction and commentary.* Cambridge University Press. (Dover reprint).

HELLEMAN, R. H. G. (Ed.) 1980. Nonlinear dynamics. *Annals of the New York Academy of Sciences* **357**, 1-507.

HENDRICKS, W. J. 1979. Multiple points for transient symmetric Lévy processes in $\mathbb{R}^d$. *Z. für Wahrscheinlichkeitstheorie* **49**, 13-21.

HERMITE, C. & STIELTJES, T. J.. 1905. *Correspondance d'Hermite et de Stieltjes.* 2 vols. Ed. B. Baillaud & H. Bourget. Paris: Gauthier-Villars.

HEYDE, C. C. & SENETA, E. 1977. *I. J. Bienaymé: statistical theory anticipated.* New York: Springer.

HILBERT, D. 1891. Über die stetige Abbildung einer Linie auf ein Flächenstück. *Mathematische Annalen* **38**, 459-460. Also in Hilbert 1932, **3**, 1-2.

HILBERT, D. 1932. *Gesammelte Abhandlungen.* Berlin: Springer (Chelsea reprint).

HILEY, B. J. & SYKES, M. F. 1961. Probability of initial ring closure in the restricted random walk model of a macromolecule. *J. of Chemical Physics* **34**, 1531-1537.

HILLE, E. & TAMARKIN, J. D. 1929. Remarks on a known example of a monotone continuous function. *American Mathematics Monthly* **36**, 255-264.

HIRST, K. E. 1967. The Apollonian packing of circles. *J. of the London Mathematical Society* **42**, 281-291.

HOFSTADTER, D. R. 1976. Energy levels and wave functions of Bloch electrons in rational and irrational magnetic fields. *Physical Review* **B14**, 2239-2249.

HOFSTADTER, D. R. 1981. Strange attractors: mathematical patterns delicately poised between order and chaos. *Scientific American* **245** (November issue), 16-29.

HOLTSMARK, J. 1919. Über die Verbreiterung von Spektrallinien. *Annalen der Physik* **58**, 577-630.

HOOGE, F. N., KEINPENNING, T. G. M. & VAN-DAMME, L. K. J. 1981. Experimental studies on 1/f noise. *Reports on Progress in Physics* **44**, 479-532.

HOPKINSON 1876. On the residual charge of the Leyden jar. *Pr. of the Royal Society of London* **24** 408-

HORN, H. 1971. *Trees.* Princeton University Press.

HORSFIELD, K. & CUMMINGS, G. 1967. Angles of branching and diameters of branches in the human bronchial tree. *Bulletin of Mathematics Biophysics* **29**, 245-259.

HORTON, R. E. 1945. Erosional development of streams and their drainage basins; Hydrophysical approach to quantitative morphology. *Bulletin of the Geophysical Society of America* **56**, 275-370.

HOSKIN, M. 1973. Dark skies and fixed stars, *J. of the British Astronomical Association*, **83**, 4-.

HOSKIN, M. A. 1977. Newton, Providence and the universe of stars. *J. for the History of Astronomy* **8**, 77-101.

HOWARD, A. D. 1971. Truncation of stream networks by headward growth and branching. *Geophysical Analysis* **3**, 29-51.

HOYLE, F. 1953. On the fragmentation of gas clouds into galaxies and stars. *Astrophysical J.* **118**, 513-528.

HOYLE, F. 1975. *Astronomy and cosmology. A modern course.* San Francisco: W.H. Freeman.

HUREWICZ, W. & WALLMAN, H. 1941. *Dimension theory.* Princeton University Press.

HURST, H. E. 1951. Long-term storage capacity of reservoirs. *Tr. of the American Society of Civil Engineers* **116**, 770-808.

HURST, H. E. 1955. Methods of using long-term storage in reservoirs. *Pr. of the Institution of Civil Engineers* Part I, 519-577.

HURST, H. E., BLACK, R. P., AND SIMAIKA, Y. M. 1965. *Long-term storage, an experimental study.* London: Constable.

HUTCHINSON, J. E. 1981. Fractals and self-similarity, *Indiana University Mathematics J.* **30** 713-747.

HUXLEY, J. S. 1931. *Problems of relative growth.* New York: Dial Press.

IBERALL, A. S. 1967. Anatomy and steady flow characteristics of the arterial system with an introduction to its pulsatile characteristics. *Mathematical Biosciences* **1**, 375-395.

JACK, J. J. B., NOBLE, D. & TSIEN, R. W. 1975. *Electric current flow in excitable cells.* Oxford University Press.

JAKI, S. L. 1969. *The paradox of Olbers' paradox.* New York: Herder & Herder.

JEANS, J. H. 1929. *Astronomy and cosmogony.* Cambridge University Press. (Dover reprint).

JERISON, H. J. 1973. *Evolution of the brain and intelligence.* New York: Academic.

JOEVEER, M., EINASTO, J. & TAGO, E. 1977. Preprint of Tartu Observatory.

JOHNSON, D. M. 1977. Prelude to dimension theory: the geometric investigation of Bernard Bolzano. *Archive for History of Exact Sciences* **17**, 261-295.

JOHNSON, D. M. 1981. The problem of the invariance of dimension in the growth of modern topology. *Archive for history of exact sciences* Part I; Part II, **25**, 85-267.

JULIA, G. 1918. Mémoire sur l'itération des fonctions rationnelles. *J. de Mathématiques Pures et Appliquées* **4**: 47-245. Reprinted (with related texts) in Julia 1968, 121-319.

JULIA, G. 1968. *Oeuvres de Gaston Julia*, Paris: Gauthier-Villars,

KAHANE, J. P. 1964. Lacunary Taylor and Fourier series. *Bulletin of the American Mathematical Society* **70**, 199-213.

KAHANE, J. P. 1968. *Some random series of functions.* Lexington, MA: D. C. Heath.

KAHANE, J. P. 1969. Trois notes sur les ensembles parfaits linéaires. *Enseignement mathématique* **15**, 185-192.

KAHANE, J. P. 1970. Courbes étranges, ensembles minces. *Bulletin de l'Association des Professeurs de Mathématiques de l'Enseignement Public* **49**, 325-339.

KAHANE, J. P. 1971. The technique of using random measures and random sets in harmonic analysis. *Advances in Probability and Related Topics*, Ed. P. Ney. **1**, 65-101. New York: Marcel Dekker.

KAHANE, J. P. 1974. Sur le modèle de turbulence de Benoit Mandelbrot. *Comptes Rendus* (Paris) **278A**, 621-623.

KAHANE, J. P. & MANDELBROT, B. B. 1965. Ensembles de multiplicité aléatoires. *Comptes Rendus* (Paris) **261**, 3931-3933.

KAHANE, J. P. & PEYRIÈRE, J. 1976. Sur certaines martingales de B. Mandelbrot. *Advances in Mathematics* **22**, 131-145.

KAHANE, J. P. & SALEM, R. 1963. *Ensembles parfaits et séries trigonométriques.* Paris: Hermann.

KAHANE, J. P., WEISS, M. & WEISS, G. 1963. On lacunary power series. *Arkiv för Mathematik, Astronomi och Fysik* **5**, 1-26.

KAKUTANI, S. 1952. Quadratic diameter of a metric space and its application to a problem in analysis. *Pr. of the American Mathematical Society* **3**, 532-542.

KANT, I. 1755-1969. *Universal natural history and theory of the heavens.* Ann Arbor: University of Michigan Press.

KASNER, E. & SUPNICK, F. 1943. The Apollonian packing of circles. *Pr. of the National Academy of Sciences U.S.A.* **29**, 378-384.

KAUFMAN, R. 1968. On Hausdorff dimension of projections. *Mathematika* **15**, 153-155.

KELLY, W. 1951. *The best of Pogo.* New York: Simon and Schuster.

KERKER, M. 1974. Brownian movement and molecular reality prior to 1900. *J. of Chemical Education* **51**, 764-768.

KERKER, M. 1976. The Svedberg and molecular reality. *Isis* **67**, 190-216.

KIRKPATRICK, S. 1973. Percolation and conduction. *Reviews of Modern Physics* **45**, 574-588.

KIRKPATRICK, S. 1979. Models of disordered materials. *Ill-condensed matter -Matière mal condensée*, Ed. R. Balian, R. Ménard & G. Toulouse, New York: North Holland, **1**, 99-154.

KLINE, S. A. 1945. On curves of fractional dimensions. *J. of the London Mathematical Society* **20**, 79-86.

KNIGHT, F. B. 1981. *Essentials of Brownian motion and diffusion.* Providence, R.I.: American Mathematical Society.

KNUTH, D. 1968–. *The art of computer programming*. Reading, MA: Addison Wesley.

KOHLRAUSCH, R. 1847. Über das Dellmann'sche Elektrometer. *Annalen der Physik und Chemie* (Poggendorf) III-12, 353-405.

KOHLRAUSCH, R. 1854. Theorie des elektrischen Rückstandes in der Leidener Flasche. *Annalen der Physik und Chemie* (Poggendorf) IV-91, 56-82 & 179-214.

KOLMOGOROV, A. N. 1940. Wienersche Spiralen und einige andere interessante Kurven im Hilbertschen Raum. *Comptes Rendus (Doklady) Académie des Sciences de l'URSS (N.S.)* 26, 115-118.

KOLMOGOROV, A. N. 1941. Local structure of turbulence in an incompressible liquid for very large Reynolds numbers. *Comptes Rendus (Doklady) Académie des Sciences de l'URSS (N.S.)* 30, 299-303. Reprinted in Friedlander & Topper 1961, 151-155.

KOLMOGOROV, A. N. 1962. A refinement of previous hypotheses concerning the local structure of turbulence in a viscous incompressible fluid at high Reynolds number. *J. of Fluid Mechanics* 13, 82-85. Original Russian text and French translation in *Mécanique de la Turbulence*, 447-458 (Colloque International de Marseille, 1961), Paris: Editions du CNRS.

KOLMOGOROV, A. N. & TIHOMIROV, V. M. 1959-1961. Epsilon-entropy and epsilon-capacity of sets in functional spaces. *Uspekhi Matematicheskikh Nauk* (N.S.) 14, 3-86. Translated in *American Mathematical Society Translations* (Series 2) 17, 277-364.

KORCAK, J. 1938. Deux types fondamentaux de distribution statistique. *Bulletin de l'Institut International de Statistique* III, 295-299.

KRAICHNAN, R. H. 1974. On Kolmogorov's inertial range theories. *J. of Fluid Mechanics* 62, 305-330.

KUO, A. Y. S. & CORRSIN, S. 1971. Experiments on internal intermittency and fine structure distribution functions in fully turbulent fluid. *J. of Fluid Mechanics* 50, 285-320.

KUO, A. Y. S. & CORRSIN, S. 1972. Experiments on the geometry of the fine structure regions in fully turbulent fluid. *J. of Fluid Mechanics* 56, 477-479.

LAMPERTI, J. 1962. Semi-stable stochastic processes. *Tr. of the American Mathematical Society* 104, 62-78.

LAMPERTI, J. 1966. *Probability: a survey of the mathematical theory*. Reading, MA: W. A. Benjamin.

LAMPERTI, J. 1972. Semi-stable Markov processes. *Z. für Wahrscheinlichkeitstheorie*, 22, 205-225.

LANDAU, L. D. & LIFSHITZ, E. M. 1953-1959. *Fluid mechanics*. Reading: Addison Wesley.

LANDKOF, N. S. 1966-1972. *Foundations of modern potential theory*. New York: Springer.

LANDMAN, B. S. & RUSSO, R. L. 1971. On a pin versus block relationship for partitions of logic graphs. *IEEE Tr. on Computers* 20, 1469-1479.

LAPLACE, P. S. DE 1878-. *Oeuvres complètes*. Paris: Gauthier-Villars.

LARMAN, D. G. 1967. On the Besicovitch dimension of the residual set of arbitrarily packed disks in the plane. *J. of the London Mathematical Society* 42, 292-302.

LAVOIE, J. L., OSLER, T. J. & TREMBLAY, R. 1976. Fractional derivatives of special functions. *SIAM Review* 18, 240-268.

LAWRANCE, A. J. & KOTTEGODA, N. T. 1977. Stochastic modelling of riverflow time series. *J. of the Royal Statistical Society* A, 140, Part I, 1-47.

LEATH, P. L. 1976. Cluster size and boundary distribution near percolation threshold. *Physical Review* B14, 5046-5055.

LEBESGUE, H. 1903. *Sur le problème des aires.* See Lebesgue 1972-, IV, 29-35.

LEBESGUE, H. 1972-. *Oeuvres scientifiques*. Genève: Enseignement Mathématique.

LEIBNIZ, G. W. 1849-. *Mathematische Schriften*. Ed. C.I. Gerhardt. Halle: H.W. Schmidt (Olms reprint).

LEOPOLD, L. B. 1962. Rivers. *American Scientist* 50, 511-537.

LEOPOLD, L. B. & LANGBEIN, W. B. 1962. The concept of entropy in landscape evolution. *U.S. Geological Survey Professional Papers* 500A.

LEOPOLD, L. B. & MADDOCK, T., JR. 1953. The hydraulic geometry of stream channels and some physiological implications. *U.S. Geological Survey Professional Papers* 252.

LEOPOLD, L. B. & MILLER, J. P. 1956. Ephemeral streams: Hydraulic factors and their relation to the drainage net. *U.S. Geological Survey Professional Papers* 282-A, 1-37.

LERAY, J. 1934. Sur le mouvement d'un liquide visqueux emplissant l'espace. *Acta Mathematica* 63, 193-248.

LÉVY, P. 1925. *Calcul des probabilités*. Paris: Gauthier Villars.

LÉVY, P. 1930. Sur la possibilité d'un univers de masse infinie. *Annales de Physique* 14, 184-189. Also in Lévy 1973- II, 534-540.

LÉVY, P. 1937-1954. *Théorie de l'addition des variables aléatoires.* Paris: Gauthier Villars.

LÉVY, P. 1938. Les courbes planes ou gauches et les surfaces composées de parties semblables au tout. *J. de l'Ecole Polytechnique,* III, **7-8,** 227-291. Also in Lévy 1973- **II,** 331-394.

LÉVY, P. 1948-1965. *Processus stochastiques et mouvement brownien.* Paris: Gauthier-Villars.

LÉVY, P. 1957. Brownian motion depending on *n* parameters. The particular case *n=5. Pr. of the Symposia in Applied Mathematics* **VII,** 1-20. Providence, R.I.: American Mathematical Society.

LÉVY, P. 1959. Le mouvement brownien fonction d'un point de la sphère de Riemann. *Circolo matematico di Palermo, Rendiconti.* II, **8,** 297-310.

LÉVY, P. 1963. Le mouvement brownien fonction d'un ou de plusieurs paramètres. *Rendiconti di Matematica* (Roma) **22,** 24-101.

LÉVY, P. 1965. A special problem of Brownian motion and a general theory of Gaussian random functions. *Pr. of the Third Berkeley Symposium in Mathematical Statistics and Probability Theory.* Ed. J. Neyman, **2,** 133-175. Berkeley: University of California Press.

LÉVY, P. 1970. *Quelques aspects de la pensée d'un mathématicien.* Paris: Albert Blanchard.

LÉVY, P. 1973-. *Oeuvres de Paul Lévy.* Ed. D. Dugué, P. Deheuvels & M. Ibéro. Paris: Gauthier Villars.

LIEB, E. H. & LEBOWITZ, J. L. 1972. The constitution of matter: existence of thermodynamics for systems composed of electrons and nuclei. *Advances in Mathematics* **9,** 316-398.

LLINAS, R. R. 1969. *Neurobiology of cerebellar evolution and development.* Chicago: American Medical Association.

LOEMKER, L. E. 1956-1969. *Philosophical papers and letters of Leibniz.* Boston: Reidel.

LORENZ, E. N. 1963. Deterministic nonperiodic flow. *J. of the Atmospheric Sciences* **20,** 130-141.

LOVE, E. R. & YOUNG, L. C. 1937. Sur une classe de fonctionnelles linéaires. *Fundamenta Mathematicae* **28,** 243-257.

LOVEJOY, S. 1982. Area-perimeter relation for rain and cloud areas. *Science* **216,** 185-187.

LUKACS, E. 1960-1970. *Characteristic functions.* London: Griffin. New York: Hafner.

LYDALL, H. F. 1959. The distribution of employment income. *Econometrica* **27,** 110-115.

MAITRE, J. 1964. Les fréquences des prénoms de baptême en France. *L'Année sociologique* **3,** 31-74.

MANDELBROT, B. B. 1951. Adaptation d'un message à la ligne de transmission. I & II. *Comptes Rendus* (Paris) **232,** 1638-1640 & 2003-2005.

MANDELBROT, B. B. 1953t. Contribution à la théorie mathématique des jeux de communication (Ph.D. Thesis). *Publications de l'Institut de Statistique de l'Université de Paris* **2,** 1-124.

MANDELBROT, B. B. 1954w. Structure formelle des textes et communication (deux études). *Word* **10,** 1-27. Corrections. *Word:* **11,** 424. Translations into English, Czech and Italian.

MANDELBROT, B. B. 1955b. On recurrent noise limiting coding. *Information Networks, the Brooklyn Polytechnic Institute Symposium,* 205-221. Ed. E. Weber. New York: Interscience. Translation into Russian.

MANDELBROT, B. B. 1956c. La distribution de Willis-Yule, relative au nombre d'espèces dans les genres taxonomiques. *Comptes Rendus* (Paris) **242,** 2223-2225.

MANDELBROT, B. B. 1956l. On the language of taxonomy: an outline of a thermo-statistical theory of systems of categories, with Willis (natural) structure. *Information Theory, the Third London Symposium.* Ed. C. Cherry. 135-145. New York: Academic.

MANDELBROT, B. B. 1956t. Exhaustivité de l'énergie d'un système, pour l'estimation de sa température. *Comptes Rendus* (Paris) **243,** 1835-1837.

MANDELBROT, B. B. 1956m. A purely phenomenological theory of statistical thermodynamics: canonical ensembles. *IRE Tr. on Information Theory* **112,** 190-203.

MANDELBROT, B. B. 1959g. Ensembles grand canoniques de Gibbs; justification de leur unicité basée sur la divisibilitié infinie de leur énergie aléatoire. *Comptes Rendus* (Paris) **249,** 1464-1466.

MANDELBROT, B. B. 1959p. Variables et processus stochastiques de Pareto-Lévy et la répartition des revenus, I & II. *Comptes Rendus* (Paris) **249,** 613-615 & 2153-2155.

MANDELBROT, B. B. 1960i. The Pareto-Lévy law and the distribution of income. *International Economic Review* **1,** 79-106.

MANDELBROT, B. B. 1961b. On the theory of word frequencies and on related Markovian models of discourse. *Structures of language and its mathematical aspects.* Ed. R. Jakobson. 120-219. New York: American Mathematical Society

MANDELBROT, B. B. 1961e. Stable Paretian random functions and the multiplicative variation of income. *Econometrica* **29**, 517-543.

MANDELBROT, B. B. 1962c. Sur certains prix spéculatifs: faits empiriques et modèle basé sur les processus stables additifs de Paul Lévy. *Comptes Rendus* (Paris) **254**, 3968-3970.

MANDELBROT, B. B. 1962e. Paretian distributions and income maximization. *Quarterly J. of Economics of Harvard University* **76**, 57-85.

MANDELBROT, B. B. 1962n. Statistics of natural resources and the law of Pareto. IBM Research Note NC-146, June 29, 1962 (unpublished).

MANDELBROT, B. B. 1962t. The role of sufficiency and estimation in thermodynamics. *The Annals of Mathematical Statistics* **33**, 1021-1038.

MANDELBROT, B. B. 1963p. The stable Paretian income distribution, when the apparent exponent is near two. *International Economic Review* **4**, 111-115.

MANDELBROT, B. B. 1963b. The variation of certain speculative prices. *J. of Business* (Chicago) **36**, 394-419. Reprinted in *The random character of stock market prices*. Ed. P. H. Cootner, 297-337. Cambridge, MA.: MIT Press).

MANDELBROT, B. B. 1963e. New methods in statistical economics. *J. of Political Economy* **71**, 421-440. Reprint in *Bulletin of the International Statistical Institute, Ottawa Session:* **40** (2), 669-720.

MANDELBROT, B. B. 1964j. The epistemology of chance in certain newer sciences. Read at *The Jerusalem International Congress on Logic, Methodology and the Philosophy of Science* (unpublished).

MANDELBROT, B. B. 1964t. Derivation of statistical thermodynamics from purely phenomenological principles. *J. of Mathematical Physics* **5**, 164-171.

MANDELBROT, B. B. 1964o. Random walks, fire damage amount, and other Paretian risk phenomena. *Operations Research* **12**, 582-585.

MANDELBROT, B. B. 1964s. *Self-similar random processes and the range* IBM Research Report RC-1163, April 13, 1964 (unpublished).

MANDELBROT, B. B. 1965c. Self similar error clusters in communications systems and the concept of conditional stationarity. *IEEE Tr. on Communications Technology* **13**, 71-90.

MANDELBROT, B. B. 1965h. Une classe de processus stochastiques homothétiques à soi; application à la loi climatologique de H. E. Hurst. *Comptes Rendus* (Paris) **260**, 3274-3277.

MANDELBROT, B. B. 1965s. Leo Szilard and unique decipherability. *IEEE Tr. on Information Theory* **IT-11**, 455-456.

MANDELBROT, B. B. 1965z. Information theory and psycholinguistics. *Scientific Psychology: Principles and Approaches*, Ed. B. B. Wolman & E. N. Nagel. New York: Basic Books 550-562.. Reprint in *Language, Selected Readings.* Ed. R. C. Oldfield & J. C. Marshall. London: Penguin. Reprint with appendices, *Readings in Mathematical Social Science.* Ed. P. Lazarfeld and N. Henry. Chicago, Ill.: Science Research Associates (1966: hardcover). Cambridge, MA: M.I.T. Press (1968: paperback). Russian translation.

MANDELBROT, B. B. 1966b. Forecasts of future prices, unbiased markets, and 'martingale' models. *J. of Business* (Chicago) **39**, 242-255. Important errata in a subsequent issue of the same Journal.

MANDELBROT, B. B. 1967b. Sporadic random functions and conditional spectral analysis; self-similar examples and limits. *Pr. of the Fifth Berkeley Symposium on Mathematical Statistics and Probability* **3**, 155-179. Ed. L. LeCam & J. Neyman. Berkeley: University of California Press.

MANDELBROT, B. B. 1967k. Sporadic turbulence. *Boundary Layers and Turbulence* (Kyoto International Symposium, 1966), *Supplement to Physics of Fluids* **10**, S302-S303.

MANDELBROT, B. B. 1967j. The variation of some other speculative prices. *J. of Business* (Chicago) **40**, 393-413.

MANDELBROT, B. B. 1967p. Sur l'épistémologie du hasard dans les sciences sociales: invariance des lois et vérification des hypothèses, *Encyclopédie de la Pléiade: Logique et Connaissance Scientifique.* Ed. J. Piaget. 1097-1113. Paris: Gallimard.

MANDELBROT, B. B. 1967s. How long is the coast of Britain? Statistical self-similarity and fractional dimension. *Science* **155**, 636-638.

MANDELBROT, B. B. 1967i. Some noises with $1/f$ spectrum, a bridge between direct current and white noise. *IEEE Tr. on Information Theory* **13**, 289-298.

MANDELBROT, B. B. 1968p. Les constantes chiffrées du discours. *Encyclopédie de la Pléiade: Linguistique*, Ed. J. Martinet, Paris: Gallimard, 46-56.

MANDELBROT, B. B. 1969e. Long-run linearity, locally Gaussian process, H-spectra and infinite variance. *International Economic Review* **10**, 82-111.

MANDELBROT, B. B. 1970p. On negative temperature for discourse. Discussion of a paper by Prof. N. F. Ramsey. *Critical Review of Thermodynamics*, 230-232. Ed. E. B. Stuart et al.

Baltimore, MD: Mono Book.

MANDELBROT, B. B. 1970e. Statistical dependence in prices and interest rates. *Papers of the Second World Congress of the Econometric Society*, Cambridge, England (8-14 Sept. 1970).

MANDELBROT, B. B. 1970y. *Statistical Self Similarity and Very Erratic Chance Fluctuations.* Trumbull Lectures, Yale University (unpublished).

MANDELBROT, B. B. 1971e. When can price be arbitraged efficiently? A limit to the validity of the random walk and martingale models. *Review of Economics and Statistics* **LIII**, 225-236.

MANDELBROT, B. B. 1971f. A fast fractional Gaussian noise generator. *Water Resources Research* **7**, 543-553.
NOTE: in the first fraction on p. 545, 1 must be erased in the numerator and added to the fraction.

MANDELBROT, B. B. 1971n. *The conditional cosmographic principle and the fractional dimension of the universe.* (Submitted to several periodicals, but first published as part of Mandelbrot 1975o.)

MANDELBROT, B. B. 1972d. On Dvoretzky coverings for the circle. *Z. für Wahrscheinlichkeitstheorie* **22**, 158-160.

MANDELBROT, B. B. 1972j. Possible refinement of the lognormal hypothesis concerning the distribution of energy dissipation in intermittent turbulence. *Statistical models and turbulence.* Ed. M. Rosenblatt & C. Van Atta. Lecture Notes in Physics **12** 333-351. New York: Springer,

MANDELBROT, B. B. 1972b. Correction of an error in "The variation of certain speculative prices (1963)." *J. of Business* **40**, 542-543.

MANDELBROT, B. B. 1972c. Statistical methodology for nonperiodic cycles: from the covariance to the R/S analysis. *Annals of Economic and Social Measurement* **1**, 259-290.

MANDELBROT, B. B. 1972w. Broken line process derived as an approximation to fractional noise. *Water Resources Research* **8**, 1354-1356.

MANDELBROT, B. B. 1972z. Renewal sets and random cutouts. *Z. für Wahrscheinlichkeitstheorie* **22**, 145-157.

MANDELBROT, B. B. 1973c. Comments on "A subordinated stochastic process model with finite variance for speculative prices," by Peter K. Clark. *Econometrica* **41**, 157-160.

MANDELBROT, B. B. 1973f. Formes nouvelles du hasard dans les sciences. *Economie Appliquée* **26**, 307-319.

MANDELBROT, B. B. 1973j. Le problème de la réalité des cycles lents, et le syndrome de Joseph. *Economie Appliquée* **26**, 349-365.

MANDELBROT, B. B. 1973v. Le syndrome de la variance infinie, et ses rapports avec la discontinuité des prix. *Economie Appliquée* **26**, 321-348.

MANDELBROT, B. B. 1974c. Multiplications aléatoires itérées, et distributions invariantes par moyenne pondérée. *Comptes Rendus* (Paris) **278A**, 289-292 & 355-358.

MANDELBROT, B. B. 1974d. A population birth and mutation process, I: Explicit distributions for the number of mutants in an old culture of bacteria. *J. of Applied Probability* **11**, 437-444. (Part II distributed privately).

MANDELBROT, B. B. 1974f. Intermittent turbulence in self-similar cascades: divergence of high moments and dimension of the carrier. *J. of Fluid Mechanics* **62**, 331-358.

MANDELBROT, B. B. 1975b. Fonctions aléatoires pluri-temporelles: approximation poissonien ne du cas brownien et généralisations. *Comptes Rendus* (Paris) **280A**, 1075-1078.

MANDELBROT, B. B. 1975f. On the geometry of homogeneous turbulence, with stress on the fractal dimension of the iso-surfaces of scalars. *J. of Fluid Mechanics* **72**, 401-416.

MANDELBROT, B. B. 1975m. Hasards et tourbillons: quatre contes à clef. *Annales des Mines* (November), 61-66.

MANDELBROT, B. B. 1975o. *Les objets fractals: forme, hasard et dimension.* Paris: Flammarion.

MANDELBROT, B. B. 1975u. Sur un modèle décomposable d'univers hiérarchisé: déduction des corrélations galactiques sur la sphère céleste. *Comptes Rendus* (Paris) **280A**, 1551-1554.

MANDELBROT, B. B. 1975w. Stochastic models for the Earth's relief, the shape and the fractal dimension of the coastlines, and the number-area rule for islands. *Pr. of the National Academy of Sciences USA* **72**, 3825-3828

MANDELBROT, B. B. 1975h. Limit theorems on the self-normalized range for weakly and strongly dependent processes. *Z. für Wahrscheinlichkeitstheorie* **31**, 271-285.

MANDELBROT, B. B. 1976c. Géométrie fractale de la turbulence. Dimension de Hausdorff, dispersion et nature des singularités du mouvement des fluides. *Comptes Rendus* (Paris) **282A**, 119-120.

MANDELBROT, B. B. 1976o. Intermittent turbulence & fractal dimension: kurtosis and the spectral exponent 5/3+B. *Turbulence and Navier Stokes Equations* Ed. R. Teman, *Lecture Notes in Mathematics* **565**, 121-145. New York: Springer.

MANDELBROT, B. B. 1977b. Fractals and turbulence: attractors and dispersion. *Turbulence Seminar Berkeley 1976/1977* Ed. P. Bernard & T. Ratiu.

Lecture Notes in Mathematics **615** 83-93. New York: Springer. Russian translation.

MANDELBROT, B. B. 1977f. *Fractals: form, chance, and dimension.* San Francisco: W. H. Freeman & Co.

MANDELBROT, B. B. 1977h. Geometric facets of statistical physics: scaling and fractals. *Statistical Physics 13*, International IUPAP Conference, 1977. Ed. D. Cabib et al. *Annals of the Israel Physical Society.* 225-233.

MANDELBROT, B. B. 1978b. The fractal geometry of trees and other natural phenomena. *Buffon Bicentenary Symposium on Geometrical Probability*, Ed. R. Miles & J. Serra *Lecture Notes in Biomathematics* **23** 235-249. New York: Springer.

MANDELBROT, B. B. 1978r. Les objets fractals. *La Recherche* **9**, 1-13.

MANDELBROT, B. B. 1978c. Colliers aléatoires et une alternative aux promenades au hasard sans boucle: les cordonnets discrets et fractals. *Comptes Rendus* (Paris) **286A**, 933-936.

MANDELBROT, B. B. 1979n. Comment on bifurcation theory and fractals. *Bifurcation Theory and Applications*, Ed. Gurel & O. Rössler. *Annals of the New York Academy of Sciences* **316**, 463-464.

MANDELBROT, B. B. 1979u Corrélations et texture dans un nouveau modèle d'Univers hiérarchisé, basé sur les ensembles trémas. *Comptes Rendus* (Paris) **288A**, 81-83.

MANDELBROT, B. B. 1980b. Fractals and geometry with many scales of length. *Encyclopedia Britannica 1981 Yearbook of Science and the Future*, 168-181.

MANDELBROT, B. B. 1980n. Fractal aspects of the iteration of $z \rightarrow \lambda z(1-z)$ for complex λ and z. *Non Linear Dynamics*, Ed. R. H. G. Helleman. *Annals of the New York Academy of Sciences*, **357**, 249-259.

MANDELBROT, B. B. 1981l. Scalebound or scaling shapes: A useful distinction in the visual arts and in the natural sciences. *Leonardo* **14**, 45-47.

MANDELBROT, B. B. 1982m. On discs and sigma discs, that osculate the limit sets of groups of inversions. *Mathematical Intelligencer*: **4**.

MANDELBROT, B. B. 1982s. The inexhaustible function z^2-m (tentative title). *Scientific American* (tentative).

MANDELBROT, B. B. & MCCAMY, K. 1970. On the secular pole motion and the Chandler wobble. *Geophysical J.* **21**, 217-232.

MANDELBROT, B. B. & TAYLOR, H. M. 1967. On the distribution of stock price differences. *Operations Research*: **15**, 1057-1062.

MANDELBROT, B. B. & VAN NESS, J. W. 1968. Fractional Brownian motions, fractional noises and applications. *SIAM Review* **10**, 422.

MANDELBROT, B. B. & WALLIS, J. R. 1968. Noah, Joseph and operational hydrology. *Water Resources Research* **4**, 909-918.

MANDELBROT, B. B. & WALLIS, J. R. 1969a. Computer experiments with fractional Gaussian noises. *Water Resources Research* **5**, 228.

MANDELBROT, B. B. & WALLIS, J. R. 1969b. Some long-run properties of geophysical records. *Water Resources Research* **5**, 321-340.

MANDELBROT, B. B. & WALLIS, J. R. 1969c. Robustness of the rescaled range R/S in the measurement of noncyclic long runstatistical dependence. *Water Resources Research* **5**, 967-988.

MANDELBROT, B. B., *see also* Apostel, M. & Morf, Berger & M., Damerau & M., Gefen, M., & Aharony, Gefen, Aharony, M. & Kirkpatrick, Gerstein & M., & Kahane & M..

MANHEIM, J. H. 1964. *The genesis of point-set topology.* New York: Macmillan.

MARCUS, A. 1964. A stochastic model of the formation and survivance of lunar craters, distribution of diameters of clean craters. *Icarus* **3**, 460-472.

MARCUS, M. B. 1976. Capacity of level sets of certain stochastic processes. *Z. für Wahrscheinlichkeitstheorie* **34**, 279-284.

MARSTRAND, J. M. 1954a. Some fundamental geometrical properties of plane sets of fractional dimension. *Pr. of the London Mathematical Society* **(3) 4**, 257-302.

MARSTRAND, J. M. 1954b. The dimension of Cartesian product sets. *Pr. of the London Mathematical Society* **50**, 198-202.

MATHERON, G. 1962. *Traité de Géostatistique Appliquée* Cambridge Philosophical Society, Tome 1, Paris: Technip.

MATTILA, P. 1975. Hausdorff dimension, orthogonal projections and intersections with planes. *Annales Academiae Scientiarum Fennicae, Series A Mathematica* I, 227-244.

MAX, N. L. 1971. *Space filling curves.* 16 mm color film. Topology Films Project. International Film Bureau, Chicago, Ill. Accompanying book (preliminary edition), Education Development Center, Newton, MA.

MAXWELL, J. C. 1890. *Scientific papers* (Dover reprint).

MCKEAN, H. P., JR. 1955a. Hausdorff-Besicovitch dimension of Brownian motion paths. *Duke Mathematical J.* **22**, 229-234.

MCKEAN, H. P., JR. 1955b. Sample functions of

stable processes. *Annals of Mathematics* **61**, 564-579.

MCKEAN, H. P., JR. 1963. Brownian motion with a several dimensional time. *Theory of Probability and its Applications* **8**, 357-378.

MCMAHON, T. A. 1975. The mechanical design of trees. *Scientific American* **233**, 92-102.

MCMAHON, T. A. & KRONAUER, R. E. 1976. Tree structures: Deducing the principle of mechanical design. *J. of Theoretical Biology* **59**, 433-466.

MEJIA, J. M., RODRIGUEZ-ITURBE, I. & DAWDY, D. R. 1972. Streamflow simulation. 2. The broken line process as a potential model for hydrological simulation. *Water Resource Research*, **8**, 931-941.

MELZAK, Z. A. 1966. Infinite packings of disks. *Canadian J. of Mathematics* **18**, 838-852.

MENGER, K. 1943. What is dimension? *American Mathematical Monthly* **50**, 2-7. Reprint in Menger 1979, Ch. 17.

MENGER, K. 1979. *Selected papers in logic and foundations, didactics and economics.* Boston: Reidel.

MENSCHKOWSKI, H. 1967. *Probleme des Unendlichen.* Braunschweig: Vieweg.

METROPOLIS, N., STEIN, M. L. & STEIN, P. R. 1973. On finite limit sets for transformations on the unit interval. *J. of Combinatorial Theory* **A15**, 25-44.

MINKOWSKI, H. 1901. Über die Begriffe Länge, Oberfläche und Volumen. *Jahresbericht der Deutschen Mathematikervereinigung* **9**, 115-121. Also in Minkowski 1911 **2**, 122-127.

MINKOWSKI, H. 1911. *Gesammelte Abhandlungen,* Chelsea reprint.

MONIN, A. S. & YAGLOM, A. M. 1963. On the laws of small scale turbulent flow of liquids and gases. *Russian Mathematical Surveys* (translated from the Russian). **18**, 89-109.

MONIN, A. S. & YAGLOM, A. M. 1971 & 1975. *Statistical fluid mechanics, Volumes 1 and 2* (translated from the Russian). Cambridge, MA: MIT Press.

MOORE, E. H. 1900. On certain crinkly curves. *Tr. of the American Mathematical Society* **1**, 72-90.

MORI, H. 1980. Fractal dimensions of chaotic flows of autonomous dissipative systems. *Progress of Theoretical Physics* **63**, 1044-1047.

MORI, H. & FUJISAKA, H. 1980. Statistical dynamics of chaotic flows. *Progress of Theoretical Physics* **63**, 1931-1944.

MUNITZ, M. K. (Ed.) 1957. *Theories of the universe.* Glencoe, IL: The Free Press.

MURRAY, C. D. 1927. A relationship between circumference and weight in trees. *J. of General Physiology* **IV**, 725-729.

MYRBERG, P. J. 1962. Sur l'itération des polynomes réels quadratiques. *J. de Mathématiques pures et appliquées* **(9)41**, 339-351.

NELSON, E. 1966. Derivation of the Schrödinger equation from Newtonian mechanics. *Physical Review* **150**, 1079-1085.

NELSON, E. 1967. *Dynamical theories of Brownian motion.* Princeton University Press.

NEUENSCHWANDER, E. 1978. Der Nachlass von Casorati (1835-1890) in Pavia. *Archive for History of Exact Sciences* **19**, 1-89.

NEWMAN, J. R. 1956. *The world of mathematics.* New York: Simon & Schuster.

NORTH, J. D. 1965. *The measure of the universe.* Oxford: Clarendon Press.

NOVIKOV, E. A. 1963. Variation in the dissipation of energy in a turbulent flow and the spectral distribution of energy. *Prikladnaya Matematika i Mekhanika* **27**, 944-946 (translation, 1445-1450).

NOVIKOV, E. A. 1965a. On correlations of higher order in turbulent motion (in Russian). *Fisika Atmosfery i Okeana* **1**, 788-796.

NOVIKOV, E. A. 1965b. On the spectrum of fluctuations in turbulent motion (in Russian). *Fisika Atmosfery i Okeana* **1**, 992-993.

NOVIKOV, E. A. 1966. Mathematical model of the intermittency of turbulent motion (in Russian). *Doklady Akademii Nauk SSSR* **168**, 1279-1282.

NOVIKOV, E. A. 1971. Intermittency and scale similarity in the structure of a turbulent flow. *Prikladnaia Matematika i Mekhanika* **35**, 266-277. English in *P.M.M. Applied Mathematics and Mechanics*

NOVIKOV, E. A. & STEWART, R.W. 1964. Intermittency of turbulence and the spectrum of fluctuations of energy dissipation (in Russian). *Isvestia Akademii Nauk SSR; Seria Geofizicheskaia* **3**, 408-413.

NYE, M. J. 1972. *Molecular reality. A perspective on the scientific work of Jean Perrin.* London: Macdonald. New York: American Elsevier.

OBUKHOV, A. M. 1941. On the distribution of energy in the spectrum of turbulent flow. *Comptes Rendus (Doklady) Académie des Sciences de l'URSS (N.S.)* **32**, 22-24.

OBUKHOV, A. M. 1962. Some specific features of atmospheric turbulence. *J. of Fluid Mechanics* **13**, 77-81. Also in *J. of Geophysical Research* **67**, 3011-3014.

OLBERS, W. 1823. Über die Durchsichtigkeit des Weltraums. *Astronomisches Jahrbuch für das*

Jahr 1826 nebst einer Sammlung der neuesten in die astronomischen Wissenschaften einschlagenden Abhandlungen, Beobachtungen und Nachrichten, **150**, 110-121. Berlin: C.F.E. Späthen.

OLDHAM, K. B. & SPANIER, J. 1974. *The fractional calculus*. New York: Academic.

OREY, S. 1970. Gaussian sample functions and the Hausdorff dimension of level crossings. *Z. für Wahrscheinlichkeitstheorie* **15**, 249-156.

OSGOOD, W. F. 1903. A Jordan curve of positive area. *Tr. of the American Mathematical Society* **4**, 107-112.

PAINLEVÉ, P. 1895. Leçon d'ouverture faite en présence de Sa Majesté le Roi de Suède et de Norwège. First printed in Painlevé 1972–**1**,200-204.

PAINLEVÉ, P. 1972-. *Oeuvres de Paul Painlevé*. Paris: Editions du CNRS.

PALEY, R. E. A. C. & WIENER, N. 1934. *Fourier transforms in the complex domain*. New York: American Mathematical Society.

PARETO, V. 1896-1965. *Cours d'économie politique*. Reprinted as a volume of *Oeuvres Complètes*. Geneva: Droz.

PARTRIDGE, E. 1958. *Origins*. New York: Macmillan.

PAUMGARTNER, D. & WEIBEL, E. 1981. Resolution effects on the stereological estimation of surface and volume and its interpretation in terms of fractal dimension. *J. of Microscopy* **121**, 51-63.

PEANO, G. 1890. Sur une courbe, qui remplit une aire plane. *Mathematische Annalen* **36**, 157-160. Translation in Peano 1973.

PEANO, G. 1973. *Selected works*. Ed. H. C. Kennedy. Toronto University Press.

PEEBLES, P. J. E. 1980. *The large-scale structure of the universe*. Princeton University Press.

PERRIN, J. 1906. La discontinuité de la matière. *Revue du Mois* **1**, 323-344.

PERRIN, J. 1909. Mouvement brownien et réalité moléculaire. *Annales de chimie et de physique* **VIII 18**, 5-114. Trans. F. Soddy, as *Brownian Movement and Molecular Reality*. London: Taylor & Francis.

PERRIN, J. 1913. *Les Atomes*. Paris: Alcan. A 1970 reprint by Gallimard supersedes several revisions that had aged less successfully. English translation: *Atoms*, by D. L. Hammick; London: Constable. New York: Van Nostrand. Also translated into German, Polish, Russian, Serbian and Japanese.

PETERSON, B. A. 1974. The distribution of galaxies in relation to their formation and evolution. *The formation and dynamics of galaxies*, Ed. Shake-shaft, J. R. IAU Symposium 58. Boston: Reidel, 75-84.

PEYRIÈRE, J. 1974. Turbulence et dimension de Hausdorff. *Comptes Rendus* (Paris) **278A**, 567-569.

PEYRIÈRE, J. 1978. Sur les colliers aléatoires de B. Mandelbrot. *Comptes Rendus* (Paris) **286A**, 937-939.

PEYRIÈRE, J. 1979. Mandelbrot random beadsets and birth processes with interaction (privately distributed).

PEYRIÈRE, J., 1981. Processus de naissance avec interaction des voisins, Evolution de graphes, *Annales de l'Institut Fourier*, **31**, 187-218.

POINCARÉ, H. 1890. *Calcul des probabilités* (2nd ed., 1912) Paris: Gauthier-Villars.

POINCARÉ, H. 1905. *La valeur de la science*. Paris: Flammarion. English tr. by G. B. Halsted.

POINCARÉ, H. 1913. *Dernières penseés*, Paris: Flammarion.

POINCARÉ, H. 1916-. *Oeuvres de Henri Poincaré*. Paris: Gauthier Villars.

PONTRJAGIN, L. & SCHNIRELMAN, L. 1932. Sur une propriété métrique de la dimension. *Annals of Mathematics* **33**, 156-162.

PRUITT, W. E. 1975. Some dimension results for processes with independent increments. *Stochastic Processes and Related Topics*, **I**, 133-165. Ed. M. L. Puri. New York: Academic.

PRUITT, W. E. 1979. The Hausdorff dimension of the range of a process with stationary independent increments. *J. of Mathematics and Mechanics* **19**, 371-378.

PRUITT, W. E. & TAYLOR, S. J. 1969. Sample path properties of processes with stable components. *Z. für Wahrscheinlichkeitstheorie* **12**, 267-289.

QUEFFELEC, H. 197. Dérivabilité de certaines sommes de séries de Fourier lacunaires. (Thèse de 3e Cycle de Mathématiques.) Orsay: Université de Paris-Sud.

RALL, W. 1959. Branching dendritic trees and motoneuron membrane resistivity. *Experimental Neurology* **1**, 491-527.

RAYLEIGH, LORD 1880. On the resultant of a large number of vibrations of the same pitch and arbitrary phase. *Philosophical Magazine* **10**, 73. Also in Rayleigh 1899 **1**, 491-.

RAYLEIGH, LORD 1899. *Scientific papers*. Cambridge University Press. Dover reprint.

RÉNYI, A. 1955. On a new axiomatic theory of probability. *Acta Mathematica Hungarica* **6**, 285-335.

RICHARDSON, L. F. 1922. *Weather prediction by numerical process*. Cambridge University Press. The Dover reprint contains a biography as part of a new introduction by J. Chapman.

RICHARDSON, L. F. 1926. Atmospheric diffusion shown on a distance-neighbour graph. *Pr. of the Royal Society of London.* **A**, **110**, 709-737.

RICHARDSON, L. F. 1960a. *Arms and insecurity: a mathematical study of the causes and origins of war.* Ed. N. Rashevsky & E. Trucco. Pacific Grove, CA: Boxwood Press.

RICHARDSON, L. F. 1960s. *Statistics of deadly quarrels.* Ed. Q. Wright & C. C. Lienau. Pacific Grove, CA: Boxwood Press.

RICHARDSON, L. F. 1961. The problem of contiguity: an appendix of statistics of deadly quarrels. *General Systems Yearbook* **6**, 139-187.

RICHARDSON, L. F. & STOMMEL, H. 1948. Note on eddy diffusion in the sea. *J. of Meteorology* **5**, 238-240.

ROACH, F. E. & GORDON, J. L. 1973. *The light of the night sky.* Boston: Reidel.

ROGERS, C. A. 1970. *Hausdorff measures.* Cambridge University Press.

ROLL, R. 1970. *Behavior of interest rates: the application of the efficient market model to U.S. treasury bills.* New York: Basic Books.

ROSE, N. J. 1981. The Pascal triangle and Sierpiński's tree. *Mathematical Calendar* 1981, Raleigh, NC: Rome Press.

ROSEN, E. 1965. *Kepler's conversation with Galileo's siderial messenger.* New York: Johnson Reprint.

ROSENBLATT, M. 1961. Independence and dependence. *Proc. 4th Berkeley Symposium Mathematical Statistics and Probability* 441-443. Berkeley: University of California Press.

ROSENBLATT, M. & VAN ATTA, C. (Eds.) 1972. *Statistical models and turbulence.* Lecture Notes in Physics **12**. New York: Springer.

ROSS, B. (Ed.) 1975. *Fractional calculus and its applications.* Lecture Notes in Mathematics **457**. New York: Springer.

RUELLE, D. 1972. Strange attractors as a mathematical explanation of turbulence. In Rosenblatt & Van Atta *Lecture Notes in Physics* **12**, 292-299. New York: Springer.

RUELLE, D. & TAKENS, F. 1971. On the nature of turbulence. *Communications on Mathematical Physics* **20**, 167-192 & **23**, 343-344.

RUSSELL, B. 1897. *An essay on the foundations of geometry* Cambridge University Press (Dover reprint).

SAFFMAN, P. G. 1968. Lectures on homogeneous turbulence. *Topics in Nonlinear Physics* Ed. N. J. Zabusky. New York: Springer.

SALEM, R. & ZYGMUND, A. 1945. Lacunary power series and Peano curves. *Duke Mathematical J.* **12**, 569-578.

SAYLES, R. S. & THOMAS, T. R. 1978. Surface topography as a nonstationary random process. *Nature* **271**, 431-434 & **273**, 573.

SCHEFFER, V. 1976. Equations de Navier-Stokes et dimension de Hausdorff. *Comptes Rendus* (Paris) **282A**, 121-122.

SCHEFFER, V. 1977. Partial regularity of solutions to the Navier-Stokes equation. *Pacific J. of Mathematics.*

SCHÖNBERG, I. J. 1937. On certain metric spaces arising from Euclidean spaces by a change of metric and their imbedding on Hilbert space. *Annals of Mathematics* **38**, 787-793.

SCHÖNBERG, I. J. 1938a. Metric spaces and positive definite functions. *Tr. of the American Mathematical Society* **44**, 522-536.

SCHÖNBERG, I. J. 1938b. Metric spaces and completely monotone functions. *Annals of Mathematics* **39**, 811-841.

SELETY, F. 1922. Beiträge zum kosmologischen Problem. *Annalen der Physik* **IV**, **68**, 281-334.

SELETY, F. 1923a. Une distribution des masses avec une densité moyenne nulle, sans centre de gravité. *Comptes Rendus* (Paris) **177**, 104-106.

SELETY, F. 1923b. Possibilité d'un potentiel infini, et d'une vitesse moyenne de toutes les étoiles égale à celle de la lumière. *Comptes Rendus* (Paris) **177**, 250-252.

SELETY, F. 1924. Unendlichkeit des Raumes und allgemeine Relativitätstheorie. *Annalen der Physik* **IV**, **73**, 291-325.

SHANTE, V. K. S. & KIRKPATRICK, S. 1971. An introduction to percolation theory. *Advances in Physics* **20**, 325-357.

SHEPP, I. A. 1972. Covering the circle with random arcs. *Israel J. of Mathematics* **11**, 328-345.

SIERPIŃSKI, W. 1915. Sur une courbe dont tout point est un point de ramification. *Comptes Rendus* (Paris) **160**, 302. More detail in Sierpiński 1974-, **II**, 99-106.

SIERPIŃSKI, W. 1916. Sur une courbe cantorienne qui contient une image biunivoque et continue de toute courbe donnée. *Comptes Rendus* (Paris) **162**, 629. More detail in Sierpiński, 1974-, **II**, 107-119.

SIERPIŃSKI, W. 1974-. *Oeuvres choisies.* Ed. S. Hartman et al. Warsaw: Éditions scientifiques.

SINAI, JA. G. 1976. Self-similar probability distributions. *Theory of Probability and Its Applications* **21**, 64-80.

SINGH, A. N. 1935-53. *The theory and construction of nondifferentiable functions.* Lucknow (India): The University Press. Also in *Squaring the Circle and Other Monographs.* Ed. E. W. Hobson, H. P. Hudson, A. N. Singh & A. B. Kempe. New York: Chelsea.

SMALE, S. 1977. Dynamical systems and turbulence. *Turbulence Seminar Berkeley 1976/1977.* Ed. P. Bernard & T. Ratiu, *Lecture Notes in Mathematics* **615** 48-70. New York: Springer.

SMITH, A. 1972. The differentiability of Riemann's function. *Pr. of the American Mathematical Society* **34**, 463-468.

SMITH, H. J. S. 1894. *Collected mathematical papers* (Chelsea reprint).

SMYTHE, R. T. & WIERMANN, J. C., (Eds.) 1978. *First-passage percolation on the square lattice. Lecture Notes in Mathematics,* **671**, New York: Springer.

SODERBLOM, L. A. 1980. The Galilean moons of Jupiter. *Scientific American,* **242**, 88-100.

SOLER, J. 1973. Sémiotique de la nourriture dans la Bible. *Annales: Economies, Sociétés, Civilisations.* English translation: The dietary prohibitions of the Hebrews. *The New York Review of Books,* June 14, 1979, or *Food and Drink in History:* Ed. R. Foster & O. Ranum. Baltimore: Johns Hopkins University Press.

STANLEY, H. E. 1977. Cluster shapes at the percolation threshold: an effective cluster dimensionality and its connection with critical-point phenomena. *J. of Physics* **A10**, L211-L220.

STANLEY, H. E., BIRGENEAU, R. J., REYNOLDS, P. J. & NICOLL, J. F. 1976. Thermally driven phase transitions near the percolation threshold in two dimensions. *J. of Physics* **C9**, L553-L560.

STAPLETON, H. B., ALLEN, J. P., FLYNN, C. P., STINSON, D. G. & KURTZ, S. R. 1980. Fractal form of proteins. *Physical Review Letters* **45**, 1456-1459. (*See also* Allen et al. 1981)

STAUFFER, D. 1979. Scaling theory of percolation clusters. *Physics Reports* **34**, 1-74.

STEIN, P. R. & ULAM, S. 1964. Non-linear transformation studies on electronic computers. *Rozprawy Matematyczne* **39**, 1-66. Also in Ulam 1974, 401-484.

STEINHAUS, H. 1954. Length, shape and area. *Colloquium Mathematicum* **3**, 1-13.

STENT, G. 1972. Prematurity and uniqueness in scientific discovery. *Scientific American* **227** (December) 84-93.

STEVENSON, B. 1956. *The home book of quotations* (8th ed.), New York: Dodd-Mead.

STONE, E. C. & MINER, E. D. 1981. Voyager I encounter with the Saturnian system. *Science* **212**, Cover & 159-163.

STRAHLER, A. N. 1952. Hypsometric (area-altitude) analysis of erosional topography. *Geological Society of American Bulletin* **63**, 1117-1142.

STRAHLER, A. N. 1964. Quantitative geomorphology of drainage basins and channel networks. In *Handbook of Applied Hydrology* sect. 4-11. Ed. V. T. Chow. New York: McGraw-Hill.

SULLIVAN, D. 1979. The density at infinity of a discrete group of hyperbolic motions. *Institut des Hautes Etudes Scientifiques. Publications Mathematiques* **50**.

SUWA, N. & TAKAHASHI, T. 1971. *Morphological and morphometrical analysis of circulation in hypertension and ischemic kidney.* Munich: Urban & Schwarzenberg.

SUWA, N., NIWA, T., FUKASAWA, H. & SASAKI, Y. 1963. Estimation of intravascular blood pressure gradient by mathematical analysis of arterial casts. *Tohoku J. of Experimental Medicine* **79**, 168-198.

SUZUKI, M. 1981. Extension of the concept of dimension—phase transitions and fractals. *Suri Kagaku (Mathematical Sciences)* **221**, 13-20.

SWIFT, J. 1733. On Poetry, a Rhapsody.

TAQQU, M. S. 1970. Note on evaluation of R/S for fractional noises and geophysical records. *Water Resources Research,* **6**, 349-350.

TAQQU, M. S. 1975. Weak convergence to fractional Brownian motion and to the Rosenblatt process. *Z. für Wahrscheinlichkeitstheorie* **31**, 287-302.

TAQQU, M. S. 1977. Law of the iterated logarithm for sums of nonlinear functions of the Gaussian variables that exhibit a long range dependence. *Z. für Wahrscheinlichkeitstheorie,* **40**, 203-238.

TAQQU, M. S. 1978. A representation for self-similar processes. *Stochastic Processes and their Applications,* **7**, 55-64.

TAQQU, M. S. 1979a. Convergence of integrated processes of arbitrary Hermite rank. *Z. für Wahrscheinlichkeitstheorie* **50**, 53-83.

TAQQU, M. S. 1979b. Self-similar processes and related ultraviolet and infrared catastrophes. *Random Fields: Rigorous Results in Statistical Mechanics and Quantum Field Theory.* Amsterdam: North Holland.

TAYLOR, G. I. 1935. Statistical theory of turbulence; parts I to IV. *Pr. of the Royal Society of London* **A151**, 421-478. Reprinted in Friedlander

& Topper 1961, 18-51.

TAYLOR, G. I. 1970. Some early ideas about turbulence. *J. of Fluid Mechanics* **41**, 3-11.

TAYLOR, S. J. 1955. The α-dimensional measure of the graph and the set of zeros of a Brownian path. *Pr. of the Cambridge Philosophical Society* **51**, 265-274.

TAYLOR, S. J. 1961. On the connection between Hausdorff measures and generalized capacities. *Pr. of the Cambridge Philosophical Society* **57**, 524-531.

TAYLOR, S. J. 1964. The exact Hausdorff measure of the sample path for planar Brownian motion. *Pr. of the Cambridge Philosophical Society* **60**, 253-258.

TAYLOR, S. J. 1966. Multiple points for the sample paths of the symmetric stable process. *Z. für Wahrscheinlichkeitstheorie* **5**, 247-264.

TAYLOR, S. J. 1967. Sample path properties of a transient stable process. *J. of Mathematics and Mechanics* **16**, 1229-1246.

TAYLOR, S. J. 1973. Sample path properties of processes with stationary independent increments. *Stochastic Analysis*. Ed. D.G. Kendall & E.F. Harding. New York: Wiley.

TAYLOR, S. J. & WENDEL, J. C. 1966. The exact Hausdorff measure of the zero set of a stable process. *Z. für Wahrscheinlichkeitstheorie* **6**, 170-180.

TENNEKES, H. 1968. Simple model for the small scale structure of turbulence. *Physics of Fluids* **11**, 669-672.

TESNIÈRE, M. 1975. Fréquences des noms de famille. *J. de la Société de Statistique de Paris* **116**, 24-32.

THOMA, R. 1901. Über den Verzweigungsmodus der Artererien. *Archiv der Entwicklungsmechanik* **12**, 352-413.

THOMPSON, D'A. W. 1917-1942-1961. *On growth and form*. Cambridge University Press. The dates refer to the first, second and abridged editions.

ULAM, S. M. 1957. Infinite models in physics. *Applied Probability*. New York: McGraw-Hill. Also in Ulam 1974, 350-358.

ULAM, S. M. 1974. *Sets, numbers and universes: selected works*. Ed. W. A. Beyer, J. Mycielski & G.-C. Rota. Cambridge, MA: M.I.T. Press.

URYSOHN, P. 1927. Mémoire sur les multiplicités cantoriennes. II: les lignes cantoriennes. *Verhandelingen der Koninglijke Akademie van Wetenschappen te Amsterdam*. (Eerste Sectie) **XIII** no. 4.

VAN DER WAERDEN, B. L. 1979. *Die Pythagoreer*.

VILENKIN, N. YA. 1965. *Stories about sets*. New York: Academic.

VON KOCH, H. 1904. Sur une courbe continue sans tangente, obtenue par une construction géométrique élémentaire. *Arkiv för Matematik, Astronomi och Fysik* **1**, 681-704.

VON KOCH, H. 1906. Une méthode géométrique élémentaire pour l'étude de certaines questions de la théorie des courbes planes. *Acta Mathematica* **30**, 145-174.

VON NEUMANN, J. 1949-1963. Recent theories of turbulence. The dates refer to publication as a report to ONR and in von Neumann, 1961- **6**, 437-472.

VON NEUMANN, J. 1961- *Collected works*. Ed. A. H. Traub. New York: Pergamon.

VON SCHWEIDLER, E. 1907. Studien über die Anomalien in Verhalten der Dielektrika. *Annalen der Physik* **(4)24**, 711-770.

VON WEIZSÄCKER, C. F. 1950. Turbulence in interstellar matter. *Problems of Cosmical Aerodynamics* (IUTAM & IAU). Dayton: Central Air Documents Office.

VOSS, R. F. & CLARKE, J. 1975. "1/f noise" in music and speech. *Nature* **258**, 317-318.

VOSS, R. F. 1978. 1/f noise in music; music from 1/f noise. *J. of the Acoustical Society of America* **63**, 258-263.

VOSS, R. F. 1982. Fourier synthesis of Gaussian fractals: 1/f noises, landscapes, and flakes (to appear).

WALLENQUIST, A. 1957. On the space distribution of galaxies in clusters. *Arkiv för Matematik, Astronomi och Fysik* **2**, 103-110.

WALSH, J. L. 1949. Another contribution to the rapidly growing literature of mathematics and human behavior. *Scientific American* (August issue) 56-58.

WEIBEL, E. R. 1963. *Morphometry of the human lung*. New York: Academic.

WEIBEL, E. 1979. *Stereological methods* (2 vols.). London: Academic.

WEIERSTRASS, K. 1872. Über continuirliche Functionen eines reellen Arguments, die für keinen Werth des letzteren einen bestimmten Differentialquotienten besitzen. Unpublished until Weierstrass 1895-, **II**, 71-74.

WEIERSTRASS, K. 1895-. *Mathematische Werke*. Berlin: Mayer & Muller.

WEYL, H. 1917. Bemerkungen zum begriff der differentialquotenten gebrochener ordnung. *Vierteljahrschrift der Naturförscher Geselschaft in*

Zürich **62**, 296-302.

WHITTAKER, E. T. 1953. *A history of the theories of aether and electricity.* New York: Philosophical Library.

WHYBURN, G. T. 1958. Topological characterization of the Sierpiński curve. *Fundamenta Mathematicae* **45**, 320-324.

WIENER, N. 1948-1961. *Cybernetics.* Paris: Hermann. New York: Wiley (1st edition). Cambridge, MA: M.I.T. Press (2d edition).

WIENER, N. 1953. *Ex-prodigy.* New York: Simon & Schuster. Cambridge, MA: M.I.T. Press.

WIENER, N. 1956. *I am a mathematician.* Garden City, N.Y.: Doubleday. Cambridge, MA: M.I.T. Press.

WIENER, N. 1964. *Selected papers.* Cambridge, MA: M.I.T. Press.

WIENER, N. 1976-. *Collected works.* Ed. P. Masani. Cambridge, MA: M.I.T. Press.

WIGNER, E. P. 1960. The unreasonable effectiveness of mathematics in the natural sciences. *Communications on Pure and Applied Mathematics* **13**, 1-14. Also in Wigner 1967, 222-237.

WIGNER, E. P. 1967. *Symmetries and reflections.* Indiana University Press. MIT Press Paperback.

WILLIS, J. C. 1922. *Age and area.* Cambridge University Press.

WILSON, A. G. 1965. Olbers' paradox and cosmology. Los Angeles, Astronomical Society.

WILSON, A. G. 1969. Hierarchical structures in the cosmos. *Hierarchical Structures,* 113-134. Ed. L. L. Whyte, A. G. Wilson & D. Wilson. New York: American Elsevier.

WILSON, K. 1975. The renormalization group: critical phenomena and the Kondo problem. *Reviews of Modern Physics* **47**, 773-840.

WILSON, K. G. 1979. Problems in physics with many scales of length. *Scientific American* **241** (August issue) 158-179.

WILSON, J. T. (Ed.) 1972. *Continents adrift.* Readings from *Scientific American.* San Francisco: W. H. Freeman.

WILSON, T. A. 1967. Design of the bronchial tree. *Nature* **213**, 668-669.

WOLF, D. (Ed.) 1978. *Noise in physical systems.* (Bad Neuheim Conference) New York: Springer.

YAGLOM, A. M. 1957. Some classes of random fields in n-dimensional space, related to stationary random processes. *Theory of Probability and Its Applications,* **2**, 273-320. Tr. R. A. Silverman.

YAGLOM, A. M. 1966. The influence of fluctuations in energy dissipation on the shape of turbulence characteristics in the inertial interval. *Doklady*

Akademii Nauk SSSR **16**, 49-52. (English trans. *Soviet Physics Doklady* **2**, 26-29.)

YODER, L. 1974. Variation of multiparameter Brownian motion. *Pr. of the American Mathematical Society* **46**, 302-309.

YODER, L. 1975. The Hausdorff dimensions of the graph and range of N-parameter Brownian motion in d-space. *Annals of Probability* **3**, 169-171.

YOUNG, W. H. & YOUNG, G. C. 1906. *The theory of sets of points.* Cambridge University Press.

YULE, G. UDNY 1924. A mathematical theory of evolution, based on the conclusions of Dr. J. C. Willis, F. R. *Philosophical Tr. of the Royal Society (London)* **213 B**, 21-87.

ZIMMERMAN, M. H. 1978. Hydraulic architecture of some diffuse-porous trees. *Canadian J. of Botany,* **56**, 2286-2295.

ZIPF, G. K. 1949. *Human behavior and the principle of least-effort.* Cambridge, MA: Addison-Wesley. (Hefner reprint.)

ZYGMUND, A. 1959. *Trigonometric series.* Cambridge University Press.

COMPUTER ILLUSTRATION CREDITS

10:V	121:V	230:L
11:V	141:H	231:L
31:L	143:N	242:N
32:H	146:L	243:N
42:H	155:H	246:H
43:H	163:H	255:H
44:H	164:H	264:V
45:HL	165:H	265:V
46:H	170:H	266:V
47:L	173:L	267:V
49:M	177:N	268:H
51:V	178:N	269:V
53:V	179:N	271:H
54:V	185:LN	285:H
55:V	187:N	286:L
56:H	188:LN	287:L
57:V	189:LN	293:L
63:H	190:LN	295:H
64:H	191:N	296:HM
66:M	192:M	297:HM
67:H	198:L	298:M
68:L	199:L	299:M
69:L	218:H	300:H
70:H	219:L	306:H
71:M	220:H	307:H
73:LM	221:H	308:H
80:H	222:H	309:H
81:M	223:H	318:M
83:H	227:H	323:L
95:H	228:V	325:L
96:H	229:L	

Sigmund W. Handelman
Richard F. Voss
Mark R. Laff
V. Alan Norton
Douglas M. McKenna

implemented most of the computer illustrations in this Essay. They are listed above in the chronological order of their first contribution.

In the itemized list on this page, which concerns the black-and-white illustrations, each plate number is followed by the initial of the author of the generating program. Illustrations that were improved upon by different hands are credited to all the direct contributors. Color illustrations are credited separately.

Other persons' kind assistance was also vital in diverse ways; they too are listed chronologically. Hirsh Lewitan contributed to Plates 296 and 297. Gerald B. Lichtenberger contributed indirectly to several plates. Plate 270 is by Jean-Louis Oneto, using a pioneering graphics package by Cyril N. Alberga. Plate 271 is a revised form of one due to Arthur Appel & Jean-Louis Oneto. Scott Kirkpatrick contributed Plate 132, and provided programs that served in the preparation of Plates 220 to 223 and 306 to 309. Peter Oppenheimer contributed to the diagrams on page 173. Peter Moldave contributed to Plates 188 to 191. David Mumford & David Wright contributed to Plate 178.

The frontispiece and end-papers are by V. A. Norton.

ACKNOWLEDGMENTS

Contrary to books that are undertaken with a precise idea of their final scope and style, the present *"macédoine de livre"* emerged gradually, in a long process. Directly or between the lines, in digressions and in the biographical and historical sketches, my main intellectual debts have already been acknowledged. Their very number and continually increasing diversity underline that none was predominant.

However, Norbert Wiener and John von Neumann have been slighted by the hazards of citation: both were kind to my work, and influenced me greatly, even more by example than by deed.

Other very different major intellectual influences that have not yet been acknowledged suitably are those of my uncle and of my brother.

A rough translation of the first (French) version was contributed by J. S. Lourie. R. W. Gosper of Stanford showed me his Peano curve before it was published. M. P. Schützenberger of Paris, J. E. Marsden of Berkeley, M. F. M. Osborne of U.S.N.R.L., Jacques Peyrière of Orsay, Y. Gefen and A. Aharony of Tel Aviv, and D. Mumford and P. Moldave of Harvard were very helpful in diverse ways.

P. L. Renz, editor at W. H. Freeman and Company, proved that his professional Guild is not entirely beyond redemption. I am grateful to him for agreeing to the idiosyncratic layout with which I wished to experiment. I am also most grateful to R. Ishikawa, at W.H. Freeman and Company.

Attractive quotes were pointed out by M. V. Berry, K. Brecher, I. B. Cohen, H. de Long, M. B. Girsdansky, A. B. Meador, J. C. Pont, M. Serres, B. L. van den Waerden, and D. Zajdenweber. Other quotes were previously used by G. Birkhoff, R. Bonola, J. Bromberg, C. Fadiman, T. Ferris, J. Gimpel, C. J. Glacken, D. M. Johnson, P. S. Stevens, and E. T. Whittaker.

M. C. Gutzwiller, P. E. Seiden, J. A. Armstrong, and P. Chaudhari, departmental directors at IBM, helped this work proceed smoothly.

D. F. Bantz smiled at our use of the color graphics equipment of his project. I. M. Cawley, C. H. Thompson, P. G. Capek, J. K. Rivlin and others on the library, text processing and graphics staffs of IBM Research were uncommonly helpful, and tolerated a deliberate policy to push every one of their devices beyond its design performance.

INDEX OF SELECTED DIMENSIONS:
EUCLIDEAN (E), FRACTAL (D), AND TOPOLOGICAL (D_T)

The letters *a* and *b* refer, respectively, to the left and right columns of a page in the text. The absence of letters means both columns. Bold numbers refer to chapters devoted to the entry. When Euclidean dimension is denoted by E, its value is an arbitrary positive integer.

I: BASIC GEOMETRIC SHAPES, AND THEIR RIGOROUS D AND D_T

	E	D	D_T	pages
Brown H fractional line-to-space fractals:				
-trail when H>1/E	E	1/H	1	252b,253a
-zeroset	1	1–H	0	252b,253a
-function	2	2–H	1	252b,253a
Brown H fractional space-to-line fractals:				
-$\mathbb{R}^2$-to-$\mathbb{R}$ function	3	3–H	2	354a
-zeroset of the $\mathbb{R}^2$-to-R function	2	2–H	1	354a
-Kolmogorov scalar turbulence isosurfaces	3	8/3	2	**30**
Lévy stable process with D<2: -trail	E	D	0	372

II: OTHER GEOMETRIC SHAPES, THEIR D_T, AND THEIR ESTIMATED D

- NONSCALING NONRANDOM FRACTAL SETS

	E	D	D_T	pages
Apollonian gasket and net (Exact bounds:1.300197<D<1.314534)	2	1.3058	1	172a

- RANDOM FRACTAL SETS

	E	D	D_T	pages
Rescaled self-avoiding random walk/polygon in $\mathbb{R}^2$	2	1.33	1	239b
Rescaled self-avoiding random walk in $\mathbb{R}^3$	3	1.67	1	239b
River in a Leopold & Langbein network	2	1.28	1	332b ff
Critical Bernoulli percolation cluster				
-full cluster in the plane	2	1.89	1	129b
-backbone in the plane	2	1.6	1	132
-backbone in $\mathbb{R}^E$, with small E	E	$\log_2(E+1)$	1	133a

III: NATURAL STANDARD (EUCLIDEAN) OBJECTS, AND THEIR D_T AND D

	E	D	D_T	pages
Very fine ball	E	0	0	18a
Very fine thread	E	1	1	18a
Empty (polished inside and out) sphere	3	2	2	18a
Polished ball (solid inside)	3	3	3	18a

IV: NATURAL FRACTAL OBJECTS, THEIR ESTIMATED D_T, AND THEIR TYPICAL D

	E	D	D_T	pages
Seacoast (Richardson exponent)	2	1.2	1	33
River network's cumulative bank	2	2	1	**7**
Individual river's outline (Hack exponent)	2	1.2	1	110a
Vascular system	3	3	2	149b ff
Pulmonary membrane on branching scales	3	2.90	2	114a,157a ff
Tree's bark	3	3	2	
Fractal errors	1	0.30	0	**8**
Galaxies in the scaling zone of sizes	3	1.23	0	**9**
Turbulence: support of dissipation	3	2.50–2.60	2	**10, 11**
Word frequencies	n.a.	0.9	n.a.	**38**

INDEX OF NAMES AND SUBJECTS

The letters *a* and *b* following a page number refer, respectively, to the left and right columns of the text. The absence of letter means "both *a* and *b*." *Passim* means "throughout the book."
   Bold numbers refer to Chapters wherein many references to the topic are included.
   The letter C preceding a page number refers to the color signature.
   Starred entries involve neologisms, like *fractal* or *squig*, or words for which this Essay introduces a nonstandard meaning, like *dust* or *trema*.

Update added in the second printing (XII 1982)

COURCHEVEL WORKSHOP: PREVIEW OF THE FORTHCOMING PROCEEDINGS

Between the delivery of this book to the publisher and its actual publication, then during the brief period before the first printing was exhausted, fractal geometry did not stand still: It moved on at increasing speed in the domains where it was already accepted, and it moved into a number of new domains.

In particular, I organized a week-long workshop on fractals in July, 1982, at Courchevel (France), and many new developments were first presented there. This update's main goal is to summarize these results and closely related ones. Some supplementary references (marked by a star *) call attention to other works presented at the workshop.

More generally, it is becoming hard to believe that only a few years ago, the fractal geometry of nature was near-exclusively my work and that of close associates. However, I can at best draw attention to some new actors, via additional supplementary references.

The topics are placed in roughly the same order as in the body of the book.

THE DEFINITION OF "FRACTAL"

This dull topic is unfortunately unavoidable, but will take mercifully little space.

To my chagrin, the term "Hausdorff dimension" has started being applied indiscriminately to either of the dimensions listed in Chapter 39, and to further variants thereof. The same is true of "Minkowski dimension," a term used once on page 164 of the 1975 *Objets fractals,* to denote Bouligand dimension. Apparently, certain foreign language articles, whose authors and topics cease to be feared as the result of my work, acquire prestige value, hence are credited—unseen!—with a variety of achievements...or crimes.

Other writers go to the opposite excess: they overstress the methods most often used to *estimate* D in practical work, such as the similarity dimension as used on pp. 130a and 214, the exponent in the mass-radius relation or a spectral exponent, and they proceed to enshrine them to define "the" fractal dimension.

It is a pity that most of these reactions to the 1977 *Fractals* manifested themselves a bit too late. They would have encouraged me to return in the present book to the well-inspired approach taken in the 1975 *Objets fractals*: to leave the term "fractal" without a pedantic definition, to use "fractal dimension" as a generic term applicable to *all* the variants in Chapter 39, and to use in each specific case whichever definition is the most appropriate.

HOMOGENEOUS FRACTAL TURBULENCE

My major conjecture on turbulence is the object of Chapter 11: it asserts that turbulence in real space is a phenomenon carried by a fractal set of dimension D~2.5 to 2.6.

Numerical work in support of this conjecture continues, witness Chorin 1982a,b.

In addition, a totally different approach has been recently advanced in Hentschel & Procaccia 1982, which handles the lengthening and folding vortices of Chapter 10 by the methods developed to handle the polymers of Chapter 36, and suggests a relation between the dimensions of turbulence and of polymers.

METAL FRACTURES AND FRACTALS (B.B.M., PASSOJA & PAULLAY 1983)

Neologisms, as mentioned in Chapter 1, demand care, and one should avoid bad conflicts of meaning. Casual examination suggested that, while broken glass surfaces are most likely *not* fractal, many stone or metal fracture surfaces *are* fractal. This informal evidence suggested that *fractal* and *fracture*

should not conflict badly.

Mandelbrot, Passoja & Paullay 1983 buttresses this informal feeling by extensive experimental evidence concerning 1040, 1095 and Cor-99 steel tensile specimens and Maraging steel impact specimens. The fractal character is tested and the value of the dimension D is estimated using methods like those Chapters 5 and 28 use for relief. These methods' success is noteworthy, because fracture surfaces are conspicuously *non*-Gaussian, and quite *unlike* the relief.

Recall that Chapters 5 and 28 proceed via island coastlines and vertical sections. Unfortunately, fractures do not naturally exhibit islands, and the definition of the vertical (as the direction such that the altitude is a single-valued function of the position in the horizontal plane) is seldom satisfied by any direction.

Nevertheless, we can define an informal vertical by the condition that the altitude is single-valued for "most" points. We then spectral-analyze the altitudes along rectilinear horizontal sections, and plot log (spectral energy above the frequency f) as a function of log f.

In addition, we find it useful to create artificial "slit islands" by "slicing" the sample along near horizontal planes (the sample is first plated with electroless nickel, and mounted in an epoxy mount by vacuum impregnation). Then we use a fixed yardstick to measure each island's area and perimeter on a digitized picture, and we plot the logarithms as suggested in Chapter 12, to test the validity of fractal dimensional analysis.

As exemplified at the bottom of page 461, very many fracture surfaces follow the fractal model admirably: both diagrams are very nearly straight and their slopes yield essentially identical D's. Furthermore, repeat of the same procedure for different samples of the same metal recovers the same D. In contrast, the traditional estimates of roughness are hard to repeat.

To echo a comment on page 112 concerning Plate 115, very few graphs in metallurgy involve all the available data and a very broad range of sizes, and are as straight as ours.

The data are so good that we can proceed immediately to a finer comparison. We observe that |D (spectral)−D (islands)| is systematically of the order of a few hundreds. A first possible cause resides in estimation bias. For example, the high frequency spectrum is overwhelmed by measurement noise, hence must be disregarded. Furthermore, we handle "lakes" and "offshore islands" the easy way: including the former and neglecting the latter because they are ill-defined.

But the discrepancy may be real. As a matter of fact, the near-identity of the D's suggested that the materials we studied were far more isotropic than expected. And for samples that *must* be anisotropic because of the way they were prepared, D(spectral) and D(island) are indeed clearly different.

An alternative explanation for conflicting D's is that the fracture may be isotropic but not self-similar, with D varying with scale (Chapter 13). Since our two methods give different weights to different ranges of scales,

they would reflect the variation of D. Indeed, for some metals we examined the slit island or spectral diagrams exhibit two clearly distinct straight zones, and for yet other metals the diagrams are even more complex.

To relate D to a metal's other characteristics, we took 300 Grade Maraging steel Charpy impact specimens, and heat treated at different temperatures. The resulting diagram, also shown on the bottom of page 461, exhibits an unmistakable relation between the impact energy and the value of D.

The facts having been established, it is worth pondering their possible causes. Our view is that fracture involves an atypical form of percolation. Let us recall that, as a specimen is pulled apart, the voids that are inevitably present around the inclusions increase in size, and eventually they coalesce into sheets that separate the specimen into parts. If the growth of a void were independent of its position, the percolation would be as in Chapter 13. Consequently, the fracture's dimension would take some universal value independent of the material. In fact, as soon as the initial void growth has coalesced into small local sheets, the strains increase on the supporting ligaments and a void grows at a rate that varies with position. There is no doubt this variability is structure dependent, hence the D need not be universal.

CLOUD AND RAIN AREA SHAPES
(LOVEJOY 1981, LOVEJOY & B.B.M. 1983)

Lovejoy's remarkable area-perimeter relation (Plate 115) is a challenge to do what Chapter 28 had done for the Earth's relief, namely, to generate fractal maps of clouds or rain areas that neither the eye nor measurement could distinguish from the meteorological maps.

A vital ingredient in the case of rain areas is provided by the finding in Lovejoy 1981, that the discontinuities in rainfall follow precisely the same hyperbolic probability distribution as the discontinuities in commodity prices according to Mandelbrot 1963b (see Chapter 37).

Lovejoy & Mandelbrot 1983 builds on this finding. Hyperbolically distributed discontinuities are shown to agree with the well-known observation that rain discontinuities occur along near rectilinear "fronts." To preserve scaling, a suitable list of exponents is introduced, reminiscent of those of the theory of critical phenomena, and even more of the turbulence exponents introduced in Mandelbrot 1976o. The outcome is extremely rewarding.

SCALING, FRACTALS & EARTHQUAKES
(KAGAN & KNOPOFF & ANDREWS)

Recall the assertions in Chapter 28, that the Earth's relief is a scaling fractal, and that it can be generated as a superposition of crude "faults." Belief in these assertions prepares one to be told that earthquakes, which are dynamic changes in the relief, are self-similar, i.e., no particular scale is connected with their time-distance-magnitude patterns, and that their geometry is fractal. These are indeed the main messages that a student of fractals retains from reading (as he is hereby advised to) Kagan & Knopoff 1978, 1980, 1981 and Andrews 1980-1981.

It is chastening to be told that Omori discovered scaling in earthquakes nearly a hundred years ago, yet the bulk of statistical work on earthquakes persisted in postulating that the occurrences are Poissonian. Again, little good can come (as I argue in Chapter 42) when a science yields to the social pressures that reward modeling and theorizing while scorning "mere" description without "theory."

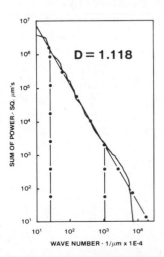

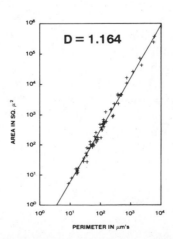

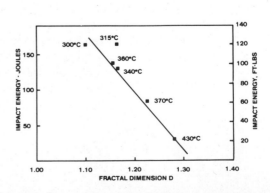

FRACTAL INTERFACES IN LITHIUM BATTERIES (A. LE MÉHAUTÉ & al.)

An electric battery is to store electricity in large amounts, and to discharge it rapidly. Everything else being fixed, storage capacity is a volume characteristic, but discharge velocity is a surface characteristic. This feature is familiar to the student of fractals (Chapters 12 and 15), and convinced Alain Le Méhauté that the balance between capacity and discharge poses a fractal problem.

A battery whose planar cross section is a Peano teragon (e.g., Plate 70) being irrealizable, Le Méhauté et al. 1982 study realistic designs theoretically, and also examine actual batteries. The effectiveness of fractal geometry is very striking.

CRITICAL PERCOLATION CLUSTERS

PERCOLATION ON LATTICES: TESTING THE MODEL OF CHAPTER 13. The specified fractal model of contact clusters in Bernoulli percolation, as proposed in Chapter 13, cries out to be verified empirically. This has now been done.

Kapitulnik, Aharony, Deutscher & Stauffer 1983 studies the number of sites in a cluster at a distance less than R from an origin, and recovers the correct $D \sim 1.9$. In addition, it recovers ξ from the crossover between the fractal region and the region of homogeneity.

PERCOLATION IN THIN FILMS OF GOLD AND LEAD. Bernoulli percolation is of course only a mathematical process. Hammersley intro-

duced it in the hope that many natural phenomena can be illustrated and clarified thereby. The applicability of the fractal geometry of Bernoulli percolation was tested for vile gold in Voss, Laibowitz & Alessandrini 1982, and for noble lead in Kapitulnik & Deutscher 1982. For example, the students of Au prepared thin films at room temperatures by electron beam evaporation onto 30 nm thick amorphous Si_3N_4 windows grown on a Si wafer frame. Sample thickness was varied to produce simultaneously a range of samples that varied from electrically insulating to conducting. The predictions of Chapter 13 are satisfied on the dot.

LOW LACUNARITY FRACTAL MODELS OF SOME FORMAL SPACES IN PHYSICS (GEFEN, MEIR, B.B.M. & AHARONY 1983)

Statistical physics finds it useful to postulate spaces of fractional dimension. Mathematicians find these spaces very upsetting, because they are nowhere constructed, and their existence and unicity are nowhere proven. Nevertheless, useful physics is achieved by assuming they do exist and possess certain strong and desirable properties: they are translationally invariant and their momentum integrals and recursion relations are obtainable by formal analytic continuations from Euclidean spaces.

These spaces puzzle the student of fractals. On the one hand, there exist *many* alternative fractal interpolated spaces, hence interpolation should have been indeterminate. On the other hand, the fractals which Gefen, Mandelbrot & Aharony 1980 applies to physics fail to be translationally invariant. In that regard, fractals may seem inferior to the postulated fractional spaces.

A response was suggested by the analogous criticism leveled against my first model of the distribution of galaxies. While it is impossible for a fractal to be *exactly* translationally invariant, Chapters 34 and 35 show that one can come as close as desired by giving a sufficiently low value to lacunarity.

In this light, Gefen, Meir, Mandelbrot & Aharony 1983 considers a certain sequence of Sierpiński carpets (Chapter 14), whose lacunarity tends to 0. Certain physical properties are computed, and the limits for zero lacunarity are shown to be identical to the properties of the postulated fractional spaces.

SIERPINSKI GASKET: PHYSICISTS' TOY

Manageable models are so attractive to physicists that every construction that promises calculations without the need for approximation will attract wide attention.

Among the ramified shapes examined in Chapter 14, the Sierpiński carpet is the more important one, but it is hard to work with. But the Sierpiński gasket is easy to manipulate. It yields fun and profit in Stephen 1981, Rammal & Toulouse 1982, 1983 and Alexander & Orbach 1982.

◁ Contrary to habit, I coined "gasket" without a French equivalent. The authors of a mathematics dictionary did not know that I had in mind the part that prevents leaks in motors, and a standard dictionary led them to ships and ropes, hence to *baderne* or *garcette*. Since the word did not fit, it was redefined to apply to the complement of what I had meant! I prefer *tamis* (sieve). ▶

CELLULAR AUTOMATA & FRACTALS

To show that global order can be generated by forces that act solely between neighbors, I cooked up the example on p. 328. Someone soon pointed out that this example involves a "cellular automaton" according to John von Neumann (Burks 1970). It had been shown by Ulam (Burks 1970) that the output of such automata can be very involved and appear random. Willson 1982, Wolfram 1983, and Vichniac 1983 observe that this output can, in fact, be fractal.

ITERATION OF $z \to z^2 - \mu$ IN COMPLEX NUMBERS: NEW RESULTS AND PROOFS

Mandelbrot 1983p includes many illustrations for which space had lacked in Chapter 19, and reports additional observations. Mandelbrot 1982s has been delayed, and is expected out in 1983.

Two major observations in Chapter 19 have now been confirmed mathematically.

Douady & Hubbard 1982, Douady 1983 prove that the closed set $\mathcal{M}$ is indeed connected. They map the exterior of $\mathcal{M}$ on that of a circle.

Ruelle 1982 proves that the Hausdorff dimension of a Julia dragon is an analytic function of the parameter μ.

THE SQUARING MAPS IN QUATERNIONS

Chapter 19 established that the properties of the map $z \to z^2 - \mu$ for real z are best understood as special cases of its properties for complex z and μ, and that iteration for complex z generates unexpected and exciting graphics. Therefore, it was natural to seek further insight and further beauty via a further generalization of z. A. Norton suggested that a next most natural environment is Hamilton's quaternions. Having been introduced in 1847, quaternions are a familiar notion in both mathematics and physics, but their role had remained peripheral. In the context of iteration, however, quaternions have proved extremely fruitful from both the mathematical and the esthetic viewpoints, as will be seen in detail in forthcoming papers by Norton and myself.

One objection often directed against quaternions is that, while complex numbers insert a space with $E=1$ into a space with $E=2$, which can be visualized, quaternions require a jump to a space with $E=4$, which cannot be visualized. A second objection is that quaternion multiplication is not commutative: in particular, the maps $z \to \lambda z(1-z)$, $z \to z^2 - \mu$, $z \to \mu z^2 - 1$, and $z \to \mu^\alpha z^2 \mu^{1-\alpha}$, differ when z is a quaternion.

To illustrate the topological interconnections of the fractal repellers of the quadratic map in quaternions, new computer graphics techniques were developed in Norton 1982. The sets of all quaternions that fail to iterate to infinity were examined in 3-dimensional sections. Their complex plane sections are, in turn, the fractal dragons of Chapter 19.

The noncommutativity of quaternion multiplication has turned out to transform into a fascinating and totally unexpected asset. To explain it, consider Plate C6. Question: do all or some of the dark yellow domains link in quaternion space? Answer: in general, each variant way of writing $z \to z^2 - \mu$ or $z \to \lambda z(1-z)$ (before moving on to quaternions) induces totally different links between the dark yellow domains. Hence, additional information is required to specify the topological interconnections.

For an example that avoids clutter, examine Plate 467, which is adapted from Norton 1982 and illustrates a simple case with a cycle of size 4. Each major segment of the dragon obtained by complex plane section is imbedded in a major segment of the spatial shape. In this instance, the major spatial sections are nearly rotationally invariant, and they are surrounded by multiple loosely fitting belts that connect the dragon's minor sections. Plate II shows a different spatial fractal obtained in roughly the same fashion. Stein 1983 reproduces further illustrations.

UNIVERSALITY AND CHAOS: $z \to \lambda(z-1/z)$ AND OTHER MAPS

A contemporary of Fatou and Julia, S. Lattès, singled out a fourth order ratio of polynomials whose iterates are "chaotic" in the whole plane, that is, not attracted to any smaller set. This example challenges us to search for chaotic behavior in lower order mappings. A second topic handled in this section is that of universality classes for the shape of islands in λ-maps.

$z \to \lambda(z-1/z)$ AND ITS λ-MAP. In the special case $\lambda=\frac{1}{2}$, $y=-iz$ follows the rule $y \to \frac{1}{2}(y+1/y)$, which also results from the application of Newton's method to the search of the roots of z^2-1. Note that one can write $z = \cotan \theta$, and $\frac{1}{2}(z-1/z)$ becomes $(\cos^2\theta-\sin^2\theta)/2\cos\theta\sin\theta = \cotan2\theta$. Thus, $z \to \frac{1}{2}(z-1/z)$ is a funny way of writing $\theta \to 2\theta$. To study other λ's, a map analogous to Plates 188 and 189 was drawn. and part of it is shown on Plate X,

We observe a very interesting form of "universality:" the "island molecules" in Plate X take precisely the same form as for the quadratic mapping. Thus, Plates X and 188-189 are built using the same "building blocks." In the open disc $|\lambda|>1$, the iteration of $z \to \lambda(z-1/z)$ converges to infinity except for points z_0 forming a dust. In the white disc $|\lambda+i/2|<1/2$, the iteration has 2 limit points. When λ falls in one of the "sprouts" of the black "corona," there is a limit cycle whose size is above 2 but not very large. As to the λ's inside the corona of the λ-map, they yield chaotic motion.

◁ The actual calculation was simplified on the following presumptions. A) When λ leads to a very large cycle, it falls within a very small atom that is not worth looking for. B) All usefully small cycles lie "near" $z=0$. Thus, any orbit that moves "far" from $z=0$ is presumed to be chaotic. The approximation lacks specific justification, but the λ-map it yields is made of familiar pieces, hence the method seems reasonable. ►

JULIA SETS OF $\lambda(z-1/z)$. When $|\lambda|>1$, infinity is an attractive point, and the Julia set is, as in Chapter 19, the boundary of the set of the z-points that do not converge to infinity. An example of Julia set defined as the boundary of the basins of attraction of $z \to \lambda(z-1/z)$ is drawn on Plate VIII, facing the foreword.

λ-MAP "UNIVERSALITY" CLASSES. In many other λ-maps, one finds the same "island molecules" as for $z^2-\mu$, except that specific constraints may create an atypical "continent."

Furthermore, the λ-maps of $z \to z^m-\lambda$ also divide into a continent and islands. However, each m induces a very characteristic shape for the atoms and for the island molecules.

When the local behavior of $z \to f(z)$ is the same near every critical z where $f'(z) = 0$, the islands' shape is locally determined. When $f(z)$ behaves differently near different critical z's, the λ-map involves more than one kind of "universal" building block. We seek a "Mendeleyev Table" for this problem. ■■

UPDATE REFERENCES, AND BRIEF SUPPLEMENTARY BIBLIOGRAPHY

ALEXANDER, S. & ORBACH, R. 1982. Density of states on fractals: "fractons". *Journal de Physique Lettres* **43**, 625-

*AGTENBERG, F. P. 1982. Recent developments in geomathematics. *Geo-processing* **2**.

ANDREWS, D. J. 1980-81. A stochastic fault model. I Static case, II Time-dependent case. *Journal of Geophysical Research* **85B**, 3867-3877 and **86B**, 10821-10834.

*BLEI, R. 1983. Combinatorial dimension: a continuous parameter. *Symposia Mathematica* (Italia), to appear.

BURKS, A. W. (Ed.) 1970. *Essays on Cellular Automata*, Urbana, IL: University of Illinois Press

*BURROUGH, P. A. 1981. Fractal dimensions of landscapes and other environmental data. *Nature* **294**, 240-242.

*CANNON, J. W. 1982. Topological, combinatorial and geometric fractals. *The 31st Earle Raymond Hedrick Lectures of the Mathematical Association of America,* delivered at the Toronto Meeting.

CHORIN, A. 1982a. The evolution of a turbulent vortex. *Communication in Mathematical Physics* **83**, 517-535.

CHORIN, A. 1982b. Numerical estimates of Hausdorff dimension. *Journal of Computational Physics* **46**.

*DEKKING, F. M. 1982. Recurrent sets. *Advances in Mathematics* **44**, 78-104.

DOUADY, A. & HUBBARD, J. H. 1982. Itération des polynomes quadratiques complexes. *Comptes Rendus* (Paris) **294I**, 123-126.

GEFEN, Y., AHARONY, A. & MANDELBROT, B. 1983. Phase transitions on fractals: I. Quasi-linear lattices. *Journal of Physics A.*

GEFEN, Y., MEIR, Y., MANDELBROT, B. & AHARONY, A. 1983. Geometric implementation of hypercubic lattices with noninteger dimensionality, using low lacunarity fractal lattices. *To appear.*

*GILBERT, W. T. 1982. Fractal geometry derived from complex bases. *Mathematical Intelligencer* **4**, 78-86.

*HATLEE, M. D. & KOZAK, J. J. 1981. Stochastic flows in integral and fractal dimensions and morphogenesis. *Proceedings of the National Academy of Sciences USA* **78**, 972-975.

HENTSCHEL, H. G. E. & PROCACCIA, I. 1982. Intermittency exponent in fractally homogeneous turbulence. *Physical Review Letters* **49**, 1158-1161.

*HENTSCHEL, H. G. E. & PROCACCIA, I. 1983. Fractal nature of turbulence as manifested in turbulent diffusion. *Physical Review A* (Rapid Communication).

*HUGHES, B. D., MONTROLL, E. W. & SHLESINGER, M. F. 1982. Fractal random walks. *Journal of Statistical Physics* **28**, 111-126.

*KAC, M. Recollections concerning Peano curves and statistical independence. *Probability, Number Theory and Statistical Physics (Selected Papers)* Cambridge, MA: M.I.T. Press, ix-xiii.

KAGAN, Y. Y. & KNOPOFF, L. 1978. Statistical study of the occurrence of shallow earthquakes. *Geophysical Journal of the Royal Astronomical Society* **55**, 67-86.

KAGAN, Y. Y. & KNOPOFF, L. 1980. Spatial distribution of earthquakes: the two-point correlation function. *Geophysical Journal of the Royal Astronomical Society* **62**, 303-320.

KAGAN, Y. Y. & KNOPOFF, L. 1981. Stochastic synthesis of earthquake catalogs. *Journal of Geophysical Research* **86B**, 2853-2862.

*KAHANE, J. P. 1976. Mesures et dimensions. *Turbulence and Navier-Stokes Equations* (Ed. R. Temam) Lecture Notes in Mathematics **565** 94-103, New York: Springer.

KAPITULNIK, A. & DEUTSCHER, G. 1982. Percolation characteristics in discontinuous thin films of Pb. *Physical Review Letters* **49**, 1444-1448.

KAPITULNIK, A., AHARONY, A., DEUTSCHER, G. & STAUFFER, D. 1983. Self-similarity and correlation in percolation. *To appear.*

*KAYE, B. H. 1983. Fractal description of fineparti-

Plate 467 ¤ ILLUSTRATES PAGE 464

cle systems. *Modern Methods in Fineparticle Characterization* (Ed. J. K. Beddow) Boca Raton, FL: CRC Press.

LE MEHAUTÉ, A. & CREPY, G. 1982. Sur quelques propriétés de transferts électrochimiques en géométrie fractale. *Comptes Rendus* (Paris) **294-II**, 685-688.

LE MEHAUTÉ, A., DE GUIBERT, A., DELAYE, M. & FILIPPI, C. 1982. Note d'introduction de la cinétique des échanges d'énergies et de matières sur les interfaces fractales. *Comptes Rendus* (Paris) **294-II**, 835-838.

LOVEJOY, S. 1981. *Preprints* 20th Conference on Radar Meteorology. A.M.S., Boston, 476-

*LOVEJOY, S. & SCHERTZER, D. 1983. Bouyancy, shear, scaling and fractals. *Sixth Symposium on Atmospheric and Oceanic Waves and Stability* (Boston).

LOVEJOY, S. & MANDELBROT, B. B. 1983. *To appear.*

MANDELBROT, B. B. 1983p. On the quadratic mapping $z \to z^2 - \mu$ for complex μ and z: the fractal structure of its $\mathcal{M}$ set, and scaling. *Order in Chaos* (Ed. D. Campbell) and *Physica D*.

MANDELBROT, B. B. & NORTON, V. A. 1983. *To appear.*

MANDELBROT, B. B., PASSOJA, D. & PAULLAY, A. 1983. *To appear.*

*MANDELBROT, B. B. 1982c. Comments on computer rendering of fractal stochastic models. *Communications of the Association for Computing Machinery* **25**, 581-583.

*MENDÈS-FRANCE, M. & TENENBAUM, G. 1981. Dimension des courbes planes, papiers pliés et suites de Rudin-Shapiro. *Bulletin de la Sociètè Mathèmatique de France* **109**, 207-215.

*MONTROLL, E. W. & SHLESINGER, M. F. 1982. On 1/f noise and other distributions with long tails. *Proceedings of the National Academy of Science of the USA* **79**, 3380-3383.

NORTON, V. A. 1982. Generation and display of geometric fractals in 3-D. *Computer Graphics* **16**, 61-67.

RAMMAL, R. & TOULOUSE, G. 1982. Spectrum of the Schrödinger equation on a self-similar struc-

ture. *Physical Review Letters* **49**, 1194-1197.

RAMMAL, R. & TOULOUSE, G. 1983. Random walks on fractal structures and percolation clusters. *Preprint.*

*ROTHROCK, D. A. & THORNDIKE, A. S. 1980. Geometric properties of the underside of sea ice. *Journal of Geophysical Research* **85C**, 3955-3963.

RUELLE, D. 1982. Analytic repellers. *Ergodic Theory and Dynamical Systems.*

*SERRA, J. 1982. *Image Analysis and Mathematical Morphology.* New York: Academic.

*SHLESINGER, M. F., HUGHES, B. D. 1981. Analogs of renormalization group transformations in random processes. *Physica* **109A**, 597-608.

STEIN, K. 1983. *Omni* (February issue).

STEPHEN, M. J. 1981. Magnetic susceptibility of percolating clusters. *Physics Letters* **A87**, 67-68.

*STEVENS, R. J., LEMAR, A. F. & PRESTON, F. H. Manipulation and presentation of multi-dimensional image data using the Peano scan.

*SUZUKI, M. 1981. *Phase transitions and fractals* (in Japanese) *Suri Kagaku* **221**, 13-20.

*TRICOT, C. 1981. Douze définitions de la densité logarithmique. *Comptes Rendus* (Paris **293I** 549-552.

VICHNIAC, G. 1983. *To appear.*

VOSS, R. F., LAIBOWITZ, R. B. & ALESSANDRINI, E. I. 1982. Fractal (scaling) clusters in thin gold films near the percolation threshold. *Physical Review Letters* **49**, 1441-1444.

WILLSON, S. J. 1982. Cellular automata can generate fractals. *Preprint.*

WOLFRAM, S. 1983. Statistical mechanics of cellular automata. *Reviews of Modern Physics.*